Functionalized Carbon Nanotubes for Biomedical Applications

Scrivener Publishing
100 Cummings Center, Suite 541J
Beverly, MA 01915-6106

Publishers at Scrivener
Martin Scrivener (martin@scrivenerpublishing.com)
Phillip Carmical (pcarmical@scrivenerpublishing.com)

Functionalized Carbon Nanotubes for Biomedical Applications

Edited by

Jeenat Aslam

Department of Chemistry, College of Science, Taibah University, Yanbu, Al-Madina, Saudi Arabia

Chaudhery Mustansar Hussain

Department of Chemistry & Environmental Science at the New Jersey Institute of Technology, Newark, NJ., U.S.A.

and

Ruby Aslam

Department of Applied Chemistry, Aligarh Muslim University, U.P., India

Scrivener Publishing

WILEY

This edition first published 2023 by John Wiley & Sons, Inc., 111 River Street, Hoboken, NJ 07030, USA and Scrivener Publishing LLC, 100 Cummings Center, Suite 541J, Beverly, MA 01915, USA

For more information about Scrivener publications please visit www.scrivenerpublishing.com.

Wiley Global Headquarters
111 River Street, Hoboken, NJ 07030, USA

For details of our global editorial offices, customer services, and more information about Wiley products visit us at www.wiley.com.

Library of Congress Cataloging-in-Publication Data

ISBN 978-1-119-90483-0

Cover image: Pixabay.Com
Cover design by Russell Richardson

Set in size of 11pt and Minion Pro by Manila Typesetting Company, Makati, Philippines

Printed in the USA

10 9 8 7 6 5 4 3 2 1

Contents

Preface

Nanotechnology suggests fascinating opportunities for a variety of applications in the biomedical field, including bioimaging and targeted delivery of biomacromolecules to cells. Numerous strategies have been recommended to functionalize carbon nanotubes with raised solubility for efficient use in these biomedical applications. Functionalized carbon nanotubes are relatively flexible, interact with cell membranes, and penetrate different biological tissues. As they possess unique arrangements and extravagant mechanical, thermal, magnetic, optical, electrical, surface, and chemical properties—the combination of which gives them widespread biomedical applications—considerable effort has been made to employ them as new materials for the development of novel biomedical tools, such as diagnostic sensors, imaging agents, and drug/gene delivery systems for both diagnostics and clinical treatment. Therefore, tremendous progress continues to be made in this field, as is reflected in the rapid growth in the amount of scattered literature available.

This book summarizes the recent progress and developments in functionalized carbon nanotubes-based biomedical devices and systems at both the experimental and theoretical model scale. It covers a broad range of topics relating to carbon nanotubes, from synthesis and functionalization to applications in advanced biomedical devices and systems. The book is logically and eloquently designed to capture an inclusive picture of functionalized carbon nanotubes-based biomedical devices and systems, thereby offering the reader a concentrated up-to-date reference source. It is divided into two parts. An overview of functionalized carbon nanotubes is presented in Part 1, in which Chapter 1 introduces functionalized carbon nanotubes and Chapter 2 describes their synthesis and characterization. Then various types of functionalization of carbon nanotubes are summarized in Chapter 3, after which their role in innovative medical technology is explored in Chapter 4. Different current and emerging biomedical applications of functionalized carbon nanotubes are covered in Part 2, beginning with Chapter 5, which discusses functionalized carbon nanotubes in biosensing. Then their use in drug delivery systems and gene therapy are covered in Chapters 6 and 7, respectively. The applications of functionalized carbon nanotubes in cancer therapy and diagnosis are discussed in Chapter 8, and recent advances in their use for biomedical imaging in Chapter 9. Functionalized carbon nanotubes for bone tissue engineering are discussed in Chapter 10, and their use in biomimetic/bioinspired systems in Chapter 11. Finally, Chapters 12 to 15 discuss the application of functionalized carbon nanotubes in tissue engineering, cell tracking, disease treatment, and antimicrobial activity, respectively.

The aim of this book is to convey information on the recent advancements in the functionalized carbon nanotubes-based biomedical devices and systems arena. It is intended for a very wide-ranging audience working in the fields of advanced materials science, chemistry, medical engineering technology, etc. Since the chapters were contributed and edited by renown researchers, scientists and experts from academia and industry, this book will be an invaluable reference addition to any library collection, including that of universities, industrial institutions, government and independent institutes, and individual research groups and scientists. Overall, it was designed to be useful for advanced undergraduate or graduate students, researchers and scientists who are searching for the functionalized carbon nanotubes-based biomedical devices and systems that modern research demands.

We are very grateful to all those whose exceptional efforts made this book possible. Special thanks go to Mr. Martin Scrivener and the editorial team at Scrivener Publishing for their wholehearted support throughout this project, and to Wiley-Scrivener for publishing the book.

The Editors:
Jeenat Aslam
Chaudhery Mustansar Hussain
Ruby Aslam

Part 1

OVERVIEW OF FUNCTIONALIZED CARBON NANOTUBES

1

Functionalized Carbon Nanotubes: An Introduction

Sheerin Masroor

Department of Chemistry, A. N. College, Patliputra University, Patna, Bihar, India

Abstract

Carbon nanotubes written in short form as, CNTs are the tubes which are made from carbon having diameters in nanometers (nm) or 10^{-9} meters. They can be considered as one of the best carbon allotropes like graphene, graphite, fullerene, diamond and amorphous carbon. Many experimental processes have been obtained to synthesize nanotubes in different sizeable quantities, such as chemical vapor deposition, arc discharge, and laser ablation methods. The blooming of technology related to nanomaterials is mostly happened in drug delivery, biomedical imaging, biosensing and designing of useful nanocomposites. While some more methods relating and realizing applications are continuing to evolve. Carbon nanotubes may be of two types single-wall carbon nanotubes (SWCNTs) having diameters in the range of a nanometers only or multi-wall carbon nanotubes (MWCNTs) possesses nest like structure of single-wall carbon nanotubes. These tubes can be allowed to functionalize via two general reactions such as esterification and amidation of nanotubes with carboxylic acids ends in it. General property like solubility of these tubes helps to know the properties of them by applying solution-based processes. A number of literatures relating functionalization of the carbon nanotubes in the modification of applications in nanocomposites and biological techniques can be seen there.

Keywords: Carbon nanotubes, allotropes, nanocomposites, drug, esterification

Email: masroor.sheerin@gmail.com

Jeenat Aslam, Chaudhery Mustansar Hussain and Ruby Aslam (eds.) Functionalized Carbon Nanotubes for Biomedical Applications, (3–20) © 2023 Scrivener Publishing LLC

1.1 Introduction

Carbon (C) is having atomic number six (6), which is non-metallic in nature and having (four electrons or tetravalency available to form covalent chemical bonds and makes about 0.025 percent of Earth's crust [1, 2]. There are three naturally occurring isotopes of carbon discovered so far, symbolized as ^{12}C and ^{13}C and ^{14}C. Out of all three first two are stable while third one is radionuclide whose half-life is about 5,730 years [3].

Also, the carbon is considered as 15th most existed element in the crust of Earth, and the fourth most abundant element in the universe in terms of mass after hydrogen (H_2), helium (He), and oxygen (O_2). Carbon is abundant element found and its unique diversity to form multiple organic compounds or collection of monomers (polymers) at the temperatures commonly encountered on Earth surface enables this element to serve as a the most common element of all known life. In addition, it is also considered as the second most abundant element found in the human body by mass of approximately 18.5% after oxygen [4].

It's a unique property of carbon atoms that they can bind together in multiple forms generating different number of carbon allotropes. Some of the well-defined allotropes of carbon include graphite, amorphous carbon, fullerenes and diamond. In turn the physical property of carbon mainly depends upon its allotropic form, like graphite is black in color and opaque, while diamond is transparent in nature. Graphite is known for its softness while the diamond is hardest material. In terms of electrical conductivity graphite always acts as good electrical conductor while diamond has diminished electrical conductivity. Recently carbon nanotubes have been also studied for best thermal conductivities along with graphite, graphene and diamond at standard temperature and pressure (STP) or under normal conditions. It can be considered as most of the carbon allotropes are solid at STP or under normal conditions, while graphite is being effectively stable thermodynamically at STP. They all are resistant chemically and generally require very high temperature and oxygen to make reaction.

The carbon in atomic form is very short-lived in nature and hence it is stabilized in different multi-atomic structures with vast molecular arrangements which are generally called as allotropes. One allotrope of carbon which is fullerene is commonly synthesized nowadays successfully and uniquely harvesting it in research in the form of carbon nanotubes, carbon nanobuds, bucky balls, and nanofibers [5–10].

Diverse variety of the exotic allotropes has also been discovered so far, such as glassy carbon, lonsdaleite, carbon nanofoam, and linear acetylenic carbon (carbyne) [11–14].

Two natural occurring crystalline forms of pure carbon are graphite and diamond. Here in diamond, the atoms of carbon show sp^3 hybridization wherein the four bonds are directly attached towards the corners of regular tetrahedron, making diamond so strong and rigid due to three-dimensional network. While in graphite we can easily see carbons hybridization in sp^2 form, in which all atoms are joined uniformly to three carbons with an angle of 120°. In 1985 a new form of carbon, Buckminster fullerene symbolized as (C_{60}) was discovered by Korto *et al.* [15]. In addition, with graphite, diamond and fullerene (C_{60}), quasi-one-dimensional nanotube was also invented in another form of carbon which was first reported in 1991 by Ijima. He reported in his finding that the soot was made by an arc-discharge method to synthesize multiwalled carbon nanotube-MWCNTs [16].

One of the allotropes of carbon is which may be tubular or rod like structure in shape fully made of graphite. The provided tubes may have at least two to many layers ranged for diameter from 3 to 30 nm in size. Later in about two years, the single-walled carbon nanotubes (SWCNTs) came into existence [17]. In and around same period of time, Dresselhaus and his co-workers synthesized single-walled carbon nanotubes by following the previous track of producing multi walled carbon nanotubes (MWCNTs) with the addition of some transition metal particles to the electrodes of carbon [18]. The expected shapes of single-walled carbon nanotubes and multiple walled carbon nanotubes can be depicted from the Figure 1.1 (a and b). Here we can easily see that wall of SWCNT are

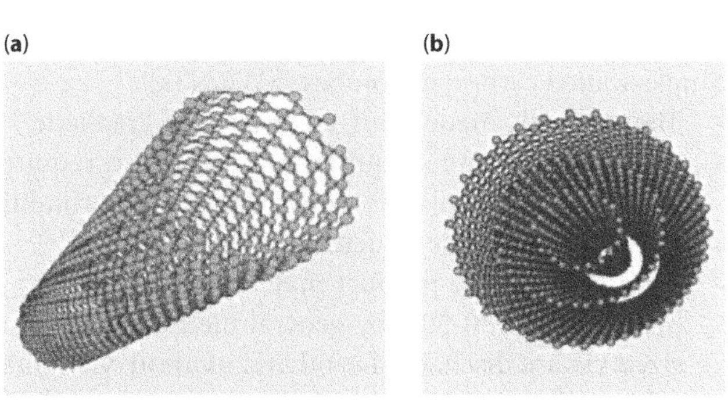

Figure 1.1 Structure of carbon nanotube (a). Single walled carbon nanotubes (b). Multiple walled carbon nanotubes.

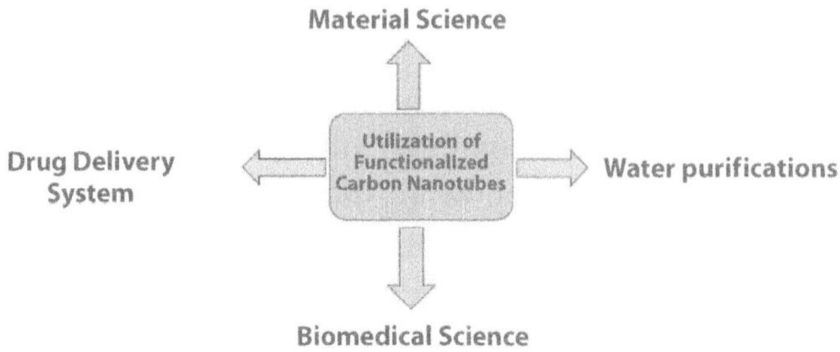

Figure 1.2 Important applications of functionalized CNTs.

generally narrower than the MWCNT, which is having a diameter in the range of 1–2 nm, and took curved shape instead to be straight.

The utmost properties of carbon nanotubes such as electrical, electrochemical, mechanical and chemical properties are extensively studied by huge amount of research that has been going on throughout since last few decades. Various reactions had been going on to find best possible results to harvest them at best possible. Nowadays the researchers have been focusing on changing/improving the quality of nanotubes those who are engaged in catalytic reactions [19]. Some important utilization of functionalized CNTs is pictorially present here in Figure 1.2.

1.2 Carbon Nanotube's Classification

Broadly classified in the given ways such as [20–23]:

a. Single-walled carbon nanotubes (SWCNTs).
b. Multiple-walled carbon nanotubes (MWCNTs).

a. Single-walled carbon nanotubes (SWCNTs).
   - They are only made from single layer of graphene.
   - The synthetic route requires catalyst, which requires an appropriate control over growth and reaction conditions.
   - Can exist in bundle structures.
   - If synthesized the product may be found with the yield percentage of 30-50% by general method but if synthesized via arc discharged synthetic method, yield may be achieved up to 80%.

   - As they are simple in nature, they are easy to characterize on synthesis.

b. Multiple-walled carbon nanotubes (MWCNTs).
   - They are made from multiple layers of graphene.
   - Synthetic route doesn't need catalyst and bulk synthesis is easy.
   - The purity is high by the synthetic route of Chemical vapor deposition (CVD) method with a yield of about 35–90%.
   - Here the unintentional defect is less especially when it is synthesized by arc-discharged method.
   - It has a convoluted structure which can't be easily twistable.

c. Armchair Carbon Nanotubes:
   It is a kind of single-walled carbon nanotube that has equal n and m indices and mainly depends on how the graphite is "rolled up" during its synthetic process. Here in this type of carbon nanotube, chiral angle is 30° and can be considered as metallic. This structure has been manually explained by the ChEBITeam [24].

d. Zigzag Carbon Nanotubes:
   The zigzag nanotubes are having n,0 indices and are metallic in nature only and only if the index n is a multiple of 3. Specifically, we can say one-third of zigzag carbon nanotubes are metallic in nature, while rest two-thirds are semiconductors [25].

e. Chiral Carbon Nanotubes:
   A carbon nanotube is said to be chiral if it has (n,m) type of indices where $m > 0$ and $m \neq n$; and its enantiomer (mirror image) has type (m,n), which is different from (n,m).

1.3 Structural and Morphological Analysis of Carbon Nanotubes

The only component present in single walled carbon nanotubes (SWCNTs) is carbon only which can be seen as tubular shells which are rolled up in the form of graphene sheets, especially benzene (Cyclohexa-1,3,5-triene) type hexagonal rings of carbon atoms. These graphene sheets look cylinders which may be borrowed from a honeycomb lattice, showing a single atomic layer of crystalline graphite. While the multiwalled carbon nanotubes are heap of graphene sheets which can be rolled up into cylinders in the concentric form. Nanotube can be considered as single molecule made up of millions of atoms and whose length can be in the range of micrometers with diameters of about 0.7 nanometers [26]. The single walled carbon nanotubes mainly contain ten (10) atoms which are present over the circumference whose thickness is only one atom thick of the whole tube. Generally, the carbon nanotubes have massive length-to-diameter ratio of approx. value of about 1000, which conclude them as one-dimensional structures [27]. The structure of multi walled carbon nanotube, mainly contains large and numerous single walled tubes which are stacked over and above each other. If the diameter of nanostructures formed have a diameter of about 15 nanometers, then it can be called as multi walled carbon nanotubes otherwise the structures are better known as carbon nanofibers, which are strands of layered-graphite sheets [28].

1.4 Synthetic Techniques of Carbon Nanotubes

There is a great demand of high-quality nanotube materials which can be used for fundamental and technological applications. The quality index can be measured on the basis of presence or absence of chemical and structural defects over a notable scale length of 1-10 microns along the axes of tube. In last few decades it can be easily seen from the literature review that research publication in the form of papers, reviews, book chapters or patents have been increasing enormously. Although there are also so many complications and challenges to think for and must be to resolve relating synthetic procedures of carbon nanotubes. On a broader vision, some existing challenges in the era of nanotube synthetic routes are: large scale production of carbon nanotubes with best quality, Controlled production over specific structure and their properties, development of distinct mechanism via thorough study of the processes of nanotube's growth. In lieu of these numerous techniques have been developed to synthesize carbon nanotubes with discrete morphology and structures in the research laboratory. At present there are specifically three synthetic routes generally used to amalgamate carbon nanotubes and these are chemical vapor deposition, arc discharge and laser ablation methods [29–35].

The important ingredients to synthesize carbon nanotubes are a pure carbon, catalyst and adequate amount of energy. The procedure followed mainly have a step of adding ample energy to a carbon which can be able to produce a fragment which then recombine in their own way to create carbon nanotube. The source of energy may be of any kind like electricity from an arc discharge or heat from a furnace with a temperature of approx. 900°C) or the high-intensity light from laser ablation. The comparison between three processes is given below:

Arc Discharge Method
In this process where there is connection two pure graphite rods which supplies the power and put them at some millimeters apart. On passing 100 amps, and at low pressure inert gas like (Helium) the carbon vaporizes and hot plasma appears there. Single walled carbon nanotubes synthesized are mainly short tubes with diameters of about 0.6–1.4 nm while multiple carbon nanotubes formed are having inner diameter of about 1–3 nm and outer diameter with value 10 nm. One disadvantage of this method is that, it involves high cost with yield of 30–90%.
Laser Ablation Method
The process involves use of blast graphite amidst intense laser pulses and in the presence of argon/nitrogen gas at 500 Torr to give rise to carbon gas from which the carbon nanotubes will be formed. Major product involves formation of single walled carbon nanotubes with long bundles of tubes having length of 5–20 microns and individual diameter with value 1–2 nm. Synthesis involves high cost with yield of about 70%.
Chemical Vapor Deposition Method
Specifically, there is use of fossil-based hydrocarbon and botanical hydrocarbons and placing them in oven at high heat, further adding slowly a carbon-bearing gas like methane in it. As soon as gas decomposes it releases the carbon atoms attached to them which recombines to form nanotubes. The temperature applied is very high with value of 500–1000°C at atmospheric pressure. The yield ranges from 20–100%.

1.5 Functionalization of Carbon Nanotubes

Upon synthesizing the nanotubes, we get them in large number differing in diameters and chirality's. The reason behind is the presence of impurities in the form of metal and amorphous materials. This all requires chemical processing protocols post synthesis, which mainly helps them to divide carbon nanotubes on the basis of diameter and specific chirality [36, 37].

The process of differentiating them depends mainly on electronic and mechanical properties of carbon nanotubes [38, 39].

The carbon nanotubes reinforcement efficiency is confined and severely confined because of its bad interfacial interaction, van der Waals interactions in between carbon nanotubes and polymer matrix. The dispersion of carbon nanotubes is very difficult from rest of the traditional fillers like as spherical particles and carbon fibers, due to their small diameter in nanoscale with high aspect ratio of greater than thousandth (1000th) and thus possessing large surface area. Furthermore, the commercialized form of carbon nanotubes is supplied in the form of heavily knotted bundles, causing difficulties in dispersion. To settle all problems, it was presumed earlier to develop some methods which can alter the surface properties of carbon nanotubes. This is called as functionalization of carbon nanotubes. These methods mainly divided into two divisions depending upon the interactions in between the carbon nanotubes and active materials:

a. Covalent (Chemical) Functionalization,
b. Noncovalent (Physical) Functionalization.

a. Covalent (Chemical) Functionalization,
It is presumed that the end caps of carbon nanotubes favored to be made up of remarkably curved fullerene-like hemispheres, so they are therefore highly reactive, comparable to side walls [40, 41]. In addition, sidewalls themselves have certain defect sites like sp^3-hybridized defects, pentagon-heptagon pairs called Stone-Walls defects, and formed void in the nanotube lattice [42].

Some defects shown in single walled carbon nanotubes are discussed here:

- Presence of five to seven membered rings in the carbon network, rather having normal six membered ring, which ultimately leads to a bent site in the tube.
- Damage caused by oxidative conditions, leaving behind a hole at -COOH groups.
- The sp^3 hybridized defects happen.
- Free/open end of single walled carbon nanotubes which mainly ends with -COOH, $-NO_2$, -OH, -H, and =O groups.

The covalent functionalization is basically a result of chemical reaction which leads to the formation of covalent bond between functional groups adjacent to carbon of nanotubes. It can be successfully performed at the end caps of nanotubes or at their sidewalls which have any defects over that place. The direct covalent sidewall functionalization can be related to the switching of hybridization from sp^2 to sp^3 and a continuous loss of p-conjugation system on graphene layer. This present process can be successfully made by reaction with some atoms or molecules of a high chemical reactivity. Some approaches for functionalization include fluorination, hydrogenation, cycloaddition (Diels-Alder reaction), chlorination, bromination, carbene, and nitrene addition and azomethineylides [43–51]. The next method to functionalize the carbon is defect functionalization. The intrinsic defects are added by oxidative damage to the nanotube network by strong acids which assent holes functionalized with oxygenated functional groups. Specifically speaking this method relates with the reaction of carbon nanotubes with strong acid such as nitric acid (HNO_3), sulfuric acid (H_2SO_4), hydrochloride (HCl) or a mixture of any of them, or using strong oxidants like potassium permanganate ($KMnO_4$), reactive plasma or ozone which tends to unlatched these tubes and to eventually generating oxygenated functional groups such as carboxylic acid or ketone or alcohol or ester groups, that serve to restrain many dissimilar types of chemical components onto the ends and defect sites of these carbon nanotubes. One advantage of functionalizing carbon nanotubes by the covalent methods is that it has good solubility in various organic solvents as the carbon nanotubes have different functional groups with polar or non-polar properties [52–62].

In addition, these approaches have two utmost drawbacks. Firstly, during the process of functionalization reaction, chiefly when applying the ruinous ultrasonication process, a numerous number of defects are mostly created on the carbon nanotubes sidewalls, and in few cases, tubes are fractured into smaller pieces. This all extremely affects the properties of carbon nanotubes and also disrupts the π-electron system in tubes affecting the thermal and electrical properties. Secondly the use of concentrated acids or strong oxidants to functionalize the carbon nanotube can create hazardous effects to the environment producing poisonous by products.

b. Non-covalent (Physical) functionalization

The positive point to note for this non-covalent functionalization is that it may not destroy the conjugated system of the nanotubes sidewalls, and hence it does not affect the basic structural properties of the provided material. This method is a secondary method for modulating the interfacial properties of carbon nanotubes. The CNTs are functionalized non-covalently by number of compounds such as surfactants, aromatic compounds, polymers, employing π-π stacking or any hydrophobic interactions for the rest of the part. In this approach the modification of nanotubes happens with the preservation of desired properties and also altering their solubilities quite exceptionally. The molecules such as polymer, biopolymers, surfactants and aromatic molecules (pyrene/porphyrin) and their derivatives on interacting with the sidewalls of nanotubes via π-π stacking interactions opened the way for the non-covalent functionalization. Number of researchers does this in their own ways like one reported a simple approach to the non-covalent functionalization of nanotubes sidewalls and the ensuing immobilization of biological molecules onto carbon nanotubes with a highest degree of control and specificity. Hecht *et al.* modified fabricated CNTs/FET devices with a molecule of zinc porphyrin derivative, which used to detect directly a photo induced electron transferring within the zinc porphyrin derivative-carbon nanotubes system [63, 64].

The conjugated polymers, are shown to serve as remarkable wrapping materials for the application of non-covalent functionalization of nanotubes which results π-π stacking and van-der Waals interactions between the conjugated polymer having aromatic rings and the exposed surfaces of carbon nanotubes [65–70].

1.6 Commercial Scale Use of Functionalized Carbon Nanotubes

The carbon nanotubes can be considered as big source to elevate the economic development as out of many of them can be effectively commercialized. Some of them have been discussed here:

a. Biomedical application.
   Here the target drugs are clubbed with carbon nanotube molecules for better results, the ultimate property of CNTs and their active counterparts to invade into deeper biological cells which further makes them interesting transportation for the transmission of small molecule drugs. Furthermore, the ability to carry one or multiple remedial agents are marvelous such as MWCNT-Taxanes [71], Hydroxyapatite–carbon nanotube (CNT) composites Hydroxyapatite drug [72] Poly-l-lactic acid matrix MWCNT and Poly-l-lactic acid matrix [73].

b. Delivery target with CNTs.
CNTs are employed for targeted delivery of proteins, nucleic acids, drugs, antibodies, and other agents to their respective sites of action. They are best applied in treatment of malignant disorders, which include choriocarcinoma, carcinoma of the cervix, breast cancer, prostate cancer, brain gliomas, and testicular tumors [74]. Furthermore, the functionalization of chitosan molecule on their surface speedup the cell attachment to the sidewalls of the carbon nanotubes, giving desired targeted release to the cells, and changing the drug absorption. Such systems have significant potential for the delivery of drugs, peptides, and nucleic acids.
c. Nerve and Tissue Regeneration via CNT.
The application of CNTs as reinforcing agents has been mainly attributed to their excellent and unique electrical and mechanical properties. CNTs may be the best tissue engineering candidate among numerous other materials of natural or synthetic origin for tissue scaffolds. In addition, they are less dense, highly flexible, and have a very high young's modulus representing good stiffness. These properties are utilized to make lighter scaffolds with very high strength. It has also been shown by various researchers that the material surface-free energies also play an important role in influencing cell adhesion, leading to greater tissue regeneration. Surface functionalization of these CNTs further imparts to them biocompatibility and biodegradability, ideal for their use in the body.
d. Vaccine delivery via CNT.
Difficulty in absorption, hypersensitivity and anaphylactic reactions are some common drawbacks related to delivery of vaccine on target molecules. Out of many approaches produced so far, CNTs have been applied to vaccine delivery and the action was found to be improved [75]. Also, with this when CNTs are allowed to conjugate with antigenic peptides, can produce new system for safe and effective delivery of synthetic vaccines.
e. Infectious Diseases Treatment via CNTs.
Some infectious diseases, like flu (bird, avian or swine), tuberculosis, leishmaniasis, and other acute respiratory

syndrome have always been a critical public health problem encountered within globe. The pathogens responsible for these diseases show's a remarkable level of resistance against numerous antivirus and antibacterial drugs. Nowadays, the functionalized CNTs are showing great results in the treatment of various diseases because of the ability which allows easy conjugation with drugs such as dapsone and amphotericin B. [76, 77]. This all conjugation mainly shows reduced toxicity and elevated antimycotic efficiency. In addition, the CNTs own their own have antimicrobial activity because bacteria, like Escherichia coli may be absorbed onto the surface of the CNTs. The CNT may cause induce oxidation of the intercellular oxidant glutathione, ensuing increased oxidative stress on the bacterial cells and thus eventually cellular death.

f. Diagnosis of Diseases via CNTs.
The biosensors have been in use since many decades in the field of diagnostics which have made a gigantic impact on basic scientific research and healthcare. These have quality to detect chemical, physical or biological quantity and transience it in the form of a signal. Properties such as fast electron transfer rate, big potential window, flexible surface chemistry, and good biocompatibility of carbon nanotubes can give them excellent property for their special use in biosensors.

Recently, the research has disclosed the fact that CNTs can also generate quantum dots. QDs, or may act like quantum dots. Because of the light-emitting property, these molecules have potential applications in taking imaging of various body parts [78].

1.7 Conclusion and Future Prospects

As in the recent years development mainly based on the advancement of material science and nanotechnology, carbon nanotubes established materials have unlocked the latest routes for the upcoming novel functional materials. In short, we can say that the nanotubes made up of carbon especially have been used as new and utmost type of material applied for electronic and thermal applications. Although variety of nanotubes are available and being investigated by researchers with equal importance to harvest them at best. Some routes have been developed to handle and to manipulate the structure and properties of carbon nanotubes at an easier level. But much research is in need in the application of supercapacitors in this field. Some peoples recently, reporting carbon nanotubes-based supercapacitors with best performance. Literally on adding carbon nanotube with specific macromolecules that may offers enhanced conductivity of the produced material is the point of interest for nowadays researchers. Especially the best electrochemical property and environment friendly nature are two important aspects to think for nowadays. Looking forward to see these functional materials in 2D or 3D forms are also making a quite big area for the utmost development.

References

1. Carbon | Facts, uses, & properties, in: *Encyclopedia Britannica*, Archived from the original on Oct. 24, 2017.
2. Carbon, in: *Britannica Encyclopedia.*
3. Carbon–naturally occurring isotopes, in: *Web Elements Periodic Table*, Archived from the original on Sept. 08, 2008. Retrieved Oct. 09, 2008.
4. Reece, J.B., *Campbell Biology*, 10 ed, Pearson, London, United Kingdom, 2013.
5. Unwin, P., *Fullerenes (An Overview)*, Archived from the original on Dec. 01, 2007. Retrieved Dec. 08, 2007.
6. Ebbesen, T.W. (Ed.), *Carbon Nanotubes—Preparation and Properties*, CRC Press, Boca Raton, Florida, 1997.
7. Dresselhaus, M.S., Dresselhaus, G., Avouris, P. (Eds.), *Carbon Nanotubes: Synthesis, Structures, Properties and Applications. Topics in Applied Physics*, Springer Berlin, Heidelberg, vol. 80, 2001.
8. Nasibulin, A.G., Pikhitsa, P.V., Jiang, H., Brown, D.P., Krasheninnikov, A.V., Anisimov, A.S., Queipo, P., Moisala, A. *et al.*, A novel hybrid carbon material. *Nat. Nanotechnol.*, 2, 3, 156–161, 2007.
9. Nasibulin, A., Anisimov, A.S., Pikhitsa, P.V., Jiang, H., Brown, D.P., Choi, M., Kauppinen, E., II, Investigations of nanobud formation. *Chem. Phys. Lett.*, 446, 1, 109–114, 2007.
10. Vieira, R., Ledoux, M.-J., Pham-Huu, C., Synthesis and characterisation of carbon nanofibers with macroscopic shaping formed by catalytic decomposition of C2H6/H2 over nickel catalyst. *Appl. Catal. A Gen.*, 274, 1–2, 1–8, 2004.
11. Frondel, C. and Marvin, U.B., Lonsdaleite, a new hexagonal polymorph of diamond. *Nature*, 214, 5088, 587–589, 1967.

12. Harris, P.J.F., Fullerene-related structure of commercial glassy carbons (PDF). *Philos. Mag.*, 84, 29, 3159–3167, 2004. Archived from the original (PDF) on 2012-03-19. Retrieved 2011-07-06.
13. Rode, A.V., Hyde, S.T., Gamaly, E.G., Elliman, R.G., McKenzie, D.R., Bulcock, S., Structural analysis of a carbon foam formed by high pulse-rate laser ablation. *Appl. Phys. A*, 69, 7, S755–S758, 1999.
14. Heimann, R.B., Evsyukov, S.E., Kavan, L., *Carbyne and Carbynoid Structures*, p. 1, Springer, NY, USA, 1999. Archived from the original on November 23, 2012. Retrieved June 06, 2011.
15. Kroto, H.W., Heath, J.R., O'Brien, S.C., Curl, R.F., Smalley, R., C60: Buckminsterfullerene. *Nature*, 318, 6042, 162–163, 1985.
16. Iijima, S., Helical microtubules of graphitic carbon. *Nature*, 354, 6348, 56–58, 1991.
17. Iijima, S. and Ichihashi, T., Single-shell carbon nanotubes of 1-nm diameter. *Nature*, 363, 6430, 603–605, 1993.
18. Dresselhaus, M.S., Dresselhaus, G., Avouris, P.H., Springer, Berlin, Germany, 2001.
19. Ajayan, P.M., Nanotubes from carbon. *Chem. Rev.*, 99, 7, 1787–1799, 1999.
20. Saifuddin, N., Raziah, A.Z., Junizah, A.R., Carbon nanotubes: A review on structure and their interaction with proteins. Hindawi Publishing Corporation, *J. Chem.*, 2013, 676815, 18, 2012. http://dx.doi.org/10.1155/2013/676815].
21. Abrahamson, J., Wiles, P.G., Rhoades, B.L., Structure of carbon fibres found on carbon arc anodes. *Carbon*, 37, 11, 1873–1874, 1999.
22. Hirlekar, R., Yamagar, M., Garse, H., Mohit, V., Kadam, V., Carbon nanotubes and its applications: A review. *Asian J. Pharm. Clin. Res.*, 2, 4, 17–27, 2009.
23. Meyyappan, M., Delzeit, L., Cassell, A., Hash, D., Carbon nanotube growth by PECVD: A review. *Plasma Sources Sci. Technol.*, 12, 2, 205–216, 2003.
24. https://www.ebi.ac.uk/chebi/chebiOntology.do?treeView=true&chebiId=CHEBI:50798.
25. Chen, Y.-R., Weng, C.-I., Sun, S.-J., Electronic properties of zigzag and armchair carbon nanotubes under uniaxial strain. *J. Appl. Phys.*, 104, 114310, 2008. https://doi.org/10.1063/1.3033167.
26. Meyyappan, M., Delzeit, L., Cassell, A., Hash, D., Carbon nanotube growth by PECVD: A review. *Plasma Sources Sci. Technol.*, 12, 2, 205–216, 2003.
27. Dresselhaus, M.S., Dresselhaus, G., Jorio, A., Unusual properties and structure of carbon nanotubes. *Annu. Rev. Mater. Res.*, 34, 247–278, 2004.
28. Lin, T., Bajpai, V., Ji, T., Dai, L., Chemistry of carbon nanotubes. *Aust. J. Chem.*, 56, 7, 635–651, 2003.
29. Ebbesen, T.W. and Ajayan, P.M., Large-scale synthesis of carbon nanotubes. *Nature*, 358, 6383, 220–222, 1992.
30. Journet, C., Maser, W.K., Bernier, P. *et al.*, Large-scale production of single-walled carbon nanotubes by the electric-arc technique. *Nature*, 388, 6644, 756–758, 1997.

31. Thess, A., Lee, R., Nikolaev, P. *et al.*, Crystalline ropes of metallic carbon nanotubes. *Science*, 273, 5274, 483–487, 1996.
32. Kong, J., Cassell, A.M., Dai, H., Chemical vapor deposition of methane for single-walled carbon nanotubes. *Chem. Phys. Lett.*, 292, 4–6, 567–574, 1998.
33. Fan, S., Chapline, M.G., Franklin, N.R., Tombler, T.W., Cassell, A.M., Dai, H., Self-oriented regular arrays of carbon nanotubes and their field emission properties. *Science*, 283, 5401, 512–514, 1999.
34. Bower, C., Zhou, O., Zhu, W., Werder, D.J., Jin, S., Nucleation and growth of carbon nanotubes by microwave plasma chemical vapor deposition. *Appl. Phys. Lett.*, 77, 17, 2767–2769, 2000.
35. Su, M., Zheng, B., Liu, J., A scalable CVD method for the synthesis of single-walled carbon nanotubes with high catalyst productivity. *Chem. Phys. Lett.*, 322, 5, 321–326, 2000.
36. Schadler, L.S., Polymer-based and polymer-filled nanocomposites, in: *Nanocomposite Science and Technology*, P.M. Ajayan, L.S. Schadler, P.V. Braun (Eds.), Wiley-VCH Verlag GmbH & Co. KGaA, Weinheim, 2004.
37. Li, H., Zhou, B., Lin, Y., Gu, L., Wang, W., Fernando, K.A.S., Kumar, S., Allard, L.F., Sun, Y.-P., Selective interactions of porphyrins with semiconducting single walled carbon nanotube. *J. Am. Chem. Soc.*, 126, 1014–1015, 2004.
38. Niyogi, S., Hamon, M.A., Hu, H., Zhao, B., Bhowmik, P., Sen, R., Itkis, M.E., Haddon, R.C., Chemistry of single-walled carbon nanotubes. *Acc. Chem. Res.*, 35, 12, 1105–1113, 2002.
39. Tasis, D., Tagmatarchis, N., Georgakilas, V., Prato, M., Soluble carbon nanotubes. *Chem. Eur. J.*, 9, 17, 4000–4008, 2003.
40. Hirsch, A., Functionalization of single-walled carbon nanotubes. *Angew. Chem. Int. Ed.*, 41, 11, 1853–1859, 2002.
41. Sinnott, S.B., Chemical functionalization of carbon nanotubes. *J. Nanosci. Nanotechnol.*, 2, 113–123, 2002.
42. Hirsch, A., Functionalization of single-walled carbon nanotubes. *Angew. Chem. Int. Ed.*, 41, 11, 1853–1859, 2002.
43. Bianco, A., Prato, M., Kostarelos, K., Bianco, A., Functionalized carbon nanotubes in drug design and discovery. *Acc. Chem. Res.*, 41, 1, 60–68, 2008.
44. Mickelson, E.T., Huffman, C.B., Rinzler, A.G., Smalley, R.E., Hauge, R.H., Margrave, J.L., Fluorination of single-wall carbon nanotubes. *Chem. Phys. Lett.*, 296, 1–2, 188–194, 1998.
45. Kelly, K.F., Chiang, I.W., Mickelson, E.T. *et al.*, Insight into the mechanism of sidewall functionalization of single-walled nanotubes: An STM study. *Chem. Phys. Lett.*, 313, 445–450, 1993.
46. Touhara, H., Inahara, J., Mizuno, T. *et al.*, Fluorination of cup-stacked carbon nanotubes, structure and properties. *Fluorine Chem.*, 114, 181–188, 2002.
47. Stevens, J.L., Huang, A.Y., Peng, H. *et al.*, Sidewall amino-functionalization of SWNTs through fluorination and subsequent reactions with terminal diamines. *Nano Lett.*, 3, 331–336, 2003.

48. Hu, H., Zhao, B., Hamon, M.A. *et al.*, Sidewall functionalization of single-walled carbon nanotubes by addition of dichlorocarbene. *J. Am. Chem. Soc*, 125, 14893–14900, 2003.
49. Unger, E., Graham, A., Kreupl, F., Liebau, M., Hoenlein, W., Electrochemical functionalization of multi-walled carbon nanotubes for solvation and purification. *Curr. Appl. Phys.*, 2, 107–111, 2002.
50. Kim, K.S., Bae, D.J., Kim, J.R., Park, K.A., Lim, S.C., Kim, J.J., Choi, W.B., Park, C.Y., Lee, Y.H., Modification of electronic structures of a carbon nanotube by hydrogen functionalization. *Adv. Mater.*, 14, 24, 1818–1821, 2002.
51. Tagmatarchis, N. and Prato, M.J., Functionalization of carbon nanotubes via 1, 3-dipolar cycloadditions. *J. Mater. Chem.*, 14, 437–439, 2004.
52. Chen, J.; Hamon, M.A.; Hu, H.; Chen, Y.; Rao, A.M.; Eklund, P.C.; Haddon, R.C., 1998. Solution properties of single-walled carbon nanotubes. Science 282, (5386), 95–98.
53. Esumi, K., Ishigami, M., Nakajima, A., Sawada, K., Honda, H., Chemical treatment of carbon nanotubes. *Carbon*, 34, 279–281, 1996.
54. Liu, J., Rinzler, A.G., Dai, H., Hafner, J.H., Bradley, R.K., Boul, P.J. *et al.*, Fullerene pipes. *Science*, 280, 1253–1256, 1998.
55. Yu, R., Chen, L., Liu, Q., Lin, J., Tan, K.-L., Ng, S.C., Chan, H.S.O., Xu, G.-Q., Hor, T.S.A., Platinum deposition on carbon nanotubes via chemical modification. *Chem. Mater.*, 10, 718–722, 1998.
56. Sham, M.-L. and Kim, J.-K., Surface functionalities of multi-wall carbon nanotubes after UV/Ozone and TETA treatments. *Carbon*, 44, 4, 768–777, 2006.
57. Wang, S.C., Chang, K.S., Yuan, C.J., Enhancement of electrochemical properties of screen-printed carbon electrodes by oxygen plasma treatment. *Electrochim. Acta*, 54, 21, 4937–4943, 2009.
58. Ma, P.C., Kim, J.K., Tang, B.Z., Functionalization of carbon nanotubes using a silane coupling agent. *Carbon*, 44, 3232–3238, 2006.
59. Sano, M., Kamino, A., Okamura, J., Shinkai, S., Self-organization of PEO-graft-single-walled carbon nanotubes in solutions and langmuir–blodgett films. *Langmuir*, 17, 17, 5125–5128, 2001.
60. Kong, H., Gao, C., Yan, D., Controlled functionalization of multiwalled carbon nanotubes by *in situ* atom transfer radical polymerization. *J. Am. Chem. Soc.*, 126, 412–413, 2003.
61. Hamon, M.A., Hui, H., Bhowmik, P., Ester-functionalized soluble single-walled carbon nanotubes. *Appl. Phys. A*, 74, 333–338, 2002.
62. Coleman, J.N., Khan, U., Gunko, Y.K., Mechanical reinforcement of polymers using carbon nanotubes. *Adv. Mater.*, 18, 689–706, 2006.
63. Hu, L., Zhao, Y.-L., Ryu, K., Zhou, C., Stoddart, J.F., Gruner, G., Light-induced charge transfer in pyrene/CdSe-SWNT hybrids. *Adv. Mater.*, 20, 939–946, 2008.
64. Hecht, D.S., Ramirez, R.J.A., Briman, M., Artukovic, E., Chichak, K.S., Stoddart, J.F., Gruner, G., Bioinspired detection of light using a

porphyrin-sensitized single wall nanotube field effect transistor. *Nano Lett.*, 6, 2031–2036, 2006.
65. Star, A., Stoddart, J.F., Steuerman, D., Diehl, M., Boukai, A., Wong, E.W., Yang, X., Chung, S.-W., Choi, H., Heath, J.R., Preparation and properties of polymer-wrapped single-walled carbon nanotubes. *Angew. Chem. Int. Ed.*, 40, 1721–1725, 20012001.
66. Steuerman, D.W., Star, A., Narizaano, R., Choi, H., Ries, R.S., Nicolini, C., Stoddart, J.F., Heath, J.R., Interactions between conjugated polymers and single walled carbon nanotubes. *J. Phys. Chem. B*, 106, 3124–3130, 2002.
67. Star, A., Liu, Y., Grant, K., Ridvan, L., Stoddart, J.F., Steuerman, D.W., Diehl, M.R., Boukai, A., Heath, J.R., Noncovalent side-wall functionalization of single walled carbon nanotubes. *Macromolecules*, 36, 553–560, 2003.
68. Star, A. and Stoddart, J.F., Dispersion and solubilization of single-walled carbon nanotubes with a hyperbranched polymer. *Macromolecules*, 35, 7516–7520, 2002.
69. Cheng, F. and Adronov, A., Noncovalent functionalization and solubilization of carbon nanotubes by using a conjugated Zn-porphyrin polymer. *Chem. Eur. J.*, 12, 5053–5059, 2006.
70. Yi, W., Malkovskiy, A., Chu, Q., Sokolov, A.P., Colon, M.L., Meador, M., Pang, Y., Wrapping of single-walled carbon nanotubes by a π-conjugated polymer: The role of polymer conformation-controlled size selectivity. *J. Phys. Chem. B*, 112, 12263–12269, 2008.
71. Kim, S., Hwang, J., Seo, J., Us, S., Production of CNT-taxol-embedded PCL microspheres using an ammonium-based room temperature ionic liquid: As a sustained drug delivery system. *J. Colloid Interface Sci.*, 442, 147–153, 2015.
72. Mukherjeea, S., Kundub, B., Chandac, A., Sen, S., Effect of functionalisation of CNT in the preparation of HAp-CNT biocomposites. *Ceram. Int.*, 41, 3766–3774, 2015.
73. Scapin, G., Salice, P., Tescari, S., Menna, E., De Filippis, V., Filippini, F., Enhanced neuronal cell differentiation combining biomimetic peptides and a carbon nanotube-polymer scaffold, 2014. https://pubmed.ncbi.nlm.nih.gov/25546847/
74. Thakare, V.S., Das, M., Jain, A.K., Patil, S., Jain, S., Carbon nanotubes in cancer theragnosis. *Nanomedicine (Lond.)*, 5, 1277–1301, 2010.
75. Bianco, A., Carbon nanotubes for the delivery of therapeutic molecules. *Expert Opin. Drug Deliv.*, 1, 57–65, 2004.
76. Mehra, N.K., Jain, K., Jain, N.K., Pharmaceutical and biomedical applications of surface engineered carbon nanotubes. *Drug Discovery Today*, 1–10, 2015. http://dx.doi.org/10.1016/j.drudis.2015.01.006.
77. Vinoth, V., Wu, J.J., Asiri, A.M., Anandan, S., Simultaneous detection of dopamine and ascorbic acid using silicate network interlinked gold nanoparticles and multi-walled carbon nanotubes. *Sens. Actuators B Chem.*, 210, 731–741, Published online 2020 Dec 16.

78. Lim, Y.T., Kim, S., Nakayama, A., Stott, N.E., Bawendi, M.G., Frangioni, J.V., Selection of quantum dot wavelengths for biomedical assays and imaging. *Mol. Imaging*, 2, 50–64, 2003.

2

Functionalized Carbon Nanotubes: Synthesis and Characterization

Neelam Sharma[1], Shubhra Pareek[2], Rahul Shrivastava[1] and Debasis Behera[3*]

[1]*Department of Chemistry, Manipal University Jaipur, Jaipur (Rajasthan), India*
[2]*Department of Materials Engineering, Indian Institute of Science, Bangalore, India*
[3]*School of Chemistry, Gangadhar Meher University, Amruta Vihar, Sambalpur, Odisha, India*

Abstract

Numerous nanomaterials have been discovered and applied with the advancement of nanotechnology. Carbon nanotubes (CNTs) are one of them due to their amazing structure and magnificent properties, such as excellent conductivity, effective electrochemical and thermal stability, and great specific surface area. Owing to such features, CNTs have a wide series of medical, environmental, and industrial applications. In this chapter, the different routes of CNT synthesis including chemical vapor deposition (CVD), arc discharge, and laser ablation/Vaporization methods have been critically analyzed for the delivery of the desired application. The selectivity in the application can be modulated with tailored surface functionalization of CNT, which is further divided broadly into three categories. (i) through covalent bonds with a π-conjugated network of CNTs; (ii) through non-covalent bonds, π–π interactions, or ionic bonds, and hydrogen bonds for the attachment of different chemical groups; (iii) Endohedral filling functionalization (Inline filling) of hollow tubes of CNTs. These functionalizations give a deep insight into oxidation and activation, heteroatom doping, sulfonation, halogenation, catalytic chemical vapor deposition (CCVD), non-covalent functionalization, and nano-particle attachment. The presence of functional groups (surface oxides, principally hydroxyl, carboxyl, and carbonyl,) affects the modification of chemical reactivity, solubility, and other physical characteristics of CNTs. Numerous characterization techniques have been employed for the structural elucidation of CNTs such as FT-IR, UV–Vis, Raman, XRD, XPS, HR-TEM, FESEM, SAED, TGA, NMR. VSM, zeta potential DLS, and pH drift.

**Corresponding author*: dbahera@gmail.com

Jeenat Aslam, Chaudhery Mustansar Hussain and Ruby Aslam (eds.) Functionalized Carbon Nanotubes for Biomedical Applications, (21–48)

***Keywords*:** Nanotechnology, carbon nanotubes, synthesis, functionalization, characterization

2.1 Introduction

In recent decades, Nanotechnology has been extensively studied, and several novel materials with various nanostructures (shapes and sizes) have emerged [1]. These nano-materials have several unique features which are not found in their micro-scale equivalents, offering up new possibilities for applications [2]. In the periodic table, one of the most adaptable elements is carbon because it can form a variety of C– C bonds (sp, sp^2, and sp^3-hybridized bonds) [3]. In recent years, the synthetic allotropes of the carbon family have developed with the discovery of nano-onions [4], carbon nanohorns [5], carbon nanotubes (CNTs) [6], graphene [7], and other synthetic allotropes of carbon. CNTs have made significant contributions to chemistry, physics, material sciences, mathematical modeling, and other fields since their discovery [8]. Carbon nanotubes (CNTs) are the most promising and extensively investigated of all the carbon allotropes. CNTs have the easiest chemical composition and atomic bonding arrangement, but their structures and structural property relationships are extremely diverse and rich [9]. These were discovered in the soot of arc discharge equipment shortly after SumioIijima of the NEC Corporation synthesized fullerene in the lab in 1991 [10]. After two years, in 1993, two companies, IBM [11], and NEC [12] disclosed CNT with a single graphene layer. CNTs are the most widely studied allotrope of carbon because it has a unique hollow structure that gives it extraordinary electrical, thermal, and mechanical properties, such as high tensile strength better than any metal. The low density, high thermal conductivity exceeding diamond, high ductility, high electrical conductivity, high chemical, and thermal stability make it suitable for a wide range of applications [13]. CNTs are one of the most innovative structures available with a high aspect ratio (* 1000), and wide surface area. All these properties are dependent on their morphology, structure, and functions [14]. In the field of hydrogen storage [15] and generation [16], supercapacitors [17], biosensors [18], fuel cells [19], lithium-ion batteries [20], molecular sieves, electrochemical sensor [21] tissue engineering [22], and removal of pollutants [23] properties, CNTs exhibit potential applications. Various approaches have been developed to synthesize functionalized CNTs as well as to increase their solubility and dispersion in many solvents [24]. Due to their specific electronic, optical, magnetic, mechanical, and chemical properties that differ from those of their respective

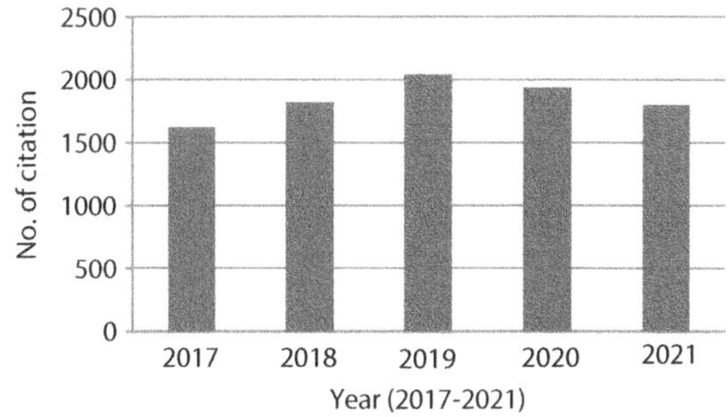

Figure 2.1 Citation trends of published articles on functionalized carbon nanotubes.

bulk materials as well as constituent molecules or atoms, nanoparticles of modest diameters (10–20 nm) have garnered a lot of attention [25]. CNTs materials decorated with metal or metal oxide nanoparticles have gained significant interest in recent years [26].

In the past few decades, functionalized CNTs have generated widespread interest in commercial applications (Figure 2.1). The right combination of properties of the functionalized group makes CNTs suitable for use in many applications like sensing, biomedical, drugs delivery, energy storage, etc. Many of which have been patented. Some patents are discussed below.

Scheinberg *et al.* have synthesized functionalized carbon nanotubes and filed a patent on May 22, 2018. In which they have explored the various application of carbon nanotubes on tumors [27]. Smith and Ghosn *et al.* have synthesized RGD-SWCNTs were used to deliver therapeutic agents for tumors and filed a patent on March 20, 2014 [28]. Myung *et al.* have synthesized metal and metal oxide co-functionalized SWCNTs and filed a patent in April 5, 2016, for the high performance of gas sensors [29]. Boday *et al.* have synthesized functionalized single-wall carbon nanotubes (FSWNTs) that were used in a printed wiring board (PWB)and filed a patent on September 5, 2017 [30]. Kobilka *et al.* discussed functionalized carbon nanotubes and filed a patent on March 19, 2019 [31]. Said *et al.* have synthesized CNTs functionalized with COOH (carboxyl) groups that are used for the solar adsorption refrigeration and filed a patent on November 1, 2016 [32]. To form macroscopic aggregates, Arnett *et al.* have synthesized DNA – functionalized SWCNTs through phosphodiester bonding and filed a patent on Jan. 7, 2020 [33].

This chapter aims to summarize (a) synthetic methods of CNTs. (b) Characterization Techniques, and (c) the systematic procedure to get the desired functionalization on CNTs. A classical balance has been presented between topics to increase readership among researchers, academicians, and students.

2.2 Synthesis Methods

Figure 2.2 is a representation of the broad classification methods employed for the synthesis of the CNTs.

2.2.1 Arc Discharge

CNTs were first produced by arc discharge between graphite electrodes [34]. Establishing a DC (direct current) between a pair of graphite electrodes under an inert atmosphere (argon or helium) at roughly 500 torr is required for this technique [35]. MWCNTs (Figure 2.3 (a)) are made by arc discharge without the use of a metal catalyst, whereas SWNCTs (Figure 2.3 (b)) require the use of mixed metal catalysts such as Co, Fe, and Ni [36]. CNTs synthesized via arc discharge method typically have a high degree of structural perfection [37], but a variety of variables (catalyst composition chamber temperature, and hydrogen present, concentration, etc.) Influence their structure and size [38]. Other electrodes [39] and compounds [40–42] have recently been used in the arc discharge synthesis

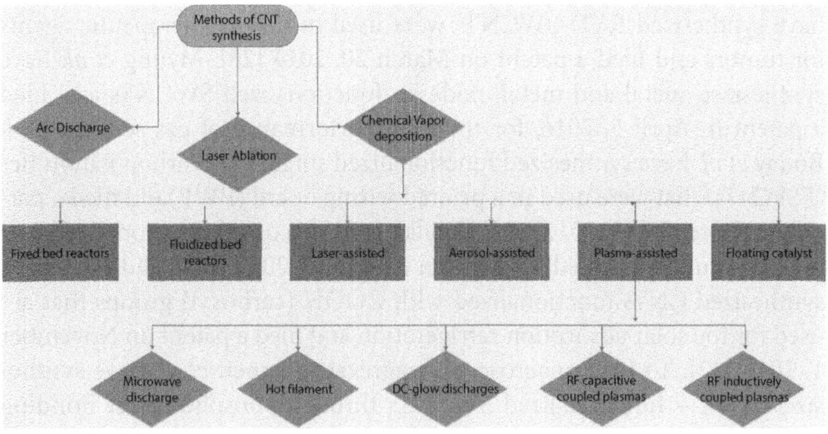

Figure 2.2 The broad classification of methods employed for the synthesis of CNTs. *Reprinted with permission [30] from Elsevier.*

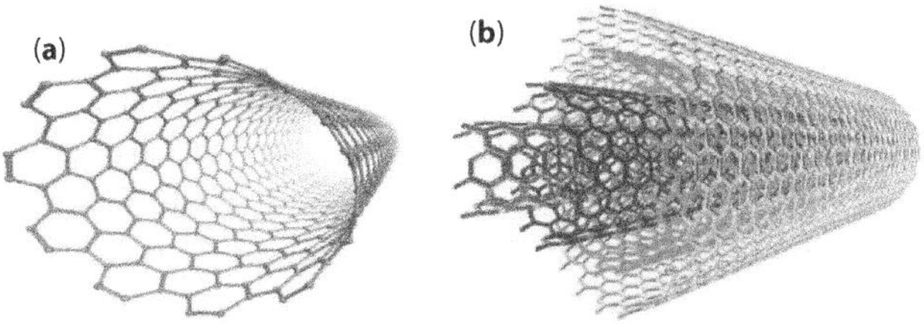

Figure 2.3 Structure of (a) single-walled carbon nanotube (SWCNT), and (b) multiwalled carbon nanotube (MWCNT).

of CNTs. Using the arc discharge approach, Belgacem *et al.* [40] were able to synthesize MWCNTs doped with boron and nitrogen. Other research groups [41] have used arc discharge in the open air to make SWCNT-SWCNT hybrids at a lower cost.

2.2.2 Laser Ablation

Guo and coworkers were the first to disclose the synthesis of CNTs utilizing laser ablation in 1995 [43]. The laser ablation technology works on the same concept as arc discharge [44]. A solid graphite target mounted in a quartz tube is vaporized by a laser beam pulse at high temperatures (800 to1500 °C) in the presence of an inert atmosphere in this method [45]. The carbon-based soot is collected from the inside of the apparatus after the target has been vaporized [46]. Because of the high temperatures used in the synthesis, the CNTs produced by this approach have a high degree of structural perfection [47]. Furthermore, SWCNTs can be made without the use of MWCNTs. Compared to the arc discharge approach, the main disadvantages of this method include the high-power consumption with the lesser numbers of CNTs production [48]. The factors that influence the volume and quality of CNTs produced by laser ablation encompasses, (i) the chemical composition of the target material, (ii) the power of the laser and wavelength, and (iii) the distance between the substrates and the target [49]. Laser ablation allows high-purity CNTs to have a specified chirality (10,10) structure. In comparison to the chemical vapor deposition (CVD) method, the arc discharge and laser ablation procedures offer the benefit of producing higher-quality CNTs. The regular replacement of graphite electrodes and targets, the need for high vacuum conditions, and temperature restrain its ability to scale up CNT manufacture economically via arc discharge and laser ablation technologies [44].

2.2.3 Chemical Vapor Deposition

Due to its simplicity, ambient pressure, low cost, and low temperature, chemical vapor deposition is the most used approach for the manufacture of carbon nanotubes [50]. The catalytic vapor phase deposition of carbon using carbon monoxide–hydrogen mixtures over Fe was first described in 1959 [51]. This procedure was used to make CNT from acetylene over Fe particles at 700°C in 1993 [52]. The CVD was first proposed as a method for large-scale CNT production in 1996 [53]. Currently, the CVD is the technology for mass synthesis of various types of CNTs that have received the most attention [54]. To accelerate the formation of nanotubes, a gaseous hydrocarbon source (ethanol, acetylene, propylene, methane, ethylene, CO, etc.) is heated to 600-1000 °C with a transition metal catalyst such as Fe, Ni, Co, etc. (Figure 2.4) [55]. The CVD enables production on large-scale, excellent alignment [56], and controlled to development of nanotube [57]. Controlling the diameter and number of shells is also possible [58]. The CNTs produced through the CVD method have a larger defect density (amorphous carbon) than that of laser ablation, and arc discharge methods [59]. These flaws harm the CNTs' thermal and electrical properties, as well as their structural properties [60]. The CVD approach is currently being used to regulate the synthesis of ultra-long CNTs with flawless structure. CNTs with good thermal electrical, and mechanical, properties can be made with this defect-free structure. The reader is directed to Zhang *et al.* [61] for a detailed explanation of the controlled synthesis of ultra-long CNTs. The integrated thermal plasma reactor and the CVD process have been employed recently to fabricate CNTs at atmospheric pressure [62].

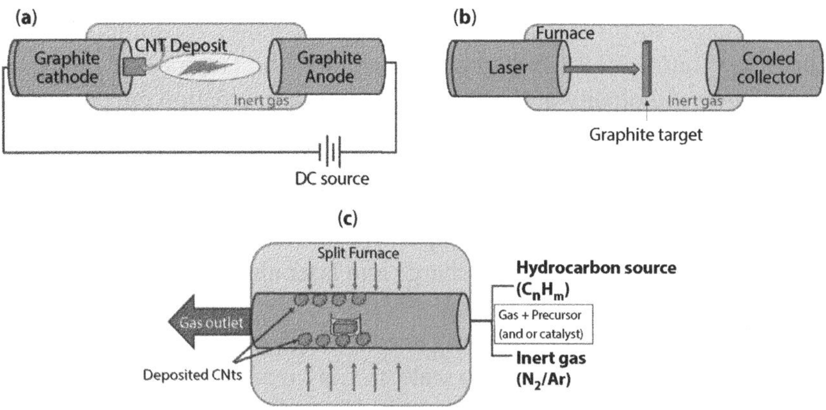

Figure 2.4 CNT preparation techniques (a) arc discharge, (b) laser ablation, and (c) CVD.

2.3 Characterization

2.3.1 Raman Spectroscopy

One of the most used approaches for identifying structural faults in CNTs is Raman spectroscopy [63]. Raman spectra of CNTs depict have peaks at 1350 cm^{-1} that correspond to the D band whereas the peak at 1580 cm^{-1} represents the G band [64]. The G band is related to sp^2 hybridization with symmetric graphite hexagonal lattices, while The D band corresponds to non-crystalline carbon (sp^3 diamond-like) [65]. In other situations, a shoulder at the G band nearby 1620 cm^{-1} (D'band) can be seen (Figure 2.5), which is related to

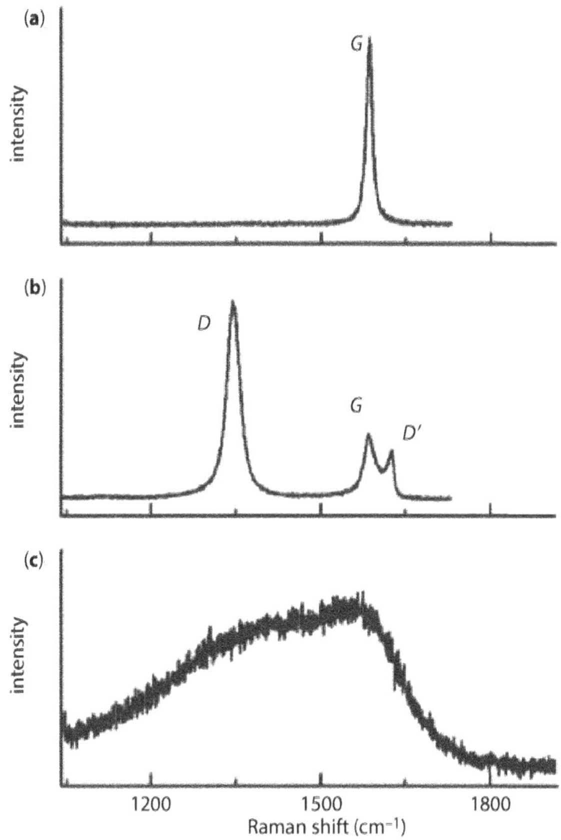

Figure 2.5 The first-order Raman spectrum of (a) symmetric hexagonal lattice, (b) defective sp3 hybridized carbon along with shoulder band D' (c) highly disordered single-layer graphene deposited on a substrate. *Reprinted with permission from [The Royal Society].*

sp^3 hybridized carbon that lacks in-plane symmetry with graphene [64, 66]. The intensity of the D band is proportional to the quantity of faulty carbon, and the ratio of the D and G bands' intensities (I_D/I_G) is commonly employed to measure structural disorder. An increase in ID/IG suggests a loss of aromaticity in CNT rings due to faults caused by the hybridization of C-C sp^2 to C-C sp^3 (*vide infra*, section 2.3.2 FTIR) [67]. Raman spectroscopy can also be used for the investigation of chirality and count CNT walls.

2.3.2 Fourier Transform Infrared Spectroscopy (FT-IR)

The surface of CNTs is evaluated qualitatively using Fourier transform infrared spectroscopy (FT-IR). In this method, infrared radiation is

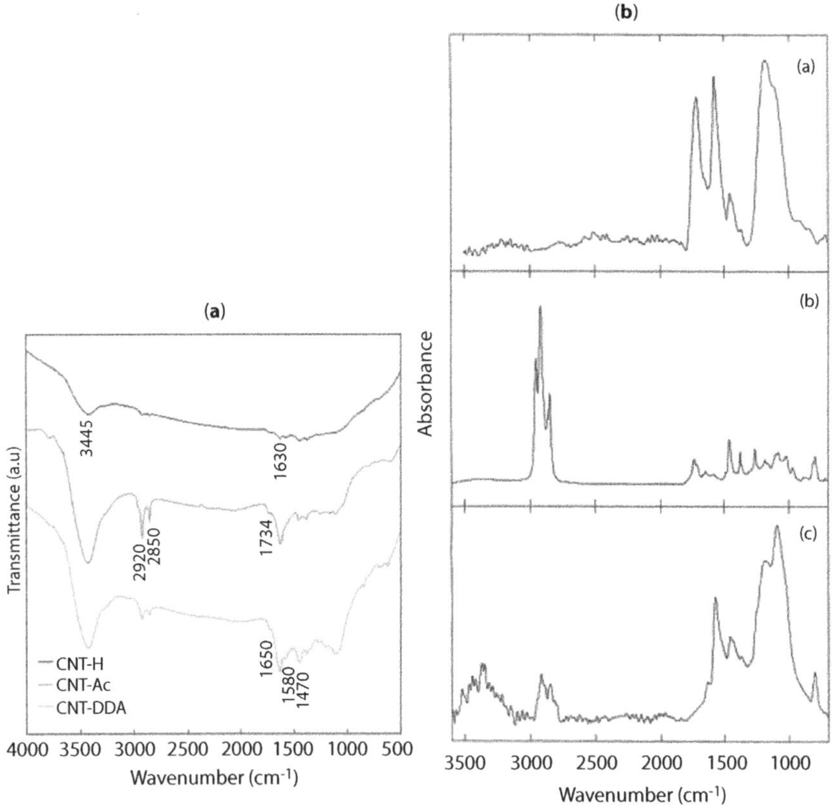

Figure 2.6 FT-IR spectra of CNT-H, CNT-Ac (CNT-COOH), and CNT-DDA (CNT-dodecylamine) (a) transmittance data, (b) absorbance data, *Reprinted with permission from [Elsevier].*

passed through the sample, and the fraction of this incident radiation absorbed at specific energy is calculated. In the spectrum, the frequency of a certain functional group is represented by each absorption peak. A distinctive band of about 3445 and 1630 cm^{-1} can be seen in the infrared spectra of raw CNTs (Figure 2.6 (a)), which is associated with O-H and carbon-carbon double bonds stretching vibrations of aromatic rings on folded CNT respectively [68]. Bands at 3800–3200 cm^{-1} (OH stretching) and 1700–1600 cm^{-1} (OH bending) have been detected in some situations. These bands are caused by water molecules being absorbed by atmospheric moisture or by a purification process [69]. The appearance of bands at 2920 and 2850 cm^{-1} (Figure 2.6 (a)) supplement the evidence to asymmetric and symmetric stretching vibrations of the methylene (CH_2) group, which are usually assigned to defects in rings of CNTs (Figure 2.6 (b)) (*vide supra*, section 2.3.1 Raman spectroscopy) [70].

2.3.3 Thermogravimetric Analysis (TGA)

The purity and thermal stability of CNTs are studied by using thermogravimetric analysis (TGA) [71]. Depending on the information needed, the analysis might be done in an oxidizing or inert atmosphere. Some important parameters are analyzed like the initiation temperature (when weight loss begins), the final temperature (when weight loss ceases), the amount of weight loss at the initiation temperature, and the residual weight exhibited at the final temperature [72]. The degree of graphitization of the CNTs walls is indicated by the initiation temperature. The initiation temperature of CNTs with faulty hexatomic rings is lower, whereas the initiation temperature of more crystalline CNTs is higher [73]. The ultimate temperature reveals the nanotubes' thermal stability: a higher temperature means more graphitizing CNTs, which means more thermal stability. The weight loss at the initiation temperature (100–300°C) represents the amount of contaminants and moisture in the sample, whereas the weight loss around 300–700°C represents the disintegration of CNT aromatic rings. The amount of contaminants, such as metal particles utilized as catalyst and oxide produced by the oxidation of these metal particles, is also indicated by the residual weight [72].

2.3.4 Scanning Electron Microscopy (SEM)

The structure and morphology of CNTs are studied using scanning electron microscopy (SEM). FEG-SEM (SEM with a field emission gun) is used to examine CNTs bundle and structural change after functionalization

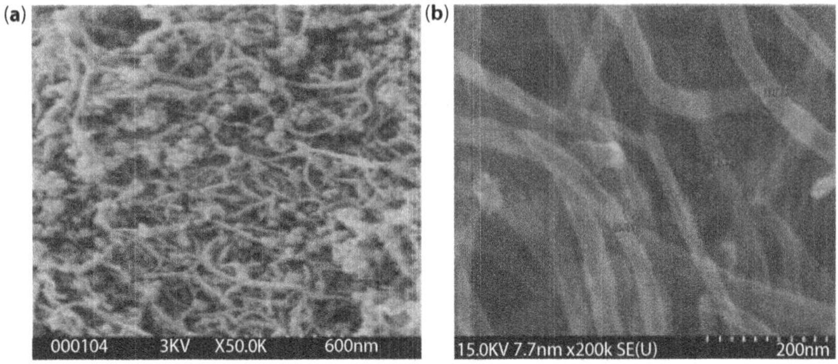

Figure 2.7 Comparison of (a) "as-prepared" MWCNTs (600 nm) and (b) oxidized MWCNTs (200 nm) by SEM. The quantities given in brackets indicate the lateral resolution applied. *Reprinted with permission from [Elsevier].*

treatments, graphitization, and purification, as well as alignment, dispersion, and homogeneity [75]. Specific features, such as tube diameter and length, CNT entanglement and curvature, and the presence of amorphous carbon or other non desired carbon forms, can be revealed by microscope examination with enhanced resolution. Stobinski *et al.* [74] have investigated MWCNTs (Figure 2.7 (a)) using SEM and compared them with oxidized MWCNTs (Figure 2.7 (a)) with a lateral resolution of 600 nm and 200 nm respectively. The surface morphology of the prepared MWCNTs and oxidized MWCNTs (Figure 2.7 (b)) reveals the bundled structure of nanorods.

2.3.5 Transmission Electron Microscopy (TEM)

Transmission electron microscopy (TEM) of CNTs enables higher magnification observation of structural properties in individual nanotubes. Metal catalysts-based nanoparticles trapped in nanotube cores, flaws in CNTs wall like opening tips, carbonaceous particles, and amorphous carbon, resulting from the non-vaporized graphitic rod can all be identified [76, 77]. Moreover, the removal of the amorphous carbon from nanotubes was also evident after the oxidation of MWCNTs (Figure 2.8). Other structural factors that can be seen with TEM include tube diameter and length, entanglement, number of walls, and curvature [78]. In addition, the bundled structure of CNT nanorods of diameters ranging from 10 to 40 nm (Figure 2.8(b)) has also been reported in previous literature [74]. Surface morphological investigation through SEM revealed no structural changes as a result of the heat treatment, whereas TEM images revealed

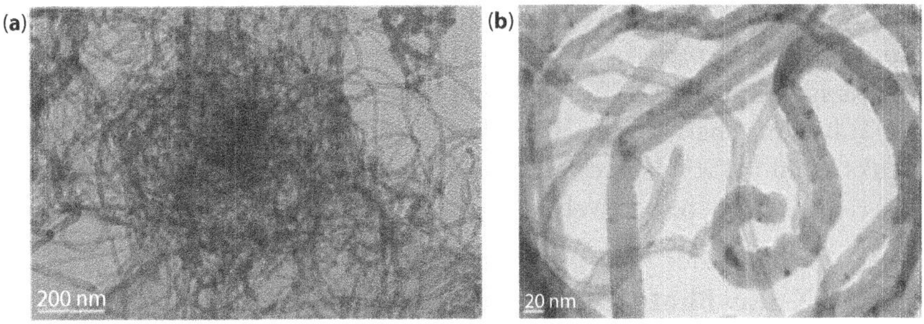

Figure 2.8 TEM image of oxidized, bundled MWCNTs. Scale bars are for (a) 200 nm and (b) 20 nm. *Reprinted with permission from [Elsevier].*

that the metal particles trapped in the nanotubes' cores had been removed [76]. The surface morphology of the prepared MWCNTs and oxidized MWCNTs reveals the bundled structure of nanorods of diameters ranging from 10 to 40 nm.

2.3.6 X-Ray Diffraction (XRD)

The crystalline structure of CNTs can be determined using X-ray diffraction (XRD) [78]. The XRD patterns of CNTs show two primary diffraction peaks: (002) hexagonal graphite structure reflection at about 2θ = 26 degrees and (100) graphite base plane diffraction roughly at 2θ = 43 degrees [79]. The XRD technique is also used to investigate the structure

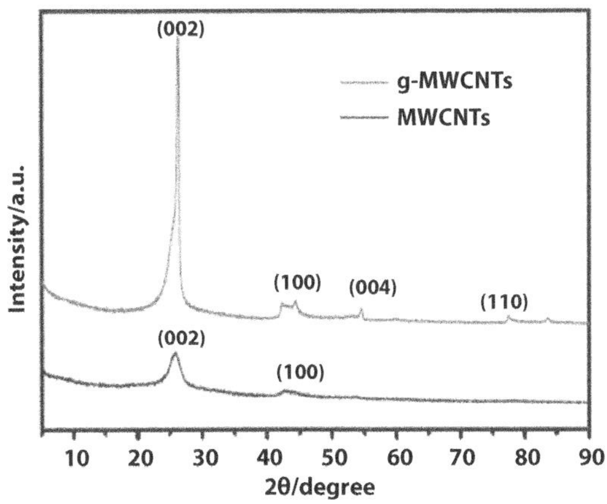

Figure 2.9 XRD pattern of graphitized multi-walled carbon nanotubes (g-MWCNTs) and pristine MWCNTs. *Reprinted with permission from [Elsevier].*

of CNTs changes following graphitization, functionalization, and purification. Using XRD patterns (Figure 2.9), Xue *et al.* [80] investigated the effect of graphitization on CNTs and found that after the treatment, the peak at (002) became sharp and narrow. Furthermore, the authors observed new peaks at 2θ = 77.5 degrees and 2θ = 54.3 degrees ascribed to (110) and (004) diffraction, respectively.

2.3.7 X-Ray Photoelectron Spectroscopy (XPS)

The components on CNT surfaces are analyzed using X-ray photoelectron spectroscopy (XPS), which provides information of the functional groups linked to CNTs wall to a 10 nm depth [78]. The O1s and C1s peaks in the XPS survey spectra of raw CNTs have binding energies of 533 and 284 eV, respectively. The peak in O1s can be due to flaws, moisture, or O_2 and CO_2 absorbed on the CNTs from the environment. Other peaks (such as N1s at 400 eV) can be seen in the CNT spectrum depending on the post-processing and fabrication method [81]. Six peaks (Figure 1.5b) can be used to deconvolute the C1s peak: C-C sp 3 (~285.2 eV), C-C sp 2 (~284.6 eV), C = O (~288.9 eV), O-C = O (~289.8 eV), C-O (~286.8 eV), and π-π* shakeup feature (~ 291.0 eV) [82]. By graphitization, functionalization, or Purification, procedures show modification on the crystalline structure of CNTs, resulting in a change in peak area (Figure 2.10) [78, 80].

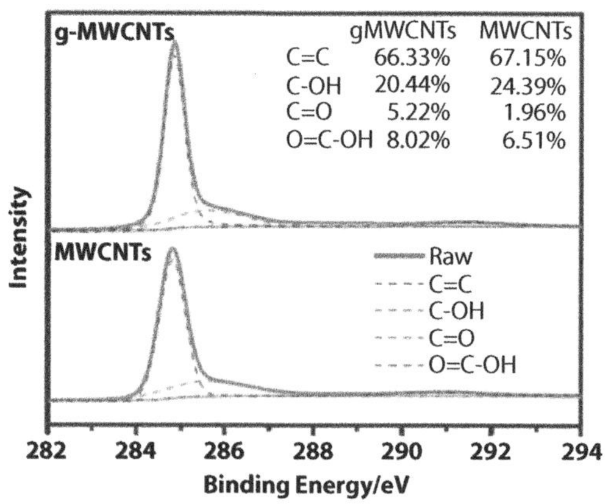

Figure 2.10 XPS spectra of C1s for graphitized multi-walled carbon nanotubes (g-MWCNTs) and pristine MWCNTs. *Reprinted with permission from [Elsevier].*

2.4 Functionalized Routes of CNTs

2.4.1 Surface Oxidation

Controlled oxidation is commonly used to introduce oxygen-containing groups like phenol, ketone, carboxyl group, lactone, acid anhydride, ether, and so on to CNT surfaces [83]. In polar solvents such as water, these oxygen-containing groups can improve the wettability or hydrophilicity of the CNT surface. These oxygen-containing groups, on the other hand, can act as anchoring sites for further functionalization via covalent, electrostatic, or hydrogen bonding. The strength of oxidants and the oxidation circumstances dictate the extent of oxidation. Hydrogen peroxide, Nitric acid permanganate, persulfate hypochlorite, chlorates, hypochlorite, dichromate, ozone, nitric oxide, and oxygen are all instances of oxidants.

Aqueous nitric acid, either concentrated or diluted, is the most commonly used oxidant for carbon materials [84]. This method is also useful for CNT purification because it removes metallic impurities and amorphous carbon [85]. Nitric acid is controllable and an efficient oxidant for the formation of surface groups by controlling acid concentration, oxidation, and temperature.

2.4.2 Doping Heteroatoms

During the one-step CNT synthesis, an effective method for doping heteroatoms like boron and nitrogen on CNTs is frequently achieved. In the case of N-doped nanotube synthesis, the carbon and nitrogen sources are a combination of ammonia [86] and hydrocarbon and or organic amine [87] and hydrocarbon. These reactants were slowly injected with argon into a heated quartz tube furnace holding quartz or ceramic boat carrying an iron-rich catalyst [88]. Organic borane is utilized instead of organic amine in the synthesis of B-doped CNTs, which uses the same CVD technique as the production of N-doped CNTs. Doping heteroatoms is an effective way of fine-tuning the structural and electrical features of nanotubes.

2.4.3 Alkali Activation

To modify the texture of carbonaceous materials, activation with alkali or CO_2 is utilized, which increases the specific surface area and pore volume [89]. This approach is also used to activate carbon nanotubes. The mass ratio of KOH to CNT was established at 4:1 before activating at 600°C, 700°C, or 800°C for 1 hour under nitrogen flow. Before filtering, the activated CNTs were washed with a weak HCl aqueous solution to eliminate alkali species, followed by drying at 100°C. The as-prepared activated CNTs have a higher specific surface area and pore volume, as well as improved hydrogen storage and migration properties [90]. Furthermore, the conductivity and electrochemical performance of alkali-activated CNTs as electrode materials for supercapacitors are shown to be superior to that of pristine CNTs [91].

2.4.4 Sulfonation

Environmentally acceptable solid–acid catalysts for esterification, rearrangement, dehydration and condensation processes have been produced using carbon compounds functionalized with sulfonic acid [92]. Sulfonation can also be accomplished by treating CNTs with ammonia sulfate. Before heating at 235 °C for 30 minutes, ammonia sulphate solution was stirred with a sufficient number of CNTs. SO_3 is formed when ammonia sulphate decomposes, forming a surface with $-SO_3H$ groups [93].

2.4.5 Halogenation

To increase the hydrophobicity of carbon surfaces, surface fluorination was explored. Several approaches were used to functionalize fluorine onto carbon surfaces. Fluorinated CNTs were directly produced for 12 hours at 350 °C in a fluorine gas environment utilizing an acceptable number of CNTs [94]. Fluorination made the exterior of the CNTs hydrophobic, allowing them to be disseminated in an epoxy matrix bromination of the CNTs is an effective approach for creating anchoring groups on the CNTs surfaces for subsequent replacement of a variety of nucleophiles such as anilines, alcohols, thiols, amines, and, among others. Bromine halogenation is accomplished by directly treating the CNTs with bromine at high temperatures over Lewis acid catalysts [95].

2.4.6 Grafting

The post grafting technique may be used to impart a wide range of surface functional groups onto CNT surfaces (Figure 2.11). The CNT surface already contains specific surface functional groups in several of these approaches, which may be further substituted by other organic groups. The pre-existing functionalities are generally produced by oxidizing CNTs with sulfuric acid/nitric acid during boiling. In some cases, CNT surfaces can also be directly coupled with functional groups.

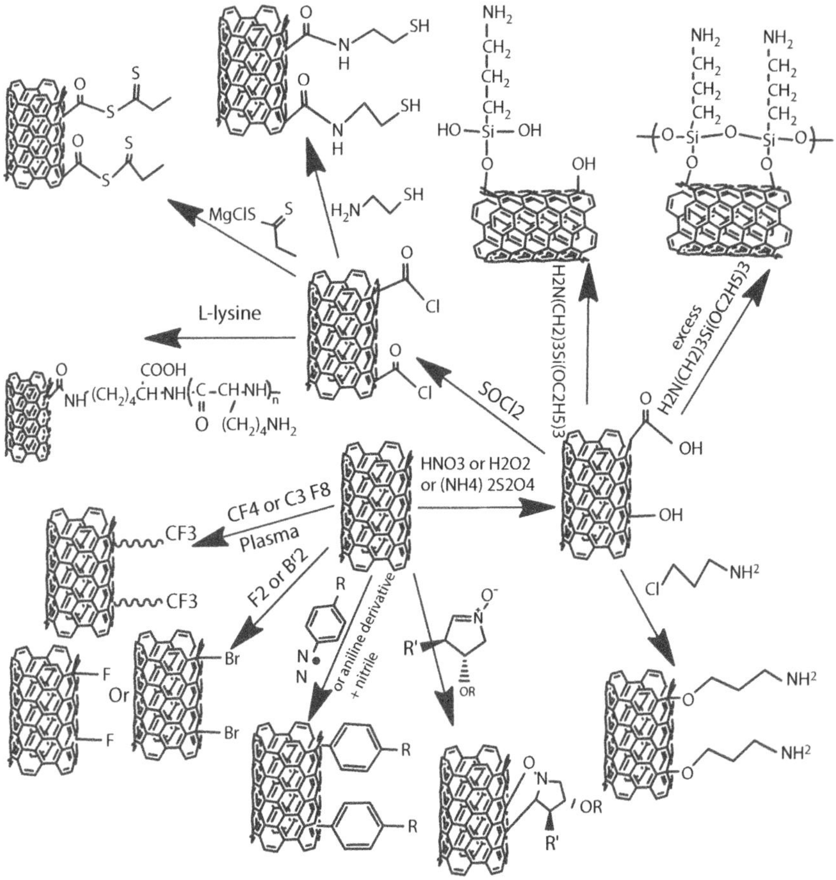

Figure 2.11 Grafting reactions to functionalized CNTs surface via covalent bonds.

2.4.6.1 *Grafting via Oxygen-Containing Groups*

On oxidant treated CNTs surface, carboxyl groups react with thionyl chloride to form acyl chloride groups, which work as an intermediate for various organic reactions. Cysteamine can react with acyl group by amidation reaction and graft thiol group. The presence of thiol group in CNT helps in attaching metal nanoparticles on the side of CNTs, which serve as catalysts and sensors [96]. The other use of acyl group can be grafting of chitosan on the surface of CNT to produce a new CNT/chitosan composite with application in the field of bone tissue engineering. Initially, 0.02 g of $K_2S_2O_8$ is mixed with 0.5 g of powdered chitosan for the duration of 20 h then lactic acid (0.055) is added gradually up to 2 h. Further, the reaction is stopped by using THF (10 mL) followed by ice drying (freeze-drying) thereafter methanol was used to rinse the dried sample. Afterward, the prepared sample is reacted with an acyl derivative of CNTs in the presence of CH_3COOH for 24 h at 75°C in vigorous stirring. Furthermore, the use of acetic acid eliminates the unreacted part of chitosan [97].

For methyl orange degradation reaction, L-lysine and acyl chloride react with each other through an amidation reaction to form poly-lysin embedded CNTs. It has been shown to have improved wettability when it comes to depositing and dispersing titanium dioxide nanoparticles. For the methyl orange degradation process, the resultant composite provided increased photocatalytic activity.

Metal-organic salts may react with acyl derivatives of CNT to create CNT polymers that can be used in reversible addition-fragmentation chain-transfer polymerization in addition to the amidation process. CNT chain transfer agents are formed when magnesium chloride di-thiopropanoate reacts with acylchloride groups on CNT surfaces [98]. A variety of groups may be grafted using the esterification depending on the carboxyl groups on CNT surfaces. For example, imidazolium functionalized carbon nanotubes have been produced, and the resulting nanohybrid composites with organometallic complexes show efficient and effective heterogeneous, which works as a hydrogen transfer catalyst for the reduction of cyclohexanone in the presence of 2-propanol as the hydrogen source [99]. The 3-aminopropyltriethoxysilane functionalized CNTs were also found to have promising CO_2 absorption capabilities and were chosen as CO_2 adsorbents for cyclic CO_2 capture study [100]. At 80°C, oxidized CNTs dispersed in water were combined with 3-aminopropyltriethoxysilane dissolved in acetone for 30 minutes under vigorous stirring. Filtration with distilled water and acetone separated the silane-functionalized CNTs before drying overnight [101].

2.4.6.2 *Grafting via Diazonium Compounds*

Diazonium compound has been used to graft aryl or aryl derivatives directly onto CNTs surface. It is a flexible grafting method that employs diazonium compounds. The interaction between diazonium salts and CNTs in a typical aqueous solution-based grafting procedure is a free radical chain reaction begun mostly by diazo-anhydride homolytic decomposition and partially by direct oxidation of CNTs by diazonium compounds. Once generated, aryl radicals can be covalently linked to carbon nanotubes (CNTs).

2.4.6.3 Other Grafting Methods

Although numerous grafting methodologies for CNT surface functionalization have been investigated, there are still certain grafting methods that are often used on carbon surfaces and can be used to functionalize CNT surfaces. A one-step procedure is used to directly functionalize carbon compounds. Under microwave irradiation, activated carbon was combined with amines, alcohols, or thiols for a solvent-free reaction, and the organic groups were directly connected to the double bonds. The microwave-assisted (MA) reactions were carried out in a controlled environment with temperatures below 200°C and duration of fewer than 100 minutes [102]. The cycloaddition process between cyclic nitrones and CNTs under sonication can covalently link cyclic nitrones to CNT surfaces, as demonstrated in Figure 2.10. By reacting with aniline precursors, nitrite, and acid, carbon black was directly functionalized with nitrophenyl, phenyl, and phenyl-azo-aniline [103].

2.4.7 Non-Covalent Functionalization of CNTs

Non-covalent functionalization approaches based on the 'p–p' interactions between CNTs and guest molecules have been used to preserve the sp^2 CNT structures and consequently their electrical properties. It is found that p–p stacking was used to functionalize the CNT sidewall with protein. Using the robust p–p interaction involving CNTs and the pyrenyl group, the CNTs were first designed and synthesized with succinimidyl ester and 1-pyrenebutanoic acid. Further, the amide bond is formed through the nucleophilic substitution of N-hydroxy succinimide on the protein [104].

2.4.8 Deposition on Functionalized CNTs

Functionalization may be described as the insertion of nanoparticles on carbon nanotubes. The addition of nanoparticles allows CNT applications to be expanded and introduces additional properties such as catalytic and electrochemical activity. Precursors attached to CNTs in the presence of surface functional groups for attaching metal ions or metal nanoparticles generate mixed metal oxides and metal oxides [105]. The most widely used functional groups on CNTs are carbonyl and carboxyl groups. This approach is also known as impregnation, and it is a simple and affordable technique. Nano-particles generated on flawless CNT surfaces by the impregnation method rely solely on 'particle–support physical contact', which can effortlessly result in leaking and degradation of the nano-particles due to the lack of surface functional groups that serves as an anchoring site. Taking MnO_2 as an example, the CNT was oxidized employing nitric acid and functionalized with oxygen-containing groups. Separately, 0.5 g of functionalized CNTs dispersed in 50 mL of toluene and a determined quantity of manganese (II) acetylacetonate dissolved in 50 mL of toluene were refluxed at 110°C for 5 h with a nitrogen flow before the two slurries were blended in a bigger flask and refluxed at 110°C overnight. The slurry was kept at room temperature for cooling, which is further filtered, washed, and desiccated before being annealed at 400°C for 6 hours in a nitrogen environment [106].

In the aerobic oxidation process of benzyl alcohol, these manganese oxides functionalized CNTs are used as catalytic substrates for palladium catalysts. Mixed metal oxides of cerium and copper were placed on carbon nanotubes pretreated with HNO_3 [107]. Impregnation was used to create highly distributed RuO_2 nanoparticles on CNTs.

2.4.9 Physiochemical Approaches

Nanoparticle-deposited CNTs may be made using a variety of physical processes. Sputtering deposition, evaporation deposition, electron, and ion-beam irradiation deposition are some of the physical processes used. The shape, size, dispersion, and homogeneity of the nanoparticles may be controlled using these strategies.

2.4.10 Electrochemical Deposition

Electrochemical deposition permits the formation of metal Nps attached on CNTs by the reduction of the metal complexes *i.e.,* platinic chloride acid, ammonium platinic chloride, and auric chloride acid, by electrons on cathodes [105]. CNTs function as conducting wires and work as a support for metal Nps deposition while exhibiting no reactivity with the metal precursors. Altering the parameters such as nucleation potential, the concentration of metal precursors, and process time indorses a difference in the homogeneity and size of metal Nps on the surfaces of CNTs. The electrochemical deposition has the benefit of producing exceptionally pure metal NPs functionalized on carbon nanotubes with robust metal support contacts in a very short process time [108]. The downside of the electrochemical deposition process, is the creation of large metal NPs on CNTs surface, typically ranging from 10 to 100 nm in size [109]. It has been reported that nickel nanoparticles can be generated on carbon nanotubes. Nickel particles were dispersed more uniformly on the surface of functionalized CNTs than on the surface of pristine CNTs after electrochemical deposition on both functionalized and pristine CNTs [110]. The shape of nickel oxide on carbon nanotubes is affected by variations in deposition time and electric current [111].

2.4.11 Electroless Deposition

Chemical reduction in the existence of reducing chemicals is required for electroless deposition of nanoparticles on CNTs. During the electroless deposition process, direct reduction occurs between the CNTs and metal precursors [105]. Nevertheless, the use of electroless deposition is limited by the fact that metal ions must have greater redox potentials than CNTs in order to be reduced to nanoparticles on CNTs support [112]. This limitation has been solved by using a 'substrate-enhanced electroless deposition' process that includes supporting CNTs with a metal substrate that has a lower redox potential than the targeted metal precursors [98]. A Pt/Mo/Ru/graphene–CNT catalyst was created by chemically reducing metal precursors at room temperature using sodium borohydride ($NaBH_4$) as the reducing agent [113].

2.5 Conclusion

This chapter gives an overview of current advances in the synthesis, characterization, and functionalization of carbon nanotubes. They are mostly employed in electronic gadgets, polymer nanocomposites, and medicinal materials for biomedical purposes as reinforcement. CNT uses are projected to grow as the understanding of synthesis methods improves, as well as future breakthroughs in surface modification. With changes in interfacial, sorption, electrical, and acid-base characteristics, heteroatom-doped CNTs can give more active sites in the meantime. Sulfonation transforms CNTs into solid acids, while diazonium functionalization allows aryl compounds to be covalently grafted onto CNT surfaces. CNTs and functionalized CNTs have been extensively used and studied in a variety of reactions, including CO/H_2 conversion to hydrocarbons or alcohols, hydrogenation, oxidation, dehydration, C–C coupling reactions, amino synthesis, and breakdown, photocatalytic reactions, CNT synthesis, and so on. With the promising inherent features of CNTs and functionalized CNTs, there are still many applications in catalysis to be investigated in depth.

In the future, we may see more designed functionalized CNTs supported catalysts with higher performance in eco-friendly processes, like plastic degradation for cellulose converted into ethanol, fuel, air pollution treatment, wastewater treatment, renewable energy conversion, and so on, to solve energy and environmental problems, which are still difficult to solve today.

References

1. Pinheiro, I.F., Ferreira, F.V., Souza, D.H.S., Gouveia, R.F., Lona, L.M.F., Morales, A.R., Mechanical, rheological and degradation properties of PBAT nanocomposites reinforced by functionalized cellulose nanocrystals. *Eur. Polym. J.*, 97, 356, 2017.
2. Sheikhpour, M., Golbabaie, A., Kasaeian, A., Carbon nanotubes: A review of novel strategies for cancer diagnosis and treatment. *Mater. Sci. Eng. C*, 76, 1289, 2017.
3. Bhakta, A.K., Kumari, S., Hussain, S., Martis, P., Mascarenhas, R.J., Delhalle, J., Mekhalif, Z., Synthesis and characterization of maghemite nanocrystals decorated multi-wall carbon nanotubes for methylene blue dye removal. *J. Mater. Sci.*, 54, 200, 2018.
4. Ugarte, D., Curling and closure of graphitic networks under electron-beam irradiation. *Nature*, 359, 707, 1992.
5. Iijima, S., Yudasaka, M., Yamada, R., Nano-aggregates of single-walled graphitic carbon nano-horns. *Chem. Phys. Lett.*, 309, 165, 1999.
6. Iijima, S., Helical microtubules of graphitic carbon. *Nature*, 354, 56, 1991.
7. Novoselov, K.S., Geim, A.K., Morozov, S.V., Electric field effect in atomically thin carbon films. *Science*, 306, 666, 2004.
8. Gupta, N., Gupta, S.M., Sharma, S.K., Carbon nanotubes: Synthesis, properties and engineering applications. *Carbon Lett.*, 29, 5, 419, 2019.
9. Dai, H., Carbon nanotubes: Synthesis, integration and properties. *Acc. Chem. Res.*, 35, 1035, 2003.
10. Iijima, S., Helical microtubules of graphitic carbon. *Nature*, 354, 56, 1991.
11. Bethune, D.S., Cobalt-catalyzed growth of carbon nanotubes with single-atomic-layer walls. *Nature*, 363, 605, 1993.
12. Iijima, S. and Ichihashi, T., Single-shell carbon nanotubes of 1-nm diameter. *Nature*, 363, 603, 1993.
13. Lan, Y., Wang, Y., Ren, Z. F., Physics and applications of aligned carbon nanotubes. *Adv. Phys.*, 60, 553, 2011.
14. Kumar, S., Rani, R., Dilbaghi, N., Carbon nanotubes: A novel material for multifaceted applications in human healthcare. *Chem. Soc. Rev.*, 46, 158, 2017.
15. Lee, Y.T., Tsai, P.J., Peterson, V.K., A microstructural and neutron-diffraction study on the interactions between microwave-irradiated multiwalled carbon nanotubes and hydrogen. *J. Mater. Sci.*, 51, 1308, 2016.

16. Wang, H., Zhou, H., Zhang, W., Yao, S., Urea-assisted synthesis of amorphous molybdenum sulfide on P-doped carbon nanotubes for enhanced hydrogen evolution. *J. Mater. Sci.*, 53, 8951, 2018.
17. Wallar, C., Luo, D., Poon, R., Zhitomirsky, I., Manganese dioxide–carbon nanotube composite electrodes with high active mass loading for electrochemical supercapacitors. *J. Mater. Sci.*, 52, 3687, 2017.
18. Morais, P.V., Gomez Jr., V.F., Silva, A.C.A., Dantas, N.O., Schoning, M.J., Siqueira Jr., J.R., Nanofilm of ZnO nanocrystals/carbon nanotubes as biocompatible layer for enzymatic biosensors in capacitive field-effect devices. *J. Mater. Sci.*, 52, 12314, 2017.
19. Adewunmi, A.A., Ismail, S., Sultan, A.S., Carbon nanotubes (CNTs) nanocomposite hydrogels developed for various applications: A critical review. *J. Inorg. Organomet. Polym. Mater.*, 26, 717, 2016.
20. Zhang, Y., Li, K., Ji, P., Silicon-multi-walled carbon nanotubes-carbon microspherical composite as high-performance anode for lithium-ion batteries. *J. Mater. Sci.*, 52, 3630, 2017.
21. Bhakta, A.K., Mascarenhas, R.J., D'Souza, O.J., Iron nanoparticles decorated multi-wall carbon nanotubes modified carbon paste electrode as an electrochemical sensor for the simultaneous determination of uric acid in the presence of ascorbic acid, dopamine and l-tyrosine. *Mater. Sci. Eng. C*, 57, 328, 2015.
22. Kouser, R., Vashist, A., Zafaryab, M., Biocompatible and mechanically robust nanocomposite hydrogels for potential applications in tissue engineering. *Mater. Sci. Eng. C*, 84, 168, 2018.
23. Naeimi, A., Saeidi, M., Baroumand, N., Carboxylated carbon nanotubes as an efficient and cost-effective adsorbent for sustainable removal of insecticide fenvalerate from contaminated solutions. *Int. Nano Lett.*, 6, 265, 2016.
24. Lata, S. and Vikas, Dispersibility of carbon nanotubes in organic solvents: Do we really have predictive models? *J. Nanopart. Res.*, 19, 211, 2017.
25. Huber, D.L., Synthesis, properties, and applications of iron nanoparticles. *Small*, 1, 482, 2005.
26. Sahebian, S., Zebarjad, S.M., Khaki, J.V., Lazzeri, A., The decoration of multi-walled carbon nanotubes with nickel oxide nanoparticles using chemical method. *Int. Nano Lett.*, 6, 183, 2016.
27. D.A. Scheinberg, Targeted self-assembly of functionalized carbon nanotubes on tumors. U.S. Patent. 9,976,137, 2018.
28. B.R. Smith and E. Ghosn, Carbon nanotubes for imaging and drug delivery. U.S. Patent Application 14/ 020,794, Leland Stanford Junior University, 2014.
29. N.V. Myung, S. Mubeen, A. Mulchandani, Metal and metal oxide co-functionalized single-walled See application file for complete search history. Carbon nanotubes for high performance gas sensors. US Patent, 2016.
30. D.J. Boday, S.R. Connor, J. Kuczynski, Propagation velocity tuning with functionalized carbon nanomaterial in printed wiring boards (pwbs) and other substrates, and design structures for same. US Patent 9,756,720, 2017.

31. B.M. Kobilka, J. Kuczynski, J.T. Porter, Functionalized carbon nanotubes. US Patent 10,233,082, 2019.
32. S.A.M. Said, S. Ahmed, N.U. Qadir, Metal organic framework adsorbent for solar adsorption refrigeration. US Patent 9,480,967, 2016.
33. C.M. Arnett, C.P. Marsh, J.H. Han, M.S. Strano, Enzyme-mediated assimilation of DNA-functionalized single-walled carbon nanotubes (swnts). US Patent, 2020.
34. Maser, W.K., Benito, A.M., Martınez, M.T., Production of carbon nanotubes: The light approach. *Carbon N. Y.*, 40, 1685, 2002.
35. Hutchison, J.L., Kiselev, N.A., Krinichnaya, E.P., Krestinin, A.V., Loutfy, R.O., Morawsky, A.P., Double-walled carbon nanotubes fabricated by a hydrogen arc discharge method. *Carbon N. Y.*, 39, 761, 2001.
36. Sano, N., Wang, H., Chhowalla, M., Alexandrou, I., Amaratunga, G.A.J., Synthesis of carbon "onions" in water. *Nature*, 414, 506, 2001.
37. Li, H., Guan, L., Shi, Z., Gu, Z., Direct synthesis of high purity single-walled carbon nanotube fibers by arc discharge. *J. Phys. Chem. B*, 108, 4573, 2004.
38. Imasaka, K., Kanatake, Y., Ohshiro, Y., Suehiro, J., Hara, M., Production of carbon nano-onions and nanotubes using an intermittent arc discharge in water. *Thin Solid Films*, 506-507, 250, 2006.
39. Sagara, T., Kurumi, S., Suzuki, K., Growth of linear Ni-filled carbon nanotubes by local arc discharge in liquid ethanol. *Appl. Surf. Sci.*, 292, 39–43, 2014.
40. Belgacem, A.B., Hinkov, I., Yahia, S.B., Brinza, O., Farhat, S., Arc discharge boron nitrogen doping of carbon nanotubes. *Mater. Today Commun.*, 8, 183, 2016.
41. Berkmans, J.A., Jagannatham, M., Reddy, R.D., Haridoss, P., Synthesis of thin bundled single walled carbon nanotubes and nanohorn hybrids by arc discharge technique in open air atmosphere. *Diam. Relat. Mater.*, 55, 12–15, 2015.
42. Su, Y. and Zhang, Y., Carbon nanomaterials synthesized by arc discharge hot plasma. *Carbon N. Y.*, 83, 90–99, 2015.
43. Guo, T., Nikolaev, P., Thess, A., Colbert, D.T., Smalley, R.E., Catalytic growth of single-walled manotubes by laser vaporization. *Chem. Phys. Lett.*, 243, 49–54, 1995.
44. Yan, Y., Miao, J., Yang, Z., Xiao, F.-X., Yang, H.B., Liu, B., Yang, Y., Carbon nanotube catalysts: Recent advances in synthesis, characterization and applications. *Chem. Soc. Rev.*, 44, 10, 3295, 2015.
45. Nagy, J.B., Bister, G., Fonseca, A., Mehn, D., Konya, Z., Kiricsi, I., On the growth mechanism of single-walled carbon nanotubes by catalytic carbon vapor deposition on supported metal catalysts. *J. Nanosci. Nanotechnol.*, 4, 326, 2004.
46. Kingston, C.T. and Simard, B., Fabrication of carbon nanotubes. *Anal. Lett.*, 36, 3119, 2003.

47. Bandaru, P.R., Electrical properties and applications of carbon nanotube structures. *J. Nanosci. Nanotechnol.*, 7, 1239, 2007.
48. Kumar, S., Rani, R., Dilbaghi, N., Tankeshwar, K., Kim, K.-H., Carbon nanotubes: A novel material for multifaceted applications in human healthcare. *Chem. Soc. Rev.*, 46, 158, 2017.
49. Dai, H., Hafner, J.H., Rinzler, A.G., Colbert, D.T., Smalley, R.E., Nanotubes as nanoprobes in scanning probe microscopy. *Nature*, 384, 147, 1996.
50. Shah, K.A. and Tali, B.A., Synthesis of carbon nanotubes by catalytic chemical vapour deposition: A review on carbon sources, catalysts and substrates. *Mater. Sci. Semicond. Process.*, 41, 67, 2016.
51. Walker, P.L., Carbon formation from carbon monoxidehydrogen mixtures over iron catalysis I. Properties of carbon formed. *J. Phys. Chem.*, 63, 133, 1959.
52. José-Yacamán, M., Miki-Yoshida, M., Rendón, L., Catalytic growth of carbon microtubules with fullerene structure. *Appl. Phys. Lett.*, 62, 657, 1993.
53. Popov, V.N., Carbon nanotubes: Properties and applications. *Mater. Sci. Eng.,: R: Reports*, 43, 61, 2004.
54. Ferreira, F.V., Cividanes, L.D.S., Brito, F.S., de Menezes, B.R.C., Franceschi, W., Simonetti, E.A.N., *Functionalizing Graphene and Carbon Nanotubes: A Review*, Springer International Publishing, Cham, 2016.
55. Jose-Yacaman, M., Miki-Yoshida, M., Rendon, L., Santiesteban, J.G., Catalytic growth of carbon microtubules with fullerene structure. *Appl. Phys. Lett.*, 62, 657, 1993.
56. Ren, Z.F., Synthesis of large arrays of well-aligned carbon nanotubes on glass. *Science*, 282, 1105, 1998.
57. Ren, Z.F., Huang, Z.P., Wang, D.Z., Wen, J.G., Xu, J.W., Wang, J.H., Growth of a single freestanding multiwall carbon nanotube on each nanonickel dot. *Appl. Phys. Lett.*, 75, 1086, 1999.
58. Hou, P.-X., Liu, C., Cheng, H.-M., Purification of carbon nanotubes. *Carbon N. Y.*, 46, 2003, 2008.
59. Collins, P.G., Engineering carbon nanotubes and nanotube circuits using electrical breakdown. *Science*, 292, 706, 2001.
60. Bandaru, P.R., Electrical properties and applications of carbon nanotube structures. *J. Nanosci. Nanotechnol.*, 7, 1239, 2007.
61. Zhang, R., Zhang, Y., Wei, F., Controlled synthesis of ultralong carbon nanotubes with perfect structures and extraordinary properties. *Acc. Chem. Res.*, 50, 179, 2017.
62. Szymanski, L., Kolacinski, Z., Wiak, S., Raniszewski, G., Pietrzak, L., Synthesis of carbon nanotubes in thermal plasma reactor at atmospheric pressure. *Nanomaterials*, 7, 45, 2017.
63. Anoshkin, I.V., Nefedova, I.I., Lioubtchenko, D.V., Nefedov, I.S., Raisanen, A.V., Single walled carbon nanotube quantification method employing the Raman signal intensity. *Carbon N. Y.*, 116, 547, 2017.

64. Ferreira, F.V., Cividanes, L.S., Brito, F.S., de Menezes, B.R.C., Franceschi, W., Simonetti, E.A.N., *Functionalizing of Graphene and Applications*, 1st ed., Springer, Cham, Zug, Switzerland, 2016.
65. Ferreira, F.V., Brito, F.S., Franceschi, W., Simonetti, E.A.N., Cividanes, L.S., Chipara, M., Functionalized graphene oxide as reinforcement in epoxy based nanocomposites. *Surf. Interfaces*, 10, 100, 2018.
66. Dresselhaus, M.S., Jorio, A., Souza Filho, A.G., Saito, R., Defect characterization in graphene and carbon nanotubes using Raman spectroscopy. *Philos. Trans. R. Soc. A Math. Phys. Eng. Sci.*, 368, 5355, 2010.
67. Varga, M., Izak, T., Vretenar, V., Kozak, H., Holovsky, J., Artemenko, A., Diamond/carbon nanotube composites: Raman, FTIR and XPS spectroscopic studies. *Carbon N. Y.*, 111, 54–61, 2017.
68. Ferreira, F.V., Franceschi, W., Menezes, B.R.C., Brito, F.S., Lozano, K., Coutinho, A.R., Dodecylamine functionalization of carbon nanotubes to improve dispersion, thermal and mechanical properties of polyethylene based nanocomposites. *Appl. Surf. Sci.*, 410, 267, 2017.
69. Zhao, Z., Yanga, Z., Hu, Y., Li, J., Fan, X., Multiple functionalization of multi-walled carbon nanotubes with carboxyl and amino groups. *Appl. Surf. Sci.*, 276, 476, 2013.
70. Vukovic, G., Marinkovic, A., Obradovic, M., Radmilovic, V., Colic, M., Aleksic, R., Uskokovic, P., Synthesis, characterization and cytotoxicity of surface amino-functionalized water dispersible multiwalled carbon nanotubes. *Appl. Surf. Sci.*, 55, 8067, 2009.
71. Chong, C.T., Tan, W.H., Lee, S.L., Chong, W.W.F., Lam, S.S., Valera-Medina, A., Morphology and growth of carbon nanotubes catalytically synthesised by premixed hydrocarbon-rich flames Catalyst-supported. *Mater. Chem.*, 197, 246, 2017.
72. Silambarasan, D., Surya, V.J., Iyakutti, K., Asokan, K., Vasu, V., Kawazoe, Y., Gamma (γ)-ray irradiated multi-walled carbon nanotubes (MWCNTs) for hydrogen storage. *Appl. Surf. Sci.*, 418, 49–55, 2016.
73. Safibonab, B., Reyhani, A., Golikand, A.N., Mortazavi, S.Z., Mirershadi, S., Ghoranneviss, M., Improving the surface properties of multi-walled carbon nanotubes after irradiation with gamma rays. *Appl. Surf. Sci.*, 258, 766, 2011.
74. Stobinski, L., Lesiak, B., Kövér, L., Tóth, J., Biniak, S., Trykowski, G., Judek, J., Multiwall carbon nanotubes purification and oxidation by nitric acid studied by the FTIR and electron spectroscopy methods. *J. Alloys Compd.*, 501, 77, 2010.
75. Herrera-Basurto, R., Lopez-Lorente, A., II, Valcarcel, M., Scanning electron microscopy of carbon nanotubes dispersed in ionic liquid: Solvent influence study. *Microchem. J.*, 122, 137, 2015.
76. Boncel, S. and Koziol, K.K.K., Enhanced graphitization of c-CVD grown multi-wall carbon nanotube arrays assisted by removal of encapsulated iron-based phases under thermal treatment in argon. *Appl. Surf. Sci.*, 301, 488, 2014.

77. Goornavar, V., Jeffers, R., Biradar, S., Ramesh, G.T., Utilization of highly purified single wall carbon nanotubes dispersed in polymer thin films for an improved performance of an electrochemical glucose sensor. *Mater. Sci. Eng. C*, 40, 299, 2014.
78. Chong, C.T., Tan, W.H., Lee, S.L., Chong, W.W.F., Lam, S.S., Medina, A.V., Morphology and growth of carbon nanotubes catalytically synthesised by premixed hydrocarbon-rich flames catalyst-supported. *Mater. Chem.*, 197, 246, 2017.
79. Silambarasan, D., Surya, V.J., Iyakutti, K., Asokan, K., Vasu, V., Kawazoe, Y., Gamma (γ)-ray irradiated multi-walled carbon nanotubes (MWCNTs) for hydrogen storage. *Appl. Surf. Sci.*, 418, 49-55, 2016.
80. Xue, Y., Zheng, S., Sun, Z., Zhang, Y., Jin, W., Alkaline electrochemical advanced oxidation process for chromium oxidation at graphitized multi-walled carbon nanotubes. *Chemosphere*, 183, 156, 2017.
81. Ferreira, F.V., Francisco, W., Menezes, B.R.C., Brito, F.S., Coutinho, A.S., Cividanes, L.S., Correlation of surface treatment, dispersion and mechanical properties of HDPE/CNT nanocomposites. *Appl. Surf. Sci.*, 389, 921, 2016.
82. Ferreira, F.V., Franceschi, W., Menezes, B.R.C., Brito, F.S., Lozano, K., Coutinho, A.R., Dodecylamine functionalization of carbon nanotubes to improve dispersion, thermal and mechanical properties of polyethylene based nanocomposites. *Appl. Surf. Sci.*, 410, 267, 2017.
83. Fanning, P.E. and Vannice, M.A., A DRIFTS study of the formation of surface groups on carbon by oxidation. *Carbon*, 31, 5, 721, 1993.
84. Klink, S., Ventosa, E., Xia, W., La Mantia, F., Muhler, M., Schuhmann, W., Tailoring of CNT surface oxygen groups by gas-phase oxidation and its implications for lithium ion batteries. *Electrochem. Commun.*, 15, 1, 10, 2012.
85. Guo, Z., Chen, Y., Li, L., Wang, X., Haller, G.L., Yang, Y., Carbon nanotube-supported Pt-based bimetallic catalysts prepared by a microwave-assisted polyol reduction method and their catalytic applications in the selective hydrogenation. *J. Catal.*, 276, 2, 314, 2010.
86. Chizari, K., Vena, A., Laurentius, L., Uttandaraman, S., The effect of temperature on the morphology and chemical surface properties of nitrogen-doped carbon nanotubes. *Carbon*, 68, 369, 2014.
87. Adjizian, J.J., Leghrib, R., Koos, A.A., Suarez-Martinez, I., Crossley, A., Wagner, P., Grobert, N., Llobet, E., Ewels, C.P., Boron-and nitrogen-doped multi-wall carbon nanotubes for gas detection. *Carbon*, 66, 662, 2014.
88. Dong, J., Qu, X., Wang, L., Zhao, C., Xu, J., Electrochemistry of nitrogen-doped Carbon Nanotubes (CNx) with different nitrogen content and its application in simultaneous determination of dihydroxybenzene isomers. *Electroanalysis (N.Y.N.Y.)*, 20, 1981, 2008.
89. Romero-Anaya, A.J., Ouzzine, M., Lillo-Ródenas, M.A., Linares-Solano, A., Pherical carbons: Synthesis, characterization and activation processes. *Carbon*, 68, 296, 2014.

90. Chen, C.-H. and Huang, C.-C., Enhancement of hydrogen spillover onto carbon nanotubes with defect feature. *Microporous Mesoporous Mater.*, 109, 1–3, 549, 2008.
91. Ko, J.M. and Kim, K.M., Electrochemical properties of MnO2/activated carbon nanotube composite as an electrode material for supercapacitor. *Mater. Chem. Phys.*, 114, 837, 2009.
92. Yu, K., Lu, G., Bo, Z., Mao, S., Chen, J., Carbon nanotube with chemically bonded graphene leaves for electronic and optoelectronic applications. *J. Phys. Chem. Lett.*, 2, 13, 1556, 2011.
93. Du, C.Y., Zhao, T.S., Liang, Z.X., Sulfonation of carbon-nanotube supported platinum catalysts for polymer electrolyte fuel cells. *J. Power Sources*, 176, 1, 9, 2008.
94. Tian, Y., Yue, H., Gong, Z., Yang, Y., Enhanced electrochemical performance of fluorinated carbon nanotube as cathode for Li–O2 primary batteries, *Electrochim. Acta*, 90, 186–193, 2013.
95. Hanelt, S., Friedrich, J.F., Orts-Gil, G., Meyer-Plath, A., Study of Lewis acid catalyzed chemical bromination and bromoalkylation of multi-walled carbon nanotubes. *Carbon*, 50, 1373, 2012.
96. Jeong, Y.T., Kim, J.S., Kim, J.T., Kim, S.H., Kumar, N.A., Functionalization of multi-walled carbon nanotubes with cysteamine for the construction of CNT/gold nanoparticle hybrid nanostructures. *Surf. Rev. Lett.*, 16, 487, 2009.
97. Carson, L., Kelly-Brown, C., Stewart, M., Oki, A., Regisford, G., Luo, Z., Bakhmutov, V., II, Synthesis and characterization of chitosan–carbon nanotube composites. *Mater. Lett.*, 63, 617, 2009.
98. Ellis, A.V., Waterland, M.R., Quinton, J., Water-soluble carbon nanotube chain-transfer agents (CNT-CTAs). *Chem. Lett.*, 36, 1172, 2007.
99. Blanco, M., Alvarez, P., Blanco, C., Jimenez, M.V., Fernandez-Tornos, J., Perez-Torrente, J.J., Oro, L.A., Menendez, R., Enhanced hydrogen-transfer catalytic activity of Iridium N-heterocyclic carbenes by covalent attachment on carbon nanotubes. *ACS Catal.*, 3, 1307, 2013.
100. Su, F.S., Lu, C.S., Chen, H.S., Adsorption, desorption, and thermodynamic studies of CO_2 with high-amine-loaded multiwalled carbon nanotubes. *Langmuir*, 27, 8090, 2011.
101. Lee, J.H. and Rhee, K.Y., Effects of moisture absorption and surface modification using 3-aminopropyltriethoxysilane on the tensile and fracture characteristics of MWCNT/epoxy nanocomposites. *Appl. Surf. Sci.*, 256, 7658, 2010.
102. Budarin, V.L., Clark, J.H., Tavener, S.J., Wilson, K., Chemical reactions of double bonds in activated carbon: Microwave and bromination methods. *Chem. Commun.*, 40, 2736, 2004.
103. Toupin, M. and Belanger, D., Thermal stability study of aryl modified carbon black by *in situ* generated diazonium salt. *J. Phys. Chem. C*, 111, 5394, 2007.

104. Chen, R.J., Zhang, Y.G., Wang, D.W., Dai, H.J., Noncovalent sidewall functionalization of single-walled carbon nanotubes for protein immobilization. *J. Am. Chem. Soc.*, 123, 3838, 2001.
105. Wu, B.H., Kuang, Y.J., Zhang, X.H., Chen, J.H., Noble metal nanoparticles/carbon nanotubes nanohybrids: Synthesis and applications. *Nano Today*, 6, 75, 2011.
106. Tan, H.T., Chen, Y.T., Zhou, C.M., Jia, X.L., Zhu, J.X., Chen, J., Rui, X.H., Yan, Q.Y., Yang, Y.H., Palladium nanoparticles supported on manganese oxide–CNT composites for solvent-free aerobic oxidation of alcohols: Tuning the properties of Pd active sites using MnOx. *Appl. Catal. B*, 119, 166, 2012.
107. Zhang, D.S., Mai, H.L., Huang, L., Shi, L.Y., Pyridine-thermal synthesis and high catalytic activity of CeO2/CuO/CNT nanocomposites, *Appl. Surf. Sci.*, 256, 6795, 2010.
108. He, Z., Chen, J., Liu, D., Zhou, H., Kuang, Y., Electrodeposition of Pt-Ru nanoparticles on carbon nanotubes and their electrolytic properties for methanol electrooxidation. *Diam. Relat. Mater.*, 13, 1764, 2004.
109. Cui, S.K. and Guo, D.J., Highly dispersed Pt nanoparticles immobilized on 1, 4-benzenediamine-modified multi-walled carbon nanotube for methanol oxidation. *J. Colloid Interface Sci.*, 333, 300, 2009.
110. Carpenter, C.R., Shipway, P.H., Zhu, Y., Electrodeposition of nickel-carbon nanotube nanocomposite coatings for enhanced wear resistance. *Wear*, 271, 2100, 2011.
111. Wen, B. and Zhang, S., Fang, H., Electrochemical growth of dispersing nickel oxide nanoparticles on carbon nanotubes. *Rare Metals*, 30, 661, 2011.
112. Qu, L. and Dai, L., Substrate-enhanced electroless deposition of metal nanoparticles on carbon nanotubes. *J. Am. Chem. Soc.*, 127, 10806, 2005.
113. Ye, F., Cao, X., Yu, L., Chen, S., Lin, W., Synthesis and catalytic performance of PtRuMo nanoparticles supported on graphene-carbon nanotubes nanocomposites for methanol electro-oxidation. *Int. J. Electrochem. Sci.*, 7, 1251, 2012.

3

Carbon Nanotubes: Types of Functionalization

Manilal Murmu[1,2], Debanjan Dey[1,2], Naresh Chandra Murmu[1,2] and Priyabrata Banerjee[1,2*]

[1]*Surface Engineering and Tribology Group, CSIR-Central Mechanical Engineering Research Institute, Mahatma Gandhi Avenue, Durgapur, India*
[2]*Academy of Scientific and Innovative Research (AcSIR), AcSIR Headquarters CSIR-HRDC Campus, Kamla Nehru Nagar, Ghaziabad, India*

Abstract

The carbon nanotubes (CNTs) have emerged as novel functional materials with the progress of materials science and nanotechnology. The CNTs are sp^2 hybridized rolled up graphene sheets having one-dimensional carbon possessing aspect ratio > 1000. The important class of CNTs *like* single walled carbon nanotubes (SWNTs), multi-walled carbon nanotubes (MWNTs) *etc.*, has been briefly outlined. This chapter covers the most significant as well as efficient chemical or physical functionalization approaches for obtaining desirable properties of mostly SWNTs and MWNTs based on covalent and non-covalent interactions. The desired properties such as chemical, physical, mechanical, electrical, transportation properties, *etc.*, of the CNTs can be achieved through its controlled functionalization. The mechanism of the functionalization of CNTs has been discussed briefly. The advantages and disadvantages of various chemical functionalization methods of CNTs have also been overviewed. The outlook of its future applications based on its functionalization has been predicted.

Keywords: Carbon nanotubes, chemical functionalization, covalent functionalization, non-covalent functionalization

**Corresponding author*: pr_banerjee@ameri.res.in; Website: http://www.priyabratabanerjee.in

Jeenat Aslam, Chaudhery Mustansar Hussain and Ruby Aslam (eds.) Functionalized Carbon Nanotubes for Biomedical Applications, (49–74) © 2023 Scrivener Publishing LLC

3.1 Introduction

The carbon nanotubes (CNTs) are hollow tubular structured carbon allotropes. The discovery of CNTs was first revealed from high resolution current arc discharge products through high resolution electron microscopy by Iijima in 1991 [1]. Later in 1993, nanotubes with single layer named as single walled carbon nanotubes (SWNTs) were revealed by the same arc process with the catalytic particles [2, 3]. The hollow concentric cylindrical carbon tubules were named as multi-walled carbon nanotubes (MWNTs). In 1991, CNTs have emerged as novel functional materials with the progress of materials science and nanotechnology.

The CNTs are synthesized using several methods such as (a) arc-discharge [4] (b) laser ablation [5], (c) chemical vapor deposition (CVD) [6], (d) catalytic gas-phase growth starting with carbon monoxide or other carbon sources (*e.g.*, HiPco methods) [7]. The first two methods involve the vaporization of carbon in an inert atmosphere, whereas CVD and HiPco are based on catalytic decomposition of hydrocarbons. The synthesis of CNTs has advanced to the point that patterns of aligned MWNTs or SWNTs can be engineered. The applications of CNTs ranges from high performance nano-electronics and re-enforced composite to hydrogen storage, field emission devises, sensors and probes [8], conductive coating, super capacitor electrodes [9], in medical sciences such as drug, gene and peptide delivery, tissue regeneration, imaging and diagnostics [10].

In this chapter, the most significant strategies used for the functionalizations of CNTs have been described briefly. Various schemes for explaining different modifications of these carbon nanostructures have also been elucidated. Since, the CNTs have diverse applications, understanding of the insight and mechanism of its functionalizations in order to impart various desired properties is very necessary so that these CNTs might find promising applications in various fields.

3.2 Carbon Nanotubes

The CNTs are hollow carbon structures with one or more concentric cylindrical walls. The CNTs are sp^2 hybridized rolled up graphene sheets having one dimensional carbon form that can be open ended or capped, having high aspect ratio (>1000) with diameter as small as 1 nm and a length of several micro-meters [11]. CNTs are designated as single walled, SWNTs double-walled carbon nanotubes DWNTs, a few-walled carbon nanotubes

(FWNTs) or multi-walled, MWNTs depending on the number of walls present in it. The structure and properties of CNTs depends upon atomic arrangement through which the graphite sheets were rolled cylindrically to form the CNTs as shown in Figure 3.1 [12]. Broadly, the CNTs are categorized as SWNTs, DWNTs, FWNTs and MWNTs based on the atomic arrangements, its diameter, tube length and nanostructure made up of several concentric shells. The structure of CNTs as a function of number of sidewalls, SWNTs as function of chirality *i.e.*, zigzag, chiral, and armchair are shown in Figure 3.1 (a), DWNTs in Figure 3.1 (b), and MWNTs in Figure 3.1(c).

The SWNTs comprised of a cylindrically rolled single graphene sheet having 1 nm diameter and lengths in centimeters [2, 3]. The arrays of

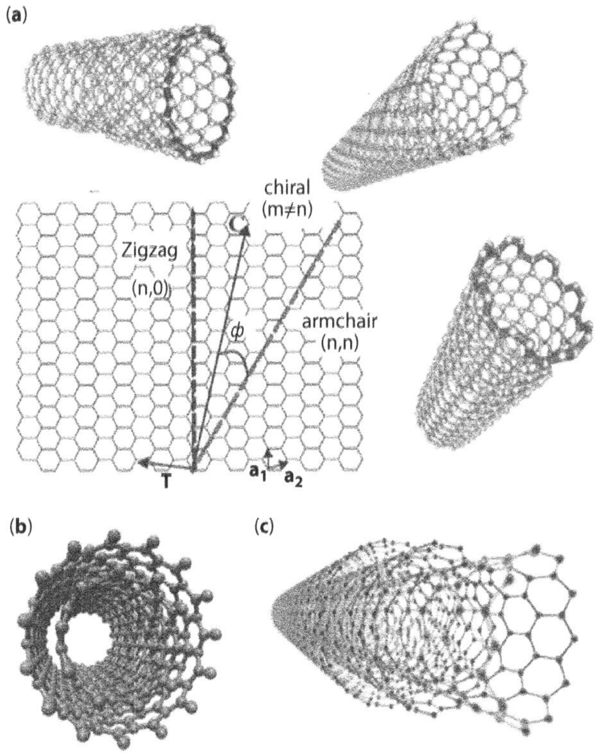

Figure 3.1 Structure of CNTs as a function of number of sidewalls (a) SWNTs as function of chirality (zigzag, chiral, and armchair), (b) DWNTs, and (c) MWNTs [Reproduced from Ref. [12]. Copyright © 2015 Tîlmaciu and Morris. This is an open-access article distributed under the terms of the Creative Commons Attribution License (CC BY)].

cylindrical tubes in MWNTs are concentrically arranged with 0.35 nm separation distance like the basal planes in graphite [13]. The cylindrical tube diameters in MWNTs are within 2 - 100 nm and tens of microns lengths [14]. The graphene sheets are rolled along (m, n) lattice vector to form SWNTs as shown in Figure 3.1 (a). Diameters as well as chirality of SWNTs are determined by the (m, n) indices. Based on the (m, n) indices, there are three types of SWNTs. When, the values of n = m, it forms an "armchair" type (= 0°) nanotube. The SWNT formed is of "zigzag" type (= 30°) when m = 0. On the other hand, a "chiral" nanotube is formed if n ≠ m, and lies within 0° and 30° as shown in Figure 3.1 (a). Chirality of CNTs is the function of (m, n) value. It affects mechanical, optical and electronic properties of the CNTs. It has been reported that the metallic nature is exhibited by all armchair and one-third of SWNTs, while, remaining CNTs with typical diameters behaves like semiconductors [15, 16].

3.3 Functionalization of Carbon Nanotubes

Curvature of graphene sheets facilitates variations of pyramidalization angle as well as it causes the misaligning of p-orbitals. These changes impart enhanced chemical reactivity of CNTs towards addition reactions. Furthermore, these changes make the sidewalls of CNTs more susceptible towards chemical reactivities compared to planar sheets of graphene.

3.3.1 Covalent Functionalization

The covalent functionalizations of CNTs are accomplished through covalent bond formation by the functional groups and carbon atoms of CNTs. This functionalization is made at the terminal boundaries of CNTs or exterior walls [17]. The CNTs are chemically modified through (a) sidewall and (b) defect functionalizations. The carbon atoms of the side walls or graphene layers of CNTs are sp^2 hybridized, as well as it possesses conjugated π-bonds. Hence, these sp^2 hybridized carbon atoms can be transformed to sp^3 system and conjugation of π-bonds can simultaneously be ceased. It can be accomplished by reacting CNTs with highly reactive chemical entities such as fluorine, *etc.*, as shown schematically in Figure 3.2 (a). Similarly, the CNTs can also be modified chemically by forming defect(s) on the graphene layer as shown schematically in Figure 3.2(a). Both the sidewall and defect functionalization give rise to form covalent bonds between the active centers of reactive species with the sp^2 hybridised carbon atoms of CNTs. Upon, initial functionalization of these CNTs, further

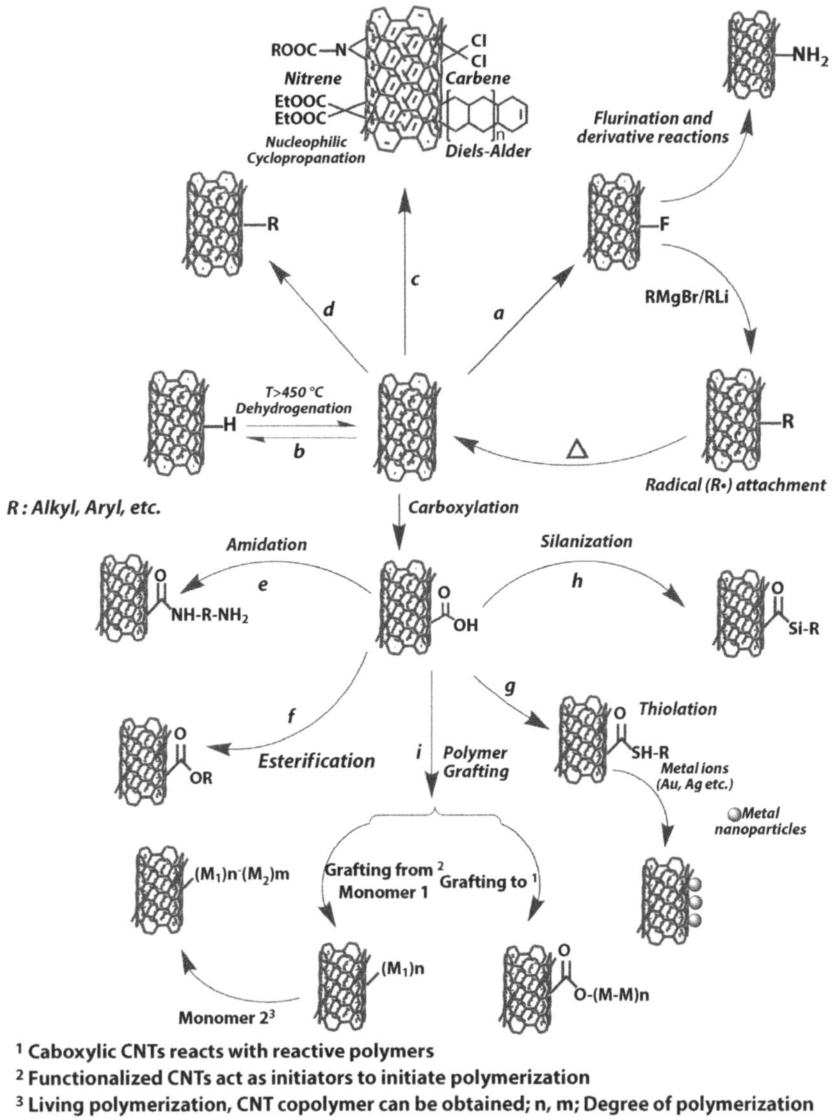

Figure 3.2 Functionalization of CNTs using different approaches.

derivatives of the CNTs can be obtained with subsequent chemical modifications. The side wall of CNTs was modified using highly reactive fluorine at 325 °C (fluorination) as shown in Figure 3.2(a). It was also reported that this fluorination process is reversible in presence of anhydrous hydrazine [18]. It was reported that the carbon fluorine *i.e.*, C-F bond formed in CNTs are weaker than that formed in the alkyl fluorides. Thus, the fluorine atoms of the C-F bonds in CNTs can be easily be substituted and furthermore additional derivatives of CNTs can be achieved [19, 20]. For instances, the alkylation of the fluorinated CNTs can be achieved by using Grignard reagent (RMgX) or alkyl lithium (RLi) as shown in Figure 3.2 (a), where R represents the alkyl or aryl group [21, 22]. Recently, several efficient approaches are executed for the functionalization of CNTs, for instances, Diels-Alder reaction, carbine, and nitrene addition like hydrogenation as shown in Figure 3.2 (b) [23], cycloaddition reactions as shown in Figure 3.2 (c) [24–26]. The alkylation or arylation of the CNTs may also be performed using radical attachment as shown in Figure 3.2 (d) [17]. Furthermore, bromination, chlorination [27], azomethine ylides [28], and so on are also adopted for functionalization of the side wall carbon atoms of the CNTs.

Another method of CNTs modification through covalent bond formation approach is the defect functionalization [17]. The defect sites such as holes in the side walls, open endings, pentagonal or heptagonal anomalies may be present in hexagonal graphene structure in the CNTs. Furthermore, the oxygenated carbon centers are also considered as defects in the CNTs. The defects on the side walls or the open ends of CNTs are chemically formed through oxidative pathway using H_2SO_4, HNO_3 and both used in combination [29]. Likewise, strong oxidants like $KMnO_4$ [30], ozone [31, 32] and reactive plasma [33, 34] are also used for defects formation on CNTs through oxidative process. The bonding of carboxylic acid or hydroxyl unit with defects formed on the CNTs stabilized it chemically (*vide* Figure 3.2). Thus, these CNTs find their applications being starting materials for additional functionalization, for instances, amidation that is shown in Figure 3.2(e) [17], esterification as shown in Figure 3.2(f) [35], thiolation and/or subsequent incorporation of metal nanoparticles as shown in Figure 3.2 (g) [36], silanization as shown in Figure 3.2(h) [32], polymer grafting as shown in Figure 3.2(i) [37]. Furthermore, the alkylation or arylation [38], and biomolecule derivatives [39] can also be prepared from this functionalization technique. In such chemical modification the polar groups are being attached in the walls of CNTs. Thus, these sorts of chemical modification transform the hydrophobic CNTs towards hydrophilic one.

Additionally, the strong bonds between the CNTs and the functional moieties in the polymers are formed at the interface which gives CNTs nanocomposites. These CNT based nanocomposites may exhibit extraordinary mechanical as well as chemical stability.

Furthermore, Diels-Alder reactions accompanied by redox polymerization has also been reported to be very efficient chemical modification technique for CNTs occurring at mild reaction environments, low temperature, stable in water and hydroxyl surface initiator. Herein, hydroxyl groups are incorporated CNTs sidewalls initially reacting it with furfuryl alcohol through Diels-Alder reaction. Subsequently, the furfuryl alcohol modified CNTs (CNTs-OH) was further modified by grafting the co-polymer which was started by Ce(IV)/HNO_3 redox system utilizing biocompatible hydrophilic poly(ethyleneglycol) methyl ether methacrylate (PEGMA) and carboxyl-rich acrylic acid (AA) as monomers as shown in Figure 3.3. The advantage of the synthetic procedure is that the reaction was accomplished at ambient condition in absence of organic solvent without involving any expensive and toxic substances [40].

Furthermore, literature also reveals that the CNTs are also functionalized using ionic liquids in which the ionic liquids exhibit strong π-cation interactions with the CNTs and thereby synergistically enhances its various properties [41]. The ionic liquids are fluids at room temperature, this property make it capable to disperse CNTs better than the traditional organic solvents. The ionic liquids undergo feeble van der Waals interactions rather than 'cation-π' interactions with CNTs. These interactions shielding mechanism also facilitates the dispersion of CNTs in ionic liquids [42]. Recently, Bains *et al.*, revealed the functionalization of MWNTs, *i.e.*, oxidized derivatives of CNTs and graphene using a series of alkylated benzimidazolium based ionic liquids in order to get super-hydrophobic and self-cleaning coatings for robust antibacterial applications in bio-medical

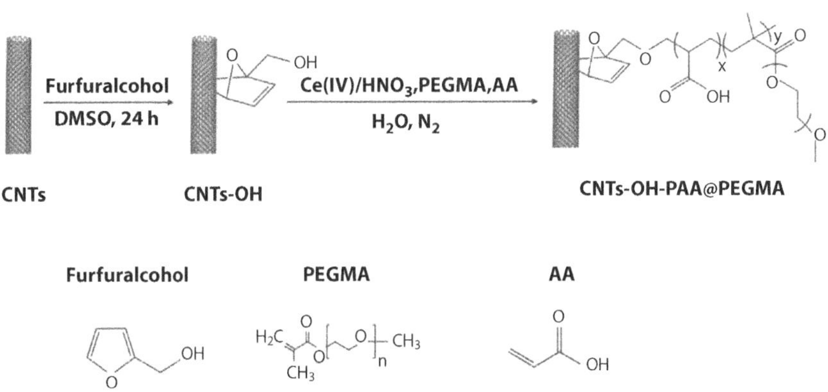

Figure 3.3 Functionalization of CNTs surface using furfur alcohol, carboxyl group and PEGMA [Adapted with permission from Ref. [40]. Copyright © 2010 Elsevier].

devices used in hospitals. These ionic liquids are designed such that the N1 position of the benzimidazole contain different chain lengths and the N3 position is left free to additional functionalization with other moieties [43]. Furthermore, poly(ionic liquids) are also used for the functionalization of CNTs in order to impart charged functional group onto CNTs surfaces for enhancing the adsorption capability of the CNTs towards environmental pollutants. This novel functionalization was performed utilizing the combined effect of mussel-inspired chemistry and subsequent surface-initiated reversible addition-fragmentation chain transfer polymerization reactions using 3-n-hexadecyl-1-vinylimidazolium bromide *i.e.*,$[C_{16}VIm^+][Br^-]$ ionic liquid as monomer [44].

Functionalization of CNTs with polymer has also gained importance with the aim to impart diverse properties such as response towards environmental stimuli, complex ability with metal ions, enhanced dispersibility in various solvents, high compatibility with polymer matrix, and so on. The polymer chains are chemically bonded to edge or surface of CNTs and give rise to highly stable CNT/polymer composite. For this purpose, the 'grafting to', 'grafting from' and 'grafting through' like chemical methods as well as the traditional free radical polymerization or controlled polymerization methods are mostly exploited [45].

The click chemistry has recently gained importance for the CNTs functionalization too [46]. He *et al.*, discussed the preparation of clicked nanohybrids from polymer coated CNTs and Fe_3O_4 nanoparticles [47]. The reagents used were (i) H_2SO_4/HNO_3, 90–133 °C, 100 minutes; (ii) $SOCl_2$, 60 °C, one day, (iii) propargyl alcohol, triethyl amine, chloroform, room temperature, one day for obtaining MWNT-C ≡ CH, which was subsequently modified by click grafting to get a polymer containing appreciable number of azide groups (MWNT-pAz), and subsequently, polymer alkyne group was incorporated (MWNT-pAlk). Fe_3O_4–CRCH and Fe_3O_4–N_3 undergo 'click' chemistry reaction on nanotubes coated with polymer and yields MWNT-pAz@Fe_3O_4 and MWNT-pAlk@Fe_3O_4 as respective magnetic nanohybrids. Similarly, Zhao *et al.*, also utilized click chemistry to synthesize ionic liquid modified MWNTs and then loaded platinum nanoparticle into it. The resulted composite film glassy carbon electron exhibited remarkable analytical performance for detecting tatrazine analytically [48]. New hybrid material of polyhedral oligomeric silsesquioxane and CNTs was prepared by Cho *et al., by* using click coupling between azide moiety-functionalized polyhedral oligomeric silsesquioxane and alkyne-functionalized MWNTs [49]. Furthermore, this approach was applied for preparing TiO_2-based nanohybrids and photo-catalytically active MWNTs [49].

Furthermore, incorporation of heteroatom into the hexagonal carbon lattice of CNTs is termed as heteroatom doping into the walls of CNTs, which is another way of CNTs functionalization. Doping of heteroatoms (*like* B, N, O, P, S *etc.*) are generally executed to impart high electronic mobility, good sodium solubility, enhances sodium storage capacity, mass transfer of oxygen, removing water, corrosion inhibition and so on [50, 51]. Hong *et al.*, reported the B and N doping by judiciously using elevated temperature and plasma treatment to obtain macroscopically dense CNTs fibers (CNTFs). It showed synergistic effect of macroscopic densification in combination with heteroatom doping as shown in Figure 3.4 (a) [52]. Hou *et al.* revealed free standing N-doped CNTs through straightforward pyrolysis of toluene and exhibited good catalytic activity for oxygen reduction reaction with superior selectivity, methanol crossover resistant and stability. Exposed C and N at surface of CNTs provide plenty of active sites [53]. There are various configurations of N doped CNTs, for instances, pyridinic N, pyrrolic N, and graphitic N. The pyridinic N and pyrrolic N are bonded to two carbon atoms

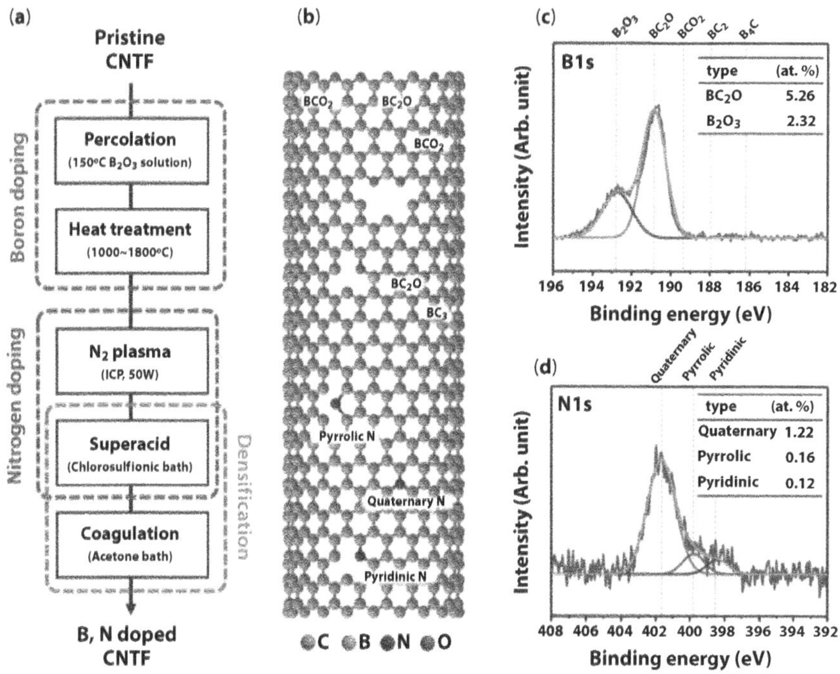

Figure 3.4 Schematic diagram of (a) boron (B) and nitrogen (N) doping on CNTFs, (b) different types of B (*e.g.*, BCO_2, BC_2O and BC_3) and N (pyrrolic N, pyridinic N and quarternary N) environment created upon doping, X-ray photoelectron spectra of (c) B 1s and (d) N 1s of boron and nitrogen co-doped CNTs. [Reproduced with permission from Ref. [52] Copyright © 2009 Elsevier].

and situated at the edges, while the quarternary or graphitic N is placed at the core of CNTs substituting the sp^2 hybridized carbon atoms as shown in Figure 3.4 (b). Figure 3.4 (c, d) showed B 1s and N 1s X-ray photoelectron spectra of B and N co-doped CNTs. These different configurations having different environments of N which changes the electronic environments of adjacent C atoms give rise to different catalytic activities [54, 55]. Recently, Iizumi *et al.*, reported simple and efficient fabrication methods of O doped CNTs employing Ultra-violet irradiation on the SWNTs thin film with a conventional Ultra-violet ozone cleaner led, which are useful for application in photonic materials owing to oxygen-induced deep trap states [56].

Furthermore, N and P co-doped CNTs are also fabricated through pyrolysis process with combination of rapidly evaporating aerosol droplets [57]. Recently, Hassani and Tavakol revealed the synthesis of S doped CNTs having high S content using chemical vapor deposition technique from sulphur powder, acetylene gas and $Fe/CaCO_3$ as catalysts [58]. Similarly, Kim *et al.*, reported S doping on CNTs (approximately 0.65 % S content) employing chemical vapor deposition technique using dimethyl sulphide as S source. The S doped CNTs exhibited excellent electrochemical performance due to enhanced electrical conductivity and pseudocapacitive property [59]. The co-doping of S and N into the structure of CNTs leads to enhance the electrochemical oxygen reduction activity [60].

3.3.2 Non-Covalent Functionalization of Carbon Nanotubes

Non-covalent functionalization deals with interactions occurring between CNTs and molecules. There are numerous approaches reported so far for non-covalent functionalization of CNTs. The host-guest interaction is one of the most significant approaches, wherein, CNTs act as host. The extended π-system of the sidewall of the carbon nanotubes are highly capable to bind with the guest molecules undergoing π-π interactions [61]. Other approaches take advantage of exploitation only through van der Waals interaction between adsorbents and nanotubes. Its advantages depend on the preservation of the properties of the CNT, while its disadvantage is the weak strength between the wrapping/pairing molecules which can reduce the load transfer in the composite [62]. CNTs also have the propensity to intertwine leading to form 3D structure by means of van der Waals interaction. After the CNTs are exfoliated, it facilitates stable dispersion formation in the organic solvents like, phenyl ethyl alcohol or N-methylpyrolidone (NMP) which are capable of efficiently interacting with the CNTs [63, 64]. During sonication, NMP was observed to penetrate the bundles and remain firmly attached with the nanotubes, leading to the enthalpy of the mixture being almost zero, just like mixing a negative value of free energy. Similarly, a relatively stable and a good dispersion of CNTs can be obtained using sonication in N, N-dimethylformamide (DMF). This clearly highlights a negligible value of hydrogen bond donation parameter. A high value of hydrogen bond acceptance, *i.e.*, basicity, and very high value for π^* (solvatochromic structure) is compulsory. However, to get a good dispersion of CNTs, this condition is not sufficient. The advantage of using non-covalent functionalization is that it doesn't hamper existing conjugation of π-electrons of the CNT sidewalls; hence, it will not alter the structural property of the material. The interfacial properties of the nanotubes can be tuned by this alternative method, non-covalent functionalization. Various substances *like,* aryl compounds, polymers, and surfactant can be used for this purpose through π-π stacking interactions and hydrophobic interactions. After this functionalization CNTs are capable of performing previous functions while the solubility factor of CNTs increases dramatically. Aromatic moieties *like,* porphyrin, pyrene and the derivatives can easily interact with π electrons of CNTs sidewalls through π-π stacking, henceforth; it can be useful for non-covalent functionalization.

Dai and co-workers have shown an attractive but general approach for CNTs functionalization using non-covalent pathway using a bifunctional molecules N-succinimidyl -1-pyrenebutanoate followed by subsequent control over immobilizing bio-molecules protein within it. In this process, specificity and high degree of control can be monitored during immobilization The succinimidyl ester get adsorbed onto sidewalls of CNTs surface through a string π-stacking interactions of highly aromatic pyrene moiety with it. The adsorbed succinimidyl ester which is highly reactive in nature, thereby, act as anchor which interacts with the primary or secondary amine groups of the proteins and facilitates the immobilization process as shown in Figure 3.5 (a) [65, 66]. Hecht *et al.*, fabricated a field effect transistor (CNTs/FET)in which the CNTs were functionalized non-covalently with the aid of zinc coated porphyrin like photoactive compound and used for detection of photo induced electron transfer as shown in Figure 3.5 (b). Such type of photosensitive device finds promising applications in photovoltaic devices [66, 67]. In 2008, Hu and co-workers revealed a uniform self-aggregation of a monolayer of highly aromatic pyrene functionalized CdSe nanoparticles over the surface of the CNTs as shown in Figure 3.5 (c). These pyrene CdSe nanoparticles system exhibit effective charge transfer towards CNTs which is the essential criteria of solar cell or heterojunction photovoltaic devices [66, 67]. Furthermore, the non-covalent functionalization of CNTs using [2] catenane tetra cations give rise to bistable monolayer. These [2] catenane

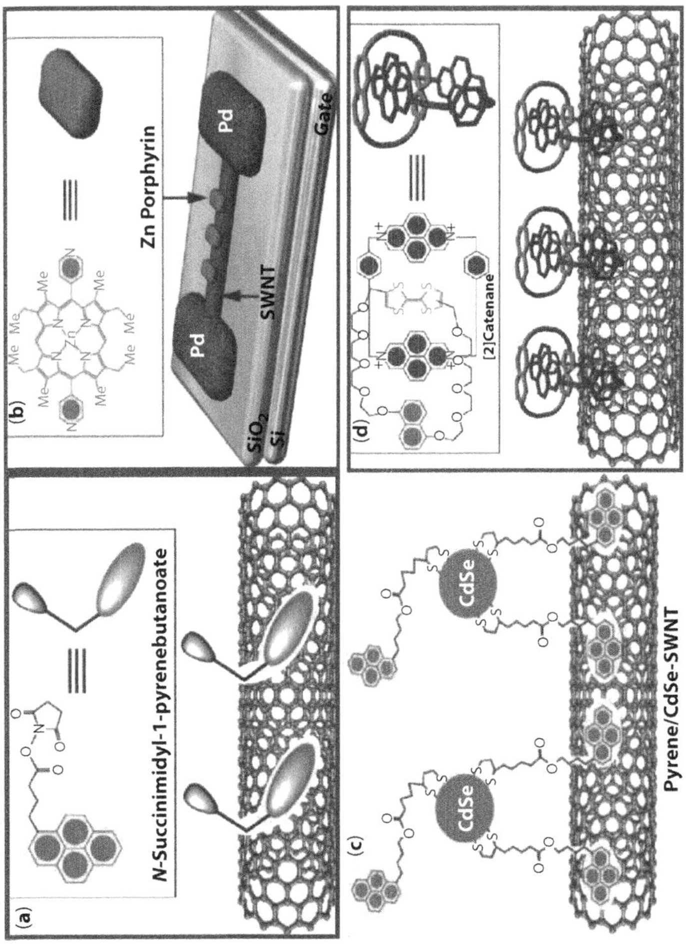

Figure 3.5 Non covalent functionalization of CNTs, (a) N-succinimidyl-1-pyrenebutanoate-decorated SWNT, (b) zinc porphyrin-coated SWNT/FET device, (c) SWNT with pyrene/CdSe nanoparticles and a SWNT FET device, (d) catenanes lined up in rows onto SWNTs surface [Adapted with permission from Ref. [66]. Copyright © 2009, American Chemical Society].

tetra cations are stabilized through self-organization by amphiphilic dimyristoyl phosphatidyl anions as shown schematically in Figure 3.5 (d) [66, 68]. Similarly, there are several organic molecules which have been used to functionalize CNTs *viz.* 1-pyrenebutanoic acid; succinimidyl ester; 1-(2-anthroquinonylaminomethyl) pyrene; [bis(2-anthaquinonyl)-aminomethyl] pyrene; 5,7,12,14-tetramethyldibenzo-1,4,8,11-tetraazacyclotetradeca-3,5,7,10,12,14-hexane; adamentane-pyrene; biotin-pyrene; nitrolotriacetic acid-pyrene; ferrecene; anthacene; anthrarobin; 9,10-dibromoanthracene; 9,10-anthracene-dicarbonitrile; 9-anthracene-methanol; thionine; methyl blue; naphthalene-1-methylphosphonic acid; 1,10-phenanthroline-5,6-dione; triphenylene; porphyrin; hydroxyferriprotoporphyrin and so on [66, 69, 70]. Conjugated polymers are proved to be excellent materials for wrapping in non-covalent functionalization of CNTs. This particularly occurs due to van der Waals as well as π-π stacking interactions occurring between surface of conjugated polymers containing aryl moieties and CNTs. Furthermore, several conjugated poly(m-phenylene)-co-(2,5-ditoxy-p-phenylene) vinylene, poly(2,6-pyridinlenevinylene)-co-(2,5-dioctoxy-p-phenylene) vinylene, stilbene *like,* dendrimers, and poly-(5-alkoxy-m-phenylenevinylene)-co-(2,5-dioctoxy-p-phenylene) vinylene, to explore CNTs functionalization using non-covalent pathway [70–74].

Surfactants material mainly polymers have also been used for functionalization of CNTs. The surfactant molecule adsorbs physically over CNTs leading to decrease surface tension of the CNTs in turn efficiently prohibit their self-agglomeration. Additionally, electrostatic/steric repulsive force helps CNTs to overcome the van der Waals attractive force. There is a dependency of this method over the strength of the surfactant, chemistry of the medium and polymer matrix.

There is a relation between CNTs and surfactants which has been studied well by several scientists for examples (i) non-ionic surfactants, *like,* polyoxoethylene 8 lauryl [75]; polyoxyethylene octylphenylether (Triton X-100) [76]; (ii) cationic surfactants, *like,* dodecyl tri-methyl ammonium bromide (DTAB); [77] cetyltrimethyl ammounium 4-vinylbenzoate; [78] and (iii) anionic surfactants, *like,* sodium dodecyl benzene sulfonate (NaDDBS), sodium dodecyl sulfate (SDS), poly(styrene sulfate) (PSS). The comparative adsorption of surfactants such as NaDDBS, SDS and Triton X-100 leading to stabilization of CNTs has been shown schematically in Figure 3.6. Herein, the parallel adsorption of surfactants onto CNTs influence the stabilization. The alkyl chain of the surfactants is adsorbed in a parallel fashion along the length of the CNTs rather than its diameter. The presence of benzene rings in NaDDBS and Triton X-100 facilitates

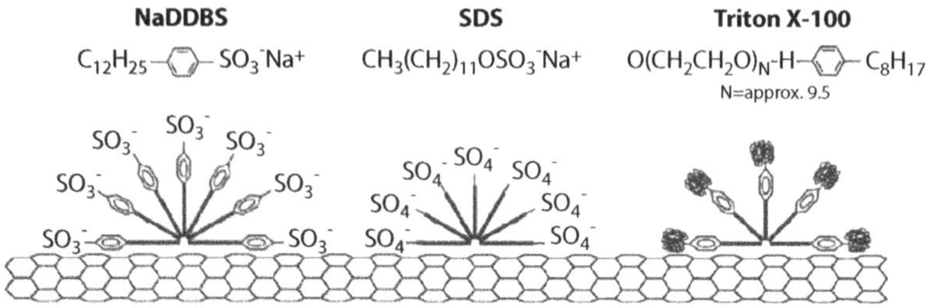

Figure 3.6 The schematic representation of how surfactants may adsorb onto the CNTs surface [Reproduced with permission from Ref. [79]. Copyright © 2009, American Chemical Society].

better adsorption which ultimately causes the better dispersion of CNTs. Furthermore, the dispersion capability of NaDDBS is greater than that of Triton X-100. This is attributed to the presence of polar and larger head group (*i.e.*, polyethylene oxide chain) in Triton X-100 which doesn't stabilize rather lowers the packing density at the CNTs compared to the smaller head group (*i.e.*, sulphur tri oxide anion) of NaDDBS (*vide* Figure 3.6). Additionally, the electrostatic interaction of sulphur tri oxide anions stabilizes the CNTs through screening Coulomb interactions [79, 80]. However, this surfactant functionalized CNTs are less useful for bio-medical applications. This is due to the toxicity fact of the surfactant, although it can efficiently solubilize the CNTs in medium. And, these surfactants are also known as permeable plasma membrane [81].

For integration of new materials, *i.e.*, biological components in living systems it is necessary to solubilize the CNTs with the component. To achieve the goal several biomacromolecules *like*, saccharides and polysaccharides, proteins, DNA, engymes, *etc.* [82–86], were used for functionalizing CNTs in a non-covalent pathway. Furthermore, various bio materials *like*, *n*-decyl-β-D maltoside, [87] η-cyclodextrin, [88] γ-cyclodextrin, pullulan, [89] chitosan, and phospholipid-dextran [90] were also exploited for non-covalent functionalization of CNTs. The saccharides and polymers of saccharides do not possess light absorption property in the UV-visible region, therefore, characterization of CNTs in presence of these macromolecules doesn't hamper its result, especially in photochemical experiments. Moreover, saccharides and polymers of saccharides, *i.e.*, polysaccharides coated CNTs are bio-compatible in nature therefore, may be applied for several bio relevant applications and medicinal purposes.

Therefore, the non-covalent functionalization through π-π stacking or van der Waals interaction. All that functionalization is exohedral derivatization *i.e.*, only outer surface will be functionalized. Considering the advantage of hollow spherical structural of CNTs, scientists have also shown great interest for endohedral functionalization. In this phenomena, the tubes are filled with atoms or small organic molecules [91–93].

Recently, it has also been seen that there are several discoveries or invention with the aim of non-covalent functionalization of CNTs. Some of the notable ones are the new non-wrapping π-stacking approach for polymer functionalization of CNTs using poly (arylene ethynylene) polymer which comprises of one functionalized moiety to the nanotube in a non-wrapping manner in organic and inorganic solvents. The aromatic or π-conjugated moieties present in the skeleton of the polymer unit undergo parallel π-stacking as the major interaction with the surface of CNTs [94]. Dai *et al.*, disclosed the hydrophilic polymer functionalized SWNTs which are useful for adsorption of amphiphilic molecules to make the SWNTs remain stable in the aqueous suspension and facilitates its drug delivery activities more facile owing to its large surface area available for supramolecular bonding, *i.e.*, π-π stacking between the targeted aromatic drug for delivery and the graphene surface on the nanotubes. Herein, the polymer poly(ethylene glycol) (PEG) and its derivatives which contains -COOH groups was covalently bonded to the sidewall of the nanotubes in order to make the them the active agents for the desired purposes [95]. Strano *et al.*, disclosed non-covalent complexes of SWNTs with biological polymers specifically proteins, polypeptides and polysaccharides which function as sensing polymers. These complexes interact selectively undergoing specific interaction with the analytes like small organic molecules such as sugars, steroids, antigens, as well as polymeric species such as enzymes, proteins, *etc.* [96]. Tour *et al.*, disclosed the functionalization of exposed sidewall of SWNTs, DWNTs, MWNTs, as well as small diameter carbon nanotubes (SDNTs) in acidic media like fuming sulphuric acid using several organic molecules in presence of $NaNO_2$, AIBN or $(t\text{-}BuO)_2$with the aim of obtaining the unbundles well dispersible derivatives of CNTs. Herein, SDNTs are CNTs having diameter of 3 nm [97].

3.3.2.1 Reversibility in Non-Covalent Functionalization

Two interesting features are hidden inside non-covalent functionalization. First one is that the functionalization does not hamper the conjugated double bond present in the sidewalls of the CNT. Second one makes this functionalization more interesting and that is the reversibility of the reaction. It is very likely to get the bare CNTs both in aqueous and organic media. Use of different and specific external stimuli is one of the intelligent pathway to prefer for non-covalent functionalization of CNTs [98]. The ability of the solvent to desorb from CNTs also depend over the used external stimuli. Many external stimuli can be applied, out of which a few of them are summarized *like,* solvent, temperature, pH, light and redox property, *etc.* Many bio molecules can also be used to get reversibility in this functionalization.

3.3.2.2 Solvent Variation in Non-Covalent Functionalization

The easier method is the changing of solvent to remove the non-covalent functionalization [99]. However, in case of some polymers, the removal of dispersion solvent is bit tough due to strong interaction between the CNTs and the solvents. A simple acetonitrile washing will result to bare CNTs by complete desorption of non-covalent functionalization [100]. When acetonitrile is added in a chlorinated solvent, it not only acts as washing solvent. Moreover, it introduces variation in structures as it possesses solvophobic effect associated with the enhance in polarity with addition of acetonitrile. Another interesting pathway to get proper reversible functionalization is the use of such molecules that can assemble surrounding the nanotubes appearing to be a supramolecular polymer. This fact occurs because hydrogen bonds are effective in keeping nanotubes dispersed with inorganic solvent. However, if the polar solvent is used it disrupts existing bonds and results in precipitation of bare CNTs.

3.3.3.3 pH of the System in Non-Covalent Functionalization

It is possible to control the aggregation and disaggregation phase of CNTs by varying the proton concentration. If 1-pyreneacetic acid is added into the solution, the system deprotonates under basic condition due to the presence of pyrene ring within the scaffold. Thereby, it turns out to be facile to go into solution and separates the CNTs along with other carbonaceous impurities in water. After that, in order to get bare CNTs, it must be rinsed thoroughly by distilled water or ethanol. Some polymers can undergo conformational changes due to protonation and deprotonation with the change in pH. Poly(acrylic acid) can be used as dispersing agents for CNTs. Change in pH leads to change in its nature, charge, *i.e.,* ability to from hydrogen bonds. CO_2 responsive polymers can also be used to tune the aggregation of CNTs. In presence of CO_2, the amidine group of the adsorbed polymer is protonated, and hence the increment of the polymer-polymer electrostatic repulsion results in bare CNTs [101].

3.3.3.4 *Temperature Responsive System in Non-Covalent Functionalization*

The temperature variation helps in controlling the precipitation of CNTs for instances like folic acid coated CNTs. Reports showed that when it is heated at 80 °C, it results in precipitation of CNTs by peeling of folic acid coating. The only drawback of this method is when it is desired to get back the dispersion, then it is required to apply an external force, for instance, sonication [102]. Wang *et al.*, have established that enthalpically favored conformation of N-(isopropylacrylamide) changes to entropically favored conformation on heating which leads to precipitation of the CNTs. The process seems reversible only by sonicating for 2 minutes at 0 °C, this regenerates the dispersion again [103]. Furthermore, the poly(N-cyclopropylacrylamide) having pendant groups like pyrene stabilizes the aqueous dispersion of CNTs. Interestingly, this dispersion rapidly changes its conformation by reducing the static layer thickness over CNTs leading to precipitation above lower critical temperature [104].

3.4 Conclusion and Future Outlook

An appropriate modification or functionalization of the skeleton of CNTs is an important criterion for its effective application in the desired emerging technologies. In this chapter, the most important strategies namely covalent and non-covalent functionalization of CNTs have been outlined and discussed briefly. The typical way of such functionalizations of CNTs helps in improving the dispersibility, stability and processability that ultimately enhances its chemical or electrochemical reactivity as well as various physical properties. The derivatives and composites of CNTs establish a foundation for its exciting and innovative applications. Furthermore, the heteroatom doping are very interesting aspects of CNTs modification. Such functionalization helps in opening new horizon to the CNTs-based composites or hybrids as well as the assembly of other functional advanced materials.

Acknowledgements

PB is very thankful to Department of Higher Education, Science & Technology and Biotechnology, Govt. of West Bengal, India for providing financial assistance *vide* sanction order no. 78(Sanc.)/ST/P/S&T/6G-1/2018

dated 31.01.2019 and project no. GAP-225612. MM acknowledges Ministry of Tribal Affairs, Govt. of India, New Delhi, India for National Fellowship for Higher Education of Schedule Tribes Students [Erstwhile known as Rajiv Gandhi National Fellowship of Schedule Tribes Students (RGNFST) from University Grants Commission, Govt. of India, New Delhi, India] *vide* reference no. F1-17.1/2014-15/RGNF-2014-15-ST-JHA-71559/(SA-III/Website) dated: February 2015. DD acknowledges Department of Science and Technology, Govt. of India, New Delhi, India for DST-INSPIRE fellowship *vide* reference no. IF160176.

Web Links

Web Link1: Bradford research group
https://sites.textiles.ncsu.edu/bradford-research-group/research
Web Link2: Max-Planck-Institut für Eisenforschung (MPIE) https://www.mpie.de/4413569/carbon-nanotubes
Web Link3: North Carolina State University https://www.mse.ncsu.edu/zhu/research/selected-carbon-nanotubes/
Web Link4: Boies Research Group https://cambridgenanoaerosol.com/research/

References

1. Iijima, S., Helical microtubules of graphitic carbon. *Nature*, 354, 56, 1991.
2. Iijima, S. and Ichihashi, T., Single-shell carbon nanotubes of 1-nm diameter. *Nature*, 363, 603, 1993.
3. Bethune, D.S., Kiang, C.H., de Vries, M.S., Gorman, G., Savoy, R., Vazquez, J., Beyers, R., Cobalt-catalysed growth of carbon nanotubes with single-atomic-layer walls. *Nature*, 363, 605, 1993.
4. Journet, C., Maser, W.K., Bernier, P., de la Chapelle Loiseau, A.L., Lefrant, A., Denart, P., Lee, R., Fischer, J.E., Large-scale production of single-walled carbon nanotubes by the electric-arc technique. *Nature*, 388, 756, 1997.
5. Rinzler, A.G., Liu, J., Dai, H., Nikolaev, P., Huffman, C.B., Rodriguez-Macias, F.J., Boul, P.J., Lu, A.H., Heymann, D., Colbert, D.T., Lee, R.S., Fischer, J.E., Rao, A.M., Eklund, P.C., Smalley, R.E., Large scale purification of single-wall carbon nanotubes: Process, product, and characterization. *Appl. Phys. A*, 67, 29, 1998.
6. Endo, M., Takeuchi, K., Kobori, K., Takahashi, K., Kroto, H.W., Sarkar, A., Pyrolytic carbon nanotubes from vapor-grown carbon fibers. *Carbon*, 33, 873, 1995.

7. Nikolaev, P., Broniko, M., Bradley, R., Rohmund, F., Colbert, D.T., Smith, K., Smalley, R.E., Gas-phase catalytic growth of single-walled carbon nanotubes from carbon monoxide. *Chem. Phys. Lett.*, 313, 91, 1999.
8. Baughman, R.H., Zakhidov, A.A., de Heer, W.A., Carbon nanotubes–the route toward applications. *Science*, 297, 787–792, 2002.
9. Maruyama, T., Carbon nanotubes, in: *Handbook of Carbon-Based Nanomaterials, Micro and Nano Technologies*, S. Thomas, C. Sarathchandran, S.A. Ilangovan, J.C. Moreno-Pirajan (Eds.), pp. 299–319, Elsevier, Amsterdam, The Netherlands, 2021.
10. Porwal, M., Rastogi, V., Kumar, A., An overview on carbon nanotubes. *MOJ Bioequiv. Bioavailab.*, 3, 5, 114, 2017.
11. Elhissi, A.M.A., Ahmed, W., Hassan, I.U., Dhanak, V.R., D'Emanuele, A., Carbon nanotubes in cancer therapy and drug delivery. *J. Drug Deliv.*, 2012, 837327, 2012.
12. Carmen-Mihaela, T. and Morris, M.C., Carbon nanotube biosensors. *Front. Chem.*, 3, 59, 2015.
13. Iijima, S., Helical microtubules of graphitic carbon. *Nature*, 354, 56, 1991.
14. Jeon, I.Y., Chang, D.W., Kumar, N.A., Baek, J.B., Functionalization of carbon nanotubes, in: *Carbon Nanotubes*, Yellampalli, S. (Ed.), pp. 91–110, 2011.
15. Hamada, N., Sawada, S., Oshiyama, A., New one-dimensional conductors: Graphitic microtubules. *Phys. Rev. Lett.*, 68, 1579, 1992.
16. Yang, S.B., Kong, B.-S., Kim, D.-W., Jung, H.-T., Comparison of the stability of surface-modified SWNTs and DWNTs network films. *J. Phys. Chem. C*, 114, 4394, 2010.
17. Ma, P.C., Siddiqui, N.A., Marom, G., Kim, J.K., Dispersion and functionalization of carbon nanotubes for polymer-based nanocomposites: A review. *Compos. A Appl. Sci. Manuf.*, 41, 10, 1345, 2010.
18. Mickelson, E.T., Huffman, C.B., Rinzler, A.G., Smalley, R.E., Hauge, R.H., Margrave, J.L., Fluorination of single-wall carbon nanotubes. *Chem. Phys. Lett.*, 296, 188, 1998.
19. Kelly, K.F., Chiang, I.W., Mickelson, E.T. *et al.*, Insight into the mechanism of sidewall functionalization of single-walled nanotubes: An STM study. *Chem. Phys. Lett.*, 313, 445, 1999.
20. Touhara, H., Inahara, J., Mizuno, T. *et al.*, Fluorination of cup-stacked carbon nanotubes, structure and properties. *Fluorine Chem.*, 114, 181, 2002.
21. Stevens, J.L., Huang, A.Y., Peng, H., Chiang, I.W., Khabashesku, V.N., Margrave, J.L., Sidewall amino-functionalization of SWNTs through fluorination and subsequent reactions with terminal diamines. *Nano Lett.*, 3, 331, 2003.
22. Zhang, Q.H. and Chen, D.J., Percolation threshold and morphology of composites of conducting carbon black/polypropylene/EVA. *J. Mater. Sci.*, 39, 1751, 2004.

23. Kim, K.S., Bae, D.J., Kim, J.R., Park, K.A., Lim, S.C., Kim, J.J., Choi, W.B., Park, C.Y., Lee, Y.H., Modification of electronic structures of a carbon nanotube by hydrogen functionalization. *Adv. Mater.*, 14, 1818, 2002.
24. Chen, J., Hamon, M.A., Hu, H., Chen, Y., Rao, A.M., Eklund, P.C., Haddon, R.C., Solution properties of single-walled carbon nanotubes. *Science*, 282, 95, 1998.
25. Hu, H., Zhao, B.M., Hamon, A., Kamaras, K., Itkis, M.E., Haddon, R.C., Sidewall functionalization of single-walled carbon nanotubes by addition of dichlorocarbene. *J. Am. Chem. Soc.*, 125, 14893, 2003.
26. Holzinger, M., Steinmetz, J., Samaille, D., Glerup, M., Paillet, M., Bernier, P., Ley, L., Graupner, R., [2 + 1] cycloaddition for crosslinking SWCNTs. *Carbon*, 42, 941, 2004.
27. Unger, E., Graham, A., Kreupl, F., Liebau, M., Hoenlein, W., Electrochemical functionalization of multiwalled carbon nanotubes for solvation and purification. *Curr. Appl. Phys.*, 2, 107, 2002.
28. Tagmatarchis, N. and Prato, M.J., Functionalization of carbon nanotubes via 1,3-dipolar cycloadditions. *J. Mater. Chem.*, 14, 437, 2004.
29. Esumi, K., Ishigami, M., Nakajima, A., Sawada, K., Honda, H., Chemical treatment of carbon nanotubes. *Carbon*, 34, 279, 1996.
30. Yu, R., Chen, L., Liu, Q., Lin, J., Tan, K.L., Ng, S.C., Chan, H.S.O., Xu, G.Q., Hor, T.S.A., Platinum deposition on carbon nanotubes via chemical modification. *Chem. Mater.*, 10, 718, 1998.
31. Sham, M.L. and Kim, J.K., Surface functionalities of multi-wall carbon nanotubes after UV/ozone and TETA treatments. *Carbon*, 44, 768, 2006.
32. Ma, P.C., Kim, J.K., Tang, B.Z., Functionalization of carbon nanotubes using a silane coupling agent. *Carbon*, 44, 2006, 3232.
33. Wang, S.C., Chang, K.S., Yuan, C.J., Enhancement of electrochemical properties of screen-printed carbon electrodes by oxygen plasma treatment. *Electrochim. Acta*, 54, 4937, 2009.
34. Ávila-Orta, C.A., Cruz-Delgado, V.J., Neira-Velázquez, M.G., Hernández-Hernández, E., Méndez-Padilla, M.G., Medellín-Rodríguez, F.J., Surface modification of carbon nanotubes with ethylene glycol plasma. *Carbon*, 47, 1916, 2009.
35. Hamon, M.A., Hui, H., Bhowmik, P., Ester-functionalized soluble single-walled carbon nanotubes. *Appl. Phys. A*, 74, 333, 2002.
36. Liu, J., Rinzler, A.G., Dai, H.J., *et al*, Fullerene pipes. *Science*, 280, 1253, 1998.
37. Liu, P., Modifications of carbon nanotubes with polymers. *Eur. Polym. J.*, 41, 2693, 2005.
38. Coleman, J.N., Khan, U., Gunko, Y.K., Mechanical reinforcement of polymers using carbon nanotubes. *Adv. Mater.*, 18, 689, 2006.
39. Coleman, J.N., Khan, U., Gunko, Y.K., Mechanical reinforcement of polymers using carbon nanotubes. *Adv. Mater.*, 18, 689, 2006.
40. He, Z., Jiang, R., Long, W., Huang, H., Liu, M., Chen, J., Deng, F., Zhou, N., Zhang, X., Wei, Y., The combination of Diels-Alder reaction and redox

polymerization for preparation of functionalized CNTs for intracellular controlled drug delivery. *Mater. Sci. Eng. C*, 109, 110442, 2020.

41. Xing, C., Zhao, L., You, J., Dong, W., Cao, X., Li, Y., Impact of ionic liquid-modified multiwalled carbon nanotubes on the crystallization behavior of poly (vinylidene fluoride). *J. Phys. Chem. B*, 116, 8312, 2012.
42. Polo-Luque, M.L., Simonet, B.M., Valcárcel, M., Functionalization and dispersion of carbon nanotubes in ionic liquids. *Trends Anal. Chem.*, 47, 99, 2013.
43. Bains, D., Singh, G., Bhinder, J., Agnihotri, P.K., Singh, N., Ionic liquid-functionalized multiwalled carbon nanotube-based hydrophobic coatings for robust antibacterial applications. *ACS Appl. Bio Mater.*, 3, 4, 2092, 2020.
44. Yang, G., Huang, Q., Gan, D., Huang, H., Chen, J., Deng, F., Liu, M., Wen, Y., Zhang, X., Wei, Y., Biomimetic functionalization of carbon nanotubes with poly(ionic liquids) for highly efficient adsorption of organic dyes. *J. Mol. Liq.*, 296, 112059, 2019.
45. Eskandari, P., Abousalman-Rezvani, Z., Roghani-Mamaqani, H., Salami-Kalajahi, M., Polymer-functionalization of carbon nanotube by *in situ* conventional and controlled radical polymerizations. *Adv. Colloid Interface Sci.*, 294, 102471, 2021.
46. Punetha, V.D., Rana, S., Yoo, H.J., Chaurasia, A., McLeskey Jr., J.T., Ramasamy, M.S., Sahoo, N.G., Cho, J.W., Functionalization of carbon nanomaterials for advanced polymer nanocomposites: A comparison study between CNT and graphene. *Prog. Polym. Sci.*, 67, 1, 2017.
47. He, H., Zhang, Y., Gao, C., Wu, J., 'Clicked' magnetic nanohybrids with a soft polymer interlayer. *Chem. Commun.*, 13, 1655, 2009.
48. Zhao, L., Zeng, B., Zhao, F., Electrochemical determination of tartrazine using a molecularly imprinted polymer–multiwalled carbon nanotubes–ionic liquid supported Pt nanoparticles composite film coated electrode. *Electrochim. Acta*, 146, 611, 2014.
49. Yadav, S.K., Mahapatra, S.S., Yoo, H.J., Cho, J.W., Synthesis of multi-walled carbon nanotube/polyhedral oligomeric silsesquioxane nanohybrid by utilizing click chemistry. *Nanoscale Res. Lett.*, 6, 122, 1, 2011.
50. Li, J.C., Hou, P.X., Liu, C., Heteroatom-doped carbon nanotube and graphene-based electrocatalysts for oxygen reduction reaction. *Small*, 13, 1702002, 2017.
51. Liu, S., Li, G., Gao, Y., Xiao, Z., Zhang, J., Wang, Q., Zhang, X., Wang, L., Doping carbon nanotubes with N, S, and B for electrocatalytic oxygen reduction: a systematic investigation on single, double, and triple doped modes. *Catal. Sci. Technol.*, 7, 4007, 2017.
52. Hong, S., Nam, J., Park, S., Lee, D., Park, M., Lee, D.S., Kim, N.D., Kim, D.Y., Ku, B.C., Kim, Y.A., Hwang, J.Y., Carbon nanotube fibers with high specific electrical conductivity: Synergistic effect of heteroatom doping and densification. *Carbon*, 184, 207, 2021.

53. Guo, Q., Zhao, D., Liu, S., Chen, S., Hanif, M., Hou, H., Free-standing nitrogen-doped carbon nanotubes at electrospuncarbon nanofibers composite as an efficient electrocatalyst for oxygen reduction. *Electrochim. Acta*, 138, 318, 2014.
54. Li, J.C., Hou, P.X., Liu, C., Heteroatom-doped carbon nanotube and graphene-based electrocatalysts for oxygen reduction reaction. *Small*, 13, 1702002, 2017.
55. Masa, N.J., Xia, W., Muhler, M., Schuhmann, W., On the role of metals in nitrogen-doped carbon electrocatalysts for oxygen reduction. *Angew. Chem. Int. Ed.*, 54, 10102, 2015.
56. Iizumi, Y., Yudasaka, M., Kim, J., Sakakita, H., Takeuchi, T., Okazaki, T., Oxygen-doped carbon nanotubes for near-infrared fluorescent labels and imaging probes. *Sci. Rep.*, 8, 6272, 2018.
57. Yang, J., Sun, H., Liang, H., Ji, H., Song, L., Gao, C., Xu, H., A highly efficient metal-free oxygen reduction electrocatalyst assembled from carbon nanotubes and graphene. *Adv. Mater.*, 28, 4606, 2016.
58. Hassani, F. and Tavakol, H., Synthesis of sulfur-doped carbon nanotubes from sulfur powder using chemical vapor deposition. *Fuller. Nanotub. Carbon Nanostructures*, 26, 8, 479, 2018.
59. Kim, J.H., Ko, Y., Kim, Y.A., Kim, K.S., Yang, C.M., Sulfur-doped carbon nanotubes as a conducting agent in supercapacitor electrodes. *J. Alloys Compd.*, 855, 1, 157282, 2021.
60. Shi, Q., Peng, F., Liao, S., Wang, H., Yu, H., Liu, Z., Zhang, B., Suc, D., Sulfur and nitrogen co-doped carbon nanotubes for enhancing electrochemical oxygen reduction activity in acidic and alkaline media. *J. Mater. Chem. A*, 1, 14853, 2013.
61. Banerjee, S., Hemraj-Benny, T., Wong, S.S., Covalent surface chemistry of single-walled carbon nanotubes. *Adv. Mater.*, 17, 17, 2005.
62. Balasubramanian, K. and Burghard, M., Chemically functionalized carbon nanotubes. *Small*, 1, 180, 2004.
63. Dumonteil, S., Demortier, A., Detriche, S., Raes, C., Fonseca, A., Rühle, M., Nagy, J.B., Nanosci, J., Dispersion of carbon nanotubes using organic solvents. *J. Nanosci. Nanotechnol.*, 6, 1315, 2006.
64. Bergin, S.D., Nicolosi, V., Streich, P.V., Giordani, S., Sun, Z.Y., Windle, A.H., Ryan, P., Niraj, N.P.P., Wang, Z.T.T., Carpenter, L., Blau, W.J., Boland, J.J., Hamilton, J.P., Coleman, J.N., Towards solutions of single-walled carbon nanotubes in common solvents. *Adv. Mater.*, 20, 1876, 2008.
65. Chen, R.J., Zhang, Y., Wang, D., Dai, H., Noncovalent sidewall functionalization of single-walled carbon nanotubes for protein immobilization. *J. Am. Chem. Soc.*, 123, 3838, 2001.
66. Zhao, Y.L. and Stoddart, J.F., Noncovalent functionalization of single-walled carbon nanotubes. *Acc. Chem. Res.*, 42, 8, 1161, 2009.
67. Hecht, D.S., Ramirez, R.J.A., Briman, M., Artukovic, E., Chichak, K.S., Stoddart, J.F., Gruner, G., Bioinspired detection of light using a

porphyrin-sensitized single wall nanotube field effect transistor. *Nano Lett.*, 6, 2031, 2006.
68. Diehl, M.R., Steuerman, D.W., Tseng, H.R., Vignon, S.A., Star, A., Celestre, P.C., Stoddart, J.F., Heath, J.R., Single-walled carbon nanotube based molecular switch tunnel junctions. *ChemPhysChem.*, 4, 1335, 2003.
69. Hu, L., Zhao, Y.L., Ryu, K., Zhou, C., Stoddart, J.F., Gruner, G., Light-induced charge transfer in pyrene/CdSe-SWNT hybrids. *Adv. Mater.*, 20, 939, 2008.
70. Zhou, Y., Fang, Y., Ramasamy, R.P., Non-covalent functionalization of carbon nanotubes for electrochemical biosensor development. *Sensors*, 19, 392, 2019.
71. Star, A., Stoddart, J.F., Steuerman, D., Diehl, M., Boukai, A., Wong, E.W., Yang, X., Chung, S.W., Choi, H., Heath, J.R., Preparation and properties of polymer- wrapped single-walled carbon nanotubes. *Angew. Chem. Int. Ed.*, 40, 1721, 2001.
72. Steuerman, D.W., Star, A., Narizaano, R., Choi, H., Ries, R.S., Nicolini, C., Stoddart, J.F., Heath, J.R., Interactions between conjugated polymers and single walled carbon nanotubes. *J. Phys. Chem. B*, 106, 3124, 2002.
73. Star, A., Liu, Y., Grant, K., Ridvan, L., Stoddart, J.F., Steuerman, D.W., Diehl, M.R., Boukai, A., Heath, J.R., Noncovalent side-wall functionalization of single walled carbon nanotubes. *Macromolecules*, 36, 553, 2003.
74. Star, A. and Stoddart, J.F., Dispersion and solubilization of single-walled carbon nanotubes with a hyperbranched polymer. *Macromolecules*, 35, 7516, 2002.
75. Gong, X., Liu, J., Baskaran, S., Voise, R.D., Young, J.S., Surfactant-assisted processing of carbon nanotube/polymer composites. *Chem. Mater.*, 12, 4, 1049, 2002.
76. Vaisman, L., Marom, G., Wagner, H.D., Dispersions of surface-modified carbon nanotubes in water-soluble and water-insoluble polymers. *Adv. Funct. Mater.*, 16, 3, 357, 2006.
77. Whitsitt, E.A. and Barron, A.R., Silica coated single walled carbon nanotubes. *Nano Lett.*, 3, 6, 775, 2003.
78. Kim, T.H., Doe, C., Kline, S.R., Choi, S.M., Water-redispersible isolated single-walled carbon nanotubes fabricated by *in situ* polymerization of micelles. *Adv. Mater.*, 19, 7, 929, 2007.
79. Islam, M.F., Rojas, E., Bergey, D.M., Johnson, A.T., Yodh, A.G., High weight fraction surfactant solubilization of single-wall carbon nanotubes in water. *Nano Lett.*, 3, 2, 269, 2003.
80. Yu, J., Grossiord, N., Koning, C.E., Loos, J., Controlling the dispersion of multi-wall carbon nanotubes in aqueous surfactant solution. *Carbon*, 45, 3, 618, 2007.
81. Klumpp, C., Kostarelos, K., Prato, M., Bianco, A., Functionalized carbon nanotubes as emerging nanovectors for the delivery of therapeutics. *Biochim. Biophys. Acta (BBA)–Biomembranes*, 1758, 3, 404, 2006.

82. Barone, P.W. and Strano, M.S., Reversible control of carbon nanotube aggregation for a glucose affinity sensor. *Angew. Chem. Int. Ed.*, 45, 8138, 2006.
83. Star, A., Steuerman, D.W., Heath, J.R., Stoddart, J.F., Starched carbon nanotubes. *Angew. Chem. Int. Ed.*, 41, 2508, 2002.
84. Chambers, G., Carroll, C., Farrell, G.F., Dalton, A.B., McNamara, M., Panhuis, M., Byrne, H.J., Characterization of the interaction of gamma cyclodextrin with single-walled carbon nanotubes. *Nano Lett.*, 3, 843, 2003.
85. Ikeda, M., Hasegawa, T., Numata, M., Sugikawa, K., Sakurai, K., Fujiki, M., Shinkai, S., Instantaneous inclusion of a polynucleotide and hydrophobic guest molecules into a helical core of cationic 1, 3-glucan polysaccharide. *J. Am. Chem. Soc.*, 129, 3979, 2007.
86. Chen, R.J., Zhang, Y., Wang, D., Dai, H., Single-walled carbon nanotubes for protein immobilization. *J. Am. Chem. Soc.*, 123, 3838, 2001.
87. Ishibashi, A. and Nakashima, N., Individual dissolution of single-walled carbon nanotubes in aqueous solution of steroid or sugar compounds and their Raman and Near-IR spectral properties. *Chem. Eur. J.*, 12, 7595, 2006.
88. Dodziuk, H., Ejchart, A., Anczewski, W., Ueda, H., Krinichnaya, E., Dolgonos, G., Kutner, W., Water solubilization, determination of the number of different types of single-wall carbon nanotubes and their partial separation with respect to diameters by complexation with η-cyclodextrin. *Chem. Commun.*, 8, 986, 2003.
89. Kim, O.K., Je, J., Baldwin, J.W., Kooi, S., Pehrsson, P.E., Buckley, L.J., Solubilization of single-walled carbon nanotubes by supramolecular encapsulation of helical amylase. *J. Am. Chem. Soc.*, 125, 4426, 2003.
90. Goodwin, A.P., Tabakman, S.M., Welsher, K., Sherlock, S.P., Prencipe, G., Dai, H., Phospholipid-dextran with a single coupling point: A useful amphiphile for functionalization of nanomaterials. *J. Am. Chem. Soc.*, 131, 289, 2009.
91. Bahr, J.L. and Tour, J.M., Covalent chemistry of single-wall carbon nanotubes. *J. Mater. Chem.*, 12, 1952, 2002.
92. Hirsch, A. and Vostrowsky, O., Functionalization of carbon nanotubes, in: *Functional Molecular Nanostructures*, p. 193, 2005.
93. Banerjee, S., Hermraj-Benny, T., Wong, S.S., Covalent surface chemistry of single-walled carbon nanotubes. *Adv. Mater.*, 1, 17, 2005.
94. J. Chen and H. Liu, Polymer for functionalizing nanotubes noncovalently and method for functionalizing nanotubes noncovalently using the polymer. Japan patent, JP4406676B2, assigned to Zyvex Performance Materials Inc, 2010.
95. H. Dai, Z. Liu, X. Li, X. Sun, Supramolecular functionalization of graphitic nanoparticles for drug delivery. US Patent US9233166B2, assigned to National Institutes of Healths (NIH), U.S. Dept. of Health and Human Services, U.S. Govt, 2016.

96. M.S. Strano, S. Baik, P. Barone, Sensors employing single - walled carbon nanotubes. US patent US10712347B2, assigned to The Board of Trustees of the University of Illinois, 2020.
97. J.M. Tour, J.L. Hudson, C.R. Dyke, J.J. Stephenson, Functionalization of carbon nanotubes in acidic media. World patent WO2005/113434A1, assigned to William Marsh Rice University, 2005.
98. Lemasson, F., Tittmann, J., Hennrich, F., Stürzl, N., Malik, S., Kappes, M.M., Mayor, M., Debundling, selection and release of SWNTs using fluorene-based photocleavable polymers. *Chem. Commun.*, 47, 7428, 2011.
99. O'Connell, M.J., Boul, P., Ericson, L.M., Huffman, C., Wang, Y., Haroz, E., Kuper, C., Tour, J., Ausman, K.D., Smalley, R.E., Reversible water-solubilization of single-walled carbon nanotubes by polymer wrapping, Chem. *Phys. Lett.*, 342, 265–271, 2001.
100. Lellouche, J.P., Piran, M., Shahar, L., Grinblat, J., Pirlot, C., A reversible decoration of multi-walled carbon nanotubes (MWCNTs) by acyclic η4-(1E,3E)-dienyl-$Fe(CO)_3$ complexes, *J. Mater. Chem.*, 18, 1093, 2008.
101. Di Crescenzo, A., Ettorre, V., Fontana, A., Non-covalent and reversible functionalization of carbon nanotubes. *Beilstein J. Nanotechnol.*, 5, 1675, 2014.
102. Ikeda, A., Totsuka, Y., Nobusawa, K., Kikuchi, J.I., Reversible solubilisation and precipitation of carbon nanotubes by temperature and pH control in water. *J. Mater. Chem.*, 19, 5785, 2009.
103. Wang, D. and Chen, L., Temperature and pH-responsive single-walled carbon nanotube dispersions. *Nano Lett.*, 7, 1480, 2007.
104. Etika, K.C., Jochum, F.D., Theato, P., Grunlan, J.C., Temperature controlled dispersion of carbon nanotubes in water with pyrene-functionalized poly (N-cyclopropylacrylamide). *J. Am. Chem. Soc.*, 131, 13598, 2009.

4

Functionalization Carbon Nanotubes Innovate on Medical Technology

Afroz Aslam[1]*, Jeenat Aslam[2], Hilal Ahmad Parray[3] and Chaudhery Mustansar Hussain[4]

[1]Department of Chemistry, Aligarh Muslim University, Aligarh, India
[2]Department of Chemistry, College of Science, Taibah University, Yanbu, Al-Madina, Saudi Arabia
[3]Department of Pathology, UT Southwestern Medical Center, Texas University Harry Hines Blvd., Dallas, TX, USA
[4]Department of Chemistry and Environmental Science, New Jersey Institute of Technology, Newark, NJ, USA

Abstract
Carbon nanotubes (CNTs) are new nanomaterials with many biomedical applications. CNTs can be utilized to deliver a variety of therapeutic compounds to disease locations, including biomolecules. Because of their exceptional optical and electrical properties they are also attractive candidates for biomedical applications. CNTs can be functionalized with various functional groups to transfer several moieties for simultaneous targeting, imaging, and therapy. The potential ability to penetrate biological membranes and the minimal toxicity is most promising. This chapter deals with the functionalized carbon nanotubes (f-CNTs) in all medical fields.

***Keywords*:** Functionalized carbon nanotubes, biomaterials, medical applications, tissue engineering, infectious diseases treatment, cancer treatment

4.1 Introduction

The most significant advantage of nanotechnology is the ability to produce advanced structures with improved ability to translocate through

**Corresponding author*: afrozaslam10@gmail.com

Jeenat Aslam, Chaudhery Mustansar Hussain and Ruby Aslam (eds.) Functionalized Carbon Nanotubes for Biomedical Applications, (75–94) © 2023 Scrivener Publishing LLC

cell membranes and higher solubilization, stability, and bioavailability of biomolecules, hence improving their delivery efficiency. Nanotechnology holds potential for various biomedical applications, including bioimaging and targeted delivery of biomacromolecules into cells. Aono and Ariga [1–3] have introduced nanoarchitectonics, an emerging new idea in nanotechnology. This innovative concept has ushered in a new era in material creation, an innovative technology that combines numerous elements and effects to create functional nanomaterials and nanosystems. Materials, mechanics, electronics, packaging, cosmetics, optical devices, textiles, and solar cells are just a few domains where this technology might be used. This technology has recently been broadened to include biomedical applications such as cultivation, detection, diagnostics, imaging, and various cancer therapies [4, 5].

The primary goal of developing nano carrier-based drug delivery structures is to increase the therapeutic efficacy of therapeutically active chemicals while lowering their toxicity. Liposomes, which are spherically shaped vesicle nanocarriers, are commonly used. CNTs, on the other hand, are cylindrical carbon atom molecules. CNTs are graphene sheets rolled into a seamless cylinder with a high aspect ratio with diameters as small as 1 nm and a length of several micrometers, which can be capped or open-ended. Single-walled carbon nanotubes (SWNTs) are generated from a single graphene sheet, whereas multi-walled carbon nanotubes (MWNTs) are made from many graphene sheets [6, 7] (Figure 4.1).

Because of their typical chemical and physical properties and possible applications in a wide range of fields, from electronic and sensors devices to nanocomposite materials of high strength and low weight, these allotropes of carbon have generated intense interest since their discovery in 1991 by Iijima [6]. Pristine CNTs are not soluble. CNTs bio applications were only possible following the invention of techniques to functionalize these molecules with organic groups and make them soluble. Due to their large surface area, they can adsorb or conjugate with a broad range of medicinal compounds. CNTs can thus be functionalized to improve their aqueous phase dispersibility or offer the proper functional groups to combine with the required target tissue or therapeutic material to elicit a therapeutic effect. CNTs may aid in the therapeutic molecule penetration into the target cell to treat diseases [8–11]. A current instance of CNTs with a range of functional groups is essential to cancer therapy [12] (Figure 4.2).

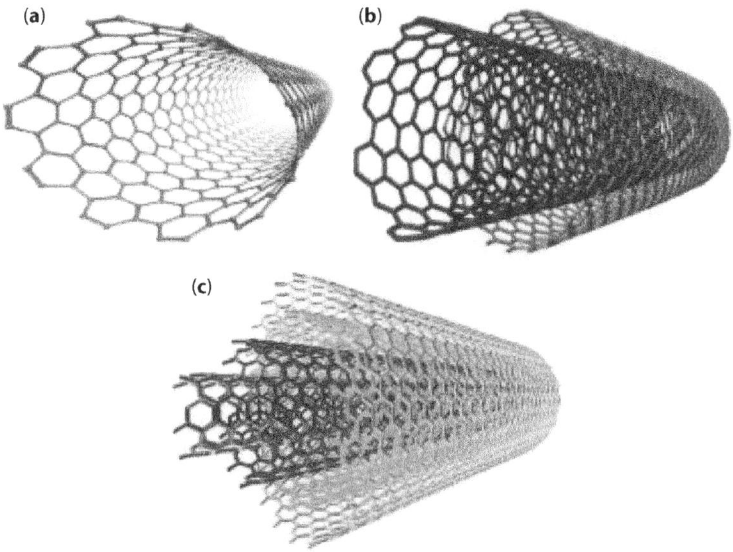

Figure 4.1 Structure of (a) a single-walled carbon nanotube (SWCNT), (b) a double-walled carbon nanotube (DWCNT), and (c) a multi-walled carbon nanotube (MWCNT) [7] (open access source).

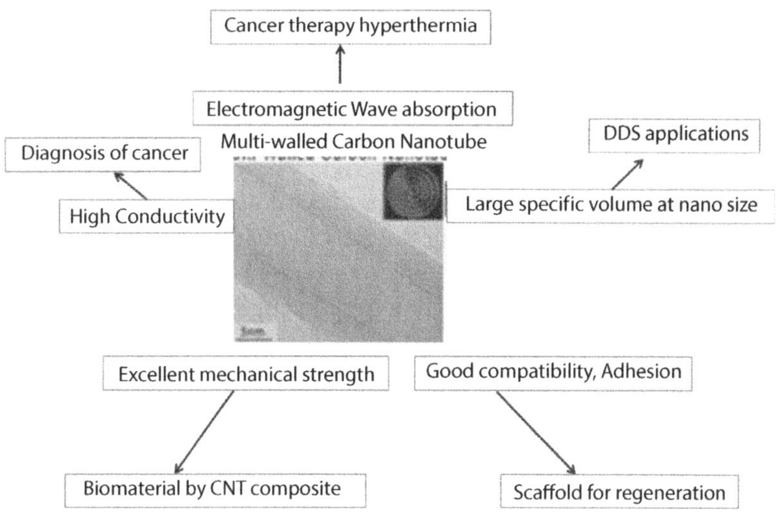

Figure 4.2 Characteristics of Carbon nanotubes and medical applications [13]. (open access source).

4.2 Functionalization CNTs for Biomedical Applications

CNTs are generally dispersible and insoluble substances; thus, it's critical to enhance their surface qualities for better solubilization, dispersion, cytotoxicity reduction, and biocompatibility. CNT surfaces could be modified to increase their biocompatibility and reduce their cytotoxicity in biological systems for biomedical applications by improving their solubility in water, serum, and other solvents [14]. The surface chemistry of CNTs and the method of purification and functionalization have a substantial impact on their biological activities and cytotoxic effects.

Based on the behavior of the biomolecule connected to CNT, the functionalization process may be categorized into two types: covalent (chemical bond formation) and noncovalent (physisorption) attachments [15, 16].

Carbon nanotubes' biocompatibility and solubility are the most significant characteristics of their utilization in biomedical applications. Purification of carbon nanotubes through covalent functionalization using a multistep acid treatment should increase solubility and lower toxicity [17]. Different purification procedures can make CNTs more soluble, exposing particular charged groups and lowering cytotoxicity [18]. Acid-oxidized, short, carboxylated CNTs with high aqueous dispersions and hydrophilic surfaces were less lethal and biocompatible in mice [19]. Human monocyte-derived macrophage cells were used to study the uptake of acid-treated, water-soluble single walled CNTs (SWCNTs).

Purified CNTs (P-CNTs) were found inside the cytoplasm and lysosomes with no effect on the structure or cell viability compared to unpurified CNTs (UP-CNTs). Functionalized SWCNTs have also been utilized on human cells, implying that eliminating hazardous toxins from carboxylated SWCNTs is essential for their progress in pharmaceutical applications [20]. The addition of chemical groups, for example –CO,-OH, and -COOH, to CNTs boosts their O_2 concentration, reducing P-CNT cytotoxicity [21, 22]. Nonviral vectors made of oxidized ultrashort SWCNTs have been employed to transfer oligonucleotide molecules to human macrophages without causing cytotoxicity [23]. It was recently discovered that scattered SWCNTs are pretty safe in cytotoxicity and that isolated and pure SWCNTs are incapable of causing acute cell death [24]. Covalent functionalization of CNTs is a better approach which depends on the degree of functionalization [21] and improves CNT biocompatibility while reducing cytotoxic effects. Noncovalent functionalization is another process to modify the CNT's surface. Surface adsorption onto the sidewalls of CNTs, hydrogen

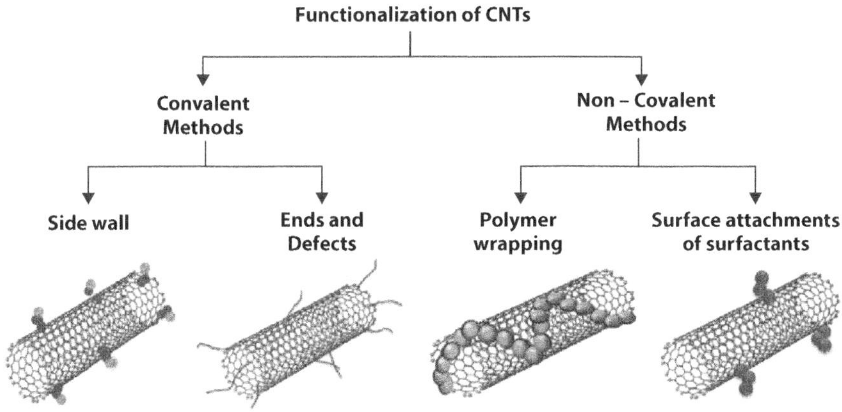

Figure 4.3 Covalent and Non-covalent functionalization methods of CNTs. "Reprinted with permission from [26]. Copyright (2018) Elsevier."

bonding, electrostatic interactions, stacking, and the van der Waals forces are all examples of noncovalent interactions between CNTs and other molecules. CNTs become more water-miscible due to noncovalent techniques, making them less poisonous [25]. To achieve biocompatibility, various biomolecules, polymers, and surfactants have been employed to functionalize CNTs noncovalently. The interaction between graphitic and pyrene surface of CNTs has also been used to coat porphyrin derivatives, and fluorescein isothiocyanate (FITC) terminated poly(ethylene glycol) (PEG) chains onto the surface of CNT, resulting in increased biocompatibility and reduced toxicity [25, 26] (Figure 4.3).

4.3 Potential Applications of CNTs in Cancer Therapy

Due to their unique features, CNTs have been comprehensively studied in delivering targeted drugs for cancer therapy. Owing to their intrinsic optical properties, CNTs can not only be used as drug carriers to deliver numerous anticancer medicines, but they can also be used as excellent phototherapy mediators. CNTs' multifunctionality allows them to be used in various cancer treatments. Many anticancer therapy techniques currently aim to target tumor cells and the environment in which they thrive. Tumor cells can be targeted directly to destroy their parenchyma. In contrast, the tumor microenvironment (TME) can be targeted to prevent tumor development and metastasis by disrupting their living conditions, indirectly killing tumor cells.

4.3.1 Anti-Tumor Immunotherapy

Many studies have indicated that CNTs utilized as carriers are effective in anticancer immunotherapy [27]. This treatment stimulates the patient's immune system to attack the malignant tumor cells. This stimulation can be induced by delivering a therapeutic antibody or a cancer vaccine as a medicine. Even though several medical trials of immunotherapy have shown potential outcomes in cancer treatment [28, 29], immune therapy's low therapeutic efficiency remains a significant concern [30]. CNTs have been approved as vaccine delivery systems due to properties such as lower immunogenicity than standard protein carriers, the capability to permeate the cell membrane without no causing damage, and a more excellent immune response when paired with an antigen [31, 32]. In a mouse model with the H22 liver tumor, a combination of multiwalled CNTs (MWCNTs) and tumor lysate protein (tumor cell vaccination) can significantly and specifically improve the efficacy of anticancer immunotherapy. Because of the high avidity of antigen on the surface and the negative charge, CNTs linked to tumor immunogens can function similarly to professional antigen-presenting cells (such as mature dendritic cells) in delivering cancer antigens to immunological effect or T cells *in vitro* [33]. CNTs are also said to activate the complement system and have adjuvant effects. The mechanism, however, is uncertain [34, 35].

4.3.2 Anti-Tumor Hyperthermia Therapy

CNT-based hyperthermia therapy has recently been recommended as a promising cancer potential treatment. SWCNTs have significant absorption in the near-infrared spectrum (NIR; 700–1100 nm) [36]. Biological systems, by coincidence, are translucent to a similar spectral window. Due to the small autofluorescence setting of tissues and cells can be employed for optical imaging of nanotubes inside living cells [37]. Its simplicity and stability make it preferable to radioactive labels or traditional fluorophores for detecting the pharmacorekinetics of SWCNTs *in vivo* [38]. Near-Infrared Light (NIR) light mixed with a laser pulse could rupture the endosome and facilitate the release of DNA carried by SWCNTs in under-regulated settings [39]. Local thermal ablation of tumor cells triggered via extreme heating of SWCNTs trapped in tumor cells is a comparable application of NIR absorbance. The photothermal effect can induce the local thermal ablation of tumor cells by excessive heating of SWCNTs shackled in tumor cells such as pancreatic cancer. Recently, some progress has been made in the technique, demonstrating clinical possibility [39–41].

4.3.3 Anti-Tumor Chemotherapy

Chemotherapy, along with surgery and radiation, is now an essential part of treating advanced cancers. Two significant difficulties have vulnerable traditional chemotherapy: low chemotherapeutic agent specificity, which results in severe systemic toxicity, and drug resistance (acquired or intrinsic to malignant cells), which results in small efficiency. CNTs have been investigated as new drug delivery vehicles with low toxicity and immunogenicity. CNTs have been used to transport a variety of medicinal compounds, with some outcomes performing those achieved with traditional carriers [42–44]. When utilized in anticancer chemotherapy, CNTs' high surface area permits the effective loading of chemotherapeutic drugs [45]. Drug-loaded CNTs (also known as macromolecular agents) can extravagate into tumor tissues over time, resulting in a concentration in the tumor that is many times higher than in the plasma due to the better permeability and retention impact in solid tumors [46]. Dai's team discovered that 13wt% of the PEGylated SWCNTs injected into cancer accumulated [47]. The same group combined paclitaxel with SWCNTs in a mouse 4T1 breast cancer model. It administered the SWCNT–paclitaxel combination *in-vivo* to obtain superior efficiency in suppressing tumor development while avoiding apparent adverse effects on normal organs [48]. SWCNTs' improved therapeutic efficiency and fewer side effects could be explained by their more extended blood circulation, slower drug release, and stronger tumors absorption (Tenfold larger than Taxol). Another broadly

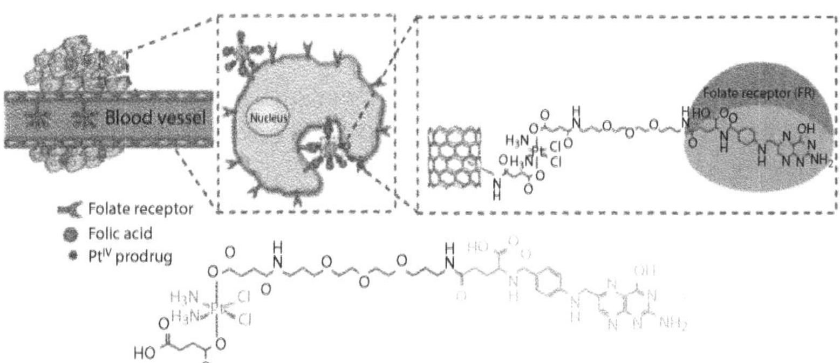

Figure 4.4 One example of CNT used as a drug carrier. Cisplatin is covalently ligated to superficially oxidized CNTs as an effective anti-tumor agent, and a folic acid molecule is further coupled to the cisplatin as a targeting molecule. The large surface area of CNTs makes it possible to carry more cisplatins into tumor cells. "Reprinted with permission from [49]. Copyright (2008) American Chemical Society."

applied anticancer drug, cisplatin, has been employed as a 'longboat delivery system' with SWCNT, resulting in increased uptake in cancerous cells [49]. Antibodies to antigens overexpressed on the cancerous cell surface or ligands that target particular receptors on the tumor cell surface can direct CNTs to the diseased surface of the cell (Figure 4.4). CNTs can be taken up by cells before chemotherapeutic medications are cleaved off, allowing for targeted administration. The previously used targeted molecule was Folic acid [49–52].

4.3.4 Other Cancer Treatment Strategies

Immunotherapy is a treatment option for incurable cancers. Because of its immense potential in targeting tumors, anticancer immunotherapy using CNTs has been intensively investigated in connection with the novel application of immunotherapy to cancer therapy. McDevitt and colleagues [53] used tumor-specific monoclonal antibodies attached to SWCNTs, a metal-ion chelate to carry or deliver a radioactive metal ion, and fluorescent chromophores to visualize their location to target lymphoma. SWCNTs were also employed by Ruggiero *et al.* [54] to bind tumor-associated angiogenesis-targeting antibodies E4G10 and radioactive metal ion chelates. Due to their dual activities in targeting and radio immune therapy, this formulation could simultaneously reduce tumor volume and raise the survival rate of LS174T tumor-bearing model mice. Using a subcutaneous injection method, Meng *et al.* [55] investigated the immunological responses generated by oxidized MWCNTs in hepatocarcinoma-bearing mice. The injection of CNTs resulted in a significant increase in the secretion of inflammatory cytokines and the stimulation of macrophage phagocytosis. These effects stimulate the synthesis of the host immune system and reduce tumor growth. As a result, these results indicate that CNTs can cause anticancer activity in the host by eliciting a tumor-specific immune response.

4.4 Treatment of Central Nervous System Disorders

Delivering medications to the central nervous system remains a crucial challenge in the anticancer drug delivery system for treating tumors of the central nervous system due to the blood-brain barrier. Acetylcholine is a cholinergic nervous system neurotransmitter linked to high-level neurological processes like thinking, memory, and learning. Acetylcholine levels in Alzheimer's brain neurons diminish due to the synthesis difficulty, causing memory, thinking, and learning impairments.

Providing acetylcholine to neurons could help patients with Alzheimer's disease demonstrate their cognitive abilities. Because acetylcholine is a molecule with high polarities, it is hard to pass by the blood-brain barrier; there has been no technique to transport it into the brain. The Raman spectrum indicated the adsorption of acetylcholine on SWCNT in a recently developed drug delivery system, albeit it remained unclear whether acetylcholine was adsorbed in the tubes or on the surfaces SWCNT. This approach effectively supplied acetylcholine into brain neurons via axoplasm neurite transformation (Figure 4.5) and notably enhanced Alzheimer's disease model animals [56, 57]. This delivery system was the first to use nanocarriers to transport medications into central nervous system neurons, opening the mode for drug delivery systems to be used to treat malignancies of the central nervous system. In the *in vivo* anticancer drug delivery system, the results of CNT-based drug delivery were compared to numerous commercially available formulations. The study indicates that carbon nanotubes had significant benefits over other nanomaterial-based drug delivery technologies. The efficiency and toxicity of doxorubicin-loaded CNTs with doxorubicin-loaded nanoliposomes were compared by Liu *et al.* [58]. Though, a significant conclusion on whether CNTs are the most excellent

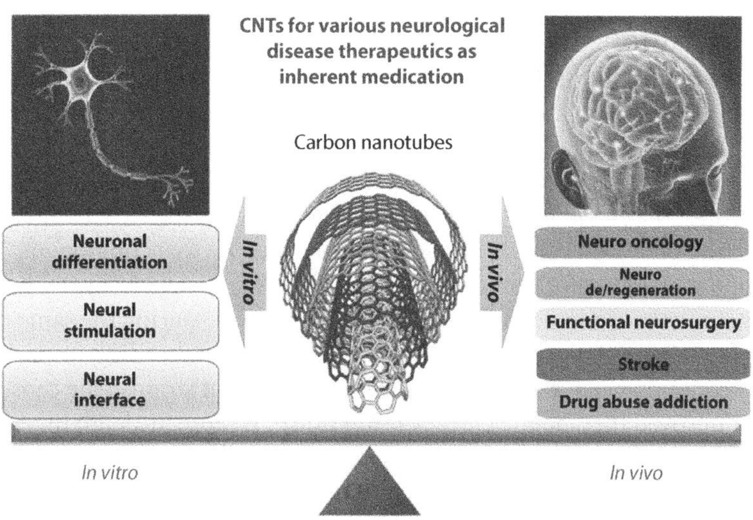

Figure 4.5 CNTs for various neurological diseases therapeutics as inherent medication [57] (open access source).

carriers in anticancer drug delivery systems has yet to be reached because several other anticancer drug delivery systems based on other nanomaterials, for example, cationic polymers [59], fibrinogens [60], and PVP/PVAL nanoparticles [61], have also demonstrated immense promise as drug carriers for cancer treatments.

Further contrast studies are needed to confirm CNTs' role as drug carriers in anticancer therapy. Moreover, the fact that the diameter, volume, and length of CNTs are not adequately managed in research, nor is there any accurate information on their dimension sharing, would not present a significant threat to the reproducibility of pharmacokinetic results. These issues will hinder the commercialization of CNT-based drug delivery technologies. To the best of our understanding, no such system has gone through a clinical study. Numerous challenges must yet be defeated before CNT-based drug delivery systems can be used in actual medicine.

4.5 Treatment of Infectious Diseases

Antibiotics could be used for f-CNTs differently. The antimycotic drug amphotericin B (AmB) is used to treat fungal strains that are very resistant [62]. It is, however, of restricted utility because, due to its low water solubility and penchant for aggregating and generating holes in the cell membrane, it is exceedingly toxic to mammalian cells. We reasoned that conjugating AmB to CNTs might alter its properties in terms of antimycotic efficiency and toxicity [63]. Cytotoxicity of f-CNT1against mammalian cells was the first issue we addressed. While AmB is indeed very poisonous at 10 g/mL, causing 40% cell death, CNT-conjugated AmB5 was not hazardous when utilized at maximum concentrations (up to 40 g/mL, equivalent to AmB concentration attached to 10 g/L tubes). The capability off-CNT**1** containing both fluorescein and AmB to penetrate cells was then studied (Figure 4.6). The detection off-CNT **1** in cells is enabled by this component. This component allows for the identification of f-CNT **1** in enclosures. The fluorescence was evident inside the cell compartments. During the same time frame, toxicity against fungi and yeasts increased. According to preliminary results, f-CNT **2** is also active on AmB-resistant strains. The cause for this improved therapeutic efficacy is unknown. A more in-depth investigation is undertaken to determine whether the action mechanism off-CNT **2** is comparable to AmB or varies in specific ways, thereby explaining the decreased toxicity to mammalian cells.

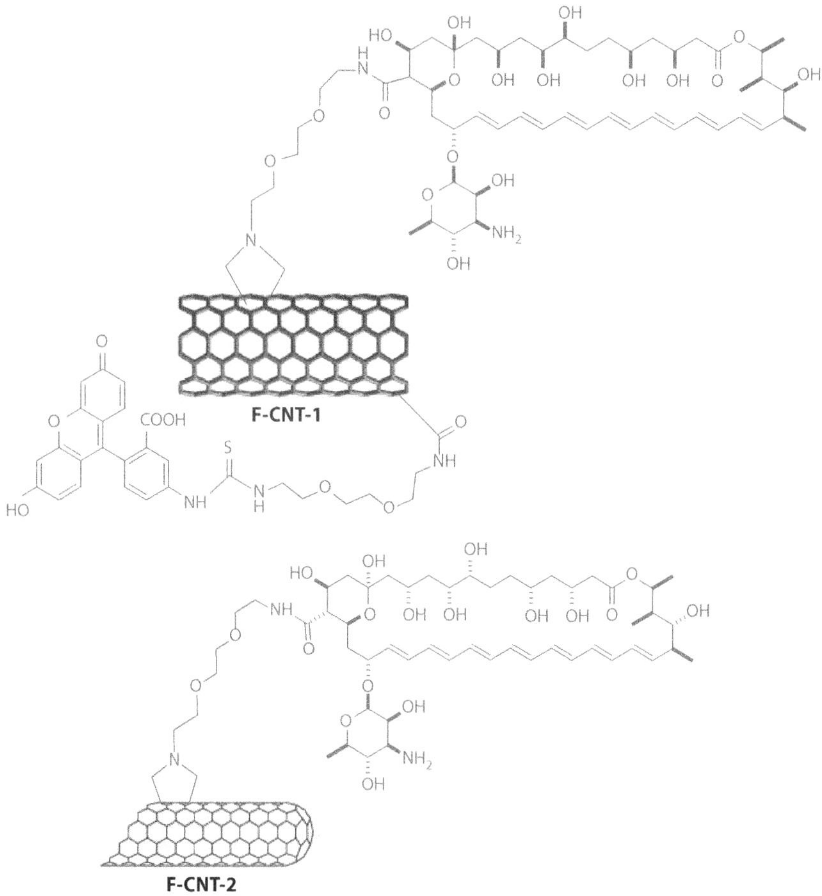

Figure 4.6 Molecular structures of the carbon nanotube conjugated with different therapeutic agents.

4.6 CNTs-Based Transdermal Drug Delivery

CNTs can successfully transfer active drug molecules through the skin because they are composed of an adsorptive material with a large surface area [64]. For hydrophobic drugs, CNTs can provide a high loading and improved transdermal penetration [65]. Because of the electro-conductive qualities of CNTs and the potential for regulated release, iontophoresis is also a viable option. The use of a CNT-based membrane can improve the application of iontophoresis. To improve the electrical sensitivity of the transdermal delivery method, MWCNTs were mixed with a matrix made of polyethylene oxide and pentaerythritol triacrylate polymers [66].

The amount of drug released rises as the electric voltage increases implies that drug release is controlled by the dissolution of polyethylene oxide in the polymer network, which CNTs support.

4.7 f-CNTs for Vaccination

Another type of CNTs-based therapy candidate uses synthetic peptides to stimulate the immune system. Adding B and T cell peptide epitopes to functionalized nanotubes can result in a multivalent system capable of generating a significant immunological response [67, 68]. Chemoselective connections can be made between peptides and tubes [69]. This methodology depends on the production off-CNTs with a maleimide group that quickly combines with peptides that include a cysteine residue on one end. The thiol group of cysteine attaches to the maleimide selectively, forming a stable covalent bond. This method can fully deprotect and characterize the peptide generated through the solid-phase synthesis before ligation to the nanotubes. By connecting a B cell epitope from a foot-and-mouth disease virus coat protein to a B cell epitope, f-CNT **3** and **4** (Figure 4.7) was created (FMDV).

The amount of peptide around the nanotubes in the two conjugates differs, doubled for f-CNT **4** utilizing a lysine branch. The antigenic and immunogenic properties of these conjugates were examined. Compared to the nonconjugated peptide, f-CNTs **3** and **4** induced significant antibody responses. Furthermore, the produced antibodies were able to neutralize the virus, demonstrating the potential of CNTs as a candidate for a synthetic vaccine—Figure 4.8 [70].

4.8 Application of f-CNTs in Tissue Engineering

CNTs have mainly been employed as reinforcement filler in polymer-based scaffolds in tissue engineering, increasing electrical and mechanical properties and permitting the production of stands for cardiac, neural, and bone tissue engineering. The addition of CNTs in the polymer matrix can improve the conductivity of nerve-related scaffolds and improve the responses of nerve cells. It also enhances the cardiac-related scaffolds' elastic strength, conductivity, and biological reactions. CNTs are employed to strengthen both the mechanical and biological aspects of bone tissue engineering. Non-toxic, greener, and safer CNTs and polymeric composites are still predicted in the future. Research into how CNTs generate toxicity is still needed to utilize better the potential factors that cause gen

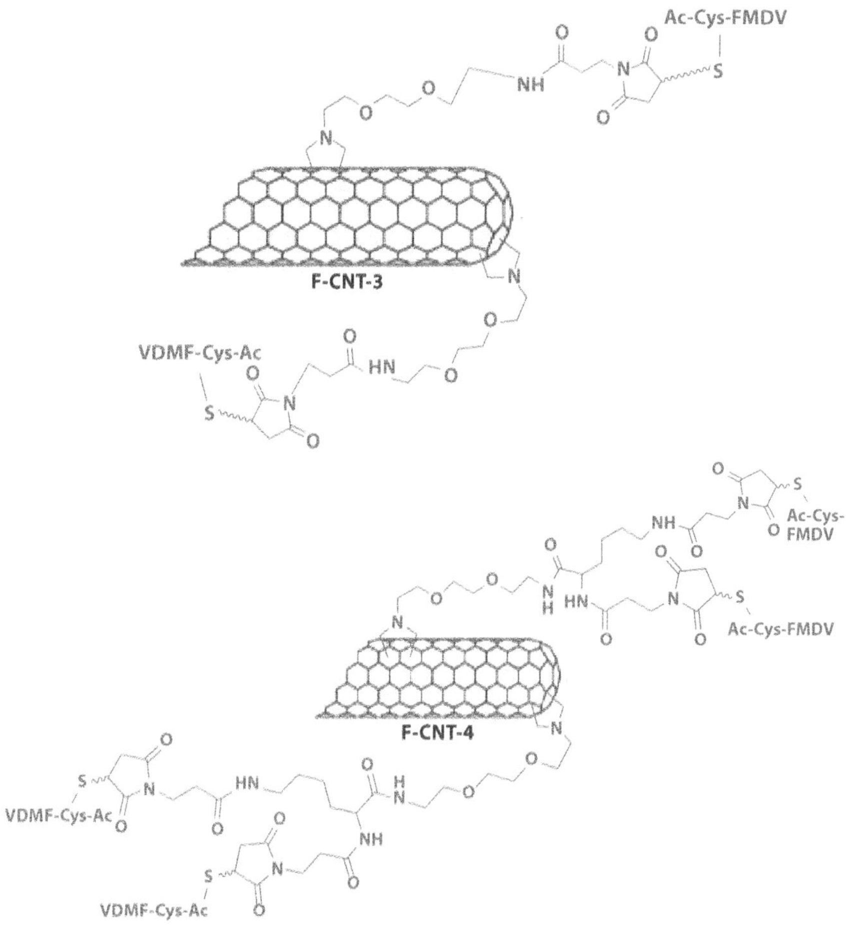

Figure 4.7 Molecular structures of the carbon nanotube conjugated with different therapeutic agents.

cytotoxicity. Many research investigations focus on the early stages of cell development *in vitro*; therefore, long-term *in vivo* studies containing the degradation of the polymeric material that leads to the CNTs release must be examined. The present research examines the small proportion of CNTs in scaffolds because of production constraints and toxicity while balancing the scaffold's electrical, mechanical, and biological capabilities. As a result, more significant CNTs concentrations in the polymer matrix are predictable. Additive manufacturing (e.g., vat-photopolymerization and extrusion-based methods) has been utilized to construct scaffolds with controlled architecture that outperformed 2D membranes or electrospun

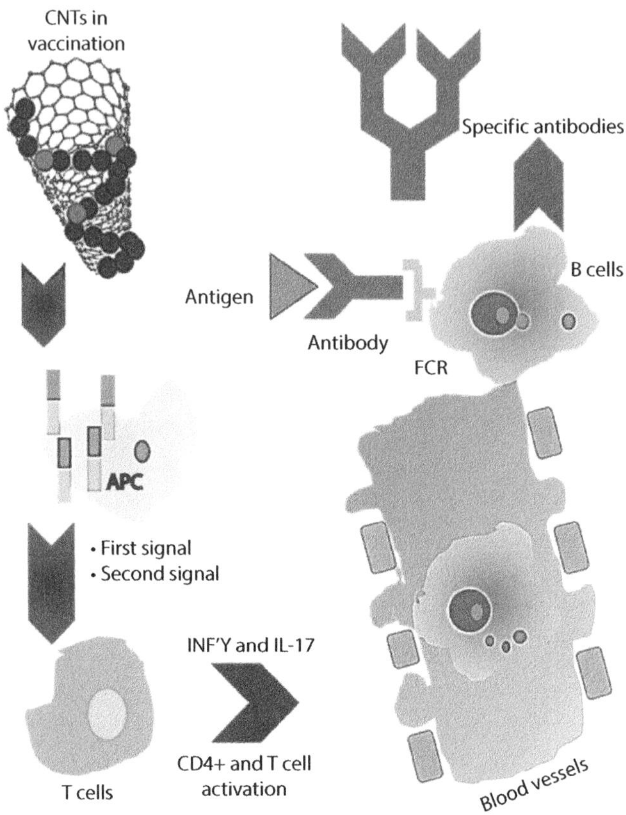

Figure 4.8 Schematic representation of CNTs mediated peptide delivery for immunization, which showed that vaccination by peptide leads to activated T cell response, which is responsible for specific antibodies. "Reprinted with permission from [70]. Copyright (2019) Elsevier."

meshes in mimicking the 3D environment of biological tissues. Different advanced manufacturing technologies will be utilized in the future to construct hierarchical and multi-material scaffolds that closely mimic the natural atmosphere for differentiation and cell proliferation to convert CNTs involving polymer composite scaffolds into clinical achievement.

4.9 Conclusion

CNTs are incredibly fascinating to medical candidates because they have chemically modifiable surfaces with variable lengths and large surface area, besides distinctive physical properties. Because of their potential as drug carriers in targeted delivery, CNTs promise to detect and treat resistant diseases, for example, central nervous system (CNS) disorders, cancer, and infectious diseases. They're fascinating to study targets in tissue engineering because of their unique mechanical properties. CNTs are currently being examined for potential toxicity; however, they are generally thought to be a safer option than other nanomaterials like quantum dots or other medicine carriers like virus carriers. As a result, CNTs still need to be tested *in vivo* for their ADME profile before being employed safely and efficiently in nanomedicine.

Important Websites

https://pubs.acs.org/doi/10.1021/acs.jmedchem.5b01770
https://books.google.com.sa/books/about/Carbon_Nanotubes_for_Biomedical_Applicat.html?id=irEuu80Rg3IC&printsec=frontcover&source=kp_read_button&hl=en&redir_esc=y#v=onepage&q&f=false
https://www.sciencedirect.com/science/article/pii/S0928493118320101
https://en.wikipedia.org/wiki/Potential_applications_of_carbon_nanotubes

References

1. Ariga, K., Minami, K., Ebara, M., Nakanishi, J., What are the emerging concepts and challenges in nano? Nanoarchitectonics, hand-operating nanotechnology, and mechanobiology. *Polym. J.*, 48, 371–389, 2016.
2. Aono, M. and Ariga, K., The way to nanoarchitectonics and the way of nanoarchitectonics. *Adv. Mater.*, 28, 989–992, 2016.
3. Ariga, K., Ji, Q.M., Nakanishi, W., Hill, J.P., Aono, M., Nanoarchitectonics: A new materials horizon for nanotechnology. *Mater. Horiz.*, 2, 406–413, 2015.
4. Nakanishi, W., Minami, K., Shrestha, L.K., Ji, Q.M., Hill, J.P., Ariga, K., Bioactive nanocarbon assemblies: Nanoarchitectonics and applications. *Nano Today*, 9, 378–394, 2014.
5. Ariga, K., Kawakami, K., Ebara, M., Kotsuchibashi, Y., Ji, Q.M., Hill, J.P., Bioinspired nanoarchitectonics as emerging drug delivery systems. *New J. Chem.*, 38, 5149–5163, 2014.
6. Iijima, S., Helical microtubules of graphitic carbon. *Nature*, 354, 6348, 56–58, 1991.
7. Ferrier, D.C. and Honeychurch, K.C., Carbon nanotube (CNT)-based biosensors. *Biosensors*, 11, 486, 2021.

8. Pantarotto, D., Briand, J.P., Prato, M., Bianco, A., Translocation of bioactive peptides across cell membranes by carbon nanotubes. *Chem. Commun.*, 10, 16–17, 2004.
9. Pantarotto, D., Singh, R., McCarthy, D., Erhardt, M., Briand, J.P., Prato, M., Kostarelos, K., Bianco, A., Functionalized carbon nanotubes for plasmid DNA gene delivery. *Angew. Chem. Int. Ed.*, 43, 39, 5242–5246, 2004.
10. Bianco, A., Kostarelos, K., Prato, M., Applications of carbon nanotubes in drug delivery. *Curr. Opin. Chem. Biol.*, 9, 6, 674–679, 2005.
11. Chen, X., Kis, A., Zettl, A., Bertozzi, C.R., A cell nanoinjector based on carbon nanotubes. *Proc. Nat. Acad. Sci.*, 104, 20, 8218–8222, 2007.
12. Bhirde, A.A., Patel, V., Gavard, J., Zhang, G., Sousa, A.A., Masedunskas, A., Leapman, R.D., Weigert, R., Gutkind, J.S., Rusling, J.F., Targeted killing of cancer cells *in vivo* and *in vitro* with EGF-directed carbon nanotube-based drug delivery. *ACS Nano*, 3, 2, 307–316, 2009.
13. Usui, Y., Haniu, H., Tsuruoka, S., Saito, N., Carbon nanotubes innovate on medical technology. *Med. Chem.*, 2, 1–6, 2012.
14. Yan, L., Zhao, F., Li, S., Hu, Z., Zhao, Y., Low-toxic and safe nanomaterials by surface-chemical design, carbon nanotubes, fullerenes, metallofullerenes, and graphenes. *Nanoscale*, 3, 362–382, 2011.
15. Zhang, Y., Bai, Y., Yan, B., Functionalized carbon nanotubes for potential medicinal applications. *Drug Discov. Today*, 15, 428–435, 2010.
16. Yang, W., Thordarson, P., Gooding, J.J., Ringer, S.P., Braet, F., Carbon nanotubes for biological and biomedical applications. *Nanotechnology*, 18, 41, 412001, 1–13, 2007.
17. Porter, A.E., Gass, M., Bendall, J.S., Muller, K., Goode, A., Skepper, J.N., Midgley, P.A., Welland, M., Uptake of noncytotoxic acid-treated single-walled carbon nanotubes into the cytoplasm of human macrophage cells. *ACS Nano*, 3, 6, 1485–1492, 2009.
18. Cui, H.F., Vashist, S.K., Al-Rubeaan, K., Luong, J.H., Sheu, F.S., Interfacing carbon nanotubes with living mammalian cells and cytotoxicity issues. *Chem. Res. Toxicol.*, 23, 7, 1131–1147, 2010.
19. Jain, S., Thakare, V.S., Das, M., Godugu, C., Jain, A.K., Mathur, R., Chuttani, K., Mishra, A.K., Toxicity of multi-walled carbon nanotubes with end defects critically depends on their functionalization density. *Chem. Res. Toxicol.*, 24, 2028–2039, 2011.
20. Wang, R., Mikoryak, C., Li, S., Bushdiecker, D., Musselman, I.H., Pantano, P., Draper, R.K., Cytotoxicity screening of single-walled carbon nanotubes: Detection and removal of cytotoxic contaminants from carboxylated carbon nanotubes. *Mol. Pharm.*, 8, 4, 1351–1361, 2011.
21. Montes-Fonseca, S.L., Orrantia-Borunda, E., Aguilar-Elguezabal, A., González Horta, C., Talamás-Rohana, P., Sánchez-Ramírez, B., Cytotoxicity of functionalized carbon nanotubes in J774A macrophages. *Nanomedicine*, 8, 853–859, 2012.

22. Cheng, C., Müller, K.H., Koziol, K.K., Skepper, J.N., Midgley, P.A., Welland, M.E., Porter, A.E., Toxicity and imaging of multi-walled carbon nanotubes in human macrophage cells. *Biomaterials*, 30, 4152–4160, 2009.
23. Crinelli, R., Carloni, E., Menotta, M., Giacomini, E., Bianchi, M., Ambrosi, G., Giorgi, L., Magnani, M., Oxidized ultrashort nanotubes as carbon scaffolds for the construction of cell-penetrating NF-κB decoy molecules. *ACS Nano*, 4, 2791–2803, 2010.
24. Holt, B.D., Short, P.A., Rape, A.D., Wang, Y.L., Islam, M.F., Dahl, K.N., Carbon nanotubes reorganize actin structures in cells and *ex vivo*. *ACS Nano*, 4, 4872–4878, 2010.
25. Yan, L., Zhao, F., Li, S., Hu, Z., Zhao, Y., Low-toxic and safe nanomaterials by surface-chemical design, carbon nanotubes, fullerenes, metallofullerenes, and graphenes. *Nanoscale*, 3, 362–382, 2011.
26. Jun, L.Y., Mubarak, N.M., Yee, M.J., Yon, L.S., Bing, C.H., Khalid, M., Abdullah, E.C., An overview of functionalized carbon nanomaterial for organic pollutant removal. *J. Ind. Eng. Chem.*, 67, 175–186, 2018.
27. Schirrmacher, V., Clinical trials of anti-tumor vaccination with an autologous tumor cell vaccine modified by virus infection: Improvement of patient survival based on improved anti-tumor immune memory. *Cancer Immunol. Immunother.*, 54, 587–598, 2005.
28. John, S.Y., Liu, G., Ying, H., Yong, W.H., Black, K.L., Wheeler, C.J., Vaccination with tumor lysate-pulsed dendritic cells elicits antigen-specific, cytotoxic T-cells in patients with malignant glioma. *Cancer Res.*, 64, 14, 4973–4979, 2004.
29. Berd, D., Sato, T., Maguire Jr., H.C., Kairys, J., Mastrangelo, M.J., Immunopharmacologic analysis of an autologous, hapten-modified human melanoma vaccine. *J. Clin. Oncol.*, 22, 3, 403–415, 2004.
30. Figdor, C.G., de Vries, I.J.M., Lesterhuis, W.J., Melief, C.J., Dendritic cell immunotherapy: Mapping the way. *Nat. Med.*, 10, 5, 475–480, 2004.
31. Salvador-Morales, C., Flahaut, E., Sim, E., Sloan, J., Green, M.L., Sim, R.B., Complement activation and protein adsorption by carbon nanotubes. *Mol. Immunol.*, 43, 3, 193–201, 2006.
32. Meng, J., Meng, J., Duan, J., Kong, H., Li, L., Wang, C., Xie, S., Chen, S., Gu, N., Xu, H., Yang, X.D., Carbon nanotubes conjugated to tumor lysate protein enhance the efficacy of an antitumor immunotherapy. *Small*, 4, 9, 1364–1370, 2008.
33. Fadel, T.R., Steenblock, E.R., Stern, E., Li, N., Wang, X., Haller, G.L., Pfefferle, L.D., Fahmy, T.M., Enhanced cellular activation with single walled carbon nanotube bundles presenting antibody stimuli. *Nano Lett.*, 8, 7, 2070–2076, 2008.
34. Dobrovolskaia, M.A. and McNeil, S.E., Immunological properties of engineered nanomaterials. *Nat. Nanotechnol.*, 2, 469–478, 2007.

35. Bottini, M., Bruckner, S., Nika, K., Bottini, N., Bellucci, S., Magrini, A., Bergamaschi, A., Mustelin, T., Multi-walled carbon nanotubes induce T lymphocyte apoptosis. *Toxicol. Lett.*, 160, 2, 121–126, 2006.
36. O'connell, M.J., Bachilo, S.M., Huffman, C.B., Moore, V.C., Strano, M.S., Haroz, E.H., Rialon, K.L., Boul, P.J., Noon, W.H., Kittrell, C., Ma, J., Band gap fluorescence from individual single-walled carbon nanotubes. *Science*, 297, 5581, 593–596, 2002.
37. Welsher, K., Liu, Z., Daranciang, D., Dai, H., Selective probing and imaging of cells with single-walled carbon nanotubes as near-infrared fluorescent molecules. *Nano Lett.*, 8, 586–590, 2008.
38. Cherukuri, P., Gannon, C.J., Leeuw, T.K., Schmidt, H.K., Smalley, R.E., Curley, S.A., Weisman, R.B., Mammalian pharmacokinetics of carbon nanotubes using intrinsic near-infrared fluorescence. *Proc. Natl. Acad. Sci.*, 103, 18882–18886, 2006.
39. Kam, N.W.S., O'Connell, M., Wisdom, J.A., Dai, H., Carbon nanotubes as multifunctional biological transporters and near-infrared agents for selective cancer cell destruction. *Proc. Natl. Acad. Sci.*, 102, 11600–11605, 2005.
40. Hirsch, L.R., Stafford, R.J., Bankson, J.A., Sershen, S.R., Rivera, B., Price, R.E., Hazle, J.D., Halas, N.J., West, J.L., Nanoshell-mediated near-infrared thermal therapy of tumors under magnetic resonance guidance. *Proc. Natl. Acad. Sci.*, 100, 13549–13554, 2003.
41. Chakravarty, P., Marches, R., Zimmerman, N.S., Swafford, A.D.E., Bajaj, P., Musselman, I.H., Pantano, P., Draper, R.K., Vitetta, E.S., Thermal ablation of tumor cells with antibody functionalized single-walled carbon nanotubes. *Proc. Natl. Acad. Sci.*, 105, 8697–8702, 2008.
42. Klumpp, C., Kostarelos, K., Prato, M., Bianco, A., Functionalized carbon nanotubes as emerging nanovectors for the delivery of therapeutics. *Biochim. Biophys. Acta Biomembr.*, 1758, 404–412, 2006.
43. Bianco, A., Kostarelos, K., Prato, M., Applications of carbon nanotubes in drug delivery. *Curr. Opin. Chem. Biol.*, 9, 674–679, 2005.
44. Pastorin, G., Crucial functionalizations of carbon nanotubes for improved drug delivery: A valuable option? *Pharm. Res.*, 26, 746–749, 2009.
45. Liu, Z., Sun, X., Nakayama-Ratchford, N., Dai, H., Supramolecular chemistry on water-soluble carbon nanotubes for drug loading and delivery. *ACS Nano*, 1, 50–56, 2007.
46. Maeda, H., Bharate, G.Y., Daruwalla, J., Polymeric drugs for efficient tumor-targeted drug delivery based on EPR-effect. *Eur. J. Pharm. Biopharm.*, 71, 409–419, 2009.
47. Liu, Z., Cai, W., He, L., Nakayama, N., Chen, K., Sun, X., Chen, X., Dai, H., *In vivo* biodistribution and highly efficient tumor targeting of carbon nanotubes in mice. *Nat. Nanotechnol.*, 2, 47–52, 2007.
48. Liu, Z., Chen, K., Davis, C., Sherlock, S., Cao, Q., Chen, X., Dai, H., Drug delivery with carbon nanotubes for *in vivo* cancer treatment. *Cancer Res.*, 68, 6652–6660, 2008.

49. Dhar, S., Liu, Z., Thomale, J., Dai, H., Lippard, S.J., Targeted single-wall carbon nanotube-mediated Pt (IV) prodrug delivery using folate as a homing device. *J. Am. Chem. Soc.*, 130, 11467–11476, 2008.
50. Kam, N.W.S., O'Connell, M., Wisdom, J.A., Dai, H., Carbon nanotubes as multifunctional biological transporters and near-infrared agents for selective cancer cell destruction. *Proc. Natl. Acad. Sci.*, 102, 11600–11605, 2005.
51. Bhirde, A.A., Patel, V., Gavard, J., Zhang, G., Sousa, A.A., Masedunskas, A., Leapman, R.D., Weigert, R., Gutkind, J.S., Rusling, J.F., Targeted killing of cancer cells *in vivo* and *in vitro* with EGF-directed carbon nanotube-based drug delivery. *ACS Nano*, 3, 2, 307–316, 2009.
52. Welsher, K., Liu, Z., Daranciang, D., Dai, H., Selective probing and imaging of cells with single-walled carbon nanotubes as near-infrared fluorescent molecules. *Nano Lett.*, 8, 586–590, 2008.
53. McDevitt, M.R., Chattopadhyay, D., Kappel, B.J., Jaggi, J.S., Schiffman, S.R., Antczak, C., Njardarson, J.T., Brentjens, R., Scheinberg, D.A., Tumor targeting with antibody-functionalized, radiolabeled carbon nanotubes. *J. Nucl. Med.*, 48, 1180–1189, 2007.
54. Ruggiero, A., Villa, C.H., Holland, J.P., Sprinkle, S.R., May, C., Lewis, J.S., Scheinberg, D.A., McDevitt, M.R., Imaging and treating tumor vasculature with targeted radiolabeled carbon nanotubes. *Int. J. Nanomed.*, 5, 783–802, 2010.
55. Meng, J., Yang, M., Jia, F.M., Kong, H., Zhang, W.Q., Wang, C.Y., Xing, J.M., Xie, S.S., Xu, H.Y., Subcutaneous injection of water-soluble multi-walled carbon nanotubes in tumor-bearing mice boosts the host immune activity. *Nanotechnology*, 21, 145104–145112, 2010.
56. Xiang, C., Zhang, Y., Guo, W., Liang, X.J., Biomimetic carbon nanotubes for neurological disease therapeutics as inherent medication. *Acta Pharm. Sin. B*, 10, 2, 239–248, 2020.
57. Yang, Z., Zhang, Y., Yang, Y., Sun, L., Han, D., Li, H., Wang, C., Pharmacological and toxicological target organelles and safe use of single-walled carbon nanotubes as drug carriers in treating Alzheimer's disease. *Nanomedicine*, 6, 427–441, 2010.
58. Liu, Z., Fan, A.C., Rakhra., K., Sherlock, S., Goodwin, A., Chen, X., Yang, Q., Felsher, D.W., Dai, H., Supramolecular stacking of doxorubicin on carbon nanotubes for *in vivo* cancer therapy. *Angew. Chem. Int. Ed. Engl.*, 48, 41, 7668–7672, 2009.
59. Nimesh, S., Gupta, N., Chandra, R., Cationic polymer-based nanocarriers for delivery of therapeutic nucleic acids. *J. Biomed. Nanotechnol.*, 7, 504–520, 2011.
60. Rejinold, N.S., Muthunarayanan, M., Chennazhi, K.P., Nair, S.V., Jayakumar, R., Curcumin loaded fibrinogen nanoparticles for cancer drug delivery. *J. Biomed. Nanotechnol.*, 7, 521–534, 2011.
61. Terence, M.C., Faldini, S.B., Miranda, L.F., Miranda, F., Munhoz Jr., A.H., Castro, P.J., Preparation and characterization of a polymeric blend of PVP/

PVAL for use in drug delivery system. *J. Biomed. Nanotechnol.*, 7, 446–449, 2011.
62. Zotchev, S.B., Polyene macrolide antibiotics and their applications in human therapy. *Curr. Med. Chem.*, 10, 211–223, 2003.
63. Wu, W., Wieckowski, S., Pastorin, G., Benincasa, M., Klumpp, C., Briand, J., Gennaro, R., Prato, M., Bianco, A., Targeted delivery of Amphotericin B to cells using functionalized carbon nanotubes. *Angew. Chem. Int. Ed.*, 44, 6358–6362, 2005.
64. Ilbasmis-Tamer, S. and Degim, I.T., A feasible way to use carbon nanotubes to deliver drug molecules: Transdermal application. *Expert Opin. Drug Deliv.*, 9, 8, 991–999, 2012.
65. Degim, I.T., Burgess, D.J., Papadimitrakopoulos, F., Carbon nanotubes for transdermal drug delivery. *J. Microencapsul.*, 27, 669–81, 2010.
66. Im, J.S., Bai, B.C., Lee, Y.S., The effect of carbon nanotubes on drug delivery in an electro-sensitive transdermal drug delivery system. *Biomaterials*, 31, 1414–19, 2010.
67. Pantarotto, D., Hoebeke, J., Graff, R., Partidos, C.D., Briand, J.-P., Prato, M., Bianco, A., Synthesis, structural characterization and immunological properties of carbon nanotubes functionalized with peptides. *J. Am. Chem. Soc.*, 125, 6160–6164, 2003.
68. Pantarotto, D., Partidos, C.D., Hoebeke, J., Brown, F., Kramer, E., Briand, J.-P., Muller, S., Prato, M., Bianco, A., Immunization with peptide-functionalized carbon nanotubes enhance virus-specific neutralizing antibody responses. *Chem. Biol.*, 10, 961–966, 2003.
69. (a) Goodman, M., Felix, A., Moroder, L., Toniolo, C., *Methods of Organic Chemistry*, vol. E22b, Houben-Weyl, Thieme, Stuttgart, 2002.
(b) Muller, S., *Laboratory Techniques in Biochemistry and Molecular Biology*, vol. 28, S. Pillai, and P.C. van der Vliet, (Eds.), pp. 79–131, Elsevier, Amsterdam, 1999.
70. Maheshwari, N., Muktika, T., Namrata, S., Piyush, G., Mukesh, C.S., Pran, K.D., Rakesh, K.T., Functionalized carbon nanotubes for protein, peptide, and gene delivery, in: *Biomaterials and Bionanotechnology*, Advances in Pharmaceutical Product Development and Research, pp. 613–637, Academic Press, 2019.

Part 2

FUNCTIONALIZED CARBON NANOTUBES: CURRENT AND EMERGING BIOMEDICAL APPLICATIONS

5

Functionalized Carbon Nanotubes: Applications in Biosensing

N. Palaniappan[1]*, Nidhi Vashistha[1] and Ruby Aslam[2]

[1]School of Chemical Sciences, Central University of Gujarat, Gandhinagar, Gujarat, India
[2]Corrosion Research Laboratory, Department of Applied Chemistry, Aligarh Muslim University, Aligarh, India

Abstract
The functionalized carbon nanotubes are used in several applications because of their excellent physicochemical properties. The current transition metal oxide decorated CNT materials have shown significant sensing. We have discussed different organic hetero atoms and nanoparticles incorporated into CNT and their sensing behaviors. However, zinc oxide functionalized CNT showed excellent glucose-sensing due to the nonbonding electron involved in the accept and donate mechanism. Hence, the porous nanoparticles decorated carbon nanotubes for sensing ammonia at room temperature. However, functionalized carbon nanotubes could be revolutionizing the biomedical sensing application at a low cost in the future.

Keywords: Sensor, functionalized carbon nanotubes, biomedical

5.1 Introduction

Since the finding of carbon nanotubes (CNTs) in 1991 by Iijima, it has gathered intense attention from the scientific community globally [1, 2]. For the past two decades, enormous advances have been made in both the understanding of the properties of CNTs and in exploring the methods for synthesis. CNTs exhibit outstanding thermal, electrical mechanical, and optical properties. CNTs can find a wide range of applications in different areas including composite materials (fillers or coatings), additives

**Corresponding author*: palaniappancecri@rediffmail.com

Jeenat Aslam, Chaudhery Mustansar Hussain and Ruby Aslam (eds.) Functionalized Carbon Nanotubes for Biomedical Applications, (97–116) © 2023 Scrivener Publishing LLC

to polymers and catalysts, flat displays, cathode rays of lighting components, lithium battery anodes, absorption and screening of electromagnetic waves, conversion of energy, hydrogen storage, telecommunication devices, nanoprobes, ultrasensitive chemical and biosensors, supercapacitors, and transparent conductive membranes, etc., [3, 4].

CNT structure can be represented as an elongated hollow cylinder constituting one or many hexagonal graphite sheets folded into a tube. Each carbon atom in hexagonal arrangement shares an sp^2 hybridization (C-C bond) with its three nearby carbon atoms [5]. Although, the circular curvature of the tube can lead to quantum confinement and rehybridization of the σ-π orbital, where three σ bonds are marginally out of the plane; to compensate for this, delocalization of the π orbital is more to the outer side of the tube. This causes nanotubes thermally and electrically to be more conductive, mechanically more robust, and enhanced chemical and biological activity than graphite. Based on the existing number of graphene sheets the CNTs can be classified into two types Single Walled and Multiple Walled. CNTs comprising only one graphene sheet are termed as SWNTs, whereas CNTs with more than one graphene sheet are defined as MWNTs. SWCNTs have a diameter of a few nanometers, whereas MWCNTs have a diameter of several tens of nanometers. It is thousands of times smaller than the diameter of human hair and can be several hundred microns in length. The interatomic bond distance is reported to be 1.44 A^0 between C-C atoms [6, 7]. Experimental and theoretical studies of CNTs have shown that they have an exceptionally high elastic modulus > 1 Tpa (diamond -1.2 TPa) and have a strength 10 to 100 times that of the strongest steel. In addition, CNTs exhibit remarkable thermal and electrical properties, including thermal stability up to 2800 °C in a vacuum, an electrical conductivity of approximately 103 Scm^{-1} with a carrying capacity several hundred times higher, and a thermal conductivity of approximately 1900 $Wm^{-1}K^{-1}$, which is two times higher than diamond [8, 9].

Since CNTs are known as potential electrode materials for devices with marked electrochemical applications due to their high surface area, and high electronic conductivity [10]. Furthermore, CNTs are ideal candidates for biosensing and electrochemical applications due to their ability to bind proteins without changing their activity [10–12]. CNTs reportedly show promising results in improving sensitivity, lowering detection limit, and reducing detection time and sample amount [13–15]. CNTs integrated with biological functionalities are anticipated to have enormous potential in future biomedical applications. This chapter outlines the recent advances and developments around biosensors which are being prepared with CNTs as the functional and prime sensing material. In the following section, we will overview and discuss various CNT-based biosensors.

5.2 CNTs-Based Biosensors

A biosensor is a chemical sensing device that integrates the active biological components like nucleic acids, enzymes, antibodies, microorganisms, tissues, etc. The detection of the analyte is based on two steps: the recognition step and transducing step. In the recognition step, the analyte can be recognized either in the solution or in the atmosphere by the biological element. Further, because of the interaction between the recognition molecule and the analyte, a signal is transmitted to the detector. Signal transduction can take place electrochemically, optically, magnetically, calorimetrically, or gravimetrically, depending on the technique used. Biosensors could find applications in drug delivery, medical diagnosis health care and food safety, military applications, and environmental monitoring [16–18]. Among the above, healthcare application persists to be of key importance for biosensors. The first biosensor used to detect oxygen was developed in 1956. Since then, numerous different types of biosensors have been developed, along with the glucose biosensor for the analysis of clinical diabetes [18].

For biosensing applications, nanomaterials offer the possibility of a significant increase in sensitivity with a lower limit of detection. This improved performance is correlated with the nano-scale dimension of nanomaterials, which bestow them with a high surface-to-volume ratio. This high specific are aallowsimmobilization of large concentrations of bioreceptorunits compared to surface/volume biosensors. Moreover, many biomaterials may act as transducers due to their intrinsic physicochemical properties. During the past decade, CNT-based hybrid systems at the nano scale have been investigated vigorously for biological interfaces. As nanotubes are structured on the surface, their entire weight is centered on their surface layers. It is because of this property that tubules possess a particularly substantial unit surface which consequently decides their electrochemical and adsorption properties [19, 20]. Nanotubes possess a strong electronic response to molecules adsorbed on their surfaces, and their unprecedented unit surfaces make them a promising tool for the development of super miniaturized biological sensors.

CNTs have been employed with a variety of biosensors, taking advantage of their useful physical and chemical properties. In general, depending on the type of transduction mechanism the biosensors can be classified as (i) electrochemical (ii) optical and (iii) field-effect sensors.

5.2.1 Electrochemical Biosensors

Electrochemical sensors characterize the aged and frequently used class of biosensors, to date. They are popular due to their low-cost, relatively fast response times, ease of use, and portability. CNTs can be highly advantageous to electrochemical sensors because of their high surface areas and exceptional electrical properties. The sensitivity and detection limit of electrochemical sensors can be improved using CNTs; this can be achieved by increasing the active surface area and enhancing the immobilization of biological recognition elements of electrodes with improved electron transfer.

5.2.1.1 Electrochemical Enzyme Sensors

An enzyme-coupled electrochemical biosensor induces an electric signal by catalyzing a reaction producing electroactive species. The biosensor usually consists of a reference electrode, a working electrode, and a counterelectrode. Analytes immobilized on the electrode which recognizes the target analyte catalyze electron transfer and contribute to the production of a voltage and current, depending on the enzymes used. Adsorption at the electrode surfaces is likely the easiest method for immobilization of enzymes on CNTs. The CNTs can be layered onto the electrodes either by drop-casting [21] or by direct synthesis of CNTs on the electrode surfaces [22]. Functionalization with enzymes onto electrodes CNTs in solution or directly onto electrodes is done via adsorption of enzymes or by covalently attaching enzymes to CNTs [23, 24].

More complex techniques involve dispersion of CNTs in ionic liquids [25] or in polymers such as Nafion [26], chitosan [27], or polymermediators [28]. Furthermore, researchers have also examined enzyme sensors using electrodes functionalized with CNTs and nanoparticles of gold [27] and platinum [29], and graphene.

It has been reported that CNTs are typically randomly distributed over a polymer film or electrode surface in most CNT-based enzyme sensors. However, many examples of vertically aligned CNTs enzyme sensors have also been revealed. This illustrates an attractive technique where vertically aligned CNTs can be formed either by the direct synthesis [22, 31] or via accumulation of CNT dispersions onto functionalized substrates of Fe^{3+} precipitated hydroxides [31]. Further, enzymes can either be immobilized through adsorption [22] or by means of covalent bonding, generally through a -COOH functionalized CNT at the end group [30, 31]. It is noteworthy that CNTs have been explored for the achievement of direct electron transfer (DET) between electrode surfaces and redox enzymes. This is technically difficult to achieve due to the inside location of redox centers of enzymes as the protein shells are shielded inside. Using functionalized SWCNTs DET has been exhibited successfully by exploiting their nano-scale size and conductivity to make 'molecular wires' between redox cores of enzymes and the electrode surfaces [32, 33].

5.2.1.2 Electrochemical Immunosensors

Immunosensor devices are based on the binding interactions between antigens and antibodies. In this approach, either the antibody or the corresponding antigen is immobilized on the sensor surface; a stable complex is formed via the binding of the corresponding antigen or antibody, which can be detected via various methods. Antibodies are frequently used in biosensing applications due to their highly specific bonding with corresponding antigens [34]. A commonly used method in CNT-immunosensors is sandwich assay, in this method, the antibodies bind at the electrode functionalized with CNTs, either directly or through another binding material present at the surface of the electrode. For detecting analyte protein binding, secondary antibodies with reporter molecules are used. The binding of reporter molecules to the immobilized analyte at the electrode surface will lead to a change in the electrical signal [35, 36].

Riberi and coworkers [36] designed an amperometry sensor (Figure 5.1) using a similar approach for detecting zearalenone (ZEA) with a detection limit of 0.15×10^{-12} g/mL. Zearalenone is a mycotoxin in maize. Here, MWCNTs were immobilized in a polyethyleneimine composite, over which deposition of AuNPs was done by drop-casting. The immunosensor

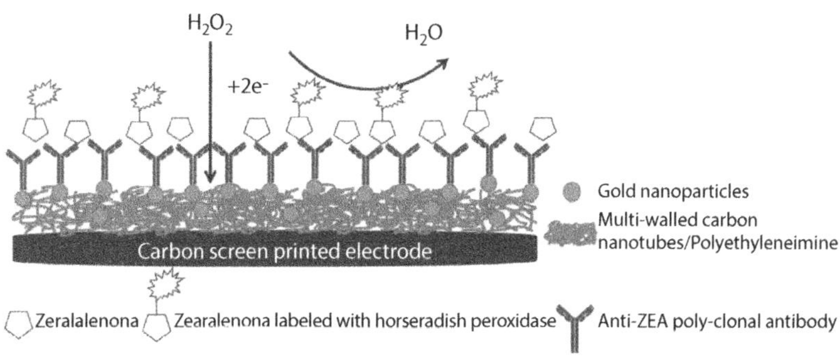

Figure 5.1 Schematic representation of the electrochemical immunosensor to determine ZEA using the ZEA-pAb/AuNPs/MWCNT/PEI/CSPE [36].

demonstrated outstanding analytical performance in terms of limit of detection (0.15 pg mL^{-1}) and sensitivity (SC_{50} = 2 pg mL^{-1}), as well as good repeatability and acceptable precision.

Similarly, Rusling *et al.* [37] Reported vertically aligned CNTs-based Immunosensors. The group has developed SWCNT forests by dipping Nafion and Fe(OH)x functionalized graphite electrodes into dispersions of SWCNT. Further, HRP-labeled antibodies are used for carrying out the sandwich assay. Thus, here an amperometric immunosensor for matrix metalloproteinase-3 (MMP-3)was developed with a 4×10^{-12} g/mL detection limit. MMP-3was a biomarker related to squamous cell carcinoma. This group further used the same technique for detecting interleukin 6 (IL-6) protein, a biomarker for oral cancer, with 0.5×10^{-12} g/mL detection limit. Additionally, Immunosensors using electrochemical impedance spectroscopy (EIS) eliminate the need for label molecules. Methods employing redox probes such as ferricyanide have also been reported [38]. In this method because of interaction between the sensor surface and the analyte, the process of electron transfer to the redox probe was retarded, causing a drop in current. By combining this method with square wave voltammetry, a detection limit of 0.163×10^{-12} g/mL was attained.

5.2.1.3 *Electrochemical DNA Sensors*

Due to its distinctly specific hybridization, DNA has been found to be a fascinating biorecognition element for sensors. DNA sensors have promising significant applications including food safety, health and medicine, and antiterrorism. A general technique for immobilizing nucleic acids onto electrodes functionalized with CNT is to covalently attach the amine end of the nucleic acids to the electrode surface. Chen and co-worker [39] developed a DNA sensor based on CNTs that covalently bound aDNA capture probe with amine-terminus to a copper oxide nanowire dispersion and carboxylated SWCNTs. By employing this sensor, the anthrax lethal factor up to a limit of 3.5 was detected using differential pulse voltammetry (DPV). Similarly, using electrophoresis, Ghrera and co-workers [40] deposited carboxylated MWCNTs onto an indium tin oxide (ITO) electrode by covalently attaching oligonucleotides to the MWCNTs. The Figure 5.2(a-d) showed various steps involved in the fabrication of pDNA/MWCNT/ITO biochip. The biosensors covalently immobilized aminated capture probe has been discovered to precisely hybridize with its complementary DNA, allowing it to detect target oligonucleotide at concentrations as low as 1 fM with a response time of 1 minute. The device's improved linear range and

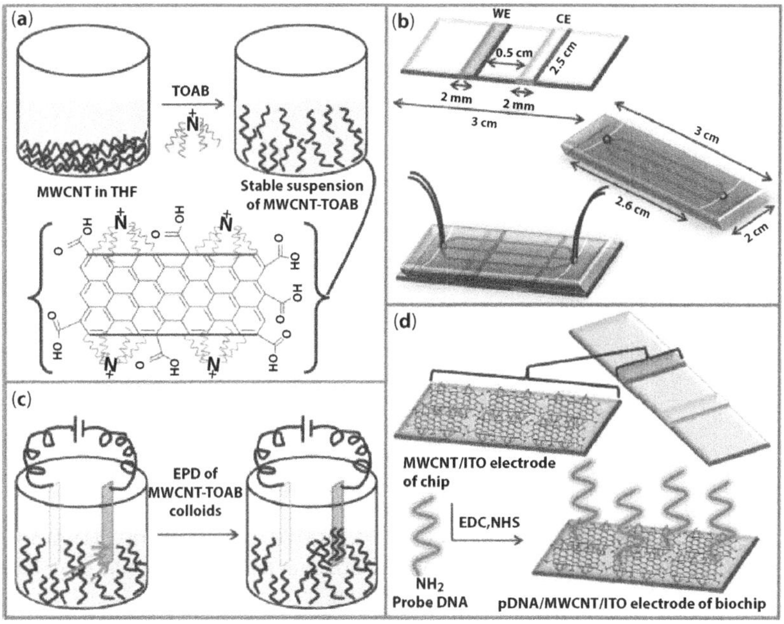

Figure 5.2 Stepwise illustration of microfluidic biochip fabrication for electrochemical detection of DNA hybridization. (a) Exfoliation of MWCNT via sonication with TOAB, (b) micropatterning of the ITO substrate and fabrication of microchannels, (c) EPD of MWCNT over the working electrode of the chip and (d) immobilization of DNA probe on the surface of MWCNT/ITO chip [40].

sensitivity can be attributed to better DNA loading thanks to MWCNT's alteration of the electrode's surface area.

A linear association was found between the Rct values and the logarithm of the complementary DNA concentrations for both biochips. In the ten-decade concentration ranges from 1 fM to 1 M, a three-fold increase in Rct value was found for the pDNA/MWCNT/ITO biochip (Figure 5.3, curve i). A considerable variance was detected with the pDNA/AS/ITO biochip (Figure 5.3, curve ii), showing that it is less suitable for sensing complementary target DNA.

Furthermore, CNT-functionalized electrodes can also be used to immobilize nucleic acids utilizing gold/thiol interactions [41, 42]. Moreover, aptamers are potential substitutes for antibodies in biosensing applications. Since they have high stability, high specificity, and selectivity in binding and can be prepared cost-effectively [43, 44]. Similar sensors based on CNTs are also an active area of research [45]. Aside from detecting DNA, RNA, and protein, metal ion has also been detected by DNA sensors based on CNTs [46].

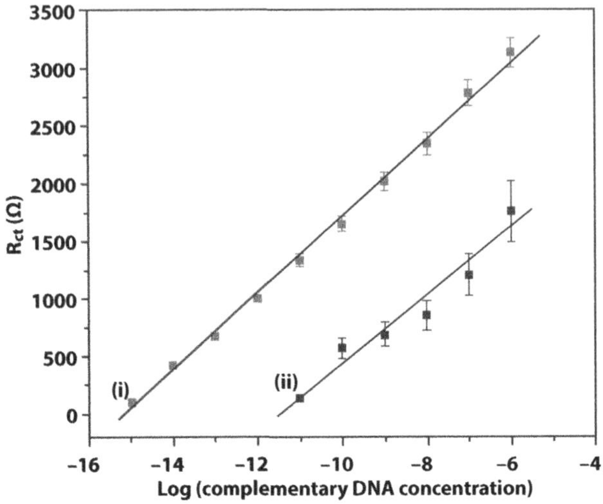

Figure 5.3 Plot of the variation in Rct of (i) pDNA/MWCNT/ITO and (ii) pDNA/AS/ITO biochips as a function of target DNA concentrations on logarithm scale. Error bars represent deviation of data points from linearity [40].

5.2.1.4 *Non-Biomolecule-Based Electrochemical Sensors*

The CNT sensors comprising biomolecules are the most used electrochemical sensors, but there are numerous examples where biomolecules are not used as recognition elements in CNT-based electrochemical sensors. In the past few years, sensors based on molecularly imprinted polymers (MIPs) have been fascinating the researchers. MIPs are synthetically analogous to antibodies. Here, polymers comprising the target molecule are synthesized, bound to the polymer structure covalently or non-covalently. This is followed by the separation of the target molecule from the polymer matrix, which in turn results in the formation of cavities (left behind the target molecule) of a certain shape, causing rebinding with the target molecule. Selective binding of MIP to the target in solution is like the binding of antibody/antigen. Compared to antibodies, MIPs may be more stable and inexpensive to mass-produce as recognition elements for sensors [47, 48]. For MIP applications, CNTs are potential candidates as they can enhance the electrical conductivity, mechanical properties, and porosity of polymer composites [49, 50]. There are many reports available, where MEPs containing functionalized MWCNTs were developed for the detection of the analyte involving epinephrine [49], myoglobin [51], and the cardiac biomarker troponin T [50]. These CNT-MIP sensors achieved detection at concentrations less than nanogram [49] or picogram [50].

Also, CNTs have been extensively used with other nanomaterials to functionalize electrode surfaces for direct electrochemical detection of a variety of analytes, resulting in greater sensitivity. Along with CNTs, nanoparticles of gold and silver are also commonly used for functionalizing electrodes for detecting analytes like bisphenol A [52], urea [53] glucose [54], and volatile organic compounds (VOCs) [55]. Nanoparticles of graphene oxide and graphene are also usually employed along with CNTs to enhance the performance of electrodes [56, 57].

5.2.2 Optical CNT Sensors

The optical biosensor detects analytes by measuring the light absorbed or emitted in catalytic and affinity reactions. Optical detection of the analyte can be achieved via absorption reflection, infrared, Raman, fluorescence, and resonance providing the nature of the biochemical reaction. The most popular optical detection techniques are fluorescence and surface plasmon resonance (SPR) because of their sensitivity and selectivity toward biochemical reactions. In fluorescent sensors and biosensors, CNTs are extensively used due to their fluorescent properties which are largely determined by their structure [58].

Further, CNTs can be functionalized with different types of biomolecules to utilize their fluorescent properties in the field of biosensing [59–62]. Along with this, fluorescent quenching can also be accomplished with CNTs. Bio-recognition molecules like oligonucleotides can be labeled using Fluorophores. Once the oligonucleotides have been attached to the CNTs non-covalently, the fluorescent quenching of fluorophores can be done by the CNTs through Förster resonance energy transfer (FRET). When oligonucleotides are present with analyte molecules, the analytes bind preferably to them, displace the CNTs, and increase the fluorescent signal significantly [63–66]. Furthermore, SPR is one of the most sensitive optical techniques universally used for chemical and biological sensing. In this technique, exploitation of surface plasmon waves is done which formed between metal surface interface and dielectric, generated via optical stimulus. The detection of binding of molecules at the surface of the metal is done via alteration in the refractive index at the surface [67–69]. Lateral flow sensors based on CNTs have also been reported in the literature for visual detection. In this technique CNTs deposition at the test site leads to the formation of a black band when the antibody or oligonucleotide capture probe is bound to the test strip and used in combination with a CNT-labelled capture probe in a sandwich assay [70, 71].

5.2.3 Field-Effect CNTs Sensors

Field-effect transistors (FETs) are microelectronic devices primarily used in electronic equipment and have three electrodes: a drain, a source, and an agate electrode. Typically, a reference electrode is used in conjunction with a liquid medium to replace the gate electrode for sensing applications. Charge carriers are influenced by changes in the electric field created by ionization or molecule binding at the surface of the channel, affecting the source/drain current. A variety of biorecognition elements have traditionally been applied to FETs in the field of biosensing. CNT-based field-effect biosensors have been a popular research field for many years because of their large surface area and high conductivity [72, 73]. Field-effect biosensors typically consist of single-walled CNTs, since SWCNTs are p-type semiconductors, making them perfect for forming semi-conductive channels [73]. There are two ways in which CNTs are dispersed in field-effect devices: either aligned between electrodes [74, 75] or arranged randomly between electrodes (the source and drain electrode) [76, 77]. To fabricate CNTs network devices, a dispersion of CNT sis made in a solvent like dimethylformamide (DMF), and further drop-casted onto the substrate. Alternatively, the electrodes can be directly fabricated on top of a substrate using standard photolithographic techniques, either between prefabricated electrodes or directly on the substrate [74]. CNTs-based FET biosensors have been widely applied to the detection of analytes based on nucleic acid and proteins [78, 79], employing various detect techniques, device configurations, and fabrication methods. These are also very frequently used in the field of immune sensors [80–82].

However, it is difficult to detect metal ions, biomolecules, and metabolites that are smaller in size, using standard FET-based detection methods. This limitation could be attributed to the fewer charged analyte, weak electric field effect due to small size, and few or single binding sites on transducer elements. These shortcomings also limit the signal amplification in sandwich assay methods. Using SWCNT-based FET biosensors [83, 84] these challenges have been overcome in many studies. This technique is based on the principle of competitive displacement, which involves biorecognition molecules bound to transducer immobilized analytes being displaced by unbound analytes present in the samples. FET-based CNTs biosensors are of great interest as new fabrication methodologies and materials provide cheaper, flexible, and mass-producible potential devices [85, 86].

5.2.4 CNT Human Strain Sensor

The alkali siloxane-modified CNT showed excellent human body movement sensing, the organic molecules mimicked in CNT and conductivity showed at 157.64 S/m due to the high surface. And the sensing movement showed 40% sensing after 500 cycles of stretch in the leg. Further, the body joint stretching such as hand, fingers, wrists, legs elbows, Knees are analyzed sensing behavior as shown in Figure 5.4(a-c) achievement symbol of different direction movement, and the finger stretching increased the

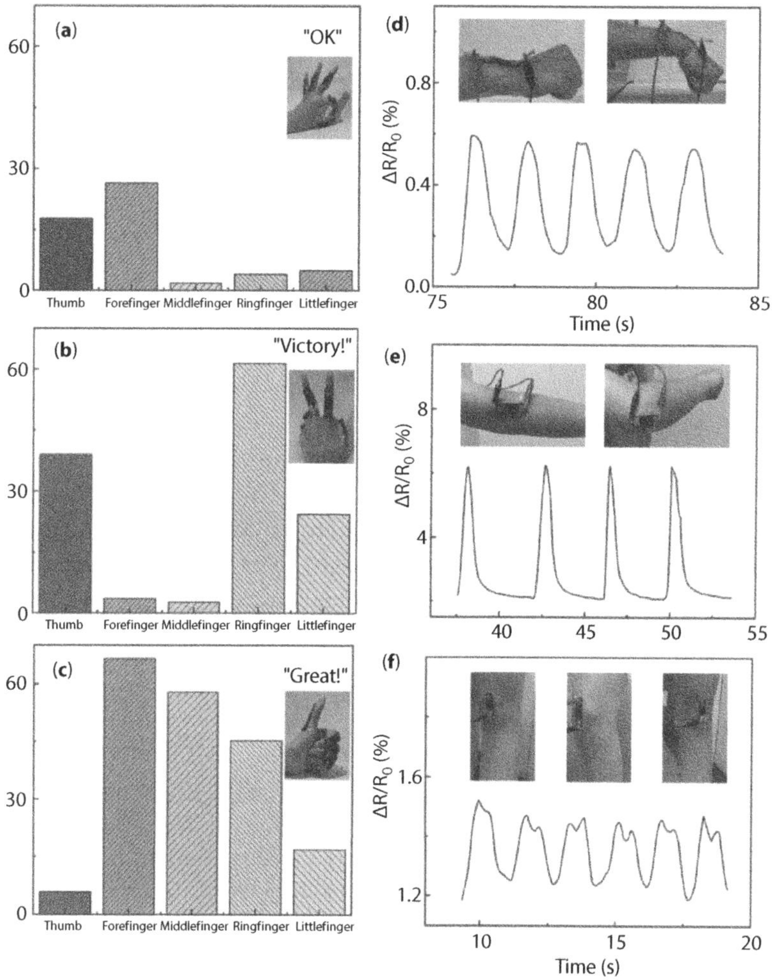

Figure 5.4 The fingers movement (a-c) and (d-f) leg knees stretching and their electrical signals [87].

conductive is decreased because of stretching resistance increased. And Figure 5.4(d-f) showed that the electrical signal is changing with respect to the movement of the knee joint.

5.3 Conclusion

CNTs offer a great platform to design a huge range of biosensors due to their distinct electronic properties like higher surface-to-volume ratio and outstanding mechanical properties. The major advantage of using CNTs in CNT-based biosensors is their enhanced detection performance of the target. Although, there are limited commercial applications of CNTs for sensing, and the reasons are the efficient growth of macroscopic size CNTs, the longevity of nanotube-based biosensors, the controllable growth of nanotubes on needed surfaces, the dispersion of CNTs in desired matrix or solution, and the control over their structure, and the scope of fabrication of biosensors using CNT. Along with this, the compatibility of biomaterial with the synthesis process must be checked. Since the synthesis of CNTs-based biosensors is done on an electrode surface or as an element of a field-effect device. For example, the requirement of high temperatures in the arc discharge and laser ablation synthesis methods or preparation of CNT scan harm many potential substrates. CNTs, therefore, account for universal and multifunctional nanostructures which imbibe the potential characteristics to be used in diagnostic and therapeutic applications. Although these may be employed as biosensors to detect cancer biomarkers with high sensitivity and cost-effective micro-fabrication, anticancer drugs could also be loaded on these and employed in oncology with therapeutic aims. They may also be applied for photoacoustic cancer cells molecular imaging and photothermic ablation of tumors.

References

1. Iijima, S., Helical microtubules of graphitic carbon. *Nature*, 354, 56–58, 1991.
2. Zhu, Z.G., Garcia-Gancedo, L., Flewitt, A.J., Xie, H., Moussy, F., Milne, W., II, Critical review of glucose biosensors Based on Carbon Nanomaterials: Carbon nanotubes and graphene. *Sensors*, 12, 5996–6022, 2012.
3. Dresselhaus, M.S., Dresselhaus, G., Avouris, P., *Carbon Nanotubes: Synthesis, Structure, Properties and Applications*, p. 464, Springer-Verlag, Berlin Heidelberg, 2008.

4. D'yachkov, P.N., *Elektronnyesvoistva Iprimenenienanotrubok [Electronic Properties and Applications of Nanotubes]*, p. 488, Binom, Laboratoriyaznanii, Moscow, 2010.
5. Forro, L., Salvetat, J.-P., Bonard, J.-M., Basca, R., Electrical and mechanical properties of carbon nanotubes, in: *Fundamental Materials Research III*, pp. 2969–2978.
6. De, B., Banerjee, S., Verma, K.D., Pal, T., Manna, P.K., Kar, K.K., Carbon nanotube as electrode materials for supercapacitors, in: *Handbook of Nanocomposite Supercapacitor Materials II*, pp. 229–43, Springer, Cham 2020.
7. Odom, T.W., Huang, J.L., Kim, P., Lieber, C.M., Atomic structure and electronic properties of single-walled carbon nanotubes. *Nature*, 391, 62–64, 1998.
8. Maruyama, B. and Alam, K., Carbon nanotubes and nanofibers in composite materials. *SAMPE J.*, 38, 59–70, 2002.
9. Collins, P.G. and Avouris, P., *Nanotubes for Electronics-Scientific American*, vol. 283, pp. 62–69, Nature Publishing Group, San Francisco, 2000.
10. Wang, J., Carbon-nanotube based electrochemical biosensors: A review. *Electroanalysis*, 17, 7–14, 2005.
11. Balasubramanian, K. and Burghard, M., Chemically functionalized carbon nanotubes. *Small*, 1, 180–192, 2005.
12. Lipomi, D.J., Vosgueritchian, M., Tee, B.C.-K., Hellstrom, S.L., Lee, J.A., Fox, C.H., Bao, Z., Skin-like pressure and strain sensors based on transparent elastic films of carbon nanotubes. *Nat. Nanotechnol.*, 6, 788–792, 2011.
13. Kong, J., Franklin, N.R., Zhou, C., Chapline, M.G., Peng, S., Cho, K., Dai, H., Nanotube molecular wires as chemical sensors. *Science*, 287, 622–625, 2000.
14. Cui, Y., Wei, Q., Park, H., Lieber, C.M., Nanowire nanosensors for highly sensitive and selective detection of biological and chemical species. *Science*, 293, 1289–1292, 2001.
15. Li, J., Ng, H.T., Cassell, A., Fan, W., Chen, H., Ye, Q., Koehne, J., Han, J., Meyyappan, M., Carbon nanotube nanoelectrode array for ultrasensitive DNA detection. *Nano Lett.*, 3, 597, 2003.
16. Pangule, R.C., Brooks, S.J., Dinu, C.Z., Bale, S.S., Salmon, S.L., Zhu, G., Metzger, D.W., Kane, R.S., Dordick, J.S., Antistaphylococcal nanocomposite films based on enzyme–nanotube conjugates. *ACS Nano*, 7, 3993–4000, 2010.
17. Kumar, A.S., Gayathri, P., Barathi, P., Vijayaraghavan, R., Improved electric wiring of hemoglobin with impure-multiwalled carbon nanotube/nafion modified glassy carbon electrode and its highly selective hydrogen peroxide biosensing. *J. Phys. Chem. C*, 116, 23692–23703, 2012.
18. Turner, A.P., Biosensors: Sense and sensibility. *Chem. Soc. Rev.*, 42, 3184–3196, 2013.
19. Akhmadishina, K.F., Bobrinetskii, I., II, Komarov, I.A., Malovichko, A.M., Nevolin, V.K., Petukhov, V.A., Golovin, A.V., Zalevskii, A.O., Flexible

biological sensors based on carbon nanotube films. *Nanotechnol. Russ.*, 8, 721–726, 2013.

20. Zhang, W.-D. and Zhang, W.-H., Carbon nanotubes as active components for gas sensors. *J. Sensors*, 2009, 16, 2009.
21. Madhurantakam, S., Babu, K.J., Rayappan, J.B.B., Fabrication of mediator-free hybrid nano-interfaced electrochemical biosensor for monitoring cancer cell proliferation. *Biosens. Bioelectron.*, 87, 832–841, 2017.
22. Wang, S.G., Zhang, Q., Wang, R., Yoon, S.F., Ahn, J., Yang, D.J., Tian, J.Z., Li, J.Q., Zhou, Q., Multi-walled carbon nanotubes for the immobilization of enzyme in glucose biosensors. *Electrochem. Commun.*, 5, 800–803, 2003.
23. Lee, Y.-M., Kwon, O.-Y., Yoon, Y.-J., Ryu, K., Immobilization of horseradish peroxidase on multi-wall carbon nanotubes and its electrochemical properties. *Biotechnol. Lett.*, 28, 39–43, 2006.
24. Ruhal, A., Rana, J.S., Kumar, S., Kumar, A., Immobilization of malate dehydrogenase on carbon nanotubes for development of malate biosensor. *Cell. Mol. Biol.*, 58, 15–20, 2012.
25. Zappi, D., Caminiti, R., Ingo, G.M., Sadun, C., Tortolini, C., Antonelli, M.L., Biologically friendly room temperature ionic liquids and nanomaterials for the development of innovative enzymatic biosensors. *Talanta*, 175, 566–572, 2017.
26. Hrapovic, S., Liu, Y., Male, K.B., Luong, J.H.T., Electrochemical biosensing platforms using platinum nanoparticles and carbon nanotubes. *Anal. Chem.*, 76, 1083–1088, 2004.
27. Magar, H.S., Ghica, M.E., Abbas, M.N., Brett, C.M.A., A novel sensitive amperometric choline biosensor based on multiwalled carbon nanotubes and gold nanoparticles. *Talanta*, 167, 462–469, 2017.
28. Dervisevic, M., Custiuc, E., Cevik, E., Senel, M., Construction of novel xanthine biosensor using polymeric mediator/MWCNT nanocomposite layer for fish freshness detection. *Food Chem.*, 181, 277–283, 2015.
29. Chen, C., Ran, R., Yang, Z., Lv, R., Shen, W., Kang, F., Huang, Z.-H., An efficient flexible electrochemical glucose sensor based on carbon nanotubes/carbonized silk fabrics decorated with Pt microspheres. *Sens. Actuators B Chem.*, 256, 63–70, 2018.
30. Lin, Y., Lu, F., Tu, Y., Ren, Z., Glucose biosensors based on carbon nanotube nanoelectrode ensembles. *Nano Lett.*, 4, 191–195, 2004.
31. Yu, X., Chattopadhyay, D., Galeska, I., Papadimitrakopoulos, F., Rusling, J.F., Peroxidase activity of enzymes bound to the ends of single-wall carbon nanotube forest electrodes. *Electrochem. Commun.*, 5, 408–411, 2003.
32. Tasca, F., Harreither, W., Ludwig, R., Gooding, J.J., Gorton, L., Cellobiosede hydrogenase aryl diazonium modified single walled carbon nanotubes: Enhanced direct electron transfer through a positively charged surface. *Anal. Chem.*, 83, 3042–3049, 2011.
33. Felix, F.S. and Angnes, L., Electrochemical immunosensors—A powerful tool for analytical applications. *Biosens. Bioelectron.*, 102, 470–478, 2018.

34. Silva, M.M.S., Dias, A.C.M.S., Silva, B.V.M., Gomes-Filho, S.L.R., Kubota, L.T., Goulart, M.O.F., Dutra, R.F., Electrochemical detection of dengue virus NS1 protein with a poly(allylsamine)/carbon nanotube layered immunoelectrode. *J. Chem. Technol. Biotechnol.*, 90, 194–200, 2015.
35. Gomes-Filho, S.L.R., Dias, A.C.M.S., Silva, M.M.S., Silva, B.V.M., Dutra, R.F., A carbon nanotube-based electrochemical immunosensor for cardiac troponin T. *Microchem. J.*, 109, 10–15, 2013.
36. Riberi, W.I., Tarditto, L.V., Zon, M.A., Arevalo, F.J., Fernandez, H., Development of an electrochemical immunosensor to determine zearalenone in maize using carbon screen printed electrodes modified with multi-walled carbon nanotubes/polyethyleneimine dispersions. *Sens. Actuators B Chem.*, 254, 1271–1277, 2018.
37. Munge, B.S., Fisher, J., Millord, L.N., Krause, C.E., Dowd, R.S., Rusling, J.F., Sensitive electrochemical immunosensor for matrix metalloproteinase-3 based on single-wall carbon nanotubes. *Analyst*, 135, 1345–1350, 2010.
38. Kalyani, T., Sangili, A., Nanda, A., Prakash, S., Kaushik, A., Jana, S.K., Bio-nanocomposite based highly sensitive and label-free electrochemical immunosensor for endometriosis diagnostics application. *Bioelectrochemistry*, 139, 107740, 2021.
39. Chen, M., Hou, C., Huo, D., Yang, M., Fa, H., An ultrasensitive electrochemical DNA biosensor based on a copper oxide nanowires/single-walled carbon nanotube nanocomposite. *Appl. Surf. Sci.*, 364, 703–709, 2016.
40. Ghrera, A.S., Pandey, C.M., Malhotra, B.D., Multiwalled carbon nanotube modified microfluidic-based biosensor chip for nucleic acid detection. *Sens. Actuators B Chem.*, 266, 329–336, 2018.
41. Liu, X., Shuai, H.-L., Liu, Y.-J., Huang, K.-J., An electrochemical biosensor for DNA detection based on tungsten disulfide/multiwalled carbon nanotube composites and hybridization chain reaction amplification. *Sens. Actuators B Chem.*, 235, 603–613, 2016.
42. Larrea, E., Sole, C., Manterola, L., Goicoechea, I., Armesto, M., Arestin, M., Caffarel, M.M., Araujo, A.M., Araiz, M., Fernandez-Mercado, M. *et al.*, New concepts in cancer biomarkers: Circulating miRNAs in liquid biopsies. *Int. J. Mol. Sci.*, 17, 627, 2016.
43. So, H.-M., Won, K., Kim, Y.H., Kim, B.-K., Ryu, B.H., Na, P.S., Kim, H., Lee, J.-O., Single-walled carbon nanotube biosensors using aptamers as molecular recognition elements. *J. Am. Chem. Soc.*, 127, 11906–11907, 2005.
44. Maehashi, K., Katsura, T., Kerman, K., Takamura, Y., Matsumoto, K., Tamiya, E., Label-free protein biosensor based on aptamer modified carbon nanotube field-effect transistors. *Anal. Chem.*, 79, 782–787, 2007.
45. Rostamabadi, P.F. and Heydari-Bafrooei, E., Impedimetric aptasensing of the breast cancer biomarker HER2 using a glassy carbon electrode modified with gold nanoparticles in a composite consisting of electrochemically reduced graphene oxide and single walled carbon nanotubes. *Microchim. Acta*, 186, 495, 2019.

46. Zhang, Z. and Yan, J., A signal-on electrochemical biosensor for sensitive detection of sliver ion based on alkanethiol-carbon nanotube-oligonucleotide modified electrodes. *Sens. Actuators B Chem.*, 202, 1058–1064, 2014.
47. Beyazit, S., Bui, B.T.S., Haupt, K., Gonzato, C., Molecularly imprinted polymer nanomaterials and nanocomposites by controlled/living radical polymerization. *Prog. Polym. Sci.*, 62, 1–21, 2016.
48. BelBruno, J.J., Molecularly imprinted polymers. *Chem. Rev.*, 119, 94–119, 2019.
49. Prasad, B.B., Prasad, A., Tiwari, M.P., Madhuri, R., Multiwalled carbon nanotubes bearing 'terminal monomeric unit' for the fabrication of epinephrine imprinted polymer-based electrochemical sensor. *Biosens. Bioelectron.*, 45, 114–122, 2013.
50. Phonklam, K., Wannapob, R., Sriwimol, W., Thavarungkul, P., Phairatana, T., A novel molecularly imprinted polymer PMB/MWCNTs sensor for highly-sensitive cardiac troponin T detection. *Sens. Actuators B Chem.*, 308, 127630, 2020.
51. Wang, Y., Han, M., Ye, X., Wu, K., Wu, T., Li, C., Voltammetric myoglobin sensor based on a glassy carbon electrode modified with a composite film consisting of carbon nanotubes and a molecularly imprinted polymerized ionic liquid. *Microchim. Acta*, 184, 195–202, 2017.
52. Messaoud, N.B., Ghica, M.E., Dridi, C., Ali, M.B., Brett, C.M.A., Electrochemical sensor based on multiwalled carbon nanotube and gold nanoparticle modified electrode for the sensitive detection of bisphenol A. *Sens. Actuators B Chem.*, 253, 513–522, 2017.
53. Kumar, T.H.V. and Sundramoorthy, A.K., Non-enzymatic electrochemical detection of urea on silver nanoparticles anchored nitrogen doped single-walled carbon nanotube modified electrode. *J. Electrochem. Soc.*, 165, 3006–3016, 2018.
54. Kangkamano, T., Numnuam, A., Limbut, W., Kanatharana, P., Thavarungkul, P., Chitosan cryogel with embedded gold nanoparticles decorated multi-walled carbon nanotubes modified electrode for highly sensitive flow based non-enzymatic glucose sensor. *Sens. Actuators B Chem.*, 246, 854–863, 2017.
55. Zhang, Y., Gao, G., Liu, H., Fu, H., Fan, J., Wang, K., Chen, Y., Li, B., Zhang, C., Zhi, X. *et al.*, Identification of volatile biomarkers of gastric cancer cells and ultrasensitive electrochemical detection based on sensing interface of Au-Ag alloy coated MWCNTs. *Theranostics*, 4, 154–162, 2014.
56. Asadian, E., Shahrokhian, S., Zad, A.I., Ghorbani-Bidkorbeh, F., Glassy carbon electrode modified with 3D graphene-carbon nanotube network for sensitive electrochemical determination of methotrexate. *Sens. Actuators B Chem.*, 239, 617–627, 2017.
57. Huang, B., Liu, J., Lai, L., Yu, F., Ying, X., Ye, B.-C., Li, Y., A free standing electrochemical sensor based on graphene foam-carbon nanotube composite coupled with gold nanoparticles and its sensing application for

electrochemical determination of dopamine and uric acid. *J. Electroanal. Chem.*, 801, 129–134, 2017.
58. Li, C. and Shi, G., Carbon nanotube-based fluorescence sensors. *J. Photochem. Photobiol. C*, 19, 20–34, 2014.
59. Heller, D.A., Pratt, G.W., Zhang, J., Nair, N., Hansborough, A.J., Boghossian, A.A., Reuel, N.F., Barone, P.W., Strano, M.S., Peptidesecondary structure modulates single-walled carbon nanotube fluorescence as a chaperone sensor for nitroaromatics. *Proc. Natl. Acad. Sci. U.S.A.*, 108, 8544–8549, 2011.
60. Zhang, J., Kruss, S., Hilmer, A.J., Shimizu, S., Schmois, Z., De La Cruz, F., Barone, P.W., Reuel, N.F., Heller, D.A., Strano, M.S., Arapid, direct, quantitative, and label-free detector of cardiac biomarker troponin T using near-infrared fluorescent single-walled carbon nanotube sensors. *Adv. Healthc. Mater.*, 14, 412–423, 2014.
61. Safee, M.M., Gravely, M., Roxbury, D., A wearable optical microfibrous biomaterial with encapsulated nanosensors enables wireless monitoring of oxidative stress. *Adv. Funct. Mater.*, 31, 2006254, 2021.
62. Zhang, J., Boghossian, A.A., Barone, P.W., Rwei, A., Kim, J.-H., Lin, D., Heller, D.A., Hilmer, A.J., Nair, N., Reuel, N.F. *et al.*, Single molecule detection of nitric oxide enabled by d(AT)15 DNA adsorbed to near infrared fluorescent single-walled carbon nanotubes. *J. Am. Chem. Soc.*, 133, 567–581, 2011.
63. Ma, H., Xue, N., Li, Z., Xing, K., Miao, X., Ultrasensitive detection of miRNA-155 using multi-walled carbon nanotube-gold nanocomposites as a novel fluorescence quenching platform. *Sens. Actuators B Chem.*, 266, 221–227, 2018.
64. Wang, X., Wang, C., Qu, K., Song, Y., Ren, J., Miyoshi, D., Sugimoto, N., Qu, X., Ultrasensitive detection of a prognostic indicator in early-stage cancer using graphene oxide and carbon nanotubes. *Adv. Funct. Mater.*, 20, 3967–3971, 2010.
65. Mattiske, S., Suetani, R.J., Neilsen, P.M., Callen, D.F., The oncogenic role of miR-155 in breast cancer. *Cancer Epidemiol. Biomarkers Prev.*, 21, 1236–1243, 2012.
66. Zhu, W., Qin, W., Atasoy, U., Sauter, E., Circulating microRNAs in breast cancer and healthy subjects. *BMC Res. Notes*, 2, 89, 2009.
67. Fan, X., White, I.M., Shopova, S.I., Zhu, H., Suter, J.D., Sun, Y., Sensitive optical biosensors for unlabeled targets: A review. *Anal. Chim. Acta*, 620, 8–26, 2008.
68. Chen, C. and Wang, J., Optical biosensors: An exhaustive and comprehensive review. *Analyst*, 145, 1605–1628, 2020.
69. Prabowo, B.A., Purwidyantri, A., Liu, K.-C., Surface plasmon resonance optical sensor: A review on light source technology. *Biosensors*, 8, 80, 2018.
70. Qiu, W., Xu, H., Takalkar, S., Gurung, A.S., Liu, B., Zheng, Y., Guo, Z., Baloda, M., Baryeh, K., Liu, G., Carbon nanotube-based lateral flow biosensor for sensitive and rapid detection of DNA sequence. *Biosens. Bioelectron.*, 64, 367–372, 2015.

71. Yao, L.T.J., Zhu, M.Z.L., Zhong, Y., Liu, G., Xue, F., Chen, W., MWCNTs based high sensitive lateral flow strip biosensor for rapid determination of aqueous mercury ions. *Biosens. Bioelectron.*, 85, 331–336, 2016.
72. Park, R.S., Hills, G., Sohn, J., Mitra, S., Shulaker, M.M., Wong, H.-S.P., Hysteresis-free carbon nanotube field-effect transistors. *ACS Nano*, 11, 4785–4791, 2017.
73. Tran, T.-T. and Mulchandani, A., Carbon nanotubes and graphene nano field-effect transistor-based biosensors. *Trends Anal. Chem.*, 79, 222–232, 2016.
74. Tlili, C., Myung, N.V., Shetty, V., Mulchandani, A., Label-free, chemiresistor immunosensor for stress biomarker cortisol in saliva. *Biosens. Bioelectron.*, 26, 4382–4386, 2011.
75. Martinez, M.T., Tseng, Y.-C., Ormategui, N., Loinaz, I., Eritja, R., Bokor, J., Label-free DNA biosensors based on functionalized carbon nanotube field effect transistor. *Nano Lett.*, 9, 530–536, 2009.
76. Oh, J., Yoo, G., Chang, Y.W., Kim, H.J., Jose, J., Kim, E., Pyun, J.-C., Yoo, K.-H., A carbon nanotube metal semiconductor field effect transistor-based biosensor for detection of amyloid-beta in human serum. *Biosens. Bioelectron.*, 50, 345–350, 2013.
77. Ramnani, P., Gao, Y., Ozsoz, M., Mulchandani, A., Electronic detection of microRNA at attomolar level with high specificity. *Anal. Chem.*, 85, 8061–8064, 2013.
78. Ohno, Y., Maehashi, K., Matsumoto, K., Graphene biosensor, in: *Frontiers of Graphene and Carbon Nanotubes*, pp. 91–103, Springer Japan, Tokyo, 2015.
79. Kim, D.J., Sohn, I.Y., Jung, J.-H., Yoon, O.J., Lee, N.-E., Park, J.-S., Reduced graphene oxide field-effect transistor for label-free femtomolar protein detection. *Biosens. Bioelectron.*, 41, 621–626, 2013.
80. Garcia-Aljaro, C., Cella, L.N., Shirale, D.J., Park, M., Munoz, F.J., Yates, M.V., Mulchandani, A., Carbon nanotubes-based chemiresistive biosensors for detection of microorganisms. *Biosens. Bioelectron.*, 26, 1437–1441, 2010.
81. Rajesh Sharma, V., Puri, N.K., Singh, R.K., Biradar, A.M., Mulchandani, A., Label-free detection of cardiac troponin-I using gold nanoparticles functionalized single-walled carbon nanotubes based chemiresistive biosensor. *Appl. Phys. Lett.*, 103, 203703, 2013.
82. Kim, J.P., Lee, B.Y., Lee, J., Hong, S., Sim, S.J., Enhancement of sensitivity and specificity by surface modification of carbon nanotubes in diagnosis of prostate cancer based on carbon nanotube field effect transistors. *Biosens. Bioelectron.*, 24, 3372–3378, 2009.
83. Cella, L.N., Chen, W., Myung, N.V., Mulchandani, A., Single-walled carbon nanotube-based chemiresistive affinity biosensors for small molecules: Ultrasensitive glucose detection. *J. Am. Chem. Soc.*, 132, 14, 5024–5026, 2010.

84. Tan, F., Saucedo, N.M., Ramnani, P., Mulchandani, A., Label-free electrical immunosensor for highly sensitive and specific detection of microcystin-LR in water samples. *Environ. Sci. Technol.*, 49, 9256–9263, 2015.
85. Liang, Y., Xiao, M., Wu, D., Lin, Y., Liu, L., He, J., Zhang, G., Peng, L.-M., Zhang, Z., Wafer-scale uniform carbon nanotube transistors for ultrasensitive and label-free detection of disease biomarkers. *ACS Nano*, 14, 8866–8874, 2020.
86. Kang, T.-H., Lee, S.-W., Hwang, K., Shim, W., Lee, K.-Y., Lim, J.-A., Yu, W.-R., Choi, I.-S., Yi, H., All-inkjet-printed flexible nanobio-devices with efficient electrochemical coupling using amphiphilic biomaterials. *ACS Appl. Mater. Interfaces*, 12, 24231–24241, 2020.
87. Demidenko, N.A., Kuksin, A.V., Molodykh, V.V., Pyankov, E.S., Ichkitidze, L.P., Zaborova, V.A., Tsymbal, A.A., Tkachenko, S.A., Shafaei, H., Diachkova, E., Gerasimenko, A.Y., Flexible strain-sensitive silicone-CNT sensor for human motion detection. *Bioengineering (Basel)*, 9, 36, 2022.

6

Applications of Functionalized Carbon Nanotubes in Drug Delivery Systems

N. Palaniappan[1], Małgorzata Kujawska[2*] and Kader Poturcu[3]

[1]*School of Chemicals Science, Central University of Gujarat, Gandhinagar, Gujarat, India*
[2]*Department of Toxicology, Poznan University of Medical Sciences, Poznan, Poland*
[3]*Department of Chemistry, Arts and Science Faculty, Suleyman Demirel University, Isparta, Çünür, Turkey*

Abstract

An inimitable and the most prominent invention in the part of nanotechnology are carbon nanotubes (CNTs). High surface areas, high aspect ratios and nano-sized stable tubes in CNTs unique structures, make them attractive materials in nanotechnology, nanomedicine, electronics, aerospace, biosensors, membranes, and capacitors applications. There are two main types of CNTs are single walled (SWNTs) and multi walled (MWNTs) carbon nanotubes. SWNTs and MWNTs consist of single long wrapped graphene sheet and multiple layers of graphite rolled over co-axially to form a tubular shape, respectively. Although one of the most important applications of CNTs is deliver drug molecules as nano containers, CNTs have disadvantages due to their hydrophobic nature which make these unique structures insoluble in organic and aqueous solvents. To overcome this challenge, synthetic functionalization process can be used. Thus, the solubility, biocompatibility enhances, and functionalized CNTs penetrate the cell and use as the potential vehicles for the drug delivery systems. This chapter covers the types, synthesis, toxic effects of CNTs and their applications in drug delivery to brain and cancer cells.

Keywords: Carbon nanotubes, drug delivery, nano particles, brain-targeted delivery, cancer

**Corresponding author*: kujawska@ump.edu.pl

Jeenat Aslam, Chaudhery Mustansar Hussain and Ruby Aslam (eds.) Functionalized Carbon Nanotubes for Biomedical Applications, (117–138) © 2023 Scrivener Publishing LLC

6.1 Introduction

In the last two decades, inorganic and organic nanostructures such as gold nanoparticles [1], SiO_2 nanoparticles [2], dendrimers [3], liposomes [4], polymeric nanomaterials [5], magnetic nanoparticles [6] and carbon-based nanocontainers [7, 8] have played an important role for targeted drug delivery studies. Particular interest is given to the clusters of pristine and functionalized carbon nanotubes and their unique structures with interesting physical and chemical properties make them attractive in drug delivery applications [6, 9]. The last five years of developments on the "use of Carbon nanotubes in drug delivery" data collected using web of Science is given in Figure 6.1.

CNTs, diamond and graphite are important allotropes of carbon atoms due to their valency. Among these allotropes, CNTs is considered the most attractive nanomaterials [10–12] due to their unique characteristics such as structure, dimension, strength, chemical stability, and electronic properties [13, 14]. CNTs, first observed in 1975 by Endo and later by S. Iijima in 1991, are basically formed of sp^2-hybridized pure C atoms and belong to the fullerenes (C_{60}) subfamily. CNTs have a cylindrical shape and are made of rolled-up graphene sheets [15–17]. Two different types of CNTs are: (a) single-walled carbon nanotubes (SWNTs), which are made up of a single graphene layer wrapped into a cylindrical structure, and (b) multi-walled carbon nanotubes (MWNTs), with a central tubule of nanometric diameter, surrounded by several graphitic layers spaced by a distance of about 0.34 nm (Figure 6.2) [18]. The properties can vary depending on

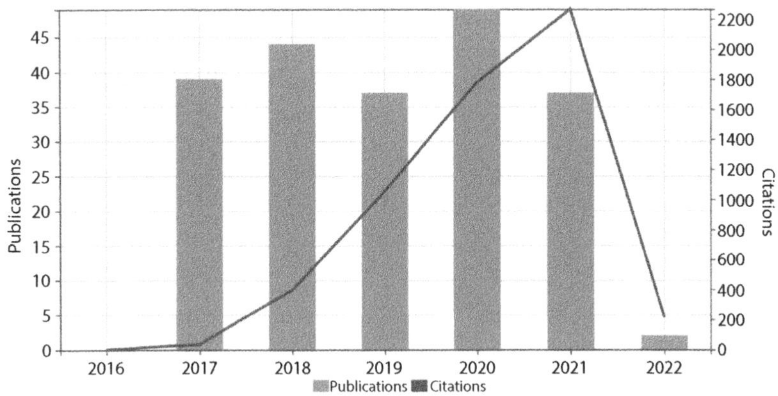

Figure 6.1 The total number of publications and citations distribution along last 5 years (2017–2022).

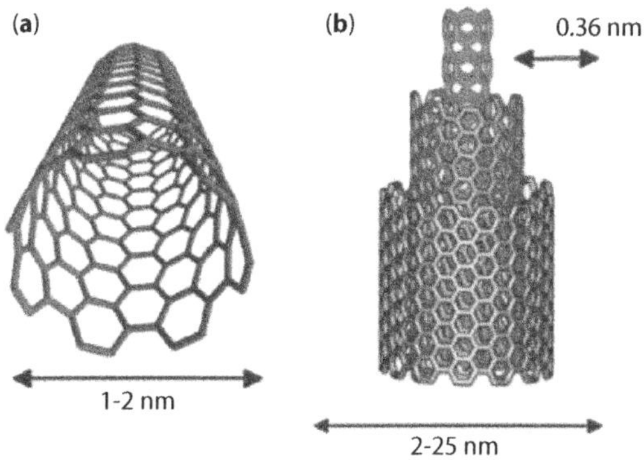

Figure 6.2 Aschematic diagram of (a) SWCNTs and (b) MWCNTs.

the types of CNTs [10, 19, 20]. SWNTs are one-dimensional nanomaterials with 1–2 nm diameters and 50nm–1cm lengths. On the other hand, MWNTs have much larger diameters (10–100 nm) [21].

Although CNTs can be synthesized naturally in a controlled flame environment by heating carbon black and graphites, produced nanotubes are mostly irregular in size, shape, mechanical strength, quality, and purity due to the uncontrollable natural environment [16, 22]. Nowadays, electric arc discharge (EAD), laser ablation technique (LA), and catalytic chemical vapor deposition (CVD) methods are widely used for synthesis of CNTs. In addition, high-pressure carbon monoxide disproportionation process (HiPCO) and plasma enhanced chemical vapored position (PE-CVD) [23] techniques are of recent interest [24].

These carbonaceous materials with their tubular shapes have novel applications in the fields of automotive industry, coatings and films, microelectronics, biotechnology, energy conversion and storage, medical and engineering applications [9]. Thanks to their specific atomic architecture coupled with high aspect ratio, characteristic configuration, excellent functionalization ability, and dynamic surface modification properties, CNTs are ideal nanomaterials for drug delivery applications [16]. In addition, CNTs with high inner volume and immobilization ability properties, are the most sought after nanomaterials in drug delivery [25] due to their versatile applications in controlled and targeted delivery, cancer treatment, bioavailability enhancement, transdermal delivery, theranostic applications, drug discovery, and delivery of proteins, peptides, RNA, DNA, siRNA [12, 16].

Generally, drug delivery systems can help the loaded drugs effectively enter the cells' interior by endocytosis through the cell membrane.

To deliver the drugs to the nucleus, the drug carrier must escape the endosomal compartment and releases loaded drug in cytosolic compartments [26, 27]. CNTs are chosen widely for their high loading of cargo molecules and excellent cell penetration ability, which enables them to deliver drug substances into neoplastic cells for selective destruction and to reduce the distribution of drugs to normal adjacent cells that aid in avoiding the toxicity of the cells, mainly in cancerous cells as chemotherapeutic and imaging agents [27]. Therefore, these nanotube systems can solve transport problems for pharmacologically relevant compounds [28, 29]. In the literature, Maleki and Soltanabadi investigated the adsorption capability of the pristine and functionalized carbon nanotubes of anticancer drug dacarbazine. They found that despite the adsorption of dacarbazine molecules on the outer sidewall of the carbon nanotubes, the spontaneous encapsulation of drug molecules into the cavity of carbon nanotubes is observed [30]. Bhatnagar *et al.* develop a new drug carrier system with SWNTs to carry cancer drug gliotoxin. They found that effective drug release was monitored against cervical and breast cancer cells [10]. Kamel *et al.* investigated the solvent and co-solvent effects on the adsorption strength between flutamide used for prostate cancer treatment and SWNTs for different configurations in the water solution based on the density calculation. The negative values of the adsorption and solvation energies demonstrate that all theoretical calculations are stable, and the interaction of the drug molecule with the nanotube sidewall is a spontaneous process [31]. Arora *et al.* evaluated cellular uptake of anticancer drug docetaxel conjugated MWNTs in human breast cancer cells (MCF-7 and MDAmb-231) and they reported that drug conjugated nanotube systems is effective against cancer cells [32]. Saberinasab *et al.* investigated the efficiency of polyethylene glycol functionalized carbon nanotubes to deliver anticancer drug temozolomide based on the density functional study [33]. Sobhani *et al.* synthesized poly citric acid functionalized MWNTs and then potent anticancer agent paclitaxel was conjugated to the carboxyl functional groups of poly citric acid. This system was suitable for the release of the drug in tumor tissues and tumor cells [34]. Anbarasan *et al.* formulated doxorubicin-loaded f-MWNTs to deliver this drug-nanotube system only to cancer cells by using pH difference. This formulated system have great potential as a drug carrier [35].

Although CNTs which are widely used in biomedical applications must be biocompatible, soluble, and non-toxic, synthesized CNTs have highly hydrophobic surfaces that make their unique structures insoluble in organic solvents, aqueous system and particularly in biologically relevant buffers [27, 36, 37]. Due to the strong van der Waals forces, these structures highly agglomerated. Therefore, obtaining a well-dispersed suspension of

CNTs in water or organic solvents without physical or chemical modifications is challenging [38, 39], and surface chemistry modifications or functionalization of CNTs is required to modify them with a wide variety of molecules [36]. There are several approaches for the functionalization of CNTs, including covalent functionalization and non-covalent coating of CNTs [37, 40]. Within the covalent strategy, the organic functionalization of SWNTs and MWNTs by 1,3-dipolar cycloaddition of azomethineylides is a powerful methodology for generating amino-functionalized carbon nanotubes with high water-soluble property [41]. The 'ends and defects' and 'side walls' functionalization [42] are two subcategories of covalent functionalization. The oxidation and cycloaddition reactions are the most common types of covalent functionalization of CNTs [18, 43–46]. The oxidation reactions is carried out with strong acid such as nitric acid [46], and various functional groups can be attached to CNTs via [42], arylatio [47], and diazonium salt chemistry [48–50] reactions. During the oxidation process, -COOH groups are formed at the ends of carbon nanotubes and the defects on the sidewalls. SWNTs are broken to very short lengths (100–300 nm) [20, 51]. Although oxidized CNTs have water soluble properties, they could not use for biological applications due to the high salt content of most biological solutions, which cause aggregation of CNTs. To overcome this limitation, further modification is performed with the attachment of hydrophilic polymers (PEG) to oxidized CNTs, yielding CNT-polymer conjugates stable in biological environments [21]. Non-covalent functionalized nanotubes do not compromise the physical properties of CNTs, but improve solubility and processability. Non-covalent functionalization mainly involves surfactants, biomacromolecules, or wrapping with polymers. CNTs can be well dispersed in water using anionic, cationic, and non-ionic surfactants [52, 53]. Sodium dodecyl sulfate (SDS) [54, 55] and sodium dodecylbenzene sulfonate (NaDDBS) are commonly preferred anionic surfactants to decrease CNT aggregation in water. Cetrimonium bromide (CTAB), 2-(methacryloyloxy)ethyl]trimethylammonium chloride (MATMAC) and Triton X-100, Triton X-305 are widely used cationic and non-ionic surfactants, respectively [20].

Nowadays, nanotechnology is revolutionizing the approaches to different fields from manufacture to health.

6.2 Nanoparticles-Doped Carbon Nanotubes

The surface functionalizing of the CNTs is a well-established approach to improving CNTs solubility and tuning the drug uptake properties of CNTs

[56, 57]. The most common two functionalization methods are: electrical doping and chemical doping. Electrical doping is a physical doping method that preserves the lattice structure or chemical composition of the graphene sheets, whereas chemical doping involves substitutional atoms (e.g., N, B, S, Si) [58–61] or adsorbed inorganic (e.g., NO_2, HNO_3) [62–64] and organic molecules (e.g., toluene) [65]. Substitution (doping) of carbon atoms of CNTs with **B** or **N** is the most effective way to alter the electronic structure and chemical reactivity of CNT because additional electronic states around the Fermi level are introduced (Figure 6.3) [66]. In biomedical applications, Nitrogen-doped CNT in rats showed low toxicity and better biocompatibility [67, 68]. Nitrogen-doped carbon nanotube (N-CNSs) sponges have good electroactive, superhydrophobic, and oil-absorbent properties, as well as good absorption capacities with a loading capacity of up to 100 times its own weight [8]. The doping of CNTs with all of these characteristics make these structures appropriate carriers for various chemotherapeutic agents for local delivery, mainly for solid tumors in which the side-effects induced by chemotherapeutics significantly reduce the quality of life of patients [69]. The B atom functions as a site for adsorption of aromatic molecules, which therapeutic molecules often are, on common anticancer drug doxorubicin [68]. Therefore, both N- and B-doped nanotubes have potential biomedical uses [70, 71].

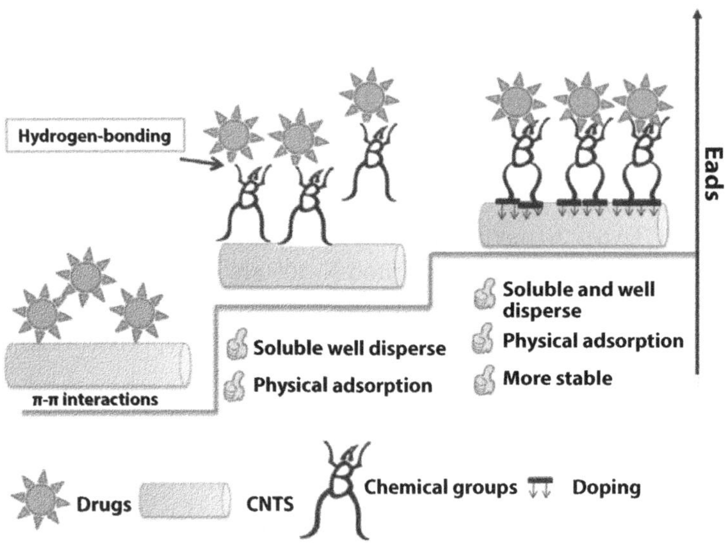

Figure 6.3 Doped CNTs for drug delivery system [66].

6.3 Brain-Targeted Delivery

The delivery of therapeutics to the central nervous system (CNS) is one of the major obstacles in treating neurological diseases. The significant drug development challenges are mainly related to crossing the blood-brain and blood-cerebrospinal fluid (CSF) barriers and systemic effects, including binding with plasma proteins, biotransformation, and excretion. The blood-brain barrier (BBB) is formed by endothelial cells, astrocytes, and pericytes. Tight junctions between endothelial cells prevent the entering of about 98% of all small molecules, and approximately 100% of large molecules cannot reach the neural tissue. The blood-CSF is a selective filter separating the blood circulation from the cerebrospinal fluid and controlling molecules' entrance into the CNS. Since this barrier is about 1000times smaller than the BBB, it is less relevant regarding CNS drug delivery [72, 73]. CNTs with outstanding cell-penetrating ability offer the potential to be developed as carriers to deliver CNS therapeutic molecules [72]. Their needle-like shape makes them suitable to cross the BBB without causing any apparent cell damage [26]. Kafa *et al.* have shown that *in vitro* "wide" MWCNTs (~35.9 nm diameter) passed a BBB co-culture model in a higher percentage than the "thin" MWCNTs (~9.2 nm diameter). *In vivo*, brain uptake of MWNTs with smaller diameters was favored after systemic injection. Conjugation of CNTs with angiopeptin-1 (ANG) improved brain distribution of wider MWCNTs [74]. Of note, CNTs can reach the CNS through the systemic, olfactory, and trigeminal pathways, which can be used as administration routes [75]. As they have the capability of delivering a vast range of cargos, including drugs, antigens, genetic materials, and biological macromolecules via the BBB [76]. CNTs represent the potential of innovative nano platforms for medicine delivery in the treatment of various neurological conditions, including infections, brain, and spinal trauma, stroke and brain cancer that has been supported in several *in vitro* and *in vivo* studies [75]. Ren *et al.* have developed the drug delivery system by modifying PEGylated oxidized MWCNTs with angiopep-2 (ANG) to target LRP receptors on BBB and glioma cells, which successfully delivered anticancer doxorubicin for glioma treatment in a mouse model. Notably, nanomedicine showed lowered cardiotoxicity compared to free doxorubicin [77]. You *et al.* have delivered oxaliplatin using carboxylated MWCNTs dual-functionalized with cell-penetrating peptidetrans-activating transcriptional activators–Polyethylenimine biotincopolymer and oxaliplatin [78]. Moreover, perspectives in neuroncology are also related to the optical feature of the CNTs, which absorb near-infrared radiation (NIR) not adsorbed by tissues. Thanks to targeting ligands, the adsorbed radiation can be converted to heat in a time and dose-dependent manner to control exclusively in the cancer cells temperature increase, i.e., hyperthermia triggering apoptosis and other death pathways [79].

In developing Alzheimer's disease (AD), acetylcholine (Ach) deficiency appears, leading to memory and cognition impairment. Due to its hydrophilic nature, Ach does not cross the BBB; therefore, its noncovalent loading into SWCNTs to distribute this neuromediator to the brain has been developed [80]. The development of berberine, an isoquinoline alkaloid used in dementia, loaded multi walled CNTs coated with polysorbate and phospholipid, has been found to be effective in managing AD by Lohan *et al.* [81]. Costa *et al.* have demonstrated that oxidized MWCNTs improved the accumulation of Pittsburgh Compound B derivatives, amyloid-β-binding molecules with low BBB permeability, in mice brains after intravenous injection [82]. Importantly, Li *et al.*, based on molecular dynamics simulation results, have suggested the ability of CNTs for direct inhibition of Aβ(16-22) and full-length Aβ fibrillation in AD [83]. Moreover, functionalized SWCNTs have been demonstrated to restore normal autophagy by reversing abnormal activation of mTOR signaling and deficits in lysosomal proteolysis, thereby improving the elimination of the mutant amyloid protein [84]. Carboxylated SWCNTs have been developed to sustain release of nanocarriers for delivering levodopa concerning Parkinson's disease (PD) [85]. Guo *et al.* loaded functionalized carboxylated SWCNT-PEG-lactoferrin with high amounts of dopamine for its targeting delivery into the striatum of parkinsonian mice after intraperitoneal injection [86]. Brain-targeted delivery of β2AR agonists downregulating α-synuclein gene expression using functionalized SWCNT or MWCNTs systems has been reviewed to overcome peripheral off-target side effects of the β2AR agonists offering potential use in managing PD [87]. Lee *et al.* have demonstrated the neuroprotective effects of amine-modified SWCNTs (a-SWNTs) in a rat model of stroke. Low levels of apoptotic, angiogenic, and inflammation markers reflected the effective blockage of ischemic damage by a-SWNTs [88]. Additionally, CNTs impregnated with subventricular zoneneural progenitor cells have been reported to repair damaged neural tissue following a stroke in rats reducing volume and area of infarct cyst that was accompanied by improved behavior [89]. It is projected that CNTs could be developed for therapeutic application in nerve regeneration. CNTs adapted to coating with nerve growth factors and neural stem cells accelerated the growth of neural networks. MWCNTs coupled with electrical stimulation can exert a synergistic effect on promoting neurite outgrowth. Although those great efforts have been made, only a few studies *in vivo* on CNTs applications aimed at accelerating the growth of neural networks have been reported so far [90].

Regarding the biocompatibility of CNTs, the potential induction of oxidative stress and related oxidative damaged products accumulation, DNA destruction, and inflammatory response are significant obstacles for their applications in neurological diseases therapy. Moreover, brain degradation and excretion are substantial concerns, a double-edged sword. Degraded and excreted too rapidly of CNTs may affect the therapeutic effect, while too high retention may cause brain damage. The rate of degradation and excretion of CNTs can be, however, tailored to the intended pharmacokinetics parameters. Moreover, contamination with byproducts of the synthesis, mainly metals and carbonaceous particles, decreasing cell viability, causing lipid peroxidation and oxygen radical formation are often responsible for the CNT toxicity. Therefore, only well-tailored, highly purified, and suitably coated CNTs can provide an acceptable safety level [75, 90].

6.4 The Organic Molecules Functionalized CNTs as Drug Delivery Vehicles

17-βEstradiol decorated MWNTs studies are performed with Doxorubicin (DOX) as drug delivery for breast cancer model. The results are showed that the DOX model has excellent inhibition efficiency in *in vitro* and *in vivo* environments. They observed that DOX along with E2- PEGMWCNT showed that specifically targeted only nuclear by estrogen receptor through due to sudden environmental changes and other genetic problems [91–95]. But still does not fulfill the drug delivery system poor materials also vary the materials physicals and chemicals properties. Hence traditional drugs delivery inorganic/organic materials failed high efficiency due to lack of chemicals properties [96–98]. In the past decades, pharmacological are used nanomaterials synthesis and delivery efficiency. As shown in Figure 6.4, nano stable store in the open-walled nanotubes along with active pharmacy materials. Nanotubes may increase functional materials released. With the varied size of nanoparticles the pharmacokinetics increased, and toxicity also decreased due to chemical properties change in molecular levels [99, 100]. However, the as shown in Figure 6.4b showed that nitrogen nonbonding electrons play a key role in transporting the active drug molecules by the support of electron exchange in the carbon nanotube moiety [101–110]. Hence the nitrogen lone pair electron could support the stable ambient environment and rapidly deliver the active drug molecules.

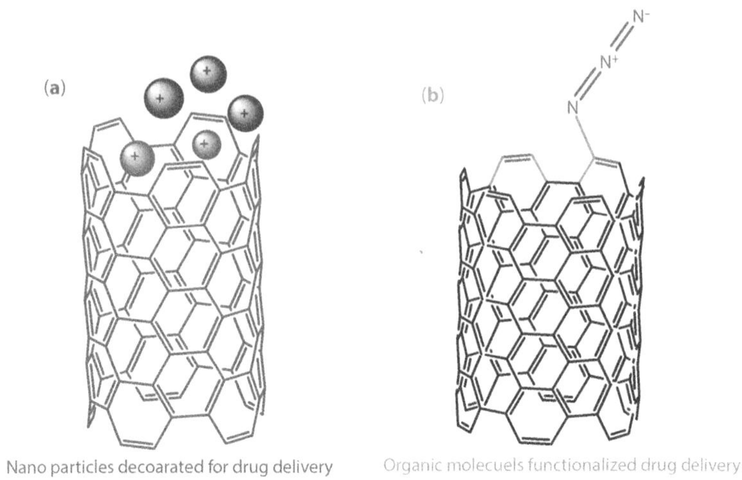

Figure 6.4 (a) Nanoparticles doped CNT as drug carrier (b) Nitrogen-based active molecules functionalized CNT.

6.5 Functionalized CNTs with Nanoparticles for Drug Active Molecular Mechanism

The CNTs have plenty of none bonding electrons in the structure. As shown in Figure 6.5, the nanoparticles as well as functionalized hetero organic molecules could be able to access the foreign particles and molecules in the microstructure [111–115]. We suggested that drug active molecules successfully release.

6.5.1 Future of Scope of Functionalized Carbon Nanotube Drug Delivery Application

Carbon based nano materials are vastly abundant in earth crust and make it new derivatives is low cost. We are proposed these functionalized carbon nano materials significantly promote the drug delivery bio science [116–119]. The carbon nanotube based organic and inorganic materials are improving the targeted delivery system by supportive high surface area and pore volume of carbon materials [120–125]. Hence carbon materials will derive several new molecules for drug deliver system by the modification SP^2 carbon structure [126–133]. However the effective carbon based molecules challenge by selective targeted molecules synthesis.

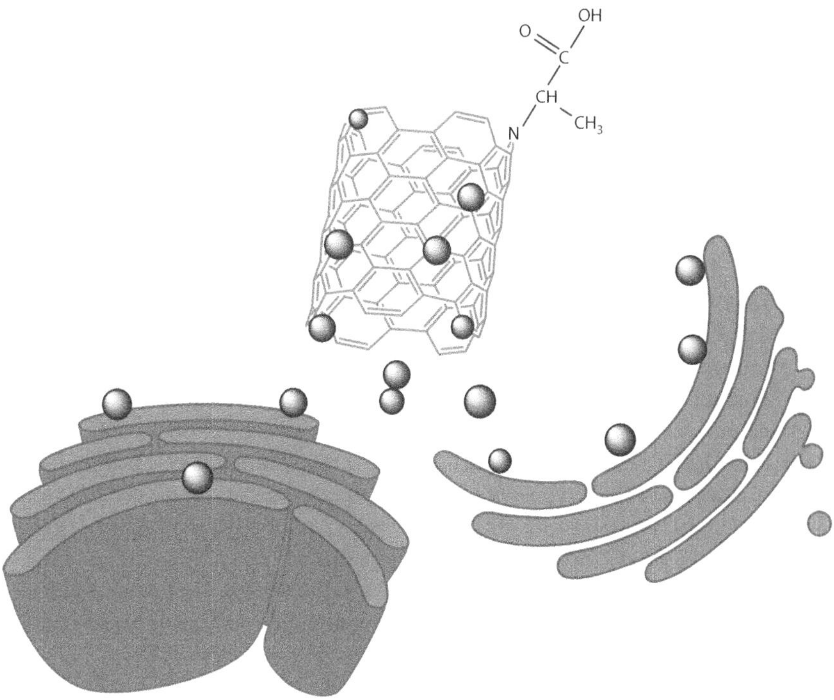

Figure 6.5 The schematic diagram for drug-loaded carbon nanoparticles releasing on respective active places.

6.6 Conclusion

We have discussed different drug delivery nano particles incorporated carbon nanotube and organic molecules functionalized materials. The nano particles decorated carbon nanotube showed improve drug deliver but toxicity issued there. Hence the biologically active organic compound showed excellent delivery efficiency the organic molecules modified carbon nanotubes will be significant drug delivery vehicle in future.

References

1. Connor, D.M. and Broome, A.M., Gold nanoparticles for the delivery of cancer therapeutics, in: *Advances in Cancer Research*, vol. 139, pp. 163–184, Academic Press Inc, 2018.

2. Li, Y., Li, N., Pan, W., Yu, Z., Yang, L., Tang, B., Hollow mesoporous silica nanoparticles with tunable structures for controlled drug delivery. *ACS Appl. Mater. Interfaces*, 9, 3, 2123–2129, 2017.
3. Chauhan, A.S., Dendrimers for drug delivery. *Molecules*, 23, 4, 938–945, 2018.
4. Pattni, B.S., Chupin, V.V., Torchilin, V.P., New developments in liposomal drug delivery. *Chem. Rev.*, 115, 19, 10938–10966, 2015.
5. Marasini, N., Haque, S., Kaminskas, L.M., Polymer-drug conjugates as inhalable drug delivery systems: A review. *Curr. Opin. Colloid Interface Sci.*, 31, 18–29, 2017.
6. Rezaei, A., Morsali, A., Bozorgmehr, M.R., Nasrabadi, M., Quantum chemical analysis of 5-aminolevulinic acid anticancer drug delivery systems: Carbon nanotube, –COOH functionalized carbon nanotube and iron oxide nanoparticle. *J. Mol. Liq.*, 340, 117182, 2021.
7. Kaur, J., Gill, G.S., Jeet, K., Applications of carbon nanotubes in drug delivery: A comprehensive review. A comprehensive review, in: *Characterization and Biology of Nanomaterials for Drug Delivery: Nanoscience and Nanotechnology in Drug Delivery*, pp. 113–135, Elsevier, 2018.
8. Karimzadeh, S., Safaei, B., Jen, T.C., Prediction effect of ethanol molecules on doxorubicin drug delivery using single-walled carbon nanotube carrier through POPC cell membrane. *J. Mol. Liq.*, 330, 115698, 2021.
9. De Volder, M.F.L., Tawfick, S.H., Baughman, R.H., Hart, A.J., Carbon nanotubes: Present and future commercial applications. *Science*, 339, 6119, 535–539, 2013.
10. Bhatnagar, I., Venkatesan, J., Kim, S.K., Polymer functionalized single walled carbon nanotubes mediated drug delivery of gliotoxin in cancer cells. *J. Biomed. Nanotechnol.*, 10, 1, 120–130, 2014.
11. Lu, X. and Chen, Z., Curved Pi-conjugation, aromaticity, and the related chemistry of small fullerenes (<C 60) and single-walled carbon nanotubes. *Chem. Rev.*, 105, 10, 3643–3696, 2005.
12. Beg, S., Rahman, M., Jain, A., Saini, S., Hasnain, M.S., Swain, S., Imam, S., Kazmi, I., Akhter, S., *Emergence in the Functionalized Carbon Nanotubes as Smart Nanocarriers for Drug Delivery Applications*, Elsevier Inc, William Andrew, 2018.
13. Lewis, G., Van Hooy-Corstjens, C.S.J., Bhattaram, A., Koole, L.H., Influence of the radiopacifier in an acrylic bone cement on its mechanical, thermal, and physical properties: Barium sulfate-containing cement versus iodine-containing cement. *J. Biomed. Mater. Res. B Appl. Biomater.*, 73, 1, 77–87, 2005.
14. Wang, F.X., The effect of HNO_3 on functionalization of CNTs. *Appl. Mech. Mater.*, 457–458, 280–283, 2014.
15. Dinadayalane, T.C. and Leszczynski, J., Fundamental structural, electronic, and chemical properties of carbon nanostructures: Graphene, fullerenes,

carbon nanotubes, and their derivatives, in: *Handbook of Computational Chemistry*, pp. 793–868, 2012.
16. Beg, S., Rizwan, M., Sheikh, A.M., Hasnain, M.S., Anwer, K., Kohli, K., Advancement in carbon nanotubes: Basics, biomedical applications and toxicity. *J. Pharm. Pharmacol.*, 63, 2, 141–163, 2011.
17. Dresselhaus, M.S., Dresselhaus, G., Charlier, J.C., Hernández, E., Electronic, thermal and mechanical properties of carbon nanotubes. *Philos. Trans. R. Soc. A Math. Phys. Eng. Sci.*, 362, 1823, 2065–2098, 2004.
18. Zare, K., Gupta, V.K., Moradi, O., Makhlouf, A.S.H., Sillanpää, M., Nadagouda, M.N., Sadegh, H., Shahryari-ghoshekandi, R., Pal, A., Wang, Z., Tyagi, I., Kazemi, M., A comparative study on the basis of adsorption capacity between CNTs and activated carbon as adsorbents for removal of noxious synthetic dyes: A review. *J. Nanostructure Chem.*, 5, 2, 227–236, 2015.
19. Schnorr, J.M., Functionalization and applications of chitosan, in: *Handbook of Chitosan Research and Applications*, G.M. Richard and M.T. Jennifer (Eds.), p. 301, Nova Science Publishers, Inc, 2011.
20. Sahoo, N.G., Rana, S., Cho, J.W., Li, L., Chan, S.H., Polymer nanocomposites based on functionalized carbon nanotubes. *Prog. Polym. Sci.*, 35, 7, 837–867, 2010.
21. Liu, Z., Tabakman, S., Welsher, K., Dai, H., Carbon nanotubes in biology and medicine: *In vitro* and *in vivo* detection, imaging and drug delivery. *Nano Res.*, 2, 2, 85–120, 2009.
22. Nasibulin, A.G., Pikhitsa, P.V., Jiang, H., Brown, D.P., Krasheninnikov, A.V., Anisimov, A.S., Queipo, P., Moisala, A., Gonzalez, D., Lientschnig, G., Hassanien, A., Shandakov, S.D., Lolli, G., Resasco, D.E., Choi, M., Tománek, D., and Kauppinen, E.I., A novel hybrid carbon material. *Nat. Nanotechnol.*, 2, 3, 156–161, 2007.
23. Awasthi, K., Srivastava, A., Srivastava, O.N., Synthesis of carbon nanotubes. *J. Nanosci. Nanotechnol.*, 5, 10, 1616–1636, 2005.
24. Bronikowski, M.J., Willis, P.A., Colbert, D.T., Smith, K.A., Smalley, R.E., Gas-phase production of carbon single-walled nanotubes from carbon monoxide via the HiPco process: A parametric study. *J. Vac. Sci. Technol. A Vac. Surf. Films*, 19, 4, 1800–1805, 2001.
25. Tsai, H.C., Lin, J.Y., Maryani, F., Huang, C.C., Imae, T., Drug-loading capacity and nuclear targeting of multiwalled carbon nanotubes grafted with anionic amphiphilic copolymers. *Int. J. Nanomedicine*, 8, 4427–4440, 2013.
26. Rahamathulla, M., Bhosale, R.R., Osmani, R.A.M., Mahima, K.C., Johnson, A.P., Hani, U., Ghazwani, M., Begum, M.Y., Alshehri, S., Ghoneim, M.M. *et al.*, Carbon nanotubes: Current perspectives on diverse applications in targeted drug delivery and therapies. *Mater. (Basel)*, 14, 21, 6707, 2021.
27. Ciobotaru, C.C., Damian, C.M., Polosan, S., Prodana, M., Iovu, H., Drug delivery study of single-wall carbon nanotubes covalent functionalized with cisplatin. *Dig. J. Nanomater. Biostructures*, 9, 2, 859–868, 2014.

28. Pantarotto, D., Briand, J.P., Prato, M., Bianco, A., Translocation of bioactive peptides across cell membranes by carbon nanotubes. *Chem. Commun.*, 4, 1, 16–17, 2004.
29. Schwarze, S.R. and Dowdy, S.F., *In vivo* protein transduction: Intracellular delivery of biologically active proteins, compounds and DNA. *Trends Pharmacol. Sci.*, 21, 2, 45–48, 2000.
30. Mirsalari, H., Maleki, A., Raissi, H., Soltanabadi, A., Investigation of the pristine and functionalized carbon nanotubes as a delivery system for the anticancer drug dacarbazine: Drug encapsulation. *J. Pharm. Sci.*, 110, 5, 2005–2016, 2021.
31. Kamel, M., Raissi, H., Morsali, A., Theoretical study of solvent and co-solvent effects on the interaction of Flutamide anticancer drug with Carbon nanotube as a drug delivery system. *J. Mol. Liq.*, 248, 490–500, 2017.
32. Arora, S., Kumar, R., Kaur, H., Rayat, C.S., Kaur, I., Arora, S.K., Srivastava, J., Bharadwaj, L.M., Translocation and toxicity of docetaxel multi-walled carbon nanotube conjugates in mammalian breast cancer cells. *J. Biomed. Nanotechnol.*, 10, 12, 3601–3609, 2014.
33. Saberinasab, A., Raissi, H., Hashemzadeh, H., Understanding the effect of vitamin B6 and PEG functionalization on improving the performance of carbon nanotubes in temozolomide anticancer drug transportation. *J. Phys. D. Appl. Phys.*, 52, 39, Article ID 395402, 2019.
34. Atyabi, F., Sobhani, Z., Adeli, M., Dinarvand, R., Ghahremani, M., Increased paclitaxel cytotoxicity against cancer cell lines using a novel functionalized carbon nanotube. *Int. J. Nanomedicine*, 6, 705–719, 2011.
35. Anbarasan, B., Babu, S.V., Elango, K., Shriya, B., Ramaprabhu, S., PH responsive release of doxorubicin to the cancer cells by functionalized multi-walled carbon nanotubes. *J. Nanosci. Nanotechnol.*, 15, 7, 4799–4805, 2015.
36. Pastorin, G., Kostarelos, K., Prato, M., Bianco, A., Functionalized carbon nanotubes: Towards the delivery of therapeutic molecules. *J. Biomed. Nanotechnol.*, 1, 2, 133–142, 2006.
37. Kostarelos, K., Bianco, A., Prato, M., Promises, facts and challenges for carbon nanotubes in imaging and therapeutics. *Nat. Nanotechnol.*, 4, 10, 627–633, 2009.
38. Abrishamchian, A., Hooshmand, T., Mohammadi, M., Najafi, F., Preparation and characterization of multi-walled carbon nanotube/hydroxyapatite nanocomposite film dip coated on Ti-6Al-4V by sol-gel method for biomedical applications: An *in vitro* study. *Mater. Sci. Eng. C*, 33, 4, 2002–2010, 2013.
39. Madani, S.Y., Mandel, A., Seifalian, A.M., A concise review of carbon nanotube's toxicology. *Nano Rev.*, 4, 1, 21521, 2013.
40. Prabhu, Y.T., Rao, K.V., Kumari, B.S., Sai, V.S., Pavani, T., Nickel and nickel oxide nanocrystals selectively grafting on multiwalled carbon nanotubes. *Nano Converg.*, 2, 2, 2015.

41. Georgakilas, V., Tagmatarchis, N., Pantarotto, D., Bianco, A., Briand, J.P., Prato, M., Amino acid functionalisation of water soluble carbon nanotubes. *Chem. Commun.*, 24, 24, 3050–3051, 2002.
42. Avilés, F., Cauich-Rodríguez, J.V., Moo-Tah, L., May-Pat, A., Vargas-Coronado, R., Evaluation of mild acid oxidation treatments for MWCNT functionalization. *Carbon N. Y.*, 47, 13, 2970–2975, 2009.
43. Raval, J.P., Joshi, P., Chejara, D.R., Carbon nanotube for targeted drug delivery, in: *Applications of Nanocomposite Materials in Drug Delivery*, pp. 203–216, Elsevier Inc, Woodhead Publishing, 2018.
44. Kazemi-Beydokhti, A., Zeinali Heris, S., Reza Jaafari, M., Nikoofal-Sahlabadi, S., Tafaghodi, M., Hatamipoor, M., Microwave functionalized single-walled carbon nanotube as nanocarrier for the delivery of anticancer drug cisplatin: *In vitro* and *in vivo* evaluation. *J. Drug Deliv. Sci. Technol.*, 24, 6, 572–578, 2014.
45. Kumar, P., Park, J.S., Randhawa, P., Sharma, S., Shin, M.S., Sekhon, S.S., A study on the effect of different chemical routes on functionalization of mwcnts by various groups (-COOH, -SO 3H, -PO 3H 2). *Nanoscale Res. Lett.*, 6, 1–9, 2011.
46. Rosca, I.D., Watari, F., Uo, M., Akasaka, T., Oxidation of multiwalled carbon nanotubes by nitric acid. *Carbon N. Y.*, 43, 15, 3124–3131, 2005.
47. Stephenson, J.J., Sadana, A.K., Higginbotham, A.L., Tour, J.M., Highly functionalized and soluble multiwalled carbon nanotubes by reductive alkylation and arylation: The billups reaction. *Chem. Mater.*, 18, 19, 4658–4661, 2006.
48. Bahr, J.L. and Tour, J.M., Highly functionalized carbon nanotubes using *in situ* generated diazonium compounds. *Chem. Mater.*, 13, 11, 3823–3824, 2001.
49. Ellison, M.D. and Gasda, P.J., Functionalization of single-walled carbon nanotubes with 1,4-benzenediamine using a diazonium reaction. *J. Phys. Chem. C*, 112, 3, 738–740, 2008.
50. Amiri, A., Maghrebi, M., Baniadam, M., Zeinali Heris, S., One-pot, efficient functionalization of multi-walled carbon nanotubes with diamines by microwave method. *Appl. Surf. Sci.*, 257, 23, 10261–10266, 2011.
51. Liu, J., Rinzler, A.G., Dai, H., Hafner, J.H., Kelley Bradley, R., Boul, P.J., Lu, A., Iverson, T., Shelimov, K., Huffman, C.B., Rodriguez-Macias, F., Shon, Y.S., Lee, T.R., Colbert, D.T., Smalley, R.E., Fullerene pipes. *Science*, 280, 5367, 1253–1256, 1998.
52. Duesberg, G.S., Muster, J., Krstic, V., Burghard, M., Roth, S., Chromatographic size separation of single-wall carbon nanotubes. *Appl. Phys. A Mater. Sci. Process.*, 67, 1, 117–119, 1998.
53. Moore, V.C., Strano, M.S., Haroz, E.H., Hauge, R.H., Smalley, R.E., Schmidt, J., Talmon, Y., Individually suspended single-walled carbon nanotubes in various surfactants. *Nano Lett.*, 3, 10, 1379–1382, 2003.

54. Yurekli, K., Mitchell, C.A., Krishnamoorti, R., Small-angle neutron scattering from surfactant-assisted aqueous dispersions of carbon nanotubes. *J. Am. Chem. Soc.*, 126, 32, 9902–9903, 2004.
55. Jiang, L., Gao, L., Sun, J., Production of aqueous colloidal dispersions of carbon nanotubes. *J. Colloid Interface Sci.*, 260, 1, 89–94, 2003.
56. Chen, J., Chen, S., Zhao, X., Kuznetsova, L.V., Wong, S.S., Ojima, I., Functionalized single-walled carbon nanotubes as rationally designed vehicles for tumor-targeted drug delivery. *J. Am. Chem. Soc.*, 130, 49, 16778–16785, 2008.
57. Zhang, Y., Bai, Y., Yan, B., Functionalized carbon nanotubes for potential medicinal applications. *Drug Discovery Today*, 15, 11–12, 428–435, 2010.
58. Zhao, L., He, R., Rim, K.T., Schiros, T., Kim, K.S., Zhou, H., Gutiérrez, C., Chockalingam, S.P., Arguello, C.J., Pálová, L., Nordlund, D., Hybertsen, M.S., Reichman, D.R., Heinz, T.F., Kim, P., Pinczuk, A., Flynn, G.W., Pasupathy, A.N., Visualizing individual nitrogen dopants in monolayer graphene. *Science*, 333, 6045, 999–1003, 2011.
59. Panchakarla, L.S., Subrahmanyam, K.S., Saha, S.K., Govindaraj, A., Krishnamurthy, H.R., Waghmare, U.V., Rao, C.N.R., Synthesis, structure, and properties of boron- and nitrogen-doped graphene. *Adv. Mater.*, 21, 46, 4726–4730, 2009.
60. Zou, Y., Li, F., Zhu, Z.H., Zhao, M.W., Xu, X.G., Su, X.Y., An *ab initio* study on gas sensing properties of graphene and Si-doped graphene. *Eur. Phys. J. B*, 81, 4, 475–479, 2011.
61. Dai, J., Yuan, J., Giannozzi, P., Gas adsorption on graphene doped with B, N, Al, and S: A theoretical study. *Appl. Phys. Lett.*, 95, 232105, 2009.
62. Zhou, S.Y., Siegel, D.A., Fedorov, A.V., Lanzara, A., Metal to insulator transition in epitaxial graphene induced by molecular doping. *Phys. Rev. Lett.*, 101, 8, 1–4, 2008.
63. Wehling, T.O., Novoselov, K.S., Morozov, S.V., Vdovin, E.E., Katsnelson, M.I., Geim, A.K., Lichtenstein, A.I., Molecular doping of graphene. *Nano Lett.*, 8, 1, 173–177, 2008.
64. Kasry, A., Kuroda, M.A., Martyna, G.J., Tulevski, G.S., Bol, A.A., Chemical doping of large-area stacked graphene films for use as transparent, conducting electrodes. *ACS Nano*, 4, 7, 3839–3844, 2010.
65. Kaverzin, A.A., Strawbridge, S.M., Price, A.S., Withers, F., Savchenko, A.K., Horsell, D.W., Electrochemical doping of graphene with toluene. *Carbon N. Y.*, 49, 12, 3829–3834, 2011.
66. Zhao, J., Liu, X., Zhu, Z., Wang, N., Sun, W., Chen, C., He, Z., Molecular insight into the enhancement of benzene-carbon nanotube interactions by surface modification for drug delivery systems (DDS). *Appl. Surf. Sci.*, 416, 757–765, 2017.
67. Carrero-Sánchez, J.C., Elías, A.L., Mancilla, R., Arrellín, G., Terrones, H., Laclette, J.P., Terrones, M., Biocompatibility and toxicological studies of carbon nanotubes doped with nitrogen. *Nano Lett.*, 6, 8, 1609–1616, 2006.

68. Saikia, N. and Deka, R.C., Density functional study on the adsorption of the drug isoniazid onto pristine and B-doped single wall carbon nanotubes. *J. Mol. Model.*, 19, 1, 215–226, 2013.
69. Salazar, A., Muñoz-sandoval, E., Chavarria, V., Potential use of nitrogen-doped carbon nanotube sponges as payload carriers against malignant glioma. *Nanomater. (Basel)*, 11, 5, 1244, 2021.
70. Mitchell, M.J., Billingsley, M.M., Haley, R.M., Wechsler, M.E., Peppas, N.A., Langer, R., Engineering precision nanoparticles for drug delivery. *Nat. Rev. Drug Discovery*, 20, 2, 101–124, 2021.
71. Malwade, C.R. and Christensen, L.P., Simple multipurpose apparatus for solubility measurement of solid solutes in liquids. *Educ. Chem. Eng.*, 16, 29–38, 2016.
72. de Lima, L.S. and Mortari, M.R., Therapeutic nanoparticles in the brain: A review of types, physicochemical properties and challenges. *Int. J. Pharm.*, 612, 121367, 2022.
73. Cao, Y. and Luo, Y., Pharmacological and toxicological aspects of carbon nanotubes (CNTs) to vascular system: A review. *Toxicol. Appl. Pharmacol.*, 385, 114801, November 2019.
74. Kafa, H., Wang, J.T.W., Rubio, N., Klippstein, R., Costa, P.M., Hassan, H.A.F.M., Sosabowski, J.K., Bansal, S.S., Preston, J.E., Abbott, N.J., Al-Jamal, K.T., Translocation of LRP1 targeted carbon nanotubes of different diameters across the blood-brain barrier *in vitro* and *in vivo*. *J. Control. Release*, 225, 217–229, 2016.
75. Facciolà, A., Visalli, G., La Maestra, S., Ceccarelli, M., D'Aleo, F., Nunnari, G., Pellicanò, G.F., Di Pietro, A., Carbon nanotubes and central nervous system: Environmental risks, toxicological aspects and future perspectives. *Environ. Toxicol. Pharmacol.*, 65, 23–30, 2019.
76. Mishra, J., Kumar, B., Pandey, M., Pottoo, F.H., Fayaz, F., Khan, F.A., Sahoo, P.K., Carbon nano tubes: Novel drug delivery system in amelioration of Alzheimer's disease. *Comb. Chem. High Throughput Screen.*, 2021, 24, 1528–1543.
77. Ren, J., Shen, S., Wang, D., Xi, Z., Guo, L., Pang, Z., Qian, Y., Sun, X., Jiang, X., The targeted delivery of anticancer drugs to brain glioma by PEGylated oxidized multi-walled carbon nanotubes modified with angiopep-2. *Biomaterials*, 33, 11, 3324–3333, 2012.
78. You, Y., Wang, N., He, L., Shi, C., Zhang, D., Liu, Y., Luo, L., Chen, T., Designing dual-functionalized carbon nanotubes with high blood-brain-barrier permeability for precise orthotopic glioma therapy. *Dalton Trans.*, 48, 5, 1569–1573, 2019.
79. Santos, T., Fang, X., Chen, M.T., Wang, W., Ferreira, R., Jhaveri, N., Gundersen, M., Zhou, C., Pagnini, P., Hofman, F.M., Chen, T.C., Sequential administration of carbon nanotubes and near-infrared radiation for the treatment of gliomas. *Front. Oncol.*, 4, 180, July 2014.

80. Wong, B.S., Yoong, S.L., Jagusiak, A., Panczyk, T., Ho, H.K., Ang, W.H., Pastorin, G., Carbon nanotubes for delivery of small molecule drugs. *Adv. Drug Deliv. Rev.*, 65, 15, 1964–2015, 2013.
81. Lohan, S., Raza, K., Mehta, S.K., Bhatti, G.K., Saini, S., Singh, B., Anti-Alzheimer's potential of berberine using surface decorated multi-walled carbon nanotubes: A preclinical evidence. *Int. J. Pharm.*, 530, 1–2, 263–278, 2017.
82. Costa, P.M., Wang, J.T.W., Morfin, J.F., Khanum, T., To, W., Sosabowski, J., Tóth, E., Al-Jamal, K.T., Functionalised carbon nanotubes enhance brain delivery of amyloid-targeting pittsburgh compound b (Pib)-derived ligands. *Nanotheranostics*, 2, 2, 168–183, 2018.
83. Li, H., Luo, Y., Derreumaux, P., Wei, G., Carbon nanotube inhibits the formation of β-sheet-rich oligomers of the Alzheimer's amyloid-β(16-22) peptide. *Biophys. J.*, 101, 9, 2267–2276, 2011.
84. Xue, X., Wang, L.R., Sato, Y., Jiang, Y., Berg, M., Yang, D.S., Nixon, R.A., Liang, X.J., Single-walled carbon nanotubes alleviate autophagic/lysosomal defects in primary glia from a mouse model of alzheimers disease. *Nano Lett.*, 14, 9, 5110–5117, 2014.
85. Tan, J.M., Foo, J.B., Fakurazi, S., Hussein, M.Z., Release behaviour and toxicity evaluation of levodopa from carboxylated single-walled carbon nanotubes. *Beilstein J. Nanotechnol.*, 6, 1, 243–253, 2015.
86. Guo, Q., You, H., Yang, X., Lin, B., Zhu, Z., Lu, Z., Li, X., Zhao, Y., Mao, L., Shen, S., Cheng, H., Zhang, J., Deng, L., Fan, J., Xi, Z., Li, R., Li, C.M., Functional single-walled carbon nanotubes "CAR" for targeting dopamine delivery into the brain of parkinsonian mice. *Nanoscale*, 9, 30, 10832–10845, 2017.
87. Sola, P., Krishnamurthy, P., Chintamaneni, P.K., Pindiprolu, S.K.S.S., Kumari, M., Novel drug delivery systems of β2 adrenoreceptor agonists to suppress SNCA gene expression and mitochondrial oxidative stress in Parkinson's disease management. *Expert Opin. Drug Deliv.*, 17, 8, 1119–1132, 2020.
88. Lee, H.J., Park, J., Yoon, O.J., Kim, H.W., Lee, D.Y., Kim, D.H., Lee, W.B., Lee, N.E., Bonventre, J.V., Kim, S.S., Amine-modified single-walled carbon nanotubes protect neurons from injury in a rat stroke model. *Nat. Nanotechnol.*, 6, 2, 121–125, 2011.
89. Moon, S.U., Kim, J., Bokara, K.K., Kim, J.Y., Khang, D., Webster, T.J., Lee, J.E., Carbon nanotubes impregnated with subventricular zone neural progenitor cells promotes recovery from stroke. *Int. J. Nanomedicine*, 7, 2751–2765, 2012.
90. Xiang, C., Zhang, Y., Guo, W., Liang, X.J., Biomimetic carbon nanotubes for neurological disease therapeutics as inherent medication. *Acta Pharm. Sin. B*, 10, 2, 239–248, 2020.
91. Gurunathan, S., Kang, M.H., Qasim, M., Kim, J.H., Nanoparticle-mediated combination therapy: Two-in-one approach for cancer. *Int. J. Mol. Sci.*, 19, 10, 1–37, 2018.

92. Blanco, E., Shen, H., Ferrari, M., Principles of nanoparticle design for overcoming biological barriers to drug delivery. *Nat. Biotechnol.*, 33, 9, 941–951, 2015.
93. Allen, T.M. and Cullis, P.R., Drug delivery systems: Entering the mainstream. *Science*, 303, 5665, 1818–1822, 2004.
94. Parveen, S., Misra, R., Sahoo, S.K., Nanoparticles: A boon to drug delivery, therapeutics, diagnostics and imaging. *Nanomed. Nanotechnol. Biol. Med.*, 8, 2, 147–166, 2012.
95. Ramezanpour, M., Leung, S.S.W., Delgado-Magnero, K.H., Bashe, B.Y.M., Thewalt, J., Tieleman, D.P., Computational and experimental approaches for investigating nanoparticle-based drug delivery systems. *Biochim. Biophys. Acta Biomembr.*, 1858, 7, 1688–1709, 2016.
96. Jadia, R., Scandore, C., Rai, P., Program, B., Nanoparticles for effective combination therapy of cancer. *Int. J. Nanotechnol. Nanomedicine*, 1, 1, 1–27, 2016.
97. Parhi, P., Mohanty, C., Sahoo, S.K., Nanotechnology-based combinational drug delivery: An emerging approach for cancer therapy. *Drug Discovery Today*, 17, 17–18, 1044–1052, 2012.
98. Poon, C., He, C., Liu, D., Lu, K., Lin, W., Self-assembled nanoscale coordination polymers carrying oxaliplatin and gemcitabine for synergistic combination therapy of pancreatic cancer. *J. Control. Release*, 201, 90–99, 2015.
99. Gad, A., Kydd, J., Piel, B., Rai, P., Sci forschen targeting cancer using polymeric nanoparticle. *Int. J. Nanomed. Nanosurg.*, 2, 3, 5–7, 2016, http://dx.doi.org/10.16966/2470-3206.116.
100. De Jong, W.H. and Borm, P.J.A., Drug delivery and nanoparticles: Applications and hazards. *Int. J. Nanomedicine*, 3, 2, 133–149, 2008.
101. Sukhanova, A., Bozrova, S., Sokolov, P., Berestovoy, M., Karaulov, A., Nabiev, I., Dependence of nanoparticle toxicity on their physical and chemical properties. *Nanoscale Res. Lett.*, 13, 44, 2018.
102. Gurturk, Z., Tezcaner, A., Dalgic, A.D., Korkmaz, S., Keskin, D., Maltodextrin modified liposomes for drug delivery through the blood-brain barrier. *Medchemcomm*, 8, 6, 1337–1345, 2017.
103. Yuan, B., Zhao, Y., Dong, S., Sun, Y., Hao, F., Xie, J., Teng, L., Lee, R.J., Fu, Y., Bi, Y., Cell-penetrating peptide-coated liposomes for drug delivery across the blood-brain barrier. *Anticancer Res.*, 39, 1, 237–243, 2019.
104. Gillies, E.R. and Fréchet, J.M.J., Dendrimers and dendritic polymers in drug delivery. *Drug Discovery Today*, 10, 1, 35–43, 2005.
105. Wijagkanalan, W., Kawakami, S., Hashida, M., Designing dendrimers for drug delivery and imaging: Pharmacokinetic considerations. *Pharm. Res.*, 28, 7, 1500–1519, 2011.
106. Bianco, A., Kostarelos, K., Prato, M., Applications of carbon nanotubes in drug delivery. *Curr. Opin. Chem. Biol.*, 9, 6, 674–679, 2005.
107. Arora, S., Kumar, V., Yadav, S., Singh, S., Bhatnagar, D., Kaur, I., Carbon nanotubes as drug delivery vehicles. *Solid State Phenom.*, 222, 145–158, 2015.

108. Ghosh, P., Han, G., De, M., Kim, C.K., Rotello, V.M., Gold nanoparticles in delivery applications. *Adv. Drug Deliv. Rev.*, 60, 11, 1307–1315, 2008.
109. Jain, T.K., Morales, M.A., Sahoo, S.K., Leslie-Pelecky, D.L., Labhasetwar, V., Iron oxide nanoparticles for sustained delivery of anticancer agents. *Mol. Pharm.*, 2, 3, 194–205, 2005.
110. Bao, G., Mitragotri, S., Tong, S., Multifunctional nanoparticles for drug delivery and molecular imaging. *Annu. Rev. Biomed. Eng.*, 15, 253–282, 2013.
111. Sun, C., Lee, J.S.H., Zhang, M., Magnetic nanoparticles in MR imaging and drug delivery. *Adv. Drug Deliv. Rev.*, 60, 11, 1252–1265, 2008.
112. Duncan, R., The dawning era of polymer therapeutics. *Nat. Rev. Drug Discovery*, 2, 5, 347–360, 2003.
113. Alanazi, F.K., Haq, N., Radwan, A.A., Alsarra, I.A., Shakeel, F., Potential of lipid nanoemulsion for drug delivery of cholesteryl-hexahydrophthaloyl-5-fluorouracil. *J. Drug Deliv. Sci. Technol.*, 24, 5, 459–463, 2014.
114. Wilson, R.J., Li, Y., Yang, G., Zhao, C.X., Nanoemulsions for drug delivery. *Particuology*, 64, 85–97, 2022.
115. Qi, L. and Gao, X., Emerging application of quantum dots for drug delivery and therapy. *Expert Opin. Drug Deliv.*, 5, 3, 263–267, 2008.
116. Xu, Z., Li, B., Tang, W., Chen, T., Zhang, H., Wang, Q., Glycopolypeptide-encapsulated Mn-doped ZnS quantum dots for drug delivery: Fabrication, characterization, and *in vitro* assessment. *Colloids Surf. B Biointerfaces*, 88, 1, 51–57, 2011.
117. Mahalunkar, S., Yadav, A.S., Gorain, M., Pawar, V., Braathen, R., Weiss, S., Bogen, B., Gosavi, S.W., Kundu, G.C., Functional design of pH-responsive folate-targeted polymer-coated gold nanoparticles for drug delivery and *in vivo* therapy in breast cancer. *Int. J. Nanomedicine*, 14, 8285–8302, 2019.
118. Manivasagan, P., Bharathiraja, S., Bui, N.Q., Jang, B., Oh, Y.O., Lim, I.G., Oh, J., Doxorubicin-loaded fucoidan capped gold nanoparticles for drug delivery and photoacoustic imaging. *Int. J. Biol. Macromol.*, 91, 578–588, 2016.
119. Lopes-Nunes, J., Agonia, A.S., Rosado, T., Gallardo, E., Palmeira-De-oliveira, R., Palmeira-De-oliveira, A., Martinez-De-oliveira, J., Fonseca-Moutinho, J., Campello, M.P.C., Paiva, A., Paulo, A., Vulgamott, A., Ellignton, A.D., Oliveira, P.A., Cruz, C., Aptamer-functionalized gold nanoparticles for drug delivery to gynecological carcinoma cells. *Cancers (Basel)*, 13, 16, 4038, 2021.
120. Li, W., Cao, Z., Liu, R., Liu, L., Li, H., Li, X., Chen, Y., Lu, C., Liu, Y., AuNPs as an important inorganic nanoparticle applied in drug carrier systems. *Artif. Cells Nanomed. Biotechnol.*, 47, 1, 4222–4233, 2019.
121. Davis, M.E., Chen, Z., Shin, D.M., Nanoparticle therapeutics: An emerging treatment modality for cancer. *Nat. Rev. Drug Discovery*, 7, 9, 771–782, 2008.
122. Bartlett, D.W., Su, H., Hildebrandt, I.J., Weber, W.A., Davis, M.E., Impact of tumor-specific targeting on the biodistribution and efficacy of siRNA nanoparticles measured by multimodality *in vivo* imaging. *Proc. Natl. Acad. Sci. U. S. A.*, 104, 39, 15549–15554, 2007.

123. Luk, B.T. and Zhang, L., Current advances in polymer-based nanotheranostics for cancer treatment and diagnosis. *ACS Appl. Mater. Interfaces*, 6, 24, 21859–21873, 2014.
124. Teirumnieks, E., Balchev, I., Ghalot, R.S., Lazov, L., Antibacterial and antiviral effects of silver nanoparticles in medicine against COVID-19-a review. *Laser Phys.*, 31, 1, 013001, 2021.
125. Eby, D.M., Luckarift, H.R., Johnson, G.R., Hybrid antimicrobial enzyme and silver nanoparticle coatings for medical Instruments. *ACS Appl. Mater. Interfaces*, 1, 7, 1553–1560, 2009.
126. Altinoğlu, E.I. and Adair, J.H., Near infrared imaging with nanoparticles. *Wiley Interdiscip. Rev. Nanomed. Nanobiotechnol.*, 2, 5, 461–477, 2010.
127. Bagalkot, V., Zhang, L., Levy-Nissenbaum, E., Jon, S., Kantoff, P.W., Langery, R., Farokhzad, O.C., Quantum dot-aptamer conjugates for synchronous cancer imaging, therapy, and sensing of drug delivery based on Bi-fluorescence resonance energy transfer. *Nano Lett.*, 7, 10, 3065–3070, 2007.
128. Kumar, R., Kulkarni, A., Nagesha, D.K., Sridhar, S., *In vitro* evaluation of theranostic polymeric micelles for imaging and drug delivery in cancer. *Theranostics*, 2, 7, 714–722, 2012.
129. Medintz, I.L., Uyeda, H.T., Goldman, E.R., Mattoussi, H., Quantum dot bioconjugates for imaging, labelling and sensing. *Nat. Mater.*, 4, 6, 435–446, 2005.
130. Murphy, C.J., Peer reviewed: Optical sensing with quantum dots. *Anal. Chem.*, 74, 19, 520A–526A, 2002.
131. Baptista, P.V., Doria, G., Quaresma, P., Cavadas, M., Neves, C.S., Gomes, I., Eaton, P., Pereira, E., Franco, R., Nanoparticles in molecular diagnostics. *Prog. Mol. Biol. Transl. Sci.*, 104, 427–88, 2011.
132. Mornet, S., Vasseur, S., Grasset, F., Duguet, E., Magnetic nanoparticle design for medical diagnosis and therapy. *J. Mater. Chem.*, 14, 14, 2161–2175, 2004.
133. Kang, S., Herzberg, M., Rodrigues, D.F., Elimelech, M., Antibacterial effects of carbon nanotubes: Size does matter! *Langmuir*, 24, 13, 6409–6413, 2008.

7

Functionalized Carbon Nanotubes for Gene Therapy

Tejas Agnihotri, Tanuja Shinde, Manoj Gitte, Pankaj Kumar Paradia, Rakesh Kumar Tekade and Aakanchha Jain*

National Institute of Pharmaceutical Education and Research (NIPER)-Ahmedabad, Palaj, Opp Air Force Station, Gandhinagar, Gujarat, India

Abstract

As conventional drug delivery system faces several shortcomings like non-specific targeting, solubility issues and low bioavailability, the focus of pharmaceutical sector has been shifted towards novel drug delivery system, of which carbon nanotubes (CNT) being one of the them. Due to their ability to transport a wide range of substances across the living cells and excellent mechanical, electrical and thermal properties, it enjoys a wide range of applications in the field of biomedicine. Though, CNTs have grabbed much of scientists' attention, some major issues like toxicity, insolubility in aqueous and organic solvents, shortened circulation half-life, and immunogenicity have marred otherwise well-suited carrier system. To overcome these shortcomings, functionalization of CNT is one of the alternatives scientists have been looking for. The present book chapter discusses about the structure, properties and methods of functionalization in detail. It also sheds light on cellular uptake process and different therapeutic classes of drugs which can be incorporated into functionalized CNT and diverse applications thereof with a special focus on gene therapy. The current book chapter gives insights into CNT applications in gene therapy as a carrier system for small interfering RNA (siRNA), microRNA (miRNA), double stranded DNA (dsDNA), aptamers, oligonucleotides, etc.

***Keywords*:** Carbon nanotubes, functionalized CNT, gene therapy, autoimmune, toxicity

**Corresponding author*: jainaakanchha83@gmail.com

Jeenat Aslam, Chaudhery Mustansar Hussain and Ruby Aslam (eds.) Functionalized Carbon Nanotubes for Biomedical Applications, (139–170) © 2023 Scrivener Publishing LLC

7.1 Introduction

The pharmaceutical industry has made great advances in the search for viable treatments for a wide range of diseases over the last three decades. Many diseases have their origins at the genetic level. Human genome projects as well as recent developments in molecular genetics and high throughput techniques have propelled to enhance our understanding of genetic basis leading to new therapeutic targets and therapies. One of the emerging and prominent therapies is gene therapy. Gene therapy mainly focuses on delivering specific genes in place of malfunctioned or defective genes without any aid of surgical or therapeutic intervention [1, 2]. According to European Medicines Agency (EMA), gene therapy product is defined as "biological medicinal product fulfilling two characteristics; (i) it contains active substance containing recombinant nucleic acid administered to human beings to regulate, repair, replace, add or delete genetic sequence and (ii) it elicits therapeutic, prophylactic or diagnostic effects which are related to nucleic acid sequence or genetic expression it induces". According to USFDA, gene therapy is defined as "products that mediate their effects by transcription and/or translation of genetic material and/or by integrating into host genome and administered as nucleic acids, viruses or genetically engineered organisms". Primarily, gene therapy has been classified into two categories: Germ line therapy concerning genetic change being reflected to next generation and somatic gene therapy wherein any change is not delivered to next generation. Out of these two, only somatic gene therapy has been approved according to legislature [3].

The history of gene therapy dates back to 1928 where scientist Griffith published the report describing transformation of nonvirulent pneumococcus into virulent one even on virtue of heat inactivation [3]. The first clinical trial that used gene therapy was in 1988 for the treatment of Gaucher's disease, but after the 1990s many adverse events were observed that led to aversion to this therapy between 1990 and 2010 [4]. It suffered major impediments in the form of suitable and effective gene delivery vectors, gene editing techniques, and safety concerns over vectors used in gene therapy [1].

Four approaches have been dealt with gene therapy:

1. Gene addition concerning the addition of plasmid DNA to produce desirable proteins
2. Gene correction concerning gene editing techniques to get rid of defective genes and translate desirable proteins by suppressing the expression of dysfunctional genes.

3. Gene silencing suppresses protein translation from binding to mRNA
4. Cell elimination specifically used to kill tumor cells by activating an immune response [4].

Genome editing techniques utilize zinc finger nuclease, transcription activator-like effector nuclease, clustered nuclease interspaced short palindromic repeats (CRISPR)/associated nuclease Cas9 [4].

Carbon nanotubes (CNT) have got a lot of interest since Japanese scientist SumioIijima discovered them in insoluble soot of arc-burned graphite rods in 1991. CNTs are considered to be one of the strongest and stiffest materials with high tensile strength and elastic modulus. CNTs also have excellent thermal and electrical properties, apart from showing photoluminescence and Raman signals. CNTs have been known for delivering a wide range of therapeutic moieties from small therapeutic molecules like antimicrobials, anticancer agents to complex vaccines, antibodies and small interfering ribonucleic acids (siRNA), etc. Depending upon outer layer chemistry and quantity, CNTs have been divided into two types: single-walled carbon nanotubes (SWCNTs) and multi-walled carbon nanotubes (MWCNTs). It is very important to know about fullerene because nanotubes have been made of the same. When a sheet of fullerene is rolled at a certain angle, carbon nanotubes with specific inner radius are obtained [5, 6].

7.2 Functionalized CNTs and Gene Therapy

A considerable attention has been paid to gene therapy as a treatment for genetic and acquired illnesses. The development of carbon nanotubes as a carrier system for protein and nucleic acid offers a great deal of potential for gene delivery in human cells as well as in plant cells. It presents low immunogenicity, increased safety, and ease of their production compared to viral vectors. However, the use of carbon nanotubes for gene delivery is still hampered by several obstacles such as poor membrane penetration capacity [7], endosomal entrapment, and lysosomal degradation [8].

The solubility of carbon nanotubes in water is essential for their use in biological systems, to permit adequate absorption, distribution, and elimination, and to render them biocompatible with reduced toxicity. Also, the CNT dispersions should be uniform and sufficiently stable to be considered as a good candidate for drug delivery applications. Therefore, the solubility of carbon nanotubes is improved by various functionalization methods [9]. Figure 7.1 represents the recent development of functionalized CNTs in the area of gene therapy.

Basically, the surface functionalization of carbon nanotubes is classified into two types, viz., covalent and noncovalent functionalization. The covalent and non-covalent functionalization make the CNTs soluble in aqueous or organic solutions especially in volatile organic solvents or strong acids and take lead role in improving dispersibility and surface interactions [10, 11]. The covalent functionalization is also known as defect functionalization which involves covalent bond formation between sidewall or at endcaps of carbon nanotubes and active functional groups [12]. The covalent functionalization is carried out either by grafting or through chemical reactions. Various biological molecules [13], polymers [14] and metals [15] are grafted onto the surface of carbon nanotubes. The covalent bond formation has been achieved through chemical reactions like cycloaddition (Diels Alder reaction), carbene and nitrene addition, chlorination, bromination and hydrogenation functionalization techniques [16]. Figure 7.2 shows the schematic structures of carbon nanotubes and functionalization methods.

Single-walled carbon nanotubes (SWCNTs) possess a needle-like shape, high aspect surface ratio, and unique surface characteristics, thus

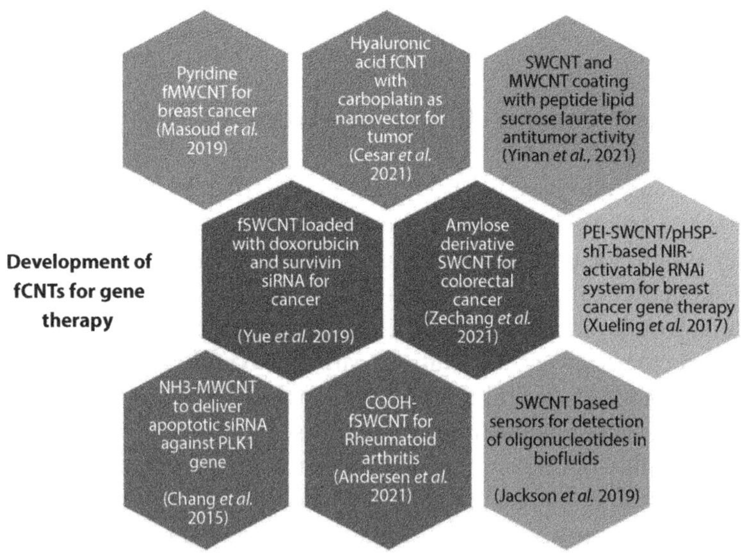

Figure 7.1 Recent development of functionalized carbon nanotubes in the area of gene therapy.

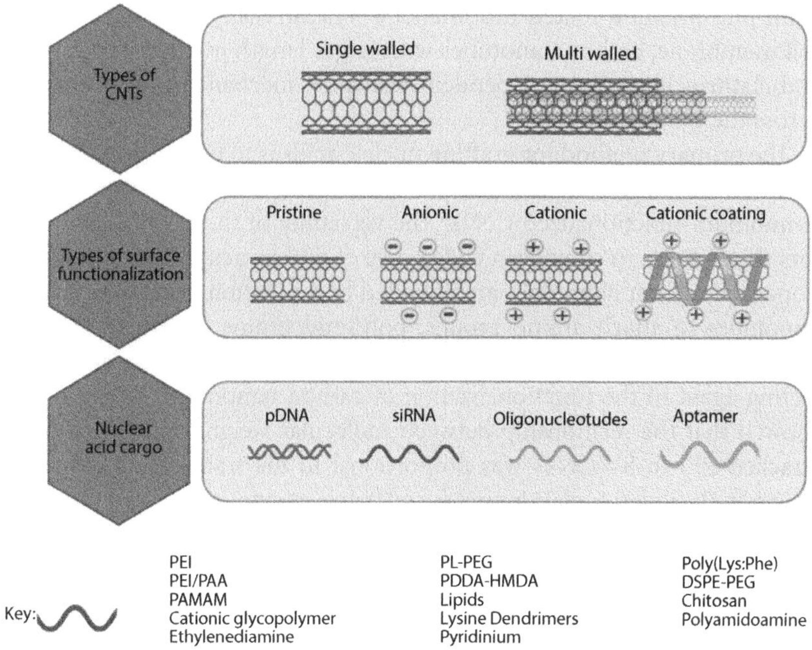

Figure 7.2 Schematic representation of carbon nanotubes and their functionalization methods.

presenting a good capability to cross cell membranes. Therefore, single walled carbon nanotubes had proven their potential of transporting several agents including single-stranded DNA, small interference RNA (siRNA), plasmid DNA into cancer cells, primary-immune cells, T cells and normal cells [17].

Since CNTs are used as promising class of gene delivery system, various surface functionalization strategies have been used to enhance the propensity of CNTs as effective candidate for the transport of nucleic acids inside the cell. The delivery of gene through functionalized carbon nanotubes is mainly dependent on several factors such as the charge ratio between positively charged fCNT and negatively charged DNA, the degree of condensation of DNA, and the functionalized surface area of CNTs. The ability of CNT to enter cells via different cell uptake mechanisms is crucial, and it depends on the length of the carbon nanotube and the type of chemical functionalization utilized, so there is a need to optimize and design CNT functionalization for efficient gene delivery. The hydrophobic affinity and small size of the carbon nanotubes are the key parameters for uniform dispersion of CNTs in the phospholipid bilayer of the cell membrane. Thus, while phospholipid coated and smaller CNTs can easily diffuse through the cell membrane, carbon nanotubes with larger length make use of caveolae and clathrin like energy dependent transport mechanism for their entry across the cell membrane.

The primary method for grafting nucleic acids is to use the ionic interaction between negatively charged phosphate groups and positively charged ammonium-functionalized CNTs. The versatility of these CNT-based vectors allows them to complex a wide range of nucleic acids. The amino functionalized carbon nanotubes are obtained by amidation of oxidized carbon nanotubes to attach amino groups, polyamidoamine groups, ethylenediamines, and polyethyleneimine (PEI) [18]. PEIis widely used as a transfecting agent in the functionalization of carbon nanotubes. Behnam *et al.* showed that the relationship between molecular weight of non-covalently attached PEI to SWCNTs was proportional to the transfection efficiency; however, the inverse relation was found between cytotoxicity and molecular weight of PEI, resulting in promising gene delivery vector, being PEI functionalized single walled carbon nanotubes had proven to be with the lack of aggregation, good dispersibility and stability in biological media [19].

The photothermal transfection (PTT) strategy utilizing near-infrared irradiation provides increased intracellular uptake of PTT agents (NGO-PEG-PEI) by enhancing membrane permeability, facilitating the escape of the photochemical agent from endosomes by generating reactive oxygen species, and promoting the release of the therapeutic gene. Kong *et al.* used the photothermal property for the functionalization of single-walled carbon nanotubes for gene therapy. They developed polyethyleneimine and cholesterol (PCS) functionalized single-walled carbon nanotubes that helped compress the DNA molecule in SWCNT via electrostatic interaction, and the results were fascinating in terms of enhanced transfection efficiency, lower toxicity, and safer intracellular transport of the gene. The PEI-DNA functionalized utilized clathrin mediated internalization and prone to undergo lysosomal degradation. This lysosomal degradation had decreased with PCS-DNA functionalized SWCNT, as it used caveolae-mediated cellular uptake in HeLa cell lines. Moreover, the PCS/DNA functionalized SWCNT exhibited a photothermal conversion ability and promoted DNA release from PCS under NIR laser irradiation [17].

Baker *et al.* carried out covalent functionalization of DNA oligonucleotides to the exterior surface of amine-terminated single walled carbon nanotubes. The evolving single-walled DNA CNT adducts selectively hybridized with the complementary sequence of DNA molecules, which allowed the formation of a more complex supramolecular CNT structure built as DNA-DNA interaction [20, 21].

On the other hand, the noncovalent functionalization method involves the physical interaction of functionalized groups on the surface of CNTs and retains the original structural integrity of CNTs without the destruction of the hybridized skeleton of the graphitic side walls sp2 compared to covalent functionalization [22]. This method of functionalization consists of weak interactions and adsorption forces such as Van der Waals forces, π-π stacking, charge transfer and hydrophobic interactions. In these, non-covalent functionalization occurs via the adsorption of surfactant molecules on the surface of carbon nanotubes or by enveloping the polymer chain around the carbon nanotubes [23]. In nanotherapeutics, the biomacromolecules like oligosaccharides [24], proteins [25], and DNA [26] are successfully attached to carbon nanotubes by non-covalent bonding. The aromatic nucleotide bases form π-stacking interaction with the sidewall of carbon nanotubes and the flexibility of nucleic acids supports to enhance these interactions. Therefore, the small interfering RNA (siRNA) is non-covalently bonded with the single-walled carbon nanotubes and delivered/targeted to cancer cells which have shown the strong specific inhibition of hypoxia-inducible factor 1 alpha activity in cancerous cells [27]. Zheng *et al.*, showed that the single-strand DNA (ssDNA) self-assembled around the carbon nanotubes in a sequence-dependent manner [28]. Both these siRNAs and ssDNAs showed firm noncovalent binding to carbon nanotubes. Recently, CNTs have been found to be the efficient carrier system for gene delivery to target cells, as it enhances DNA amplification in the polymerase chain reaction [29]. Lizeng Gao *et al.* carried out covalent functionalization of amino, carboxyl, and alkyl group to the multi-walled carbon nanotubes to which negatively charged DNA got attached to positively charged multi-walled carbon nanotubes and these electrostatic interactions had been confirmed by Kaiser assay, which successfully delivered the green fluorescent protein to cells [30].

The functionalization of CNTs has been patented with their utilities in gene therapy. For instance, Ensysce Biosciences filed patent (Patent No: US2016/0228544 A1) entitled, 'single walled carbon nanotube polynucleotide complexes and methods related thereto' which encompasses delivery of polynucleotides encoding proteins of therapeutic and immunologic in nature, and methods of manufacturing SWCNTs. In the other patent entitled, 'Novel microcapsule nanotube devices for targeted delivery of therapeutic molecules' (Patent No: WO 2008/066507 A2) filed by McGill university, carbon nanotube with incorporation of DNA in gel matrix was proposed to be given orally through microcapsule. Microcapsules protect DNA from harsh gastrointestinal environment and this carbon nanotube device claim to have various biomedical applications. Patent entitled, 'dispersion and debundling of carbon nanotubes using gemini surfactant compounds' (Patent No: US 2011/0300126A1) talks about the method of solubilization of CNTs have been claimed using gemini surfactant and has got clinical applications in carrier system for biological agents. The delivery of DNA encoding EGFP protein through functionalized carbon nanotubes has been patented by Shyam Mohapatra and Arun Kumar (Patent No: US 8536324B2). Carbon nanotubes were complexed with modified chitosan and showed enhanced transfection efficiency *in vivo.*

7.3 Cellular Uptake of CNT

CNTs have been found to have remarkable properties, which allows them to be taken up by various cells. Therefore, it is possible to deliver various drugs and genes using the CNTs. Various studies have been carried; however, there are still few questions about the mechanism and cellular pathways for drug or gene delivery that are utilized by CNT [31].

Various mechanisms have been proposed for the internalization of CNTs:

1. Direct penetration through the cell membrane
2. Passive or active mechanism (endocytosis)

Cellular internalization of CNTs through a passive or nanoneedle mechanism allows diffusion across the cell membrane (CM). The cylindrical shape of CNT, the high aspect ratio, and the elongated form help in overcoming barriers to cross the CM. Another way of internalization includes endocytosis, which is further divided into five subcategories: Phagocytosis, pinocytosis (mainly micropinocytosis), clathrin-mediated endocytosis or receptor-mediated endocytosis, caveolin-mediated endocytosis and independent endocytosis with clathrin/caveolae [32]. CNTs of up to 100 nm size are taken by caveolin- and clathrin-mediated endocytosis, and CNTs of greater than 100 nm size are taken by macropinocytosis [33]. The mechanism used by the cells depends upon the material properties of CNTs such as surface charge and functionalization, dimension of CNTs, degree of agglomeration and aggregation and the types of CNT (SW/MWCNTs) involved [31]. Figure 7.3 represents the internalization mechanism of carbon nanotubes.

On the CNT surface (Carboxylated or amidated CNT), biomacromolecules such as proteins, antibodies, or DNA can form tiny bundles that are ingested by energy-dependent endocytosis. In a recent study, it was

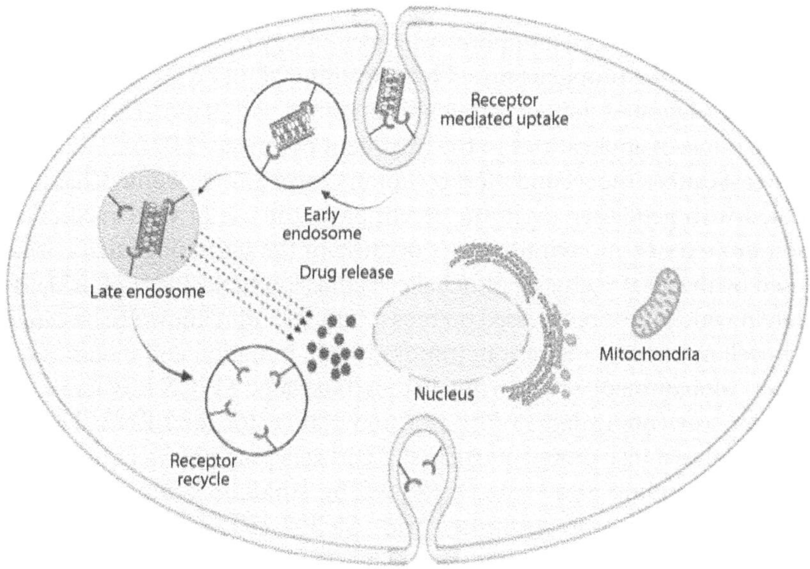

Figure 7.3 Cellular uptake of carbon nanotubes. Receptor mediated endocytosis of CNTs loaded with drug causes formation of early endosomes followed by late endosomes and then drug release from the system.

discovered that SWCNTs have a different cell absorption mechanism than MWCNT bundles. Direct penetration is more common for SWCNTs, whereas endocytosis is more common for MWCNTs. The consequences of the coating on the surface of CNTs have been investigated in some studies. Collagen, polymers, proteins, ssDNA PEGylated lipid or block copolymers, and pluronic F108 copolymer-functionalized SWCNTs can be absorbed into mammalian cells. Cai *et al.* conducted a theoretical investigation of penetration routes. They demonstrated that endocytosis is the route for amphiphilic nanotubes to pass through artificial lipid bilayers [7].

7.4 Functionalized Carbon Nanotubes and Cancer

Cancer, the second leading cause of mortality worldwide, is defined by uncontrolled cell division leading to metastasis when cancerous cells migrate from one part of the body to another. Various types of therapies exist to address the issue, yet some of them still fall short. Current preferential therapies include surgical removal of tumor, chemotherapy, and radiation therapy. However, drug delivery remains to be inadequate in terms of lacking specificity, off-target side effects and insufficient drug concentration to affected tumors. Nanotechnology has been shown to alleviate some of the problems posed by conventional therapies. The reported literature suggested that gene therapy combined with nanotechnology has a wide range of applications in the treatment of cancer [34].

Breast cancer, the second most common cancer among women, has been treated with gene therapy in the recent past. The use of viral vectors has been limited over nonviral vectors because of the immunogenicity issues posed by the viral vectors. Epithelial cell adhesion molecule (EpCAM) is a protein which is overexpressed on breast tumors and is known to be cancer stem cell marker, taking part in tumor growth, initiation and metastasis of tumor. Mohammadi *et al.* formulated carboxylated SWCNT, wherein the aptamer was conjugated to PEI as a nonviral vector for siRNA delivery. Using Malvern instrument, particle size and zeta potential were measured. The DNA binding assay was performed by ethidium bromide. Further, cytotoxicity assay was performed with MCF-7 cells and MDA-Mb-231 cells. The results indicated that the aptamer/siRNA conjugation was stable with a size of around 245 nm along with the stabilized SWCNT/polyethyleneimine/piperazine conjugate with a particle size of 85 nm. The transfection efficiency was also found to be higher than PEI alone. Besides, greater potential of inducing apoptosis was observed for aptamer/siRNA complex on MCF-7 cells for which EpCAM responded positively in contrast to MDA-MB231 cells for which EpCAM responded negatively. This has been further validated by protein immunoblotting wherein, BCL9 expression was lowered in case of MCF-7 cells, thus providing good platform for functionalized CNT for siRNA targeted delivery [35].

In order to have efficient gene delivery for the treatment of breast cancer, SWCNTs have been conjugated with polyethylene glycol and PEI. To impart lipophilicity, PEI was also derivatized by alkyl carboxylation. The results showed a particle size of less than 150 nm that could effectively condense DNA with a charge of up to 31 mV. Scientists further developed 5TR1 aptamer and conjugated to SWCNT to increase tumor efficacy. *In vitro* studies had been carried out on MCF-7 cells and transfection efficiencies were determined by Renilla luciferase. It was found that by conjugation with derivatized PEI-PEG-SWCNT, the transfection efficiency increased 6-8 times with a C/P ratio of 6 compared to PEI alone with a C/P ratio of only 0.8, indicating the importance of SWCNTs in gene delivery [36].

Functionalization of SWCNTs and MWCNTs with PEI was performed for the delivery of siRNA and tested on cervical cancer cells by Huang *et al.* Functionalization assisted in electrostatic interaction of positively charged CNTs with negatively charged siRNA. HeLa-S3, human cervical cancer cells were used for the study and incubated with prepared formulations for 48h. It was found that 5.08% w/w and 5.28% w/w of PEI was grafted on SWCNT and MWCNT, respectively. Through an electrophoretic assay, it was reported that siRNA was associated with PEI-SWCNT and PEI-MWCNt at PEI-CNT-/siGAPDH at a mass ratio of 80: 1 and 160:1, respectively. Finally, it was concluded that PEI functionalized SWCNT and MWCNT were effective in delivering siRNA to cervical cancer cells and could be used as nonviral vectors for gene delivery [37].

CNTs have also found applications in the field of transdermal drug delivery. Siu *et al.* developed functionalized single-walled carbon nanotubes with succinated polyethyleneimine. The formulation containing siRNA/PEI SWCNT was analyzed by transmission electron microscopy (TEM), zeta potential, *in vitro* and *in vivo* gene silencing activity, and *in vivo* study on tumor-bearing melanoma mouse models. The results demonstrated that after topical administration of CNT complex, greater uptake was observed for Cy3-labeled siRNA. The *in vivo* study also showed a gradual reduction in tumor growth over a 25-day period along with gene silencing. This delivery provided the foundation for siRNA delivery through CNT complexation by transdermal route [38].

One of the genes involved in the cancer pathogenesis is Polo like kinase-1 (PLK-1), a serine/threonine kinase, playing a key role in cell cycle regulation and overexpressed on tumors. Thus, gene silencing could be achieved through the use of functionalized CNTs by delivering siRNA. The results demonstrated that PLK1 gene when delivered through multi-walled CNT and tested on Calu6 tumor xenograft animal models showed improved therapeutic efficacy and increased *in vivo* cellular uptake in tumor by utilizing fluorescently labelled, non-coding siRNA. It was further validated by doing biological analysis, TUNEL staining, RT-PCR and western blotting. The complex was compared with that of cationic liposomes (lipoplexes). The study showed that *in vivo* activity (tumor growth inhibition) was not observed in lipoplexes while *in vitro* activity was observed. This could be due to the quick washing of the lipoplexes due to poor retention inside the tumors. In case of CNT formulation, it was retained for more than 24 h [39].

In the study conducted by Li *et al.*, single-walled CNTs were functionalized with 1,2-dioleoyl-3-trimethylammonium-propane (DOTAP) for effective siRNA delivery. It was characterized for particle size, zeta potential, morphology, and stability of siRNA by gel electrophoresis. *In vitro* activity was further evaluated on PC-3 prostate carcinoma cells and transfection efficiency was assessed by flow cytometry. The results indicated effective delivery system in form of functionalized CNT with particle size of 195 nm, zeta potential of 45 mV and stable siRNA in the system. On PC-3 cells, the efficacy of gene silencing was found to be 83% and the efficacy of transfection was around 92%. Therefore, functionalized SWCNT has been shown to be a better gene carrier system in the treatment of cancer [40].

7.5 Miscellaneous Diseases and Gene Delivery Through Functionalized CNT

Cystic fibrosis is a respiratory tract genetic disorder that is characterized by a defective gene of the cystic fibrosis transmembrane conductance regulator (CFTCR). Generally, the mutation in the CFTCR gene consists of deletion of a phenylalanine moiety at position 508.The primary feature of CF is a failure in ion transport in the apical membrane of most secretory cells, resulting in abnormal epithelial mucus production in the airway epithelia, the digestive system, the pancreas, the reproductive tract, and the liver. In most secretory cells, particularly respiratory tract cells, mutations in the CFTR gene result in a defective CFTR protein that is unable to efficiently transport chloride anions, resulting in an imbalance of ion regulation that depletes the water content and acidifies the pH in the airway surface liquid. This, together with the increased mucin polymer cross-links, increases the viscosity and quantity of mucus, making ciliary clearance more difficult.

Conventional therapy deals with correcting the consequences that occur due to the defective gene. Recently, gene therapy is becoming a hallmark for the treatment of cystic fibrosis, which particularly restores the activity of the mutated CFTCR gene and overcomes the underlying causes of cystic fibrosis compared to conventional treatment therapy. Basically, the gene therapy is targeted through vectors, that is, viral vector and non-viral vector. Viruses are used as vectors for the transfer of genetic material into human cells, as they easily get access to enter into human cells due to their transduction ability. However, the viral vectors are associated with major limitations that is they induce immune response against host cells. On the contrary, nonviral vectors are gaining much attention in gene and drug delivery systems [41]. The nonviral vectors such as carbon nanotubes have been efficiently used for the treatment of cystic fibrosis because they are safer, present a lack of immune response, and can be easily tailored with polymers and biological agents for the required drug delivery applications.

Liu *et al.* used modified single walled carbon nanotubes which were functionalized with the previously grafted phospholipids on the amine-terminated polyethylene glycol by adsorbing it on the surface of single walled CNT. Further resulting functionalized single walled CNTs was attached to thiol linked siRNA moiety for the antiviral therapy to target the CD4 and CXCR4 coreceptor. Here, the advantage of two alkyl groups present on the phospholipid molecule was that they impart firm binding to single-walled carbon nanotubes without detachment in phosphate buffer even after exposure to heating at 700°C for one week. Further the *in vitro* cell line study was carried out on three cell lines such as T cell line, MAGI cell line and primary human peripheral blood mononuclear cells. The *in vitro* cell line studies suggested that the reported siRNA grafted single walled carbon nanotubes were proved to be the generic molecular transporters for T cells and primary cells. The outcomes of these modified SWCNTs were found to be enhanced solubility, stability of CNTs in biological media and maximum silencing of gene expression on the surface of cells [42].

In addition, the functionalized CNTs have also been used for the diagnosis of human immunodeficiency virus (HIV) infection. Rapid and quantitative detection of viral oligonucleotides without the need for purification and amplification has played a crucial role in the early diagnosis of HIV virus in biological fluids such as serum and urine sample. Harvey *et al.* found a unique phenomenon wherein the carbon nanotube's optical response (photoluminescence) to nucleic acids enhanced by sodium dodecyl sulphate denatured protein present in HIV virus and thereafter the hydrophobic residue released from the SDS-denaturated protein binds to newly exposed nanotubes surface after grafting of the DNA on the carbon nanotubes, which resulted in increase in blue shift or fluorescence. Here, the miR-19 DNA grafted carbon nanotube sensor showed response to the denaturated viral genetic material. The mechanistic study confirmed that hydrophobic patches of the denaturated protein chain interact with the freed nanotube surface after hybridization, resulted in enhanced blue shift of nanotube emission. This mechanism has proved to be useful for the diagnosis of HIV infection [43].

Autoimmune disorders are characterized as a condition in which the immune system of the host mistakenly attacks itself. It occurs when the immune system of host is incapable to differentiate between normal healthy tissues and antigens that may be harmful. In usual scenario, the immune system of the host will attack foreign antigens and produce an immune response against antigens. However, the immune system fails in the event of autoimmune disorders. These disorders force the immune system to attack several organs, causing a systemic reaction, or they can be localized, affecting only one organ. The actual mechanism of several autoimmune diseases is unknown; however, the assumed mechanism varies between disorders [44]. Rheumatoid arthritis (RA) is a chronic disorder that primarily affects joints and is also associated with serious cartilage destruction. However, other body tissues such as the skin, eyes, and kidneys might be damaged. It is a type of T cell-mediated disease in which the body's immune system attacks its tissues. Rheumatoid factors are secreted in response to immunological reactions, which cause cartilage and bone degradation as the disease progresses. The auto-reaction of immune modulators such as effector cells and cytokines causes joint injury. It begins by attacking synovial membranes, then progresses to surrounding structures. The most important element in RA pathology is persistent activation of synovial macrophages which causes the development of inflammation in RA [45]. The pathophysiology of RA has been linked to genetic factors such as class II major histocompatibility antigens/human leukocyte antigens (HLA-DR) and non-HLA genes. Citrullinated protein autoimmunity (ACPA [anti-citrullinated protein antibody]) and smoking have been linked to HLA and various non-HLA genes [46]. B-lymphocytes produce autoantibodies such as rheumatoid factor (RF) and antibodies against the cyclic citrullinated peptide (anti-CCP) [47]. In summary, smoking, as well as other environmental and lifestyle factors, may stimulate ACPA production and the development of seropositive ACPA RA [46]. *In vivo* and *ex vivo* delivery strategies are commonly used to deliver therapeutic genes. These methods can be used to transfer both local and systemic genetic material. Gene transfer through intraarticular route is generally done in the case of *in vivo* strategies but *ex vivo* strategies include three steps:

1. Removal of synoviocytes from the diseased location, which is usually a joint
2. *In vitro* gene transduction
3. Injecting transduced cells back into the joints [45].

The mechanistic approach behind delivering genes through vectors can be via blocking the production and activity of inflammatory agents that are involved in RA pathogenesis or by expressing the proteins that can further block the production of these agents. In two clinical trials which were carried out in RA and osteoarthritis models, retroviral vectors were used to deliver IL-1Ra cDNA to arthritic joints, and it was found that it successfully reduced the joint damage. The sequence of specific genes includes RNA interference (RNAi) and small interfering RNA (siRNA) that silence or down-regulated proteins that are present in the cytoplasm. Silencing or down-regulation of proteins associated with RA, such as Notch1 and TNF-α, could be beneficial in the treatment of RA. Protection of therapeutic genes should have been done to avoid nuclease degradation and it should be delivered intact to the nucleus of the target cells by avoiding entering the plasma and nucleus membrane without giving any adverse effects to the surrounding tissues [48]. Activation and expression of the Notch signaling pathway have been seen in multiple cell types and it is one of the targets for gene therapy. Furthermore, it was found that Notch inhibitor when administered reduced arthritis in collagen-induced arthritis mouse. Another study utilizing siRNA to block the secretase essential for Notch1 activation discovered a decreased autoimmune response which was followed by an increase in regulatory T cells. In this study, two types of single-wall carbon nanotubes (SWCNTs) high-pressure carbon monoxide synthesized SWCNT (HiPco-SWCNTs) and COOH functionalized SWCNT (carboxyl-SWCNTs) were linked with siRNA or methotrexate (MTX), or both. These conjugates targeted the Notch1 signaling pathway involved in arthritis. MTX attachment to SWCNTs has the potential to improve efficacy while reducing MTX toxicity [49].

Atherosclerosis is a cardiovascular disease in which there is narrowing of blood vessels due to deposition of fatty plaques. The most widely used interventional therapy to treat the atherosclerosis is percutaneous trans vascular balloon angioplasty and stent technology. However long-term therapy with these techniques has been associated with certain limitations such as in-stent restenosis wherein the obstruction of dilated arteries occurs due to incomplete endothelial recovery and smooth muscle overgrowth at the site of stent implantation. The implication of this event leads to serious effects such as heart attack. Various literature reports suggested the use of gene therapy by naked plasmids, viral vectors and non-viral nano delivery systems. However, these approaches are associated with several shortcomings such as high cytotoxicity, inefficient *in vivo* gene transfer, strong inflammatory reactions and risk of viral DNA integration to host genome. The insufficient vascular endothelial growth factor cytokine concentration at the stent site combined with their short half-life leads to suboptimal biological effects to trigger vascular re-endothelialization. Paul *et al.* developed nanoparticles from nano biohybrid stent platform, containing layers of functionalized single walled carbon nanotubes and bio hydrogel. These bio hydrogels consisted of fibrin matrices which were assembled in a layer by layer on stent surface with alternate layers carrying endosomolytic Tat peptide or DNA nanoparticles or nanoparticles hybridized to polyacrylic acid wrapped single-walled carbon nanotubes. These single walled carbon nanotubes nanohybrid hydrogel delivered proangiogenic vascular endothelial growth factor, plasmid DNA and angiopoietin-1 gene to the target site and tuned the bioactivity of the stents, reduced the loss of therapeutics while traversed through the vessels and *in vivo* experiments in balloon injured canine femoral artery demonstrated enhanced re-endothelialization and attenuated stenosis thus showing promise to address the existing shortcomings of current stent technologies [50].

Type II diabetes is involved in impaired glucose transportation and also shows insulin resistance. The said causes are due to impairment in the insulin signaling pathway and as a result of glucose transporter 4 (GLUT4) translocation. Johanna and co-workers tested the involvement of Ca^{2+} through canonical transient receptor potential 3 channels (TRPC3) in insulin-mediated glucose uptake in adult skeletal muscles. To prove the above hypothesis, they conducted two sets of experiments. In the first, they tested the anti-TRPC3 antibodies, which have earlier been shown to impede Nonselective cation currents (NSCC) mediated through TRPC3 channels and were used in patch pipette to detect OAG (1-oleyl-2-acetyl-sn-glycerol)-induced NSCC in skeletal muscle cells. In the second, siRNA was linked to f-CNTs and was used to knock down the TRPC3. Results were promising and knockdown of TRPC3 showed reduction of approximately 70% in OAG induced Ca^{2+} influx and also reduction in insulin-mediated glucose uptake. Through this study, TRPC3 has been proposed to be a therapeutic target for insulin resistance [51].

Al Jamal *et al.* prepared functionalized MWCNTs by the addition of an ammonium group to transport anticaspase-3 siRNA into the central nervous system (CNS). Effective delivery of siRNA directly to the CNS corrected characteristics of various neurological disorders that have been shown in animal studies. Using an endothelin-1 induced stroke model, the complex reduced neurodegeneration and improved functional preservation before and after focal ischemia injury to the rodent motor cortex. This is the first instance to show that delivering siRNA into the CNS via CNTs had biological and functional (motor recovery) effects in an animal model of endothelin-1 induced stroke. As a result, this system can be considered as an effective carrier for siRNA molecules implicated in neurological diseases, as well as a system capable of inducing functional recovery [52].

Joshua *et al.* fabricated functionalized SWCNTs for vascular diseases. CNTs were functionalized with lipids and dendrimers (lysine based) (TOT) and finally complexed with siRNA. The advantage of using lipid was that it helped in cell binding and also masked the hydrophilicity of siRNA. The complex (TOT- siRNA) was used for the delivery of anti-ApoB siRNA. The dose administered to the mice was 0.96 mg/kg. The complex was effective in reduction of Apo mRNA and protein levels. Results showed a reduction in total cholesterol level without any signs of toxicity [53].

Table 7.1 shows the summary of biomedical applications of functionalized carbon nanotubes in different therapeutic area.

Table 7.1 Applications of functionalized carbon nanotubes in different therapeutic area.

Sr. no.	Functionalized carbon nanotube	Target	Therapy	Remark	Reference
1	Phospholipid grafted on amine terminated polyethylene glycol which adsorbed on the surface of single walled carbon nanotube	siRNA	Antiviral therapy	Increased solubility and stability of functionalized carbon nanotubes in biological media and achieved maximum silencing of gene expression on the surface of cells	[42]
2	miR-19DNA grafted carbon nanotube sensor	miR-19DNA	Diagnosis of HIV Infection	Enhanced optical response to denaturated viral genetic material	[43]
3	Carbon monoxide and carboxyl group functionalized single walled carbon nanotube	siRNA, Methotrexate	Treatment of rheumatoid arthritis	Enhanced efficacy and decreased toxicity of methotrexate	[49]

(Continued)

Table 7.1 Applications of functionalized carbon nanotubes in different therapeutic area. (*Continued*)

Sr. no.	Functionalized carbon nanotube	Target	Therapy	Remark	Reference
4	SWCNT nanohybrid -fibrin matrix biohydrogel/Tat peptide or DNA nanoparticle hybridized to polyacrylic acid wrapped SWCNT containing fibrin matrix biohydrogel	Vascular growth factor, plasmid DNA and angiopoietin-1 gene	Treatment of atherosclerosis	Reduced loss therapeutics from SWCNT during traversed through the blood vessels	[50]
5	Functionalized CNT with polyethylene glycol 2000	siRNA	Type 2 diabetes	Knockdown of TRPC3 leads near about 70% reduction in OAG(1-oleyl-2acetyl-sn-glycerol) induced nonselective cation current (NSCC) in skeletal muscles	[51]

(*Continued*)

Table 7.1 Applications of functionalized carbon nanotubes in different therapeutic area. (*Continued*)

Sr. no.	Functionalized carbon nanotube	Target	Therapy	Remark	Reference
6	Ammonium group functionalized multiwalled carbon nanotube	Anticaspase-3 siRNA	Neurological disorder	Promising biological and functional (motor recovery) effects which implicated proposed fMWCNT effective carrier for siRNA molecule in neurological disorders	[52]
7	Lipids and dendrimers functionalized SWCNT	Anti-Apo siRNA	Vascular disease	Functionalized SWCNT with lipid provide promising cell binding and masked the hydrophilicity of siRNA and found significant reduction in total cholesterol level with lack of toxicity	[53]

7.6 Toxicology and Environmental Aspects of Functionalized CNT

Functionalization of CNT has major impact on human health as it can alter the morphological and physicochemical characteristics of carbon nanotubes. In an aqueous environment, fresh CNTs tend to cluster owing to the presence of Van der Waals forces near the surface. Because of high hydrophobicity and high aspect ratio, CNTs can pass biological membranes easily. When functionalization of CNTs is carried out, it reduces the hydrophobicity of graphene surfaces which allows it to disperse more easily in aqueous solutions. It also allows the graphene surface to link together to form composite materials. After undergoing internalization, CNTs accumulate in diverse subcellular compartments, owing to their physicochemical characteristics, functionalization and other factors such as the cytosol, endosomes, perinuclear area, mitochondria, and nucleus [54].

The mechanism involved in CNTs toxicity is represented in Figure 7.4.

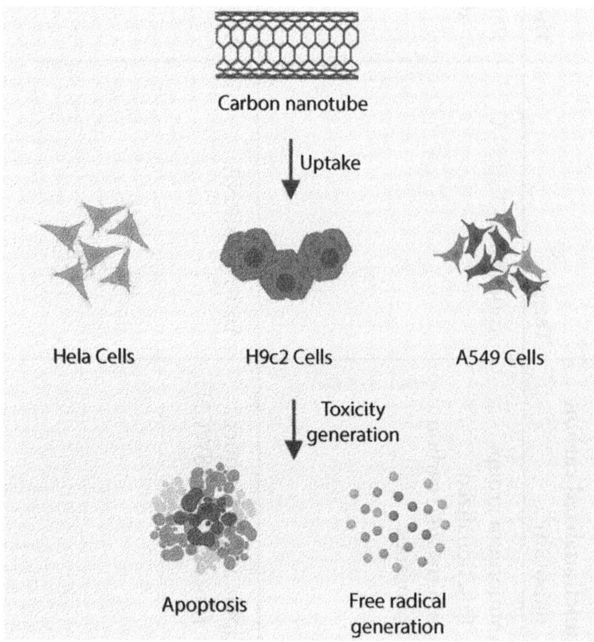

Figure 7.4 The mechanism of toxicity caused by CNTs; Through cell lines study (on Hela, H9c2 and A549 cells), free radicals are generated which subsequently generate toxicity to cells and ultimately cause cell apoptosis.

7.6.1 Cellular Toxicity

When CNTs penetrate the cellular membrane, it results in the generation of oxidative stress which causes inflammation to the cell. A foreign body response (FBR) is induced to get rid of CNTs as it behaves as a foreign body to cells. In response to CNT-induced oxidative damage, the major signaling factors like protein kinases and nuclear factor kappa B regulate cytokines. Free radicals are also one of the reasons for the generation of oxidative stress which in turn causes cytotoxicity. The excess free radicals are responsible for the activation of inflammatory responses like activating transcription factors and activator protein-1 causes oxidation of proteins, lipids, and DNA [55].

Requardt *et al.* investigated *in vitro* cytotoxicity studies on lung A549 epithelial cells and liver HepG2 hepatocytes. For these, they used two types of multi-walled carbon nanotubes of which one with iron that was used as a catalyst while the other had an iron that was removed in post-production heat treatment. The MWCNT without iron had lesser surface defects. For testing cell viability, WST-8 assay was used. They used the lactate dehydrogenases assay (LDH) to examine the cell membrane integrity and measured the cellular generation of reactive oxygen species (ROS). The result found that both MWCNTs and cell types demonstrated a dose- and time-dependent decrease in cell viability. Furthermore, LDH leakage increased in a dose- and time-dependent manner, indicating a loss of membrane integrity. The heat-treated MWCNTs had greater adverse effects and ROS production than non-treated MWCNT. Also, cell cycle arrest was induced in A549 cells with heat-treated MWCNTs [56].

7.6.2 Liver Toxicity

Liver is the most important organ in the body, and it plays an important role in the body's metabolic pathways. As a result, any substance administered to the body is metabolized by the liver. So, there is a need to study the toxic effects of CNTs on liver tissues [55]. A toxicity study carried out by Principi *et al.* used MITO-Luc bioluminescence reporter mice and CD1 mice. SWCNTs were exposed in these models for nine weeks to check systemic biodistribution in the body. SWCNTs when accumulated in the body induced cell proliferation, inflammation with an increased amount of enzyme production. Different biomarkers were used for liver damage such as aspartate aminotransferase, alanine aminotransferase, and bilirubinemia. When histological studies were carried out, concentration of these biomarkers was found to be elevated [57].

In vitro cytotoxicity of PEG-modified carbon nanotubes on hepatoma cells (HepG2) was carried out. The HepG2 hepatocytes were cultivated with varying concentrations of PEG-CNTs for 24, 48, and 72 h. The colorimetric assay method was used to check the survival rate and effect of PEG-CNTs concentration. It was found that necrotic cells were aggregated in HepG2 when the concentration was greater than 100 μg/mL i.e., grade II (middle toxic) and less aggregated when the concentration was less than 100 μg/mL (Grade I; non-toxic) according to the scoring of ISO2109932-5 cytotoxicity standard. This result showed that as the concentration of water-soluble PEG-CNTs increased, the cytotoxicity also increased [58].

A study was conducted to compare the effects of carboxylated multiwalled carbon nanotubes (MWCNTs-COOH) and pristine multiwalled carbon nanotubes (p-MWCNTs) on human normal liver cell line L02. Nuclear morphology changes were observed to check apoptosis induced by MWCNTs. It was observed that p-MWCNTs induced apoptosis at a half concentration as compared to (MWCNTs-COOH) at 24h. Lactate dehydrogenase is a cytoplasmic stable enzyme present in all kinds of cells which is released as a result of oxidative stress and causes damage to the cell membrane integrity. This result was demonstrated by lactate dehydrogenase leakage and it was higher in the case of p-MWCNTs. Also, the L02 cell line showed higher G0/G1 cell cycle arrest in the case of p-MWCNTs. Furthermore, caspase-8, a key mediator of the extrinsic apoptotic pathway when activated got resulted in cleavage of downstream caspases such as caspase-3 leading to cell apoptosis. MWCNTs-COOH had reduced the activation of caspase-8 as compared with p-MWCNTs. This result showed that for *in vivo* applications, MWCNTs-COOH might be safer as compared to p-MWCNTs [59].

7.6.3 Central Nervous System Toxicity

The therapeutic use of CNTs for drug delivery is extremely appealing in the treatment of several neurological disorders, infections, brain and spinal trauma, stroke, and breast cancer. Through cortical stereotactic injection in the mouse brain, Bardi *et al.* studied the neuronal toxicity of various types of CNTs. In particular, oxMWCNTs-NH^{3+} activate cytokines and glial cells, implying that oxidation can contribute to a long-term inflammatory response in healthy brains, whereas MWCNTs-NH^{3+} was well tolerated, causing only a local and temporary inflammatory response. Even though the two engineered NPs had equal uptake levels in different brain tissue cells (astrocytes, microglia, and neurons), this distinct effect was seen. The release of the proinflammatory cytokines TNF- and IL-1,

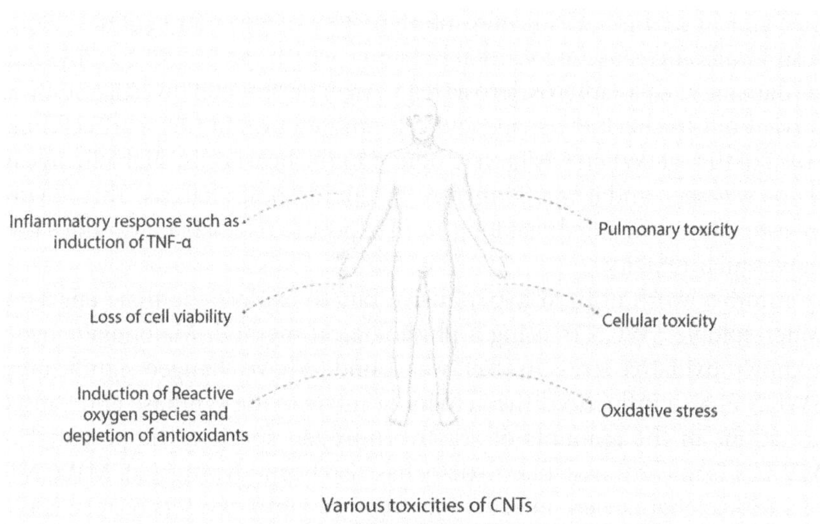

Figure 7.5 Toxicities caused by carbon nanotubes.

as well as microglia activation, was considerably higher after injection of ox-MWCNTs-NH^{3+} compared to MWCNTs-NH^{3+} [54]. A study was conducted to check the biological impact of MWCNTs on glial and neuronal cells. These cells were isolated from the frontal cortex and striatum of the fetal rat. Modified LDL assay result showed that neurons were not affected when MWCNTs were exposed to neurons of both regions. Cytotoxicity was unaffected by MWCNT type or dose, implying that CNTs have a developed sensitivity. Before nanotube exposure, cell populations in mixed glial cultures were characterized, and the ST-derived mixed glial cultures had a higher number of CD11b/c positive cells. CNT was taken up more effectively by CD11b/c positive cells (microglia) after exposure to MWCNTs than by GFAP positive cells (astrocytes). ST-derived glial cultures released more NO than FCO-derived cells when exposed to conditioned media from microglia enriched cultures treated to MWCNTs. These findings suggested that the increased number of microglial cells in this brain region was related to the higher cytotoxic response observed from ST-derived mixed glia cultures [60]. Figure 7.5 represents the summary of different toxicities caused by CNTs.

7.6.4 Cardiovascular Toxicity

When CNTs get circulated after administration into the body, it may cause cardiovascular damage. Because of the properties like hardness, mechanical strength, CNTs may interfere with the formation of new blood vessel walls and cause myocardial ischemia, which can lead to cardiovascular damage such as atherosclerosis [61]. An experimental investigation was carried out to check the shape and cell proliferation in the heart cell lines H9c2 of the rat. When CNTs were bound to the heart cells, slight change in shape was seen and it was confirmed by microscopic analysis. This *in vitro* investigation concluded that long-term exposure to CNTs may induce cardiovascular toxicity.

Another work utilized hypertensive rats to explore the toxic effects of water-soluble SWCNTs using a physiological method. Malondialdehyde, a known oxidative stress marker, was found to have changed significantly. SWCNTs also influenced the activity of nitric oxide synthase and caused variations in the amounts of reactive nitrogen species. The potential of MWCNTs in cardiovascular toxicity has also been investigated. MWCNTs of various lengths were injected intravenously into two sets of rats: spontaneous hypertensive (SH) and Wistar-Kyoto (WKY) rats. In the SH group of rats, the data demonstrated reduced blood pressure and a consistent decrease in the heart rate. These studies showed findings that were similar to those obtained with SWCNT. The toxicity of CNTs in cardiovascular organs suggested an increase in oxidative stress and inflammation as a result of CNT accumulation in cardiac muscle and vascular tissues [55].

7.7 Regulatory Concerns Over Functionalized Carbon Nanotubes

All the nanomaterials including the carbon nanotubes have been linked to health issues. Inflammatory reactions in the lungs have been observed in several investigations of pulmonary toxicity after inhalation, and specific forms of MWCNTs have been found to cause mesothelioma after intraperitoneal injection. As a result of these findings, the use of CNTs, in general, has sparked controversy [62].

Previously, it was unclear if CNTs would be controlled as carbon or graphite, or whether they would be treated as a new substance due to their unique structure. CNTs are regulated as a new material in the United States and the European Union, according to recent statements from the US Environmental Protection Agency ("EPA") and the European Commission ("EU"). The EPA issued a notice on CNT regulation under the Toxic Substances Control Act ("TSCA") on Friday, October 31, 2008. According to this view, everyone who manufactures or imports CNTs must file the required documents with the EPA at least 90 days before doing so [63].

A number of national bodies in the United States, including the Environmental Protection Agency (EPA), the Food and Drug Administration (FDA), and the Consumer Product Safety Commission (CPSC), have issued nano-regulations. The Toxic Substances Control Act (TSCA) and the Federal Insecticide, Fungicide, and Rodenticide Act (FIFRA) regulate the EPA's policy on nanomaterials. The EPA has issued Significant New Use Rules (SNURs) of premanufacture notices (PMNs) for 13 substances, including carbon nanotubes and fullerenes, as part of its regulation. The Registration Evaluation Authorization and Restriction of Chemicals (REACH) committee of the European Union is in charge of registering compounds that are created or imported in quantities more than one ton per year. Nanomaterials, as well as carcinogenic, mutagenic, and reproductive (CMR) and Persistent, Bio accumulative, and Toxic (PBT) compounds, are among these substances [64].

The nanomaterials and nanostructured materials are classified into four categories.

1. Carbon-based nanomaterials
2. Inorganic-based nanomaterials
3. Organic-based nanomaterials
4. Composite-based nanomaterials

Our concern of category is type 1 which includes nanomaterials that contain carbon in the form of hollow tubes, ellipsoids and spheres. Examples include fullerene, carbon nanotubes, carbon nanofibers, carbon black, graphene and carbon onions [65].

In 2011, the European Commission recommended the definition of nanomaterials. "Nanomaterial means a natural, incidental or manufactured material containing particles, in an unbound state or as an aggregate or as an agglomerate and where, for 50% or more of the particles in the number size distribution, one or more external dimensions is in the size range 1–100 nm". In specific cases and where warranted by concerns for the environment, health, safety or competitiveness, the number size distribution threshold of 50 % may be replaced by a threshold between 1 and 50%. By derogation from the above point, fullerenes, graphene flakes and single-wall carbon nanotubes with one or more external dimensions below 1 nm should be considered as nanomaterials [66].

Apart from that, a collaboration of US domestic and international advocacy groups issued a book "Principles for the Oversight of Nanotechnologies and Nanomaterials," which was recognized by 70 groups from six continents. This paper calls for robust and complete oversight of nanomaterial derived products. This includes a precautionary foundation for specific nanomaterial rules, as well as public and worker health and safety, transparency, public involvement, environmental preservation, and the consideration of larger implications and manufacturer accountability [65].

According to the US EPA, CNTs are a key regulatory concern in terms of toxicity and environmental safety. Another concern for CNTs is bioaccumulation. Since CNTs show high bioavailability, bioaccumulation also needs to be taken into account [67].

7.8 Conclusion and Future Prospects

Due to increased burden of genetic disorders, gene therapy has been considered as a viable treatment option over conventional drug delivery system. The safety and efficacy of gene therapy is mainly dependent on the choice of transfection vector mainly viral or nonviral, each having its own advantages and disadvantages. Carbon nanotubes have been considered as one of the efficient drug delivery systems for a wide range of synthetic and biological molecules. However, due to drawbacks posed by CNTs such as toxicity, environmental safety, biosafety profiles as a carrier system, various approaches such as functionalization of CNTs are sought. Functionalization of CNTs for delivery of proteins and nucleic acids have been found to improve the safety, efficacy, and formulation related aspects such as stability, solubility and is coupled with advantages such as low immunogenicity compared to viral vectors. Using functionalization, various biomacromolecules (DNA, oligosaccharides, proteins) are attached to CNTs. Gene delivery through functionalized CNTs has been found to have applications in a wide range of therapeutic areas (cancers, autoimmune disorders, cardiovascular diseases) and diagnosis. Though CNTs carry certain risks and norms to be followed during CNTs designing as per regulatory agencies, further improvement in functionalization strategies will overcome these issues and will improve the potential of delivered therapeutics.

Prominent Groups Working in the Field of Carbon Nanotubes Especially in Gene Therapy

Mahdi Karimi and his teammates are working on carbon nanotubes in the Department of Medical Nanotechnology at Iran University of Medical Sciences. In University of California Berkeley, scientist Landry and colleagues are also working on carbon nanotubes and their applications in gene therapeutic area. Professor Hongjie Dai is another leading figure in the world of carbon nanotubes. His team was pioneered in vapor deposition for carbon nanotubes growth and their applications in nanomedicines.

Arkema group, Nanocyl, Nanoshell, Carbon solutions Inc, Hyperion catalysis International, Showa Denko KK, Klean commodities Inc are some of the notable companies working on the development, production and commercialization of carbon nanotubes for biomedical applications.

Important Website

https://pubmed.ncbi.nlm.nih.gov/33688185/
https://news.stanford.edu/news/2004/january21/dai-121.html
https://dailab.stanford.edu/
https://www.fortunebusinessinsights.com/carbon-nanotubes-cnt-market-102700

References

1. Zhou, W. and Wang, X., Human gene therapy: A patent analysis. *Gene*, Elsevier B.V., 803, 145889, 2021 [Internet].
2. Mirón-Barroso, S., Domènech, E.B., Trigueros, S., Nanotechnology-based strategies to overcome current barriers in gene delivery. *Int. J. Mol. Sci.*, 22, 2021.
3. Wirth., T., Parker, N., Ylä-Herttuala, S., History of gene therapy. *Gene*, Elsevier B.V., 525, 162–9, 2013 [Internet].
4. Alhakamy, N.A., Curiel, D.T., Berkland, C.J., The era of gene therapy: From preclinical development to clinical application. *Drug Discovery Today*, Elsevier Ltd, 26, 1602–19, 2021 [Internet].
5. Wong, B.S., Yoong, S.L., Jagusiak, A., Panczyk, T., Ho, H.K., Ang, W.H., Pastorin, G., Carbon nanotubes for delivery of small molecule drugs. *Adv. Drug Deliv. Rev.*, Elsevier B.V., 65, 1964–2015, 2013 [Internet].
6. Rathinavel, S., Priyadharshini, K., Panda, D., A review on carbon nanotube: An overview of synthesis, properties, functionalization, characterization, and the application. *Mater. Sci. Eng. B Solid-State Mater. Adv. Technol.*, Elsevier B.V., 268, 115095, 2021 [Internet].
7. Karimi, M., Solati, N., Ghasemi, A., Estiar, M.A., Hashemkhani, M., Kiani, P., Mohamed, E., Saeidi, A., Taheri, M., Avci, P., Aref, A., Amiri, M., Baniasadi, F., Hamblin, M., Carbon nanotubes part II: A remarkable carrier for drug and gene delivery. *Expert Opin. Drug Deliv.*, 12, 1089–105, 2015.

8. Wang, T., Upponi, J.R., Torchilin, V.P., Design of multifunctional non-viral gene vectors to overcome physiological barriers: Dilemmas and strategies. *Int. J. Pharm.*, Elsevier B.V., 427, 3–20, 2012.
9. Sajid, M.I., Jamshaid, U., Jamshaid, T., Zafar, N., Fessi, H., Elaissari, A., Carbon nanotubes from synthesis to *in vivo* biomedical applications. *Int. J. Pharm.*, Elsevier B.V., 501, 278–99, 2016.
10. Tasis, D., Tagmatarchis, N., Georgakilas, V., Prato, M., Soluble carbon nanotubes. *Chem. Eur. J.*, 9, 4000–8, 2003.
11. Khan, M.U., Gomes, V.G., Altarawneh, I.S., Synthesizing polystyrene/carbon nanotube composites by emulsion polymerization with non-covalent and covalent functionalization. *Carbon N Y.*, Elsevier Ltd, 48, 2925–33, 2010.
12. Cha, J., Jin, S., Shim, J.H., Park, C.S., Ryu, H.J., Hong, S.H., Functionalization of carbon nanotubes for fabrication of CNT/epoxy nanocomposites. *Mater. Des.*, Elsevier Ltd, 95, 1–8, 2016.
13. Huang, W., Taylor, S., Fu, K., Lin, Y., Zhang, D., Hanks, T.W., Rao, A.M., Sun, Y., Attaching proteins to carbon nanotubes via diimide-activated amidation. *Nano Lett.*, 2, 311–4, 2002.
14. Sun, Y., Fu, K., Lin, Y., Huang, W., Functionalized carbon nanotubes. *Acc. Chem. Res.*, 35, 1096–104, 2002.
15. Lordi, V., Yao, N., Wei, J., Method for supporting platinum on single-walled carbon nanotubes for a selective hydrogenation catalyst. *Chem. Mater.*, 13, 733–7, 2001.
16. Voiry, D., Goswami, A., Kappera, R., Silva, C.D.C.C.E., Kaplan, D., Fujita, T., Chen, M., Asefa, T., Chhowalla, M., Covalent functionalization of monolayered transition metal dichalcogenides by phase engineering. *Nat. Chem. Nat.* Publishing Group, 7, 45–9, 2015.
17. Kong, F., Liu, F., Li, W., Guo, X., Wang, Z., Zhang, H., Li, Q., Luo, L., Du, Y., Jin, Y., You, J., Smart carbon nanotubes with laser-controlled behavior in gene delivery and therapy through a non-digestive trafficking pathway. *Small*, 12, 6753–66, 2016.
18. Caoduro, C., Hervouet, E., Girard-Thernier, C., Gharbi, T., Boulahdour, H., Delage-Mourroux, R., Pudlo, M., Carbon nanotubes as gene carriers: Focus on internalization pathways related to functionalization and properties. *Acta Biomater*, 49, 36–44, 2017.
19. Behnam, B., Shier, W.T., Nia, A.H., Abnous, K., Ramezani, M., Non-covalent functionalization of single-walled carbon nanotubes with modified polyethyleneimines for efficient gene delivery. *Int. J. Pharm.*, Elsevier B.V., 454, 204–15, 2013 [Internet].
20. Baker, S.E., Cai, W., Lasseter, T.L., Weidkamp, K.P., Hamers, R.J., Covalently bonded adducts of deoxyribonucleic acid (DNA) oligonucleotides with single-wall carbon nanotubes: Synthesis and hybridization. *Nano Lett.*, 2, 1413–7, 2002.

21. Dwyer, C., Guthold, M., Falvo, M., Washburn, S., Superfine, R., Erie, D., DNA-functionalized single-walled carbon nanotubes. *Nanotechnology*, 13, 601–4, 2002.
22. Chen, J., Liu, H., Weimer, W.A., Halls, M.D., Waldeck, D.H., Walker, G.C., Noncovalent engineering of carbon nanotube surfaces by rigid, functional conjugated polymers. *J. Am. Chem. Soc.*, 124, 9034–5, 2002.
23. Mittal, V., Carbon nanotubes surface modifications: An overview, in: *Surface Modification of Nanotube Fillers*, pp. 1–23, 2011.
24. Gorityala, B.K., Ma, J., Wang, X., Chen, P., Liu, X.W., Carbohydrate functionalized carbon nanotubes and their applications. *Chem. Soc. Rev.*, 39, 2925–34, 2010.
25. Oliveira, S.F., Bisker, G., Bakh, N.A., Gibbs, S.L., Landry, M.P., Strano, M.S., Protein functionalized carbon nanomaterials for biomedical applications. *Carbon N Y.*, Elsevier Ltd, 95, 767–79, 2015.
26. Sanz, V., Borowiak, E., Lukanov, P., Galibert, A.M., Flahaut, E., Coley, H.M., Silva, S.R., McFadden, J., Optimising DNA binding to carbon nanotubes by non-covalent methods. *Carbon N Y.*, Elsevier Ltd, 49, 1775–81, 2011.
27. Bartholomeusz, G., Cherukuri, P., Kingston, J., Cognet, L., Lemos, R., Leeuw, T.K., Russo, L., Weisman, R., Powis, G., *In vivo* therapeutic silencing of hypoxia-inducible factor 1 alpha (HIF-1α) using single-walled carbon nanotubes noncovalently coated with siRNA. *Nano Res.*, 2, 279–91, 2009.
28. Zheng, M., Jagota, A., Semke, E.D., Diner, B.A., McLean, R.S., Lustig, S.R., Richardson, R., Tassi, N., DNA-assisted dispersion and separation of carbon nanotubes. *Nat. Mater.*, 2, 338–42, 2003.
29. Cui, D., Tian, F., Kong, Y., Titushikin, I., Gao, H., Effects of single-walled carbon nanotubes on the polymerase chain reaction. *Nanotechnology*, 15, 154–7, 2004.
30. Gao, L., Nie, L., Wang, T., Qin, Y., Guo, Z., Yang, D., Yan, X., Carbon nanotube delivery of the GFP gene into mammalian cells. *ChemBioChem*, 7, 239–42, 2006.
31. Zare, H., Ahmadi, S., Ghasemi, A., Ghanbari, M., Rabiee, N., Bagherzadeh, M., Karimi, M., Webster, T., Hamblin, M., Mostafavi, E., Carbon nanotubes: Smart drug/gene delivery carriers. *Int. J. Nanomedicine*, 16, 1681–706, 2021.
32. Cui, X., Xu, S., Wang, X., Chen, C., The nano-bio interaction and biomedical applications of carbon nanomaterials. *Carbon N Y.* Elsevier B.V., 138, 436–50, 2018.
33. Singh, B., Lohan, S., Sandhu, P.S., Jain, A., Mehta, S.K., Functionalized carbon nanotubes and their promising applications in therapeutics and diagnostics, in: *Nanobiomaterials in Medical Imaging Applications of Nanobiomaterials*, pp. 455–78, Elsevier Inc, 2016.
34. Mahajan, S., Patharkar, A., Kuche, K., Maheshwari, R., Deb, P.K., Kalia, K. *et al.*, Functionalized carbon nanotubes as emerging delivery system for the treatment of cancer. *Int. J. Pharm.*, 548, 540–58, 2018 [Internet].

35. Mohammadi, M., Salmasi, Z., Hashemi, M., Mosaffa, F., Abnous, K., Ramezani, M., Single-walled carbon nanotubes functionalized with aptamer and piperazine-polyethylenimine derivative for targeted siRNA delivery into breast cancer cells. *Int. J. Pharm.*, Elsevier B.V., 485, 50–60, 2015 [Internet].
36. Taghavi, S., HashemNia, A., Mosaffa, F., Askarian, S., Abnous, K., Ramezani, M., Preparation and evaluation of polyethylenimine-functionalized carbon nanotubes tagged with 5TR1 aptamer for targeted delivery of Bcl-xL shRNA into breast cancer cells. *Colloids Surf. B Biointerfaces*, Elsevier B.V., 140, 28–39, 2016 [Internet].
37. Huang, Y.P., Lin, I.J., Chen, C.C., Hsu, Y.C., Chang, C.C., Lee, M.J., Delivery of small interfering RNAs in human cervical cancer cells by polyethylenimine-functionalized carbon nanotubes. *Nanoscale Res. Lett.*, 8, 1–11, 2013.
38. Siu, K.S., Chen, D., Zheng, X., Zhang, X., Johnston, N., Liu, Y. *et al.*, Noncovalently functionalized single-walled carbon nanotube for topical siRNA delivery into melanoma. *Biomaterials*, Elsevier Ltd, 35, 3435–42, 2014 [Internet].
39. Guo, C., Al-Jamal, W.T., Toma, F.M., Bianco, A., Prato, M., Al-Jamal, K.T. *et al.*, Design of cationic multiwalled carbon nanotubes as efficient siRNA vectors for lung cancer xenograft eradication. *Bioconjug. Chem.*, 26, 1370–9, 2015.
40. Li, H., Hao, Y., Wang, N., Wang, L., Jia, S., Wang, Y. *et al.*, DOTAP functionalizing single-walled carbon nanotubes as non-viral vectors for efficient intracellular siRNA delivery. *Drug Deliv.*, 23, 840–8, 2016.
41. Villate-Beitia, I., Zarate, J., Puras, G., Pedraz, J.L., Gene delivery to the lungs: pulmonary gene therapy for cystic fibrosis. *Drug Dev. Ind. Pharm.*, Taylor & Francis, 43, 1071–81, 2017.
42. Liu, Z., Winters, M., Holodniy, M., Dai, H., siRNA delivery into human T cells and primary cells with carbon-nanotube transporters. *Angew. Chem. Int. Ed.*, 46, 2023–7, 2007.
43. Harvey, J.D., Baker, H.A., Ortiz, M.V., Kentsis, A., Heller, D.A., HIV detection via a carbon nanotube RNA sensor. *ACS Sens., Am. Chem. Soc.*, 4, 1236–44, 2019.
44. Angum, F., Khan, T., Kaler, J., Siddiqui, L., Hussain, A., The prevalence of autoimmune disorders in women: A narrative review. *Cureus*, 12, 2020.
45. Kesharwani, D., Paliwal, R., Satapathy, T., Paul, S.D., Rheumatiod Arthritis: An updated overview of latest therapy and drug delivery. *J. Pharmacopuncture*, 22, 210–24, 2019.
46. Kurkó, J., Besenyei, T., Laki, J., Genetics of rheumatoid arthritis—A comprehensive review. *Clin. Rev. Allergy Immunol.*, 45, 2, 170–9, 2013.
47. Nemtsova, M.V., Zaletaev, D.V., Bure, I.V., Mikhaylenko, D.S., Kuznetsova, E.B., Alekseeva, E.A. *et al.*, Epigenetic changes in the pathogenesis of rheumatoid arthritis. *Front. Genet.*, 10, 1–13, 2019.
48. Journal, A.I., Chegini, S.P., Varshosaz, J., Taymouri, S., Recent approaches for targeted drug delivery in rheumatoid arthritis diagnosis and treatment. *Artif.*

Cells Nanomed. Biotechnol., Informa UK Limited, trading as Taylor & Francis Group, 1–13, 2018.

49. Andersen, C.K., Khatri, S., Hansen, J., Slott, S., Parvathaneni, R.P., Mendes, A.C. *et al.*, Carbon nanotubes-potent carriers for targeted drug delivery in rheumatoid arthritis. *Pharmaceutics*, 13, 4, 453, 2021.
50. Paul, A., Shao, W., Shum-Tim, D., Prakash, S., The attenuation of restenosis following arterial gene transfer using carbon nanotube coated stent incorporating TAT/DNAAng1+Vegf nanoparticles. *Biomaterials*, Elsevier Ltd, 33, 7655–64, 2012.
51. Lanner, J.T., Bruton, J.D., Assefaw-Redda, Y., Andronache, Z., Zhang, S.-J., Severa, D. *et al.*, Knockdown of TRPC3 with siRNA coupled to carbon nanotubes results in decreased insulin-mediated glucose uptake in adult skeletal muscle cells. *FASEB J.*, 23, 1728–38, 2009.
52. Al-Jamal, K.T., Gherardini, L., Bardi, G., Nunes, A., Guo, C., Bussy, C. *et al.*, Functional motor recovery from brain ischemic insult by carbon nanotube-mediated siRNA silencing. *Proc. Natl. Acad. Sci.*, 108, 27, 10952–10957, 2011.
53. McCarroll, J., Baigude, H., Yang, C.S., Rana, T.M., Nanotubes functionalized with lipids and natural amino acid dendrimers: A new strategy to create nanomaterials for delivering systemic RNAi. *Bioconjug. Chem.*, 21, 56–63, 2010.
54. Facciolà, A., Visalli, G., La, S., Ceccarelli, M., Aleo, F.D., Nunnari, G. *et al.*, Carbon nanotubes and central nervous system: Environmental risks, toxicological aspects and future perspectives. *Environ. Toxicol. Pharmacol.*, Elsevier, 65, 23–30, 2020.
55. Mohanta, D., Patnaik, S., Sood, S., Das, N., Carbon nanotubes : Evaluation of toxicity at biointerfaces. *J. Pharm. Anal.*, Elsevier Ltd, 9, 293–300, 2019.
56. Requardt, H., Braun, A., Steinberg, P., Hampel, S., Hansen, T., Surface defects reduce carbon nanotube toxicity *in vitro*. *Toxicol. In Vitro*, Elsevier, 60, 12–8, 2019.
57. Principi, E., Girardello, R., Bruno, A., Manni, I., Gini, E., Pagani, A. *et al.*, Systemic distribution of single-walled carbon nanotubes in a novel model: Alteration of biochemical parameters, metabolic functions, liver accumulation, and inflammation *in vivo*. *Int. J. Nanomedicine*, 11, 4299–316, 2016.
58. Yu, S., Su, X., Du, J., Wang, J., Gao, Y., Zhang, L. *et al.*, The cytotoxicity of water-soluble carbon nanotubes on human embryonic kidney and liver cancer cells. *New Carbon Mater.*, Institute Coal Chemistry, Chin. Acad. Sci., 33, 36–45, 2018.
59. Liu, Z., Liu, Y., Peng, D., Carboxylation of multiwalled carbon nanotube attenuated the cytotoxicity by limiting the oxidative stress initiated cell membrane integrity damage, cell cycle arrestment, and death receptor mediated apoptotic pathway. *J. Biomed. Mater. Res. A*, 103, 8, 2770–7, 2015.
60. Bussy, C., Al-jamal, K.T., Boczkowski, J., Prato, M., Bianco, A., Kostarelos, K., Microglia determine brain region-specific neurotoxic responses to

chemically functionalized carbon nanotubes microglia determine brain region-specific neurotoxic responses to chemically functionalized carbon nanotubes. *ACS Nano*, 9, 8, 7815–7830, 2015.

61. Yan, H., Xue, Z., Xie, J., Dong, Y., Ma, Z., Sun, X. *et al.*, Toxicity of carbon nanotubes as anti-tumor drug carriers. *Int. J. Nanomedicine*, 14, 10179–94, 2019.
62. Amenta, V. and Aschberger, K., Carbon nanotubes: Potential medical applications and safety concerns. *Wiley Interdiscip. Rev. Nanomed. Nanobiotechnol.*, 7, 371–86, 2015.
63. Thomas, J., Uvesato, D., Bergkamo, L., Herbat, N., One bird with two stones-international regulation of carbon nanotubes. *Client Alert*. 2008, www.huntonak.com (accessed on 16/02/22).
64. Park, H.G. and Yeo, M.K., Nanomaterial regulatory policy for human health and environment. *Mol. Cell Toxicol.*, 12, 223–36, 2016.
65. Jeevanandam, J., Barhoum, A., Chan, Y.S., Dufresne, A., Danquah, M.K., Review on nanoparticles and nanostructured materials: History, sources, toxicity and regulations. *Beilstein J. Nanotechnol.*, 9, 1050–74, 2018.
66. Commission recommendation of 18 October 2011 on the definition of nanomaterial (Text with EEA relevance) (2011/696/EU. www.eur-lex.europa.eu (accessed on 16/02/22).
67. Beg, S., Rizwan, M., Sheikh, A.M., Hasnain, M.S., Anwer, K., Kohli, K., Advancement in carbon nanotubes: Basics, biomedical applications and toxicity. *J. Pharm. Pharmacol.*, 63, 141–63, 2011.

8

Applications of Functionalized Carbon Nanotubes in Cancer Therapy and Diagnosis

Irshad Ahmad[1,2]*, Talat Parween[2], Lina Khandare[3], Aafaq Tantray[4] and Weqar Ahmad Siddiqi[1]

[1]*Department of Applied Science & Humanities, Faculty of Engineering & Technology, Jamia Millia Islamia University, New Delhi, India*
[2]*Department of Biosciences, Jamia Millia Islamia University, New Delhi, India*
[3]*Department of Physics, Savatribai Phule Pune University, Pune, (M.S. India)*
[4]*Department of Chemistry, Savatribai Phule Pune University, Pune, (M.S. India)*

Abstract

Carbon has a large aspect ratio, low mass density, and distinct physicochemical properties, carbon nanotubes (CNTs), one-dimensional allotropes of carbon, concern significant research attention in recent decades, providing exciting possibilities for nano-scale applications such as versatile functionalization chemistry and biological compatibility. Carbon nanotubes (CNTs) are a revolutionary carrier technology for both tiny and big medicinal compounds that were recently presented, CNTs maybe functionalized with different techniques. CNTs have been employed as nano-carriers for different types of applications like-biomedical, cancer diagnosis and treatment. Chemotherapy still utmost functioning treatment regimen for cancer therapy by eliminating free cancer cells and eradicating undetected tumors micro-focuses the carbon nanotubes.

CNTs' enormous superficial extent and capacity towards variations with their shallows and sensual dimensions were reconnaissance's to be employed as a photo-thermal killing of neoplastic cells, in addition to their potential to broadly function as nano-carriers for therapeutic compounds. CNTs can efficiently carry various compounds to cancer cells, including peptides, antigens, and nucleic acids. Researchers have enormous work on development in cancer theranostics using

**Corresponding author*: irshadahmad85@gmail.com; irshadalig2005@yahoo.com

Jeenat Aslam, Chaudhery Mustansar Hussain and Ruby Aslam (eds.) Functionalized Carbon Nanotubes for Biomedical Applications, (171–196) © 2023 Scrivener Publishing LLC

CNT-based drug delivery systems, with a focus mainly on intracellular components of lump cells and extracellular components of the tumor microenvironment.

Keywords: CNTs, cancer cells, drug delivery systems, target chemotherapy, cancer theranostics, tumor microenvironment

8.1 Introduction

Cancer is considered as one of the most common cause of mortality and disability on earth. It is characterized as unregulated cell proliferation that eventually spreads to other sections of the body and destroys normal tissues. Various kinds of treatment strategies and therapy like phototherapy, radiotherapy, immunotherapy, gene therapy, endocrine therapy etc. play a ray of hope for cancer patient and extended the patients' survival time. These treatments are highly specific because of their inherent nature [1]. One of the improving clinical outcomes for the treatment of cancer includes Immunotherapy, though patients experienced allergic reaction and other side effects due to complex interaction within immune system [2]. In recent decades, nanotechnology has gained as a novel treatment for malignant disease like cancer. The development of a nano-based drug delivery system (DDS) has opened up new possibilities for combining several anticancer drugs. This approach offers a good remedial consequence with few side effects and better targeting ability [3–5]. In this area, several inorganic nanomaterials have been broadly explored clinically and pre-clinically in anticancer drug delivery. Numerous nanocarriers have been investigated for treatment, including micelles, reconstituted high-density lipoprotein, liposomes, metallic and hybrid NPs, and carbon linked nanomaterials [6–8]. The multifunctional application possibilities of carbon nanotubes (CNTs) in cancer treatment and detection have piqued the interest of numerous researchers.

Researchers recently proposed that carbon nanotubes (CNTs) can be used as a versatile new carrier for medication delivery and diagnostics. Their adaptable physicochemical properties allow for the covalent and non-covalent interaction of a variety of pharmaceutically relevant substances, as well as the logical creation of innovative applicant microscale constructions for drug discovery. CNTs may be functionalized with several functional groups to carry many moieties for targeting, imaging, and treatment at the same time. The primary purpose of developing nano-carrier drug delivery systems is to improve the therapeutic effectiveness or reduce the toxicity of therapeutically active drugs. Carbon nanotubes are made up of graphite sheets that have been wound into cylindrical shapes. CNTs have a diameter of about 100 nanometers and are measured in micrometers. Carbon nanotubes (CNTs) are a cross between carbon fibers and fullerene, with 60 carbon atoms grouped in a muted tube structure. CNTs are more versatile than other nanomaterials in biomedical activities, and they are one of the most exciting nanocarriers in scientific research. As new delivery systems, carbon nanotubes have demonstrated their ability to cross biological barriers. Carbon nanotubes (CNTs) can be utilized for medicine delivery and thermal ablation, unlike quantum dots, which are usually used for cancer cell imaging [9]. They can infiltrate cells, regardless of the cell type or functional group present on their surface [10–12]. CNTs have a large surface area, which allows for many locations for various molecules to bind, allowing for polyvalent derivatization. According to *in vitro* and *in vivo* evidence collected from several research groups, a variety of chemically functionalized CNTs has the potential to be biocompatible with the biological environment. These studies revealed how the degree and kind of functionalization *in vivo* might affect the material's behavior, and how these two elements must be carefully balanced [13]. CNTs have certain distinguishing characteristics, such as a large interfacial area with the cellular membrane and the ability to include numerous functionalizations. CNTs, on the other hand, are compatible with and transportable in biological fluids. Functionalized carbon nanotubes with water solubility and biocompatibility may pass cell membranes and transport a variety of physiologically active substances into cells, including medicines, proteins, DNA, and RNA [14–16]. Carbon nanotube cytotoxicity is primarily determined by their surface functionalization, with fully functionalized, serum-stable nanotubes having little toxic effects [17–19]. These characteristics make them an effective tool for a wide range of diagnostic, therapeutic, and drug delivery applications [20]. We detailed the process and attributes of CNT functionalization, along with examples of its use in cancer detection and treatment. These nanotubes are made of graphene sheets rolled up into cylinders with high aspect ratios, diameters as small as 1nm and lengths of few micrometers, which can be open-ended or capped. In contrast to single-walled carbon nanotubes (SWNTs), multi-walled carbon nanotubes (MWNTs) are made from multiple graphene sheets [21, 22] (as shown in Figure 8.1). Carbon nanotubes with single and multilayer walls have previously been shown to be safer and more effective than earlier medication delivery systems. Since their discovery in 1991 by Iijima [21]. Due to their unusual physical and chemical properties and possible uses in a wide range of disciplines, from electrical devices and sensors to high-strength, low-weight nanocomposite materials, these allotropes of carbon

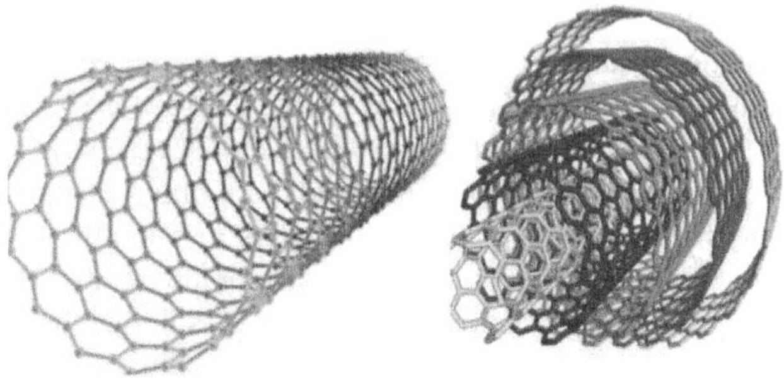

Figure 8.1 (a) SWNT: 0.2-2nm. (b) MWNTs: 2-100nm [10, 22]. Reproduced with permission from Ref. [22]. Copyright 20121 @ Elsevier.

have attracted a lot of attention. CNTs may be used in a multifunctional approach for cancer therapy and diagnostics because of their superior optical property, simplistic functionalization facility, thermal and electrical conductivity, and high drug loading ability. They can travel across membranes, delivering medicinal medicines, vaccinations, and nucleic acids to previously unreachable parts of the cell. They're also good non-toxic carriers that, in certain situations, boost the solubility of the linked medicine, consequential in improved effectiveness and protection. General, current CNT research has provided a very hopeful view into the future of medicine. Methotrexate, an anticancer medication, is released by enzymatic cleavage *in vitro* breast cells with the help of multi-walled carbon nanotubes. They can traverse membranes, allowing them to carry medications, injections, and nucleic acids to regions of the cell that were formerly inaccessible. They are good non-toxic transporters that raise the solubility of the associated medicament in specific settings, resulting in better efficacy and safety. Overall, present CNT research offers a very optimistic outlook on the future of medicine. Methotrexate, an anticancer drug, is released *in vitro* breast cells via enzymatic cleavage with the aid of multi-walled carbon nanotubes. Dendrimer-modified multiwalled carbon nanotubes have also been utilized to deliver the medication doxorubicin. Because the enzymes on the interior of the cell surface quickly degrade peptides, proteins, and genes, multi-walled carbon nanotubes are perfect transporters of these macromolecules. Because of their high surface area, they can adsorb or conjugate with a wide spectrum of therapeutic substances. Surface engineering (i.e., functionalization) of CNTs can improve their aqueous phase dispersibility or provide the appropriate functional groups to bind to the desired therapeutic material or target tissue to induce a therapeutic effect. CNTs may help the connected therapeutic chemical penetrate the target cell and heal ailments [23–27]. CNTs are increasingly becoming more relevant in the developing area of nanomedicine; nevertheless, we must ensure that the exciting possibilities they provide are converted into viable and safe constructions that can be used in drug discovery and development pipelines. An overview of the therapeutic applications of carbon nanotubes is presented here, with special emphasis on cancer treatment. A recent study discusses the use of carbon nanotubes in cancer therapy and diagnostics.

8.2 Characteristic Properties of CNTs and Their Performance

The orderly production of carbon atoms in graphene cylinders results in a variety of features [28]. Laser ablation, electric arc discharge, chemical vapor deposition are now the most common methods for manufacturing [29]. Carbon nanotubes are enormous cylindrical molecules made up of a hexagonal arrangement of sp_2 hybridization (with bond distance 1.4). Carbon atoms are produced in graphene cylinders in an ordered manner, resulting in a range of properties [28–31]. Three key properties of CNTs have been used in the medical field:

- Compact size;
- Large surface area to volume ratio;
- Chemical-containment capacity.

Figure 8.1 shows the structures of these two CNT. MWCNTs consists of concentric sheets of rolled graphite having internal diameters ranging from 1-3 nm to external diameters ranging from 2-100 nm [32]. SWCNTs have a single cylindrical carbon layer that is 0.2-2 nm thick [33]. MWCNTs have an average interlayer separation of 0.34 nm, with each graphene layer creating an independent tube, having a wider outer diameter (2.5 to 100 nm) than SWCNTs (0.6–2.4 nm). The length-to-diameter ratio of SWCNTs is larger than that of conventional carbon nanotubes. When compared to SWCNTs, MWCNTs have a larger inner diameter, which allows for additional drug loading areas. It has also been claimed that MWCNTs' outer shells can be functionalized without causing harm to the side wall [34]. CNTs may be improved in solubility using efficient approaches that can be used in a variety of biological applications [35]. Strong acids can oxidize carbon nanotubes, causing them to lose length and produce carboxylic groups. In addition, these modifications can help them dissolve more

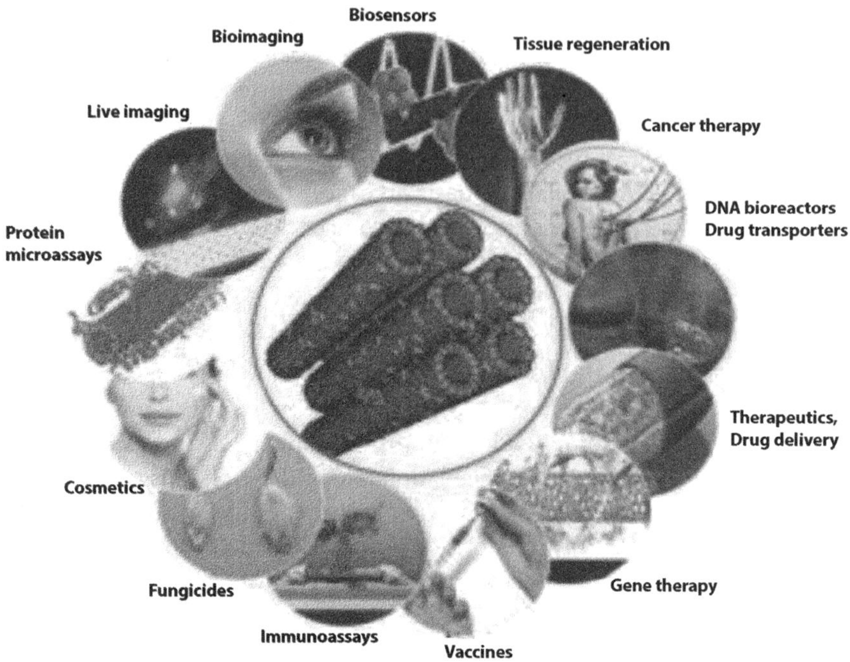

Figure 8.2 Bioactive compounds conjugated through carbon nanotubes [50]. Reproduced with permission from Ref. [50]. Copyright 2021 @ Elsevier.

easily in water [36]. Alternatively, the additional reactions to the CNTs might improve the solubility of the exterior walls and tips [37]. Solubility is one of the most important properties that provide biocompatibility under physiological settings [35]. Furthermore, functionalized carbon nanotubes (f-CNTs) may be connected to nearly any active molecule, including therapeutic agents, proteins and peptides. CNTs feature a variety of physicochemical properties, including as a surface area, high aspect ratio, mechanical strength, electrical and thermal conductivity, and ultra-lightweight properties [38]. CNTs are a promising candidate material with significant biological potential because of these benefits [40]. Because of their toxicity, CNTs must be functionalized to make them soluble and biocompatible, allowing them to be incorporated into biological systems. CNTs may bind a wide range of compounds (Figure 8.2) [39].

8.2.1 Physicochemical Properties of CNTs

Depending on the synthetic technique used, the length and diameter of CNTs might vary greatly. In contrast, SWCNTs have a more defined wall, whereas MWCNTs are more likely to have structural defects, resulting in a

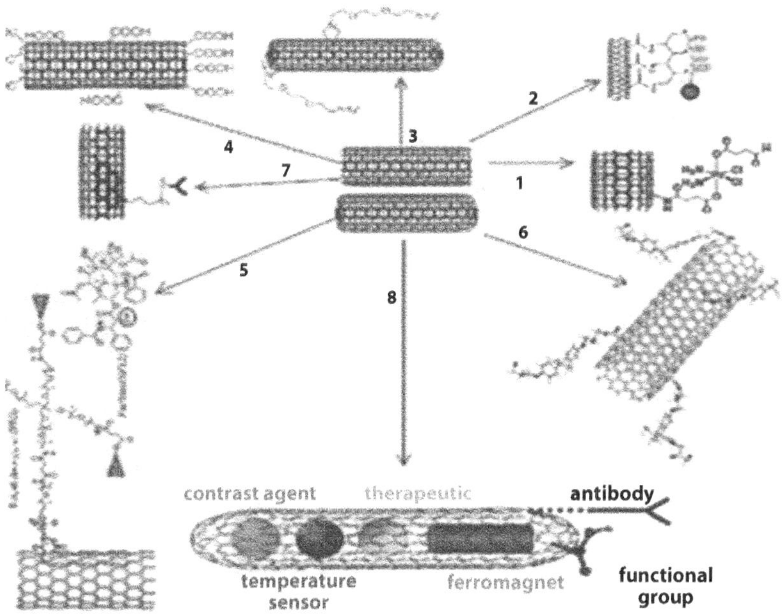

Figure 8.3 Schematic illustration for the preparation of CNT-based drug delivery systems [39]. Reproduced with permission from Ref. [39]. Copyright 2021 @ Elsevier.

less stable nanostructure. Despite this, MWCNTs continue to receive wide attention due to their simple processing; they continue to be highlighted in numerous publications. SWCNTs have no clear benefits over MWCNTs in terms of drug delivery; their specified smaller diameter may be helpful for quality control. Due to the van der Waals forces, SWCNTs and MWCNTs have a robust tendency to pack together in ropes. Nanotube bundles are made of a variety of nanotubes, and their size and width can differ significantly from their originators. CNTs come in a variety of shapes and sizes, depending on the hexagonal orientation of the graphene sheet, having huge surface area and high aspect ratio. The usable surface area is affected by the length, diameter, and degree of bundling. CNTs can be chemically or physically reduced to make them appropriate for drug carriers by opening their two ends with favorable wall defects for intratube drug loading and chemical functionalization (Figure 8.3).

8.3 The Techniques of CNTs Functionalization

Carbon nanotubes' intrinsic restrictions, such as poor solubility, aggregation, and cytotoxicity, severely limit their range of applications because they are initially insoluble and poorly dispersible. Enhancing the natural features of biological systems by surface modification is crucial for achieving increased solubility, enhanced biocompatibility, increased penetration, and reduced cytotoxicity [41]. For gastrointestinal absorption, blood transportation, secretion, and biocompatibility, CNTs must be soluble in an aqueous solvent, among other things. As a result, CNT composites used in therapeutic delivery systems must fulfill this essential criterion. CNTs offer several benefits, but their biological applications have been limited. CNT purification is still in its early stages. Mean-while. CNTs are difficult to dissolve in aqueous conditions, which restricts their biological applications [42]. However, this drawback can be solved by functionalization. Similarly, to acquire correct concentration data, such CNT dispersions must be homogeneous and stable to a sufficient degree. For successful CNT dispersion, the medium should be capable of wetting the hydrophobic tube surfaces as well as changing the tube surfaces to reduce tube bundle formation. Foldvari *et al.* identified four key strategies for achieving optimal dispersion:

(1) Surfactant dispersion,
(2) Solvent dispersion,
(3) Functionalization of sidewalls and
(4) Bimolecular dispersion.

Functionalization has shown to be the most successful of the procedures outlined above. Furthermore, functionalization has been found to reduce cytotoxicity, improve biocompatibility, and allow for the development of delivery systems using appendage molecules of medicines, proteins, or genes [20].

Furthermore, CNTs are frequently undissolved and thus have poor biocompatibility. There are two functionalization strategies for CNTs to boost their bio solubility i.e. Covalent, and non-covalent. Different chemical groups can be used to covalently or non-covalently functionalize CNTs [43]. The surface of CNTs is slightly damaged by non-covalent functionalization via Vander Waals interactions, – interactions, and hydrophobic interactions [44]. Non-covalent functionalization of CNTs by Vander Waals interactions, – interactions, and hydrophobic interactions causes minimal harm to the surface of CNTs, whereas covalent modification of CNTs such as oxidation and carboxylation provides a stable platform for drug administration. A variety of bioactive molecules, such as antibodies and various polymers and surfactants, have been utilized to non-covalently functionalize CNTs to obtain improved biocompatibility and lower toxicity

Figure 8.2 [45]. Furthermore, because of their hydrophobic interaction, raw CNTs have a low biodegradability; however, their biodegradability can be improved through functionalization, which increases their solubility and creates defect sites that provide preferred enzyme binding sites and promote enzymatic degradation [46]. For example, defective sites on the carbon walls and ends of carboxylated- or nitrogen-doped MWNTs are more likely to bind to horseradish peroxidase (HRP), facilitating enzymatic degradation [47]. As a result, the next sections are mostly focused with carbon nanotube modification utilizing various functionalized chemicals.

Polyethylene glycol (PEG), a commonly used polymer, was licensed for pharmaceutical use by the FDA because of its flexibility and safety and has been utilized as an adjuvant material in drug delivery for many years [48]. As a result, PEG functionalization is an effective technique to improve CNT physicochemical properties, as well as their solubility and biocompatibility. Taghavi *et al.* have reported PEGylated SWNTs for gastric cancer treatment, and the synergistic therapeutic impact resulted in increased cancer cell death [49]. To facilitate active targeting, the PEGylated SWNTs were covalently linked to the nucleolin ligand AS1411 aptamer. PEG modification improved SWNT solubility, reduced particle aggregation, and reduced the deleterious effects of DDS. According to *in vitro* experiments, this combination of shRNA-based gene silencing with the chemotherapeutic agent DOX has a high tumoricidal effectiveness. In addition to PEG, another often-used polymer for gene transfer is polyethylenimine (PEI). PEI, as a polycation, may bond to CNTs, enhancing their ability to mix with nucleic acid through a stacking and electrical response [50]. Huang *et al.* used amination to modify both single walled and multiwalled nano tubes with PEI to produce PEI-NH-based CNT [51]. Furthermore, because of the positive charge on the surface of CNTs, they were able to engage with siRNAs via an electrical response, successfully transporting siRNAs into cells and subsequently increased transfection and gene silencing efficiency. Hyaluronic acid (HA) is a biodegradable and biocompatible linear macromolecular mucopolysaccharide made up of alternatingly connected two saccharide units of N-acetylglucosamine and glucuronic acid [52].

Chitosan and polylactic-co-glycolic acid (PLGA) are two more polymers that can be utilized to modify CNTs [53, 54]. Chitosan is a bioactive polymer derived from the natural polysaccharide chitin, whereas PLGA is a synthetic polymer with a high biodegradability and low toxicity that has been approved for therapeutic use by the FDA [55]. Both changes have the potential to improve the intrinsic properties of CNTs, resulting in increased anticancer treatment efficacy. Because of their functional qualities, carbon nanotubes are a useful tool for biological applications (as shown in Figure 8.4).

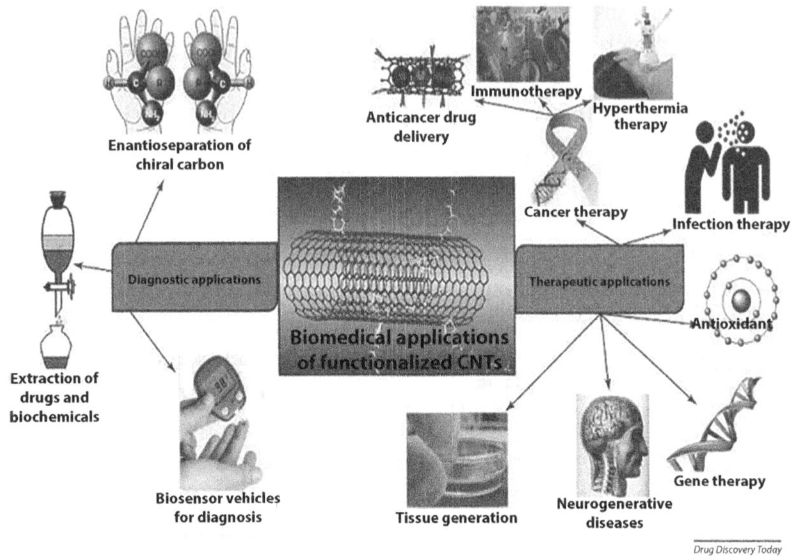

Figure 8.4 Biomedical applications of functionalized CNTs [50]. Reproduced with permission from Ref. [50]. Copyright 2021 @ Elsevier.

8.4 Application of Carbon Nanotubes in Cancer Therapy and Diagnostic

8.4.1 The Use of Carbon Nanotubes in Cancer Treatment

CNTs are being studied intensively in targeted medication delivery for cancer treatment because of their unique features. They also serve as effective phototherapy mediators due to their intrinsic optical properties. Several anticancer treatment methods are now aimed against tumor cells and their environment.

8.4.2 Intracellular Targeting Using Carbon Nanotubes

Intracellular drug delivery, in which anticancer drugs are given to specific intracellular targeting areas, has recently received a lot of interest in cancer therapy. Its job is to increase their therapeutic efficacy while lowering their associated toxicity [56]. The nucleus, cytoplasm, and mitochondria are the most significant intracellular targets. The nucleus is a regulatory hub for cellular genetics and plays an important role in metabolism, growth, and cell differentiation [57]. The cytoplasm is the prime location for the activity of cell, while Mitochondria are fundamental intracellular organelles that important functions involved in cellular enzymatic pathways. As a result, these cellular particles might be exploited as remedial targets in the treatment of certain cancers [58].

8.4.2.1 *Nucleus Targeting*

The cellular nucleus acts as the regulating center for controlling cell proliferation and metabolism, as well as managing the cell cycle, making it a desirable therapeutic target for the treatment of malignant tumors. The use of this technology in cancer treatment is predicted to be more successful. Lee *et al.* developed a unique method for nanotube that combined SWNTs with the EGFR inhibitor cetuximab and the topoisomerase I inhibitor SN38 to treat EGFR over-expressing colorectal cancer precisely [59]. For breast cancer therapy, Karmakar *et al.* employed functionalized SWNTs as the vector and the oncogene suppressor protein 53 (p53) genes that participates in fixation of DNA or triggers cellular death when DNA damage is irreversible [60, 61]. All of the aforementioned investigations together showed that f-SWNTs are potential gene vectors capable of efficiently transporting the particular gene to the nucleus.

8.4.2.2 *Cytoplasm Targeting*

Due to the diversity of its elements, the cytoplasm is a jelly-like structure made of numerous cellular structure with a highly ordered and diverse design [62]. Cytoplasm plays a crucial role in a variety of cellular processes. Chen *et al.* built a nano DDS based on carbon nanotubes to successfully administer the 5-FU for the treatment of gastric cancer [63]. SWNTs were employed by Bartholomeusz *et al.* to transfer hypoxia-inducible factor 1 alpha (HIF-1) siRNA into pancreatic cancer cells, inhibiting HIF-1 expression [64]. Tumor hypoxia is being considered a possible therapeutic target, and many studies have already shown that inhibiting HIF-1 can slow down the growth of tumor by decreasing its potential [65]. This nano-complex was able to effectively achieve the purpose of RNAi and reduce tumor development, highlighting the importance of CNTs. To summarize, nano DDS based on carbon nanotubes is a potential technique for delivering siRNA into the cytoplasm for effective gene therapy.

8.4.2.3 *Mitochondria Targeting*

Mitochondria are recognized to have a role in energy metabolism, including ATP production [66]. Furthermore, they are important cellular organelles involved in the formation of reactive oxygen species (ROS) as well as the control of apoptosis, autophagy, and necroptosis [67]. Mitochondrial bioenergetics, biosynthesis, and signaling have all been shown to play a role in carcinogenesis, and mitochondria are an essential source of energy for fast proliferating cancer cells. As a result, mitochondria are being viewed as key therapeutic targets in cancer treatment [68]. CNTs, fortunately, may be internalized by a variety of cancer cells and can be targeted by mitochondrial targeting ligands such as mitochondrio tropics [69]. Chemotherapy is still the most extensively used treatment for a variety of malignancies, however, because of its non-specificity, it frequently produces unfavorable side effects, and its use is still restricted [70]. PtBz entrapped inside the cores of hydrophobic MWNTs may be converted into platinum (II) by glutathione, a hydrophilic species that could be released from the MWNTs [71]. Kim *et al.* developed a PEGylated CNT-based nano DDS (PEG-CNTs-ABT737) with improved mitochondrial targetability against lung cancer cells [72, 73]. Mitochondria are heat sensitive and can be used as PTT targets [74]. *In vitro* tests revealed that following modest hyperthermia induced by SWNTs-mediated PTT, this nano-system could raze the mitochondria of MB49 cancer cells, resulting in a rupture of ROS. Although PTT is one of the newer therapeutic techniques for cancer therapy that is non-invasive or very marginally invasive, it can unavoidably cause some skin damage.

8.4.3 CNTs for Immunotherapy

TME, as previously stated, is a complex ecosystem made up of malignant parenchymal cells and non-cancerous stromal components. Immune cells, which make up the majority of TME cells, play an important role in TME-based targeted treatment [75]. These immune cells interact with other suppressive qualities. Thus, cancer immunotherapy relies heavily on targeting particular immune cells or altering the immunosuppressive milieu of TME.

Because of CNTs' ability to be modified, a variety of immune-modulators can be directly associated outside into their inner structure for effective delivery. Furthermore, s.c. injection in HCC tumor-bearing mice was reported to increase systematic immune activity, indicating the promising application prospect of CNTs in cancer immunotherapy [76]. As a result, CNT-mediated DDS has a lot of research promise in terms of precision immunotherapy. Luo *et al.* recently developed a nanosystem for triple-negative breast cancer (TNBC) treatment made up of MWNTs-loaded ginsenoside Rg3 (Rg3-CNT), which first established the novel role of Rg3-CNT through immunotherapy [77]. Intratumoral injection resulted in a significant reduction in cancer development *in vivo*, as well as an increase in CD4+ and CD8+ T cell population.

8.4.4 Cancer Stem Cell Inhibition

CSCs are a kind of cancer cell that is extremely tumorigenic and drug-resistant by nature.

They have the ability to self-renew and differentiate into several kinds of tumor cells [78]. Increasing evidence suggests that CSCs play a critical role in modifying TME, which is involved in several phases of carcinogenesis, including tumor development, metastasis, and recurrence [79, 80]. Furthermore, mounting evidence suggests that CSCs can withstand traditional remediation radiotherapy through an array of methods, including increased drug transporter expression, enhanced DNA repairability [81]. Furthermore, because of their tumorigenic and survival qualities, CSCs can likely survive traditional tumor healing procedures and start new tumors in next generation, resulting in a rapid cancer relapse [82]. As a result, CSCs are thought to be potential cancer therapeutic targets inside the TME. Because CNTs combine molecular targeting and PTT into a single platform, CNT-based therapeutic techniques are potential for destroying CSCs and inducing apoptosis by hyperthermia. Osteosarcoma stem cells (OSCs) are a kind of CSC that are intimately linked to osteosarcoma metastasis and recurrence [83]. TGF1 (transforming growth factor type I) is a cytokine that has a role in osteosarcoma development and carcinogenesis. TGF1 signaling was found to be important in sustaining the OSC population, and a high amount of TGF1 in the osteosarcoma microenvironment caused osteosarcoma cells to differentiate into OSCs, according to Zhang *et al.* [84]. As a result, inhibiting TGF1 signaling might be a way to change the condition of OSCs and lower their number. Intriguingly, Miao *et al.* revealed that both raw SWNTs and carboxyl-modified SWNTs block TGF1-induced differentiation of osteosarcoma cells, prevent phenotypic changes in OSCs, and reduce OSC viability in osteosarcoma microenvironment-mimetic settings [85]. All of these findings pointed to CNTs having an unanticipated biological effect on OSCs, providing plausible references for CSC-targeted therapy by the deployment of CNT-mediated treatment.

8.5 Carbon Nanotubes in Cancer Diagnosis

Cancer diagnosis is one of an important procedure for the examination of various kind of cancer through latest technologies. Due to unexceptional properties as a kind of inorganic material, CNTs has got awareness for the diagnosis of this malignant tumor.

8.5.1 CNTs in Cancer Imaging

Due to various unique properties, CNTs are regarded as one of the multifunctional bio-probes. In the recent year, this technique got attention in other types of tumor analysis.

8.5.1.1 Raman Imaging

It is the technique in which there is an excitement of light with wavelength of photons. CNTs show various prominent Raman peaks, detected by Raman microscope [86]. For study of fluorescence imagery of colon-26 cancer cells, Sekiyama, *et al.*, prepared SWNTs as a currently researched over-thousand-nanometer (OTN)-NIR [87]. To study the o-SWNTs-PEG's distribution in colon-26 cells, this technique was used which proved that the signals were elevated that proves the suitability of applying Raman imaging. The most recognizable method for the improvement of Raman signals is covering metal all around the CNTs [88]. To prove this function, Wang, *et al.*, tailored CNTs with Au or Ag to attain the better effect of nanocomposites [89].

8.5.1.2 Nuclear Magnetic Resonance Imaging

The benefit of MRI is that it prevents the human body from being harmed by ionizing radiation. Medical imaging can be enhanced with its ability to capture the original 3D cross-section images [90]. Combining NGR (asparagine-glycinearginine) peptide with DOX and MR on CNTs. This nanosystem exhibited both outstanding anticancer efficacy and MRI function, providing exceptional synergy in tumor diagnosis and treatment [91, 92]. MWNTs were also used in MRI research by Zhang, *et al.* [93]. A type of functionalized MWNTs was synthesizes by modifying gadolinium nanoparticles (GdN), magneto-fluorescent CQDs and folic acid (FA). With these changes, the MWNTs were stable and water tight, which are both crucial for MRI.

8.5.1.3 Ultrasonography

Because of low detection value and built-in safety, Ultrasonography is an fundamental diagnostic imaging approach [94]. As in contrast to others, CNTs are regarded as the most applicable nanomaterials for ultrasonic imaging due to the excessive sign produced by means of them at some point of ultrasonography. Saghatchi *et al.* developed type of functionalized MWNTs, acquiring mf-MWNT/AuNPs for most cancers remedy [95]. These NPs confirmed excessive distinction beneath all applied concentrations, exhibiting the opportunity to turn out to be a kind of characteristic means. Delogu, *et al.*, learn about on MWNTs utilized in ultrasonography [96]. They tailor-made MWNTs with azomethine ylides to polish their properties. According to more than a few test results, these geared up functionalized MWNTs (ox-MWNT-NH3+) confirmed the ultrasound signal.

8.5.1.4 Photoacoustic Imaging

With new methodology, Photoacoustic imaging is viewed as a technology offers *in vivo* pathology [97]. As detected with the aid of the sensors, the captivated radiant strength can be transformed into ultrasonic [98]. The use of PA science may additionally enhance the tissue diffusion successfully as in contrast to different strategies. Because CNTs have a powerful NIR penetration, it is considered a suitable PA component agent [99, 100]. In addition, Wang and colleagues reported silica-coated, Au nanorods-functional MWNTs with the ability to detect *in vivo* PA imaging of stomach cancer [101]. This work first synthesized Au nanorods (sGNR) synthetic silica into MWNTs. After i.v. Management of these findings, the treated mice were all exposed with this system, proving a strong effect of this system [102–106].

8.5.1.5 Near-Infrared Fluorescence Imaging

NIRF imaging is an electrifying imaging tool, applied to deep tissue and cancerous body. CNTs shows potential feature used in NIRF thinking during treatment and these images are often obtained with *in vivo* internal imagery [107, 108]. In comparison to MWNTs, SWNTs are highly suited to NIR imaging due to their high absorption structures with strong visual absorption, better E11 optical modification, and low image output [109]. Ghosh, *et al.* designed a stable MN SWNTs investigation that could accurately identify hidden, acidic and rich cysteines (SPARCs) [110]. NIR-II used in this study as a source of fluorescence to avoid detection and detection of deep tissue infiltration during NIRF imaging. As a result of the diagnostic, these NIR2-releasing SWNTs show high specificity to tumors in the *in situ* and ovarian tumors Another study of the adoption of SWNTs in the NIRF model was conducted by Welsher, *et al.* [111]. Nanobiosensors are a vital biosensors comprising both biological detection elements and nanomaterials with a magnitude of 1 to 100 nm [112–116].

8.6 Future Prospects

A wide range of CNT applications has been demonstrated in anticancer drug, including intracellular targeting spots, TME components, single treatment methods, and combinations of techniques, indicating CNTs' potential future use. Several cancer imaging methods can utilize carbon nanotubes, including PA imaging, MRI, radionuclide and near-infrared fluorescence imaging. Together with other diagnostic agents, CNTs can work as excellent nanosensors to detect a wide range of cancer types early, including prostate cancer, cervical cancer and pancreatic cancer. Prior to their use in clinics, CNT-based nanomedicine must be proven safe in the human body over the long term.

It is therefore vital to examine and improve the precision with which cancer cells can be targeted or TME components targeted using carbon-based therapies, thereby minimizing administration doses and reducing toxicity. Graphene and its derivatives have amazing electrical characteristics in terms of diagnosis, making them more attractive in cancer biosensing. Despite the benefits indicated above, there are still several feral roadblocks to CNT-based treatments being used in clinical settings. For example, the safety of carbon nanotubes in the human body has not been well investigated. *In vitro* experiments using a variety of cell lines have shown that CNTs are safe, but *in vivo* studies using animal models are generally short. Preclinical studies based on CNT-based nanocarriers will become more sophisticated as science and technology advance, providing more relevant information for studying their clinical potential.

8.7 Conclusion

This chapter summarizes the most recent breakthroughs in cancer theranostics using carbon nanotubes. Nanotechnology has advanced rapidly in the biomedical area in recent decades. CNTs are used as a drug delivery system and also play a vital role in radiation-based phototherapy cancer treatment due to their distinct physiochemical properties, such as large superficial zones, with modified surface possibilities, better drug delivery capacity, also excellent photosensitive properties. Furthermore, CNTs exhibit unrivaled electrical and thermal conductivity properties. These features allow them to be used in therapeutic settings. In fact, carbon nanotubes shown enormous gravity for cancer treatment, we can't ignore their non-biodegradability and cytotoxicity in clinical application. We anticipate that shortly, this revolutionary CNT-based system will give a new opportunity to us for better and compatible cancer therapy.

CNTs can be functionalized to allow molecules to adhere to their surfaces through covalent and non-covalent interactions. These nano-carriers have several benefits, including a high drug delivery capacity, strong absorptivity, photothermally accessible, and built-in analytical capabilities. The above characteristics of functionalized CNTs have shown crucial in making them suitable nano-carriers for tumor therapy. Furthermore, many researchers shown functional applications of carbon nanotubes in antineoplastic drug delivery system, ranging from intracellular elements to tumor microenvironment system, easy therapeutic modality to multi-functional treatment approaches, all pointing to CNTs targeted aspects in cancer treatment. CNTs show promise as needle-like carriers for both tiny medicinal molecules and macromolecules like genes and proteins. CNTs can puncture cellular sheaths and transfer therapeutic chemicals to cellular components which are present inside the cancer cells because of their needle-like structure. Endocytosis is considered to be involved in this process. CNTs offer unique features that make them ideal for medicinal applications, including the capacity to absorb harmful germs and transfer heat.

Carbon nanotubes have only been used in drug delivery studies for a few years, a lot more study is planned to be done in the next years to fully understand their potential. Functionalized CNTs to be employed in a different cancer imaging technique, like that Postero-Anterior radiograph, Magnetic Resonance imaging, Raman imaging, radionuclide imaging, and near-infrared fluorescence imaging. CNTs may also be used as nano-bio sensors for the primary diagnosis of several cancer forms such as pancreatic, prostate, and cervical neoplastic cells have excellent specificity when combined with other diagnostic agents. As a result, functionalized carbon nanotubes (CNTs) have versatile nano-materials which shown enormous potential in biological applications and cancer therapy including diagnosis.

CNTs are a fascinating nano-tool with a lot of therapeutic and scientific potential in cancer therapy and diagnostics. More mature preclinical trials for CNT-based nano-carriers will be designed as science and technology advance, providing us with more relevant data to examine their clinical potential.

Important Websites

https://www.cheaptubes.com/product-category/industrial-carbon-nanotubes-products/

https://www.alfa-chemistry.com/product/double-walled-crabon-nanotubes-dwnts-2-4nm-cas-1333-86-4-282359.html?gclid=CjwKCAjwi-6WSBhA-EiwA6Niok_apP5IykmfDHfIF2mSM1Kaf3KZ-p-PskbqKy-4QsV1qdVb6dvRQj3RoCCisQAvD_BwE

- https://www.ctimaterials.com/product-category/industrial-carbon-nanotubes-products/
- https://www.sigmaaldrich.com/IN/en/product/aldrich/652482?gclid=CjwKCAjwi6WSBhA-EiwA6N-iok-O1ig3bUnfW5ksMdJP61axrRmgIt2nBSUq6bD00d1jS-G6HcEAVxORoC3voQAvD_BwE
- https://nanointegris.com/
- http://www.timesnano.com/

References

1. Su, Y., Liu, Y., Xu, X., Zhou, J., Xu, L., Xu, X. *et al.*, On-demand versatile prodrug nanomicelle for tumor-specific bioimaging and photothermal-chemo synergistic cancer therapy. *ACS Appl. Mater. Interfaces*, 10, 45, 38700–14, 2018.
2. Yan, S., Luo, Z., Li, Z., Wang, Y., Tao, J., Gong, C. *et al.*, Improving cancer immunotherapy outcomes using biomaterials. *Angew. Chem. Int. Ed. Engl.*, 59, 40, 17332–43, 2020.
3. Bockamp, E., Rosigkeit, S., Siegl, D., Schuppan, D., Nano-enhanced cancer immunotherapy: Immunology encounters nanotechnology. *Cells*, 9, 9, 2102, 2020.
4. Tang, L., Li, J., Zhao, Q., Pan, T., Zhong, H., Wang, W., Advanced and innovative nano-systems for anticancer targeted drug delivery. *Pharmaceutics*, 13, 8, 1151, 2021.
5. Liu, H., Mei, Y., Zhao, Q., Zhang, A., Tang, L., Gao, H. *et al.*, Black phosphorus, an emerging versatile nanoplatform for cancer immunotherapy. *Pharmaceutics*, 13, 9, 1344, 2021.
6. Mei, Y., Tang, L., Xiao, Q., Zhang, Z., Zhang, Z., Zang, J. *et al.*, Reconstituted high density lipoprotein (rHDL), a versatile drug delivery nanoplatform for tumor targeted therapy. *J. Mater. Chem. B*, 9, 3, 612–33, 2021.
7. Su, Y., Wang, T., Su, Y., Li, M., Zhou, J., Zhang, W. *et al.*, A neutrophil membrane-functionalized black phosphorus riding inflammatory signal for positive feedback and multimode cancer therapy. *Mater Horiz.*, 7, 2, 574–85, 2020.
8. Mitchell, M.J., Billingsley, M.M., Haley, R.M., Wechsler, M.E., Peppas, N.A., Langer, R., Engineering precision nanoparticles for drug delivery. *Nat. Rev. Drug Discovery*, 20, 2, 101–24, 2021.

9. Utreja, P., Jain, S., Tiwary, A.K., Novel drug delivery systems for sustained and targeted delivery of anti-cancer drugs: Current status and future prospects. *Curr. Drug Deliv.*, 7, 2, 152–161, 2010.
10. Kostarelos, K., Lacerda, L., Pastorin, G. *et al.*, Cellular uptake of functionalized carbon nanotubes is independent of functional group and cell type. *Nat. Nanotechnol.*, 2, 2, 108–113, 2007.
11. Kam, N.W., Liu, Z., Dai, H., Carbon nanotubes as intracellular transporters for proteins and DNA: An investigation of the uptake mechanism and pathway. *Angew. Chem.*, 118, 4, 591–595, 2006.
12. Fabbro, C., Ali-Boucetta, H., Ros, T.D., Kostarelos, K., Bianco, A., Prato, M., Targeting carbon nanotubes against cancer. *Chem. Commun.*, 48, 33, 3911–3926, 2012.
13. Singh, R., Pantarotto, D., Lacerda, L. *et al.*, Tissue bio distribution and blood clearance rates of intravenously administered carbon nanotube radiotracers. *Proc. Natl. Acad. Sci. U. S. A.*, 103, 9, 3357–3362, 2006.
14. Lacerda, L., Ali-Boucetta, H., Herrero, M.A. *et al.*, Tissue histology and physiology following intravenous administration of different types of functionalized multiwalled carbon nanotubes. *Nanomedicine*, 3, 2, 149–161, 2008.
15. Lacerda, L., Herrero, M.A., Venner, K., Bianco, A., Prato, M., Kostarelos, K., Carbon-nanotube shape and individualization critical for renal excretion. *Small*, 4, 8, 1130–1132, 2008.
16. Mc Devitt, M.R., Chattopadhyay, D., Jaggi, J.S. *et al.*, PET imaging of soluble yttrium-86-labeled carbon nanotubes in mice. *PloS One*, 2, 9, e907, 2007.
17. Liu, Z., Davis, C., Cai, W., He, L., Chen, X., Dai, H., Circulation and long-term fate of functionalized, biocompatible single-walled carbon nanotubes in mice probed by Raman spectroscopy. *Proc. Natl. Acad. Sci. U. S. A.*, 105, 5, 1410–1415, 2008.
18. Kostarelos, K., Bianco, A., Prato, M., Promises, facts and challenges for carbon nanotubes in imaging and therapeutics. *Nat. Nanotechnol.*, 4, 10, 627–633, 2009.
19. Ruggiero, A., Villa, C.H., Bander, E. *et al.*, Paradoxical glomerular filtration of carbon nanotubes. *Proc. Natl. Acad. Sci. U. S. A.*, 107, 27, 12369–12374, 2010.
20. Faria, P.C.B.D., Santos, L., II, Coelho, J.P. *et al.*, Oxidized multiwalled carbon nanotubes as antigen delivery system to promote superior CD8+ T Cell response and protection against cancer. *Nano Lett.*, 14, 9, 5458–5470, 2014.
21. Iijima, S., Helical microtubules of graphitic carbon. *Nature*, 354, 6348, 56–58, 1991.
22. Zhang, M. and Li, J., Carbon nanotube in different shapes. *Mater. Today*, 12, 6, 12–18, 2009.
23. Pantarotto, D., Briand, J.P., Prato, M., Bianco, A., Translocation of bioactive peptides across cell membranes by carbon nanotubes. *Chem. Commun.*, 10, 1, 16–17, 2004.

24. Pantarotto, D., Singh, R., McCarthy, D. *et al.*, Functionalized carbon nanotubes for plasmid DNA gene delivery. *Angew. Chem. Int. Ed.*, 43, 39, 5242–5246, 2004.
25. Bianco, A., Kostarelos, K., Prato, M., Applications of carbon nanotubes in drug delivery. *Curr. Opin. Chem. Biol.*, 9, 6, 674–679, 2005.
26. Chen, X., Kis, A., Zettl, A., Bertozzi, C.R., A cell nanoinjector based on carbon nanotubes. *Proc. Natl. Acad. Sci. U. S. A.*, 104, 20, 8218–8222, 2007.
27. Bhirde, A.A., Patel, V., Gavard, J. *et al.*, Targeted killing of cancer cells *in vivo* and *in vitro* with EGF-directed carbon nanotube-based drug delivery. *ACS Nano*, 3, 2, 307–316, 2009.
28. Jemal, A., Devesa, S.S., Hartge, P., Tucker, M.A., Recent trends in cutaneous melanoma incidence among whites in the United States. *J. Natl. Cancer Inst.*, 93, 9, 678–683, 2001.
29. Yan, Y., Chan-Park, M.B., Zhang, Q., Advances in carbon nanotube assembly. *Small*, 3, 1, 24–42, 2007.
30. Kesharwani, P., Mishra, V., Jain, N.K., Validating the anticancer potential of carbon nanotube-based therapeutics through cell line testing. *Drug Discovery Today*, 20, 9, 1049–1060, 2015.
31. Bethune, D.S., Kiang, C.H., de Vries, M.S. *et al.*, Cobalt catalysed growth of carbon nanotubes with single-atomic-layer walls. *Nature*, 363, 6430, 605–607, 1993.
32. Bekyarova, E., Ni, Y., Malarkey, E.B. *et al.*, Applications of Carbon Nanotubes in Biotechnology and biomedicine. *J. Biomed. Nanotechnol.*, 1, 1, 3–17, 2005.
33. Klumpp, C., Kostarelos, K., Prato, M., Bianco, A., Functionalized carbon nanotubes as emerging nanovectors for the delivery of therapeutics. *Biochim. Biophys. Acta Biomembr.*, 1758, 3, 404–412, 2006.
34. Madani, S.Y., Naderi, N., Dissanayake, O., Tan, A., Seifalian, A.M., A new era of cancer treatment: Carbon nanotubes as drug delivery tools. *Int. J. Nanomedicine*, 6, 2963–2979, 2011.
35. Bianco, A., Kostarelos, K., Prato, M., Applications of carbon nanotubes in drug delivery. *Curr. Opin. Chem. Biol.*, 9, 6, 674–679, 2005.
36. Liu, J., Rinzler, A.G., Dai, H. *et al.*, Fullerene pipes. *Science*, 280, 5367, 1253–1256, 1998.
37. Tasis, D., Tagmatarchis, N., Georgakilas, V., Prato, M., Soluble carbon nanotubes. *Chem. Eur. J.*, 9, 17, 4000–4008, 2003.
38. Georgakilas, V., Kordatos, K., Prato, M., Guldi, D.M., Holzinger, M., Hirsch, A., Organic functionalization of carbon nanotubes. *J. Am. Chem. Soc.*, 124, 5, 760–761, 2002.
39. Hossen, S., Hossain, M.K., Basher, M.K., Mia, M.N.H., Rahman, M.T., Uddin, M.J., Smart nanocarrier-based drug delivery systems for cancer therapy and toxicity studies: A review. *J. Adv. Res.*, 15, 1, 1–18, 2019.
40. Bianco, A., Kostarelos, K., Partidos, C.D., Prato, M., Biomedical applications of functionalized carbon nanotubes. *Chem. Commun.*, 5, 571–577, 2005.

41. Abousalman-Rezvani, Z., Eskandari, P., Roghani-Mamaqani, H., Salami-Kalajahi, M., Functionalization of carbon nanotubes by combination of controlled radical polymerization and "grafting to" method. *Adv. Colloid Interface Sci.*, 278, 102126, 2020.
42. Kesharwani, P., Ghanghoria, R., Jain, N.K., Carbon nanotube exploration in cancer cell lines. *Drug Discovery Today*, 17, 17-18, 1023–1030, 2012.
43. Rosca, I.D., Watari, F., Akasaka, T., Oxidation of multiwalled carbon nanotubes by nitric acid. *Carbon*, 43, 15, 3124–3131, 2005.
44. Eskandari, P., Abousalman-Rezvani, Z., Roghani-Mamaqani, H., Salami-Kalajahi, M., Polymer-functionalization of carbon nanotube by *in situ* conventional and controlled radical polymerizations. *Adv. Colloid Interface Sci.*, 294, 102471, 2021.
45. Raphey, V.R., Henna, T.K., Nivitha, K.P., Mufeedha, P., Sabu, C., Pramod, K., Advanced biomedical applications of carbon nanotube. *Mater. Sci. Eng. C Mater. Biol. Appl.*, 100, 616–30, 2019.
46. Chen, M., Qin, X., Zeng, G., Biodegradation of carbon nanotubes, graphene, and their derivatives. *Trends Biotechnol.*, 35, 9, 836–46, 2017.
47. Zhou, H., Ge, J., Miao, Q., Zhu, R., Wen, L., Zeng, J. *et al.*, Biodegradable inorganic nanoparticles for cancer theranostics: insights into the degradation behavior. *Bioconjug. Chem.*, 31, 2, 315–31, 2020.
48. Veronese, F.M. and Pasut, G., PEGylation, successful approach to drug delivery. *Drug Discovery Today*, 10, 21, 1451–8, 2005.
49. Taghavi, S., Nia, A.H., Abnous, K., Ramezani, M., Polyethylenimine functionalized carbon nanotubes tagged with AS1411 aptamer for combination gene and drug delivery into human gastric cancer cells. *Int. J. Pharm.*, 516, 1–2, 301–12, 2017.
50. Kesharwani, P., Mishra, V., Jain, N.K., Validating the anticancer potential of carbon nanotube-based therapeutics through cell line testing. *Drug Discovery Today*, 20, 9, 1049–1060, 2015.
51. Huang, Y.-P., Lin, I.J., Chen, C.-C., Hsu, Y.-C., Chang, C.-C., Lee, M.-J., Delivery of small interfering RNAs in human cervical cancer cells by polyethylenimine-functionalized carbon nanotubes. *Nanoscale Res. Lett.*, 8, 1, 267, 2013.
52. Huang, G.L. and Huang, H.L., Application of hyaluronic acid as carriers in drug delivery. *Drug Deliv.*, 25, 1, 766–72, 2018.
53. Dong, X., Liu, L.X., Zhu, D.W., Zhang, H.L., Leng, X.G., Transactivator of transcription (TAT) peptide-chitosan functionalized multiwalled carbon nanotubes as a potential drug delivery vehicle for cancer therapy. *Int. J. Nanomedicine*, 10, 3829–41, 2015.
54. Cheng, Q., Blais, M.-O., Harris, G., Jabbarzadeh, E., PLGA-carbon nanotube conjugates for intercellular delivery of caspase-3 into osteosarcoma cells. *PloS One*, 8, 12, e81947, 2013.

55. Mirakabad, F., Nejati-Koshki, K., Akbarzadeh, A., Yamchi, M.R., Milani, M., Zarghami, N. *et al.*, PLGA-based nanoparticles as cancer drug delivery systems. *Asian Pac. J. Cancer Prev.*, 15, 2, 517–35, 2014.
56. Harisa, G.I. and Faris, T.M., Direct drug targeting into intracellular compartments: Issues, limitations, and future outlook. *J. Membr. Biol.*, 252, 6, 527–39, 2019.
57. Afanzar, O., Buss, G.K., Stearns, T., Ferrell Jr., J.E., The nucleus serves as the pacemaker for the cell cycle. *Elife*, 9, e59989, 2020.
58. Bradshaw, P.C., Cytoplasmic and mitochondrial NADPH-coupled redox systems in the regulation of aging. *Nutrients*, 11, 3, 504, 2019.
59. Lee, P.C., Chiou, Y.C., Wong, J.M., Peng, C.L., Shieh, M.J., Targeting colorectal cancer cells with single-walled carbon nanotubes conjugated to anticancer agent SN-38 and EGFR antibody. *Biomaterials*, 34, 34, 8756–65, 2013.
60. Karmakar, A., Bratton, S.M., Dervishi, E., Ghosh, A., Mahmood, M., Xu, Y. *et al.*, Ethylenediamine functionalized-single-walled nanotube (f-SWNT)-assisted *in vitro* delivery of the oncogene suppressor p53 gene to breast cancer MCF-7 cells. *Int. J. Nanomedicine*, 6, 1045–55, 2011.
61. Kern, S.E., Kinzler, K.W., Bruskin, A., Jarosz, D., Friedman, P., Prives, C. *et al.*, Identifcation of p53 as a sequence-specific DNA-binding protein. *Science*, 252, 1708–11, 1991.
62. Etienne-Manneville, S., Cytoplasmic intermediate filaments in cell biology. *Annu. Rev. Cell Dev. Biol.*, 34, 1–28, 2018.
63. Chen, W., Yang, S., Wei, X., Yang, Z., Liu, D., Pu, X. *et al.*, Construction of aptamer-siRNA chimera/PEI/5-FU/carbon nanotube/collagen membranes for the treatment of peritoneal dissemination of drug resistant gastric cancer. *Adv. Healthc. Mater.*, 9, 21, e2001153, 2020.
64. Bartholomeusz, G., Cherukuri, P., Kingston, J., Cognet, L., Lemos, R., Leeuw, T.K. *et al.*, *In vivo* therapeutic silencing of hypoxia-inducible factor 1 alpha (HIF-1alpha) using single-walled carbon nanotubes noncovalently coated with siRNA. *Nano Res.*, 2, 4, 279–91, 2009.
65. Wang, B., Zhao, Q., Zhang, Y., Liu, Z., Zheng, Z., Liu, S. *et al.*, Targeting hypoxia in the tumor microenvironment: a potential strategy to improve cancer immunotherapy. *J. Exp. Clin. Cancer Res.*, 40, 1, 24, 2021.
66. Liang, Z., Currais, A., Soriano-Castell, D., Schubert, D., Maher, P., Natural products targeting mitochondria: Emerging therapeutics for age-associated neurological disorders. *Pharmacol. Ther.*, 221, 107749, 2021.
67. Suomalainen, A. and Battersby, B.J., Mitochondrial diseases: the contribution of organelle stress responses to pathology. *Nat. Rev. Mol. Cell Biol.*, 19, 2, 77–92, 2018.
68. Mani, S., Swargiary, G., Singh, K.K., Natural agents targeting mitochondria in cancer. *Int. J. Mol. Sci.*, 21, 19, 6992, 2020.
69. Szabo, I., Zoratti, M., Biasutto, L., Targeting mitochondrial ion channels for cancer therapy. *Redox Biol.*, 42, 101846, 2021.

70. Liu, Y.Q., Wang, X.L., He, D.H., Cheng, Y.X., Protection against chemotherapyand radiotherapy-induced side effects: A review based on the mechanisms and therapeutic opportunities of phytochemicals. *Phytomedicine*, 80, 153402, 2021.
71. Li, J., Yap, S.Q., Chin, C.F., Tian, Q., Yoong, S.L., Pastorin, G. *et al.*, Platinum (iv) prodrugs entrapped within multiwalled carbon nanotubes: selective release by chemical reduction and hydrophobicity reversal. *Chem. Sci.*, 3, 6, 2083–7, 2012.
72. Niazi, A.K., Mileshina, D., Cosset, A., Val, R., Weber-Lotf, F., Dietrich, A., Targeting nucleic acids into mitochondria: progress and prospects. *Mitochondrion*, 13, 5, 548–58, 2013.
73. Kim, S.W., Lee, Y.K., Lee, J.Y., Hong, J.H., Khang, D., PEGylated anticancer-carbon nanotubes complex targeting mitochondria of lung cancer cells. *Nanotechnology*, 28, 46, 465102, 2017.
74. Li, C., Zhang, W., Liu, S., Hu, X., Xie, Z., Mitochondria-targeting organic nanoparticles for enhanced photodynamic/photothermal therapy. *ACS Appl. Mater. Interfaces*, 12, 27, 30077–84, 2020.
75. Guedan, S., Ruella, M., June, C.H., Emerging cellular therapies for cancer. *Annu. Rev. Immunol.*, 37, 145–71, 2019.
76. Meng, J., Yang, M., Jia, F., Kong, H., Zhang, W., Wang, C. *et al.*, Subcutaneous injection of water-soluble multi-walled carbon nanotubes in tumor bearing mice boosts the host immune activity. *Nanotechnology*, 21, 14, 145104, 2010.
77. Luo, X., Wang, H., Ji, D., Carbon nanotubes (CNT)-loaded ginsenosides Rb3 suppresses the PD-1/PD-L1 pathway in triple-negative breast cancer. *Aging (Albany NY)*, 13, 13, 17177–89, 2021.
78. Han, J., Won, M., Kim, J.H., Jung, E., Min, K., Jangili, P. *et al.*, Cancer stem cell targeted bio-imaging and chemotherapeutic perspective. *Chem. Soc. Rev.*, 49, 22, 856–78, 2020.
79. El-Sahli, S. and Wang, L., Cancer stem cell-associated pathways in the metabolic reprogramming of breast cancer. *Int. J. Mol. Sci.*, 21, 23, 9125, 2020.
80. de Andres, J., Grinan-Lison, C., Jimenez, G., Marchal, J.A., Cancer stem cell secretome in the tumor microenvironment: A key point for an effective personalized cancer treatment. *J. Hematol. Oncol.*, 13, 1, 136, 2020.
81. Shen, S., Xu, X., Lin, S., Zhang, Y., Liu, H., Zhang, C. *et al.*, A nanotherapeutic strategy to overcome chemotherapeutic resistance of cancer stem-like cells. *Nat. Nanotechnol.*, 16, 1, 104–13, 2021.
82. Walcher, L., Kistenmacher, A.K., Suo, H., Kitte, R., Dluczek, S., Strauss, A. *et al.*, Cancer stem cells-origins and biomarkers: Perspectives for targeted personalized therapies. *Front. Immunol.*, 11, 1280, 2020.
83. Brown, H.K., Tellez-Gabriel, M., Heymann, D., Cancer stem cells in osteosarcoma. *Cancer Lett.*, 386, 189–95, 2017.
84. Zhang, H., Wu, H., Zheng, J., Yu, P., Xu, L., Jiang, P. *et al.*, Transforming growth factor beta1 signal is crucial for dedifferentiation of cancer cells to cancer stem cells in osteosarcoma. *Stem Cells*, 31, 3, 433–46, 2013.

85. Miao, Y., Zhang, H., Pan, Y., Ren, J., Ye, M., Xia, F. *et al.*, Single-walled carbon nanotube: One specific inhibitor of cancer stem cells in osteosarcoma upon downregulation of the TGFbeta1 signaling. *Biomaterials*, 149, 29–40, 2017.
86. Liu, Z., Tabakman, S., Welsher, K., Dai, H., Carbon nanotubes in biology and medicine: *In vitro* and *in vivo* detection, imaging drug delivery. *Nano Res.*, 2, 2, 85–120, 2009.
87. Sekiyama, S., Umezawa, M., Iizumi, Y., Ube, T., Okazaki, T., Kamimura, M. *et al.*, Delayed increase in near-infrared fluorescence in cultured murine cancer cells labeled with oxygen-doped single-walled carbon nanotubes. *Langmuir*, 35, 3, 831–7, 2019.
88. Chen, Y.-C., Young, R.J., Macpherson, J.V., Wilson, N.R., Silver-decorated carbon nanotube networks as SERS substrates. *J. Raman Spectrosc.*, 42, 6, 1255–62, 2011.
89. Wang, X., Wang, C., Cheng, L., Lee, S.T., Liu, Z., Noble metal coated single walled carbon nanotubes for applications in surface enhanced Raman scattering imaging and photothermal therapy. *J. Am. Chem. Soc.*, 134, 17, 7414–22, 2012.
90. Dai, Y., Su, J., Wu, K., Ma, W., Wang, B., Li, M. *et al.*, Multifunctional thermosensitive liposomes based on natural phase-change material: near-infrared light-triggered drug release and multimodal imaging-guided cancer combination therapy. *ACS Appl. Mater. Interfaces*, 11, 11, 10540–53, 2019.
91. Ghaghada, K.B., Starosolski, Z.A., Bhayana, S., Stupin, I., Patel, C.V., Bhavane, R.C. *et al.*, Pre-clinical evaluation of a nanoparticle-based blood-pool contrast agent for MR imaging of the placenta. *Placenta*, 57, 60–70, 2017.
92. Yan, C., Chen, C., Hou, L., Zhang, H., Che, Y., Qi, Y. *et al.*, Single-walled carbon nanotube-loaded doxorubicin and Gd-DTPA for targeted drug delivery and magnetic resonance imaging. *J. Drug Targeting*, 25, 2, 163–71, 2017.
93. Zhang, M., Wang, W., Cui, Y., Zhou, N., Shen, J., Magnetofuorescent carbon quantum dot decorated multiwalled carbon nanotubes for dual-modal targeted imaging in chemo-photothermal synergistic therapy. *ACS Biomater. Sci. Eng.*, 4, 1, 151–62, 2018.
94. Sanginario, A., Miccoli, B., Demarchi, D., Carbon nanotubes as an effective opportunity for cancer diagnosis and treatment. *Biosensors (Basel)*, 7, 1, 9, 2017.
95. Saghatchi, F., Mohseni-Dargah, M., Akbari-Birgani, S., Saghatchi, S., Kaboudin, B., Cancer therapy and imaging through functionalized carbon nanotubes decorated with magnetite and gold nanoparticles as a multimodal tool. *Appl. Biochem. Biotechnol.*, 191, 3, 1280–93, 2020.
96. Delogu, L.G., Vidili, G., Venturelli, E., Menard-Moyon, C., Zoroddu, M.A., Pilo, G. *et al.*, Functionalized multiwalled carbon nanotubes as ultrasound contrast agents. *Proc. Natl. Acad. Sci. U. S. A.*, 109, 41, 16612–7, 2012.
97. Zhang, J., Yang, S., Ji, X., Zhou, Q., Xing, D., Characterization of lipid-rich aortic plaques by intravascular photoacoustic tomography: *Ex vivo* and

in vivo validation in a rabbit atherosclerosis model with histologic correlation. *J. Am. Coll. Cardiol.*, 64, 4, 385–90, 2014.

98. Ji, Z., Ding, W.Z., Ye, F.H., Lou, C.G., Xing, D., Shape-adapting thermoacoustic imaging system based on flexible multi-element transducer. *Appl. Phys. Lett.*, 107, 9, 2015.
99. De La Zerda, A., Zavaleta, C., Keren, S., Vaithilingam, S., Bodapati, S., Liu, Z. *et al.*, Carbon nanotubes as photoacoustic molecular imaging agents in living mice. *Nat. Nanotechnol.*, 3, 9, 557–62, 2008.
100. Avti, P.K., Hu, S., Favazza, C., Mikos, A.G., Jansen, J.A., Shroyer, K.R. *et al.*, Detection, mapping, and quantification of single walled carbon nanotubes in histological specimens with photoacoustic microscopy. *PloS One*, 7, 4, e35064, 2012.
101. Wang, C., Bao, C., Liang, S., Fu, H., Wang, K., Deng, M. *et al.*, RGD-conjugated silica-coated gold nanorods on the surface of carbon nanotubes for targeted photoacoustic imaging of gastric cancer. *Nanoscale Res. Lett.*, 9, 1, 264, 2014.
102. Tao, Y., Sun, Y., Shi, K., Pei, P., Ge, F., Yang, K. *et al.*, Versatile labeling of multiple radionuclides onto a nanoscale metal-organic framework for tumor imaging and radioisotope therapy. *Biomater. Sci.*, 9, 8, 2947–54, 2021.
103. Hernandez, R., Grudzinski, J.J., Aluicio-Sarduy, E., Massey, C.F., Pinchuk, A.N., Bitton, A.N. *et al.*, Lu-177-NM600 targeted radionuclide therapy extends survival in syngeneic murine models of triple-negative breast cancer. *J. Nucl. Med.*, 61, 8, 1187–94, 2020.
104. Gong, H., Peng, R., Liu, Z., Carbon nanotubes for biomedical imaging: The recent advances. *Adv. Drug Deliv. Rev.*, 65, 15, 1951–63, 2013.
105. Wang, J.T., Fabbro, C., Venturelli, E., Menard-Moyon, C., Chaloin, O., Da Ros, T. *et al.*, The relationship between the diameter of chemically-functionalized multi-walled carbon nanotubes and their organ biodistribution profiles *in vivo*. *Biomaterials*, 35, 35, 9517–28, 2014.
106. Zhao, H., Chao, Y., Liu, J., Huang, J., Pan, J., Guo, W. *et al.*, Polydopamine coated single-walled carbon nanotubes as a versatile platform with radionuclide labeling for multimodal tumor imaging and therapy. *Theranostics*, 6, 11, 1833–43, 2016.
107. Mandal, A.K., Wu, X., Ferreira, J.S., Kim, M., Powell, L.R., Kwon, H. *et al.*, Fluorescent sp (3) defect-tailored carbon nanotubes enable NIR-II single particle imaging in live brain slices at ultra-low excitation doses. *Sci. Rep.*, 10, 1, 5286, 2020.
108. Dang, X., Bardhan, N.M., Qi, J., Gu, L., Eze, N.A., Lin, C.-W. *et al.*, Deep-tissue optical imaging of near cellular-sized features. *Sci. Rep.*, 9, 3873, 2019.
109. Silindir-Gunay, M., Sarcan, E.T., Ozer, A.Y., Near-infrared imaging of diseases: A nanocarrier approach. *Drug Dev. Res.*, 21532, 1–14, 2019. https://doi.org/10.1002/ ddr.21532.
110. Ghosh, D., Bagley, A.F., Na, Y.J., Birrer, M.J., Bhatia, S.N., Belcher, A.M., Deep, noninvasive imaging and surgical guidance of submillimeter tumors

using targeted M13-stabilized single-walled carbon nanotubes. *Proc. Natl. Acad. Sci. U.S.A.*, 111, 38, 13948–53, 2014.

111. Welsher, K., Liu, Z., Sherlock, S.P., Robinson, J.T., Chen, Z., Daranciang, D. *et al.*, A route to brightly fluorescent carbon nanotubes for near-infrared imaging in mice. *Nat. Nanotechnol.*, 4, 11, 773–80, 2009.
112. Clark Jr., L.C. and Lyons, C., Electrode systems for continuous monitoring in cardiovascular surgery. *Ann. N. Y. Acad. Sci.*, 102, 29–45, 1962.
113. Gupta, S., Murthy, C.N., Prabha, C.R., Recent advances in carbon nanotube based electrochemical biosensors. *Int. J. Biol. Macromol.*, 108, 687–703, 2018.
114. Moon, D., Cha, Y.K., Kim, S.O., Cho, S., Ko, H.J., Park, T.H., FET-based nanobiosensors for the detection of smell and taste. *Sci. China Life Sci.*, 63, 8, 1159–67, 2020.
115. Bellan, L.M., Wu, D., Langer, R.S., Current trends in nanobiosensor technology. *Wiley Interdiscip. Rev. Nanomed. Nanobiotechnol.*, 3, 3, 229–46, 2011.
116. Yoosefan, M. and Etminan, N., Leucine/Pd-loaded (5, 5) single-walled carbon nanotube matrix as a novel nanobiosensors for in silico detection of protein. Amino Acids, 50, 6, 653–61, 2018.

9

Functionalized Carbon Nanotubes for Biomedical Imaging: The Recent Advances

Alina Abbas[1], Saman Zehra[2*], Ruby Aslam[2†], Mohammad Mobin[2] and Shahidul Islam bhat[2]

[1]Department of Chemistry, Jamia Hamdard (Hamdard University), New Delhi, India
[2]Corrosion Research Laboratory, Department of Applied Chemistry, Faculty of Engineering and Technology, Aligarh Muslim University, Aligarh, India

Abstract
The chapter summarizes recent advancements in the applicability of Functionalized Carbon Nanotubes (fCNTs) in biomedical imaging. In the ever-expanding area of imaging techniques, the development of molecular contrast agents enabling visual assessment of biological, structural, and functional information is an ongoing process. These non-destructive and non-invasive methods protect the specimen while also carrying a plethora of information. Various Nano-technological instruments are inexplicably employed as molecular probes, opening the door for smart and Theragnostic action modules. Carbon nanotubes, both single-walled (SWCNT) and multi-walled (MWCNT), are a key class of synthetic nanostructures sought in virtually every conceivable field of application because of their remarkable characteristics. It has advanced to the point that it is now a promising imaging tool with extensive medical research. fCNTs have the potential to be used in a wide range of biological applications. This chapter describes a method for imaging biological subjects *in vitro* and *in vivo* utilizing the inherent photo physical characteristics of fCNTs, which involves a large number of components to match the current situation of multiplexed and multimodal imaging and a brief review of different optical methods that use them as contrast agents is provided with regards to the recent advances in the field.

Corresponding author: samanzehra2050@gmail.com
†*Corresponding author*: drrubyaslam@gmail.com

Jeenat Aslam, Chaudhery Mustansar Hussain and Ruby Aslam (eds.) Functionalized Carbon Nanotubes for Biomedical Applications, (197–224) © 2023 Scrivener Publishing LLC

Keywords: Carbon nanotubes, single and multi-walled carbon nanotube, biomedical applications, biomedical imaging, and photocaustic imaging

9.1 Introduction

The exceptional properties of the Carbon Nanotubes (CNTs), like high mechanical strength, adaptability, and outstanding electrical properties, make it an appealing and possible contender for different applications [1–5]. The blend of these attributes presents them with broad biomedical applications. Carbon nanotubes, similar to diamond, graphite, graphene, fullerene, and indeterminate carbon, are some of the carbon allotropes. The one-layered carbon structure, on the other hand, can have a viewpoint proportion greater than 1000, which is remarkable. CNTs offer a unique set of physical and chemical characteristics, which some have potential uses in biomedicine [6].

Nonetheless, the vast majority of the inherent characteristics are unsuitable for direct application, e.g., CNTs have been seen to display infrared discharges; however, the fluxes are so weak that *in vivo* imaging is impossible [7]. The usage of radiant CNTs in conjunction with fluorescent materials has attracted a lot of attention. Another theory states that carbon nanotubes can diminish water protons' crossover unwinding, making them T2-weighted MRI contrast experts. However, the MRI signal is simply too weak to provide any meaningful information. CNTs need to be functionalized with attractive materials, such as superparamagnetic iron oxide nanoparticles (SPIO). The deliberate connection of various SPIO nanoparticles on the outer layer of the CNTs in a one-layered course can upgrade the attractive impact because of the attractive coupling between SPIO nanoparticles.

The novel actual characteristics of CNTs, especially SWNTs have a lot of potential in the field of biological imaging. The narrow band of around 1 eV in 1D semi-conducted SWNTs allows fluorescence discharge in the near-infrared (NIR) region, which includes both NIR-I (700-900 nm) and NIR-II (700-900 nm) (1100-1400 nm). SWNTs have a wide dispersion cross-segment and show solid Raman scattering, making them ideal for Raman testing in biological sensing and imaging.

CNTs are the most mysterious material, yet they have a high absorption in the near-infrared (NIR) area and may therefore be utilized as photoacoustic (PA) imaging contrast specialists [8, 9]. Metal nanoparticle contamination discovered during CNT testing can be used to give solid T2-weighted image contrast in magnetic resonance imaging (MRI) [10]. Apart from embracing, radionuclides can be created or even implanted in CNTs to give alternative imaging modalities, such as Positron emission tomography (PET) and Single-photon emission computed tomography (SPECT) [11]. In the NIR-II region, fluorescent imaging of SWNTs has made enormous progress in recent years, making it a promising imaging technique with enormous potential in biomedical research. By coating SWNTs with noble metal nanoparticles, surface-enhanced Raman scattering (SERS) may significantly improve Raman scattering of organic samples, allowing them to be scanned much faster. Multiplexed photoacoustic imaging *in vivo* has been produced by layering different colors with NIR absorbance at varying frequencies. It has also been produced SWNTs with suitable surface functionalization. It has been employed for immature microbe tagging and *in vivo* multimodal following thanks to Raman imaging, magnetic resonance imaging, and photoacoustic imaging. The usage of carbon nanotubes in biomedical imaging is likely to be updated in this audit area. Conclusively, this chapter focuses on application of functionalized carbon nanotubes (fCNTs) in the area of various biomedical imaging fields, with little or no cytotoxicity created by functionalization procedures, which is the fundamental need for CNT applications in biomedical imaging.

9.2 CNT-Based Imaging Methods

In the imaging sector, there are a variety of advances based on CNT. As an example, photoluminescent imaging uses the fluorescence of activated SWCNT in the NIR-I and II regions, at which tissues and water are easily distinguished, allowing for more penetration depth. Welsher *et al.* employed this characteristic to continually monitor the distribution of implanted SWCNTs in deep tissues and arteries of mice [12, 13] SWCNT fluorescence is developing as a feasible optical imaging method for vascular perception [14, 15], and it has recently been employed as a cost-effective alternative to expensive Positron emission tomography-computed tomography for earthy-colored fat identification [16].

Another intriguing way of exploiting CNT NIR intake is the photoacoustic effect. The ultrasounds supplied by tissues developing surrounding heat-producing CNT subjected to NIR excitation are estimated using photoacoustic imaging. De La Zerda *et al.* acquired better differentiation on specified growth cells due to SWCNT synthesized with Arg-Gly-Asp peptides, as well as an increased awareness and discovery edge, when compared to virgin SWCNT [17, 18].

Because of the G band produced by SWCNT about 1580 cm-1, Raman dissipating is also an imaging option. Liu *et al.* demonstrated five-color multiplexed Raman imaging utilizing a combination of peak moving for SWCNT based on the 13C/12C isotopic fraction and marking by variable ligand focusing [19]. Further, Raman imaging's acquiring time, one of its biggest limitations, has been significantly improved through the use of Raman hyperspectral instruments with Bragg tunable channels, dramatically improving output speed and filtering quality [20]. In comparison to SWCNT and double-walled CNT (DWCNT) MWCNT has a lower signal/clamor ratio; therefore, its responsiveness is based on the concept of the natural grid.

Hong *et al.* radio's tests, which were built of metal halides exemplified inside CNT and used for Single-Photon Emission Computed Tomography imaging (SPECT), demonstrate another use of CNT in bioimaging [21]. A further advantage to the 1,3-dipolar cycloaddition of CNT is that it achieves functionalization without opening the CNT closed ends, preventing radioactivity from reaching the thyroid and stomach. This functionalization technique was also used to create oxidized MWCNT with gadolinium, a component commonly employed as a contrast specialist in attractive reverberation imaging (MRI). When compared to actual business ones, Gd-MWCNT demonstrated better MRI contrast [22].

9.2.1 Fluorescence Imaging

Fluorescence imaging is important in logical inquiry and clinical analysis. In any event, the limited depth of light entry becomes a substantial barrier to the continued use of fluorescence imaging [23]. Experts have concentrated on inventing and performing fluorescence experiments with excitation and outflow frequencies that fall inside the naturally simple NIR window to counteract this difficulty [24]. The classic NIR window, also known as the NIR I window, is defined as a range of wavelengths between 700 and 900 nm in which neither hemoglobin nor water exhibits significant light absorption. Various tests, such as natural colors and semiconducting quantum dots (QDs), are currently available in this field [25–28]. In addition to being consumed by water, light with a longer wavelength of 1100-1400 nm also shows decreased tissue absorbance and, more importantly, dramatically reduced dissipation by organic tissues, which allows *in vitro* examination [29] as well as *in vivo* fluorescence imaging with much better spatial resolution and infiltration [30–34].

The quantum repression effect causes SWNTs to be semi-one-layered quantum wires with segregated energy levels. Van Hoff singularities (VHSs) are used to explain the thickness of states, which indicate that the DOS force is greatest within a limited energy range [35, 36]. The band-hole between each semiconducting SWNT is one eV in request, accounting for fluorescence in the NIR-II region under illumination in the NIR-I region [36]. Further, the significant Stoke-shift between the excitation wavelength of 550-850 nm and the outflow wavelength of 900-1600 nm reduces organic tissue autofluorescence during imaging, resulting in improved *in vivo* imaging [37].

Welsher *et al.* studied the course of SWNTs through mouse anatomy using SWNTs as an NIRII contrast agent in high frame rate fluorescent video imaging of mice i.v. injected with SWNTs [13]. SWNTs first reached the lungs several seconds after injection, followed by the spleen and liver at subsequent time points, as recorded in real time (Figure 9.1a). The anatomic resolution of organs was greatly improved by principal component analysis (PCA). After PCA analysis of NIR-II fluorescence images of mice

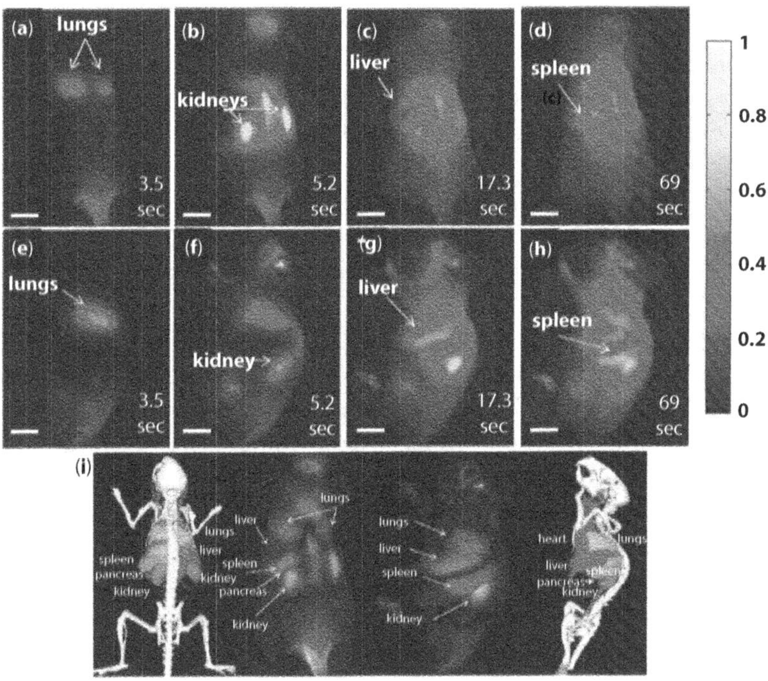

Figure 9.1 In a live mouse, video-rate NIR-II imaging of SWNTs. (a–h) At various time intervals after injection, frames from video imaging of a mouse injected with PEGylated SWNTs were collected. I Mouse injected with SWNTs by PCA with dynamic contrast-enhanced imaging. Every 150 equally spaced frames from the 2000-frame dataset were used to create PCA pictures throughout the first 130 seconds after injection. The lungs, liver, kidney, spleen, and even the pancreas have prominent characteristics [13].

with SWNT injection, even the pancreas, which could not be resolved from real-time raw images, could be distinguished (Figure 9.1b). As a result, this research showed that NIR-II fluorescence imaging combined with PCA could give powerful tools for a wide range of applications ranging from biological research to disease diagnostics.

SWNTs' low quantum yield (QY) was a major stumbling block to their use in *in vivo* imaging at the time, despite their promising findings in NIR fluorescence imaging. The formation of SWNTs would be disrupted by covalent functionalization, and their NIR fluorescence would be wholly lost. Despite the high QY of SWNTs suspended by tiny surfactant atoms, such as sodium cholate, these particles showed all the signs of harming a natural environment. PEGylation of SWNTs can result in remarkably water-soluble and biocompatible nanotubes, however, it is often associated with a significant decrease in SWNT quality.

In 2009, Dai and colleagues devised a novel covering trade strategy for obtaining biocompatible SWNTs with a high QY. DE packed SWNTs were solubilized in a sodium cholate solution before being uprooted with phospholipid-PEG (PL-PEG). Unlike the previous adjustment method, which simply sonicated SWNTs in PL-PEG arrangements for a long time (N15 min), our comparatively careful covering trading methodology avoids SWNT QY deficit. PEGylated SWNTs with high QY arranged using this approach were used for *in vivo* total body NIR-II fluorescence imaging of mice with intravenously (i.e.) injected SWNTs. Following SWNT injection, high-resolution intravital microscopy images revealed small tumor arteries beneath the barrier [38].

Using PEGylated SWNTs with high QY, we imaged total body NIR-II fluorescence *in vivo* of mice injected with SWNTs intravenously (i.e., intravenously injected). Further, high-resolution intravital microscopy imaging of the SWNT-injected tumor revealed small tumor arteries. The anatomic target of organs was greatly enhanced using head part examination (PCA). Indeed, following PCA assessment of NIR-II fluorescence photos of mice with SWNT injection, the pancreas, which could not be separated from the crude photographs, could be isolated. As a consequence, our study showed that NIR-II fluorescence imaging paired with PCA could be beneficial in a variety of applications, from biological inspection to infection detection [39].

Robinson *et al.* [40] employed poly(maleic anhydride-alt-1-octadecene)-polyethylene glycol (C18PMH-PEG) to produce a well-functionalized SWNT plan with a long blood course (half-life of 30 h) *in vivo* to reach ultrahigh accumulation of 30% infused part (percent ID/g) in 4T1 bosom disease in mice. To screen the fluorescent indications in malignancies and diverse organs, high-fluorescence video-rate imaging and PCA were used for the first time. They discovered visible luminous indicators in the growth 20 minutes after injection, and the fluorescent indication lasted for 72 hours. Furthermore, 3D modelling of NIR-II fluorescence signals inside cancerous growths revealed a positive correlation between SWNTs and cancer vasculatures, suggesting EPR may play a crucial role in preventing cancerous nanotube accumulation. High-fluorescence video-rate imaging and PCA analysis were used to perform vascular imaging *in vivo* in addition to cancer imaging.

Hong *et al.* [37] used a substantial difference enhanced NIR-II imaging approach to clearly identify blood vessels and venous vessels based on their distinct hemodynamics. As compared to the tiny CT image of the mouse's distal hind leg, the NIR-II image revealed far more small vessels. NIR-tiniest II's vessel had a maximum width of only 35.4 m, while tiny CT couldn't see any vessels smaller than 100 m. The blood speed inside the veins can also be calculated in both ischemia and normal appendages, even at very low blood speeds, which are beyond the resolution of ultrasonography. As a consequence of their findings, they concluded that NIR-II fluorescence imaging of vasculatures using SWNTs as Nano-tests has a few advantages over Micro-CT and ultrasound imaging approaches.

Currently, SWNTs are all different in their chiralities, and each chirality relates to a specific combination of excitation and discharge frequencies. This resulted in only a small percentage of SWNTs being excited by the laser to produce fluorescence, whereas most were not in resonance with the excitation laser frequency, appearing dim during fluorescence imaging. This results in significantly reduced nanotube injection into animals when using chirality-decontaminated SWNTs for NIR-II fluorescence imaging.

Dielectrophoresis, thickness slope centrifugation [41], DNA wrapping chromatography [42], and gel filtration [43–45] have all been used to obtain SWNTs with pure chirality. As part of their ongoing research, Diao *et al.* are creating a fundamental gel detachment strategy to increase semiconducting (12,1) and (11,3) SWNTs with indistinguishable resonance assimilation at 808 nm and emanation at 1200 nm. On a for every mass basis, chirality-organized SWNTs showed a 5-fold greater fluorescence output under total excitation at 808 nm than unsorted SWNTs, reducing the SWNT infusion dosage required for *in vivo* imaging by six times [46].

In a nutshell, NIR-II fluorescence imaging with SWNTs has shown tremendous promise in biological imaging, offering several intriguing advantages over previous imaging approaches. Continued progress on this route necessitates that better SWNT tests with high QY and sanitized chirality are required for further progress along this road, which will result in significantly more brilliant fluorescence. Furthermore, unique SWNTs with filtered single chirality would show isolated excitation and discharge frequencies, which might be relevant for multicolor NIR-II fluorescence imaging, which is still being researched. Other fluorescence enhancement methods can be combined, such as gold substrate-based surface resonance amplification of SWNT fluorescence [47].

9.2.2 Raman Imaging

Raman Scattering, like fluorescence, is a photon dissipating process rather than photoluminescence, as it involves the emission of photons with different frequencies in response to light stimulation. Atoms' inborn Raman dissipating indications are normally quite weak without the use of enhancement equipment such as SERS. The dispersing productivity rises when the laser intensity matches the energy required for the electron to migrate from the valence to the conduction groups, which is known as resonance Raman dissipating [48]. The extreme breathing mode (RBM, 100 cm1-300 cm1) and the digressive G band (1580 cm1), which correlate to the vibration of carbon particles in the extreme and extraneous directions, respectively, are found in SWNTs. The Raman resonance scattering is not fixed in stone by the DOS accessible for optical change (e.g., E11 and E22 advances), which are still up in the air due to their chirality files and breadths [49]. Among SWNTs with resonance E11, those with solid resonance Raman scattering would be useful for natural detecting and imaging.

For the first time, Heller *et al.* employed the Raman properties of SWNTs for live-cell imaging in 2005 [50]. SWNTs wrapped in DNA oligonucleotides were used as living cell identifiers in their study. The RBM apex of SWNTs was observed in live 3T3 cells under the activation of a 785-nm laser without any additional fluorescent names. In contrasting and natural colors, SWNTs demonstrated rather strong Raman dispersion signals with little to no critical extinguishing or photograph fading, as well as quantum dabs that could be progressively extinguished or image dyed over time. This allowed them to be used for long-term following in organic examples over months [51]. The inborn Raman G band of carbon nanotubes allowed Lamprecht *et al.* to direct and track the circulation of carbon nanotubes inside cancer cells [52]. Recently, Kang *et al.* used fast confocal Raman imaging to study the uptake of SWNTs by cells. In RAW 264, they created films with two cell-inborn and nine nanotube-determined Raman signals. In seven macrophages, we studied SWNTs of different sizes and aggregation phases, as well as their placement inside cells. Raman spectroscopy was shown to be an effective technique for atomic imaging of live cells in this study [53]. Using ligand-formed SWNTs as a focus, Raman imaging of tumors in mice has also been demonstrated [54, 55]. Raman scattering, in general, produces sharp tops with fluctuating regions for different atoms, making it ideal for multiplexed detection and imaging [56–58]. In the G band Raman pinnacle, the carbon iota mass does not entirely settle the occurrence of carbon-carbon bond vibration. Shifting the carbon isotope from 12C to 13C might change the G band pinnacles of SWNTs in a similar way [59]. Liu *et al.* were the first to discover multiplexed Raman imaging of live cells [60]. Three types of SWNTs with varying 13C doping fractions and therefore discrete Raman G-band tops were generated with different ligand concentrations to identify the respective cell lines. Unmixing and multiplexed Raman imaging were both acceptable over the range at the same time. By constructing SWNTs with varied atom focusing on the surface of the surface, this group was able to go a step further in understanding five-shading Raman imaging of cells, which could be applied both to *in vitro* multiplexed cell imaging as well as to *ex vivo* tissue labelling.

Despite the fact that SWNTs are likely to have the most grounded inherent Raman signals of any atom, the procurement time to obtain a suitable Raman image for natural instances marked with SWNTs is still lengthy (for a picture with 100100 pixels). Because the Raman signs of Raman labels do not totally determine the procurement time in Raman imaging, improving the Raman signs of those labels is the best strategy to reduce the imaging time.

By placing substantial metal nanoparticles on the nanotube surface [61–64], surface-enhanced Raman scattering (SERS) may fill in as a promising strategy for improving the Raman signals of SWNTs. There has been much discussion about the use of metal nanoparticles on SWNTs for SERS effect despite the fact that the recommendations have been widely publicized by various groups [65, 66]. In addition, gold nanoparticles have been attached to covalently functionalized SWNTs, which generally exhibit rather low Raman signals due to the modification of the nanotube structure during functionalization (for example, by oxidation) [67, 68], as well as improvements of metal-covered SWNTs with conductive coatings. The use of single-strand DNA to functionalize perfect SWNTs prevented solid Raman dissipation. The nanotubes were attached with contrarily charged gold Nano-seeds after being covered with forcefully charged poly(allyl amine hydrochloride) (PAH), resulting in the formation of gold or silver Nano-shells on the outer layer of SWNTs (Figures 9.2a–d). During the arrangement stage, the noble metal-covered SWNTs had significant SERS effects, with a maximal upgrade variable of more than 20 (Figures 9.2e–g). As a result of the uniquely improved Raman dispersion, gold-coated

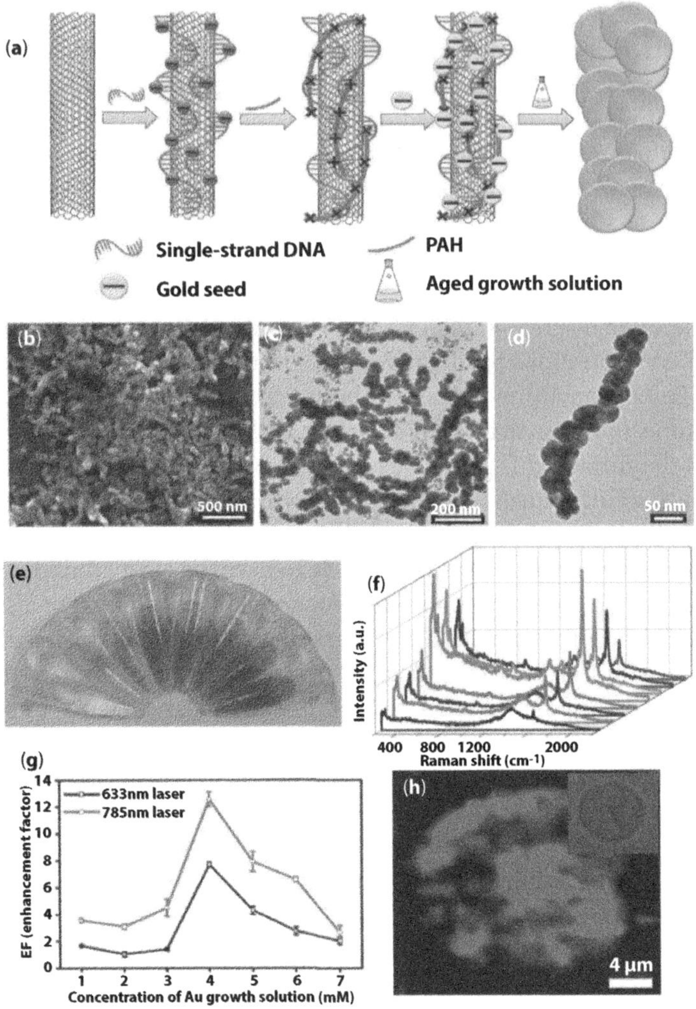

Figure 9.2 The SERS of SWNTs was improved by using noble metals. (a) A schematic representation of the SWNT–Au (or SWNT–Ag) nanocomposite fabrication process. A SEM picture of the SWNT–Au–PEG nanocomposite (b). TEM pictures of the SWNT–Au–PEG nanocomposite (c,d). (e) A shot of SWNT–Au–PEG solutions made by varying the growth solution concentrations. (f) Raman spectra of SWNT–Au–PEG samples prepared with various amounts of growth solutions. The spectra were collected using a laser with a wavelength of 785 nm. (g) Under the 633-nm and 785-nm excitations, enhanced factors of SWNT–Au–PEG as a function of increased gold growth solution concentrations. (h) High-resolution Raman picture of a KB cell tagged with SWNT–Au–PEG–FA. A bright field picture of this cell was included in the inset. The American Chemical Society owns the rights to this work [61].

SWNTs were created with folic corrosive for specific cell naming and Raman imaging. The imaging duration of the Raman images of organic cells was reduced by one important degree, allowing faster planning of organic instances (see Figure 9.2h).

SWNTs have a few advantages over other natural Raman hues as an original Raman label. First, SWNTs have a simple, restricted, and remarkable Raman peak, with a complete width at half-limit of less than 2 nm, making them very easy to distinguish from the auto fluorescent foundation.

SWNTs' Raman signals are also powerful and do not fade or die over time, enabling for long-term monitoring and imaging. Finally, the Raman movements may be changed by changing the isotope arrangement inside SWNTs, resulting in different "colors" and multiplexed Raman imaging. Lastly, SERS can be used to increase identification responsiveness and reduce imaging time by employing the resonance Raman signals of SWNTs.

9.2.3 Photoacoustic Imaging

Photoacoustic (PA) imaging is a recently developed imaging technique that is widely used in the biological area [69–71]. In the natural world, light-engrossing particles, such as endogenous atoms or differentiation specialists, produce heat, prompting transient thermoelastic development and thus prompting wideband ultrasonic emanation, which can be distinguished by an ultrasound amplifier to build 2D or 3D pictures, can retain the laser beats. By avoiding the absorption and dissipation of the outflow light in fluorescence imaging, PA imaging identifies sound rather than light and thus exhibits various evident benefits, such as vastly improved tissue infiltration and improved spatial objective. Different nanomaterials with strong absorbance in the near-infrared region are useful PA imaging differentiation specialists [72–74].

Due to their solid NIR absorbance [75–77], both MWNTs and SWNTs have been hired as photo thermal experts. Nanotubes, on the other hand, are perfect differentiation specialists for photoacoustic imaging due to their strong NIR absorbance. Avti *et al.* [78] used photoacoustic microscopy to detect, map, and quantify the SWNT follow sum (from nanograms to micrograms) in a variety of histological tissue samples for cell imaging. The results revealed that the commotion same location response was as low as seven pictograms, allowing for future tissue inspection applications. Gambhir *et al.*, for example, used RGD-formed SWNTs as PA contrast specialists *in vivo* PA imaging [79].

In their analysis, the development of the SWNT-RGD infused bunch showed robust PA signals, but the pure SWNT infused bunch showed very faint indicators. A gold layer or a few natural atoms with NIR absorbance were coupled with SWNTs to increase their absorbance in the NIR region and improve the responsiveness of the photoacoustic sign. To increase the inborn PA indications of SWNTs, Zharov and colleagues generated "golden nanotubes" (GNTs) by coating CNTs with gold [80]. The fragile coating of gold embedded on the outer layer of SWNTs improved their optical thickness in the NIR district, and the GNTs were then produced with antibodies to explicitly sense the endothelium of murine lymphatic arteries, according to their findings. Using incredibly low laser familiarity levels of a few mJ/cm^2, the obtained GNTs provided increased NIR PA imaging contrast (102 overlap) for focusing on lymphatic arteries in mice. The lymphatic endothelial receptor was also planned using neutralizer-formed GNTs. As the photoacoustic signal and the photograph warm sign in the counteracting agent formed gathering outperformed endogenous foundation and were specifically located in the dividers of lymphatic vessels, whereas the GNTs without formation with antibodies showed only irregular signs and no signs in the lymphatic mass of vessels. A subsequent study by the same group expanded on the use of GNTs to differentiate circular cancer cells (CTCs) under PA imaging. Folate-formed GNTs were used as the PA contrast specialist to image CTCs *in vivo* after being trapped by an outside attractive field, utilizing the solid PA indications of GNTs [81].

Gambhir *et al.* [82] used pi-pi stacking to stack indocyanine green (ICG) particles atop PEGylated SWNTs for color enhanced PA imaging using SWNTs, which increased the optical thickness of the Nano-test by 20 times at 780 nm. When compared to the awareness of plain SWNTs, SWNT-ICG showed sub-Nano molar recognition response in PA imaging (50 nM). They also produced multiplexed PA imaging tests based on a similar concept by layering five different NIR hues with moved absorbance tops atop PEGylated SWNTs (Figure 9.3a and b). SWNTs stacked with QSY21 (SWNT-QSY) or indocyanine green (SWNT-ICG) showed stable and isolated absorbance spectra, allowing them to be used *in vivo* for multicolor PA imaging with exceptional responsiveness (Figure 9.3c and d). As a result, CNTs with strong NIR absorbance are prospective photoacoustic imaging difference experts. CNTs might be used as a flexible Nano-stage by linking them with other light-retaining Nano-constructions or atoms for better or multiplexed photoacoustic imaging, in addition to utilising their distinctive optical absorbance. Although SWNTs are currently used in the

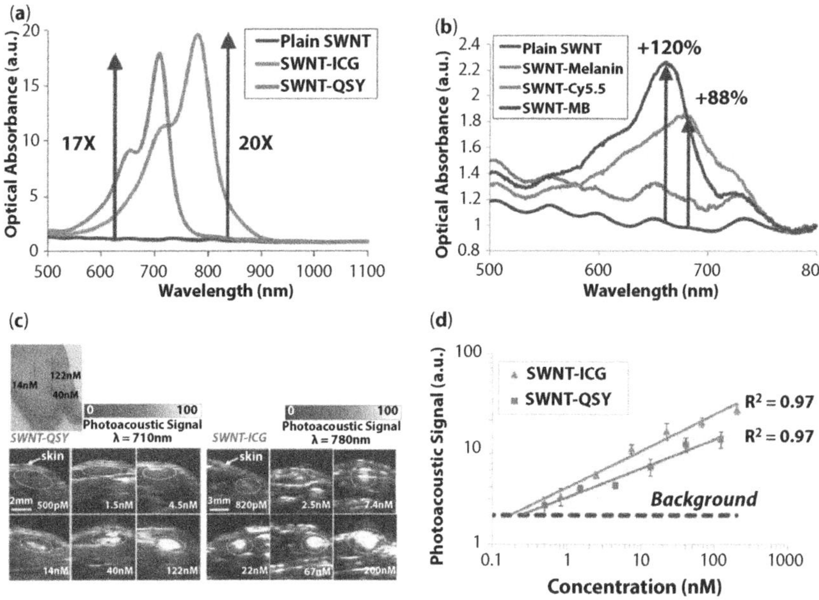

Figure 9.3 Photoacoustic imaging using multiplexing. SWNT–QSY (red) and SWNT–ICG (blue) optical absorbance spectra (green). At their peak absorption wavelengths of 710 and 780 nm, respectively, the QSY and ICG dye-enhanced SWNTs displayed 17- and 20-fold greater optical absorption than plain SWNTs. Plain SWNTs (black), SWNT-melanin (purple), SWNT-Cy5.5 (brown), and SWNT-MB optical absorbance spectra (blue). (c) PA imaging of mice injected subcutaneously with SWNT–QSY or SNWT–ICG at doses ranging from 0.5 to 122 nanometers. For SWNT–QSY and SWNT–ICG, PA pictures were taken at = 710 nm (red) and 780 nm (green) excitation, respectively. Ultrasound pictures (black/white) were used to view the skin and inclusion area. (d) PA signals from SWNT–ICG and SWNT–QSY *in vivo* at various SWNT doses. Standard error (n = 3 mice) was shown by the error bars. According to linear regression of the two PA signal curves (R2 = 0.97 for both particles), a concentration of 450 pM for SWNT–QSY or 170 pM for SWNT–ICG produced the equivalent background signal of tissues. The American Chemical Society owns the rights to this work [83].

majority of thorough CNT-based photoacoustic imaging studies, MWNTs should also be useful in this type of imaging technology.

9.2.4 Magnetic Resonance Imaging

Magnetic resonance imaging (MR imaging) is one of the most impressive and safe imaging modalities [84]. Throughout 60 million MR imaging cases are performed in centers around the world each year, with contrast specialists [85] being used in 30% of them. T1-shortening specialists (containing Gd3+, Mn2+, and so on) and T2 shortening specialists (containing Gd3+, Mn2+, and so on) are the most common types of experts (for example iron oxide nanoparticles). There has been a lot of research into the potential use of CNTs in MR imaging [86–100].

Sitharaman *et al.* originally discovered the nanoscale stacking and suppression of aquated Gd3+-particle groups within super short SWNTs (US-tubes) for T1-weighted MR imaging, resulting in Gd3+@US-tube with solid T1-contrast useful in MR imaging [94]. Richard *et al.* discovered a noncovalent technique for forming MWNTs from amphiphilic gadolinium (III) chelate (GdL), which makes them useful for T1-contrast in MR imaging [95].

Strano and colleagues previously shown that SWNTs with iron oxide nanoparticles attached at the nanotube closures may be used as T2 contrast specialists for MR imaging without the need for further naming [96]. A solvothermal method was used by Wu *et al.* to merge MWNTs with cobalt ferrite. In T2-weighted MR imaging of disease cells, the acquired nanocomposite revealed a significant negative differential upgrade.

During the amalgamation of CNTs, metal catalysts (such as Fe, Co) are commonly used. Faraj *et al.* [96] considered the bio dispersion of SWNTs by evaluating MR signals *in vivo*, using the pollutions of iron present in SWNTs. Indeed, the improved SWNTs with a follow measure of iron material have been recognized as nonetheless compelling to give robust T2-contrast in MR imaging, even after cleaning to remove most of the impulse metal nanoparticles in the nanotube test [101].

Immature microorganisms have recently attracted a lot of attention for their potential in regenerative medicine [102, 103]. As a result, delicate and reliable procedures for fundamental microbe identification and *in vivo* tracking are essential. Carbon nanotubes have also been used by a few groups for designating and tracking immature cells. According to Vittorio *et al.*, MWNT-labeled mesenchymal foundational microorganisms (MSC) could be adequately tracked after being administered to animals using T2-weighted MR imaging [104, 105]. The presence of SWCNTs in systemic organs, a gradient-echo imaging sequence sensitive to changes of relaxation time values was used [104]. CNRs measured in regions of interest, such as the liver, spleen, and kidneys, were chosen as an index of SWCNT deposition (for systemic proton MRI protocol, see Supporting Information). ROIs were made around visible vascular structures in the liver, and these regions were removed from the map to leave only liver parenchyma. For signal measurement, ROIs comprising the entire spleen and both kidneys were manually selected. Instilled groups showed no difference in axial proton imaging of the various organs of interest (kidneys, spleen, and liver) in systemic MRI. CNR readings in raw- and SP-SWCNT-exposed rats were statistically identical during the course of the 2-week study. The CNR levels were comparable to those seen in the control group. Up to 24 hours after injection, the raw-SWCNT injected group showed a 2- to 3-fold drop in CNR (corresponding to proton NMR signal attenuation) in the spleen and kidneys. This impact fades over time, and two weeks after instillation, CNR levels are equivalent to control groups (Figure 9.4). CNR levels had not changed statistically.

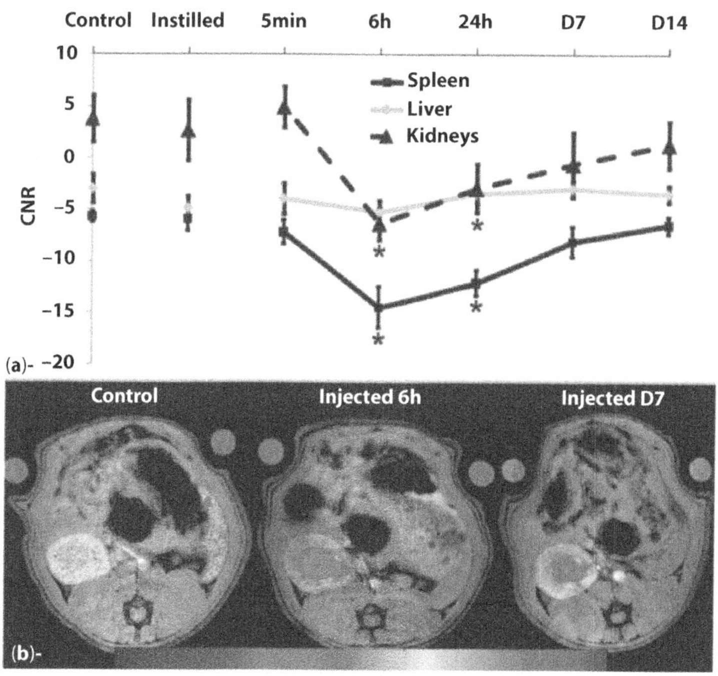

Figure 9.4 (a) Graph showing CNR variation of control, instilled (averaged over the 2-week investigation) and injected animals 6 h, 24 h, 1 week, and 2 weeks after raw-SWCNT injection in spleen, liver, and kidneys. Asterisks indicate statistically different values from the control group ($p < 0.05$)> (b) Proton MR images of an injected rat: (i) before, (ii) 6 h after, and (iii) 7 days after injection showing signal variation in the spleen and kidneys. The superimposed signal intensity color maps over kidneys and spleen indicate the MR signal intensity drops post injection. The American Chemical Society owns the rights to this work [104].

9.2.5 Nuclear Imaging

Outside markings, such as radioisotopes, can also be used to increase the versatility of SWNT-based imaging studies, in addition to embracing the natural properties of SWNTs. Wang *et al.* discovered that the 125I may be used to track the bio distribution of SWNTs in living things [105]. The same group that had been following the long-term bio distribution of MWNTs had also figured out a 14C-based alternative to 125I [106]. By labelling CNTs with 111In, the methodology of tiny single-photon Emission computerized scanning tomography (miniature SPECT) was successfully acquired [107]. As a result, McDevitt and colleagues used 86Y to label SWNTs for Positron Emission Tomography (PET) imaging [108]. Several studies have used radionuclide-labelled SWNTs for *in vivo* cancer imaging in mice models [98, 99]. Through a 1,4,7,10-tetraazacyclododecane-1,4,7,10-tetraacetic corrosive (DOTA) chelator, Liu *et al.* [109] used 64Cu to label PEGylated SWNTs with RGD peptide synthesis in 2007. The specific restriction between RGD peptide and integrin av3 overexpressed on U87MG growth cells as well as growth vasculatures was recognized *in vivo* growth designated PET imaging of U87MG tumors using 64Cu/DOTA-named SWNT-PEG-RGD. In addition to the RGD peptide, hostile to CD20 antibodies were generated using 111In-named SWNTs and utilized to target human Burkitt lymphoma [100]. Radioisotopes might be incorporated into nanotubes for radiolabeling instead of using traditional chelation technology to obtain radiolabeled CNTs [101].

A 125I-marked SWNT can be imaged using SPECT/CT after Na125I is inserted into the vacant design of SWNTs, according to Hong *et al.* The advantages of radio-marked SWNTs over other methods are that they have no depth limitation in tissue infiltration and are highly responsive. The use of CNTs for nuclear imaging has slowed down considerably in the last couple of years, with little significant progress, according to our best knowledge. Overall, combining nuclear imaging with other imaging processes might be extremely beneficial to the advancement of CNT-based bio-imaging assays.

9.3 Prospects and Challenges

Different imaging methods for SWNTs have their own set of advantages and disadvantages. Because of the low native tissue dissipation and much-reduced photon scattering in the NIR-II district compared to the apparent (400-700 nm) and traditional NIR-I region, fluorescence imaging in the NIR-II natural, straightforward window with SWNTs enables *in vivo* optical imaging with profound tissue entrance and high spatial goal (700-900 nm). Despite this, the QY of SWNTs is still insufficient. The goal of this project is to develop new approaches to NIR-II imaging using SWNTs with multi-shaded fluorescence, which will require the development of simple union and filtration techniques to get SWNTs with pure chirality. Solid resonance Raman scattering is seen in SWNTs, which have a large dissipating cross-segment. Raman imaging using SWNT-tests has a high sensitivity and is unaffected by image fading and quenching.

Noble metal-covered SWNTs display also enhanced Raman signals as a result of the SERS effect, allowing for much faster Raman planning. In any case, even with the SERS augmentation, acquiring a Raman picture using standard Raman planning takes roughly 10 minutes. SWNT-based Raman imaging may become a feasible technique for genuine natural exploration as technology improves. In addition, mass-based Raman signals of SWNTs will be greatly enhanced when chirality-cleaned SWNTs are used with the majority of nanotubes.

Other CNT-based imaging approaches have shown considerable potential in biological imaging, such as PA imaging, MR imaging, and atomic imaging. Coordination of many imaging modalities might boost each imaging technology's strengths while overcoming its inherent limits. Carbon nanotubes' multifunctionality is the actual benefit of employing them as an imaging test. As opposed to many other nanomaterials used in bioimaging, CNTs possess increased optical and appealing properties, which eliminate the need to design complex nanostructures in order to accomplish diverse capabilities. Furthermore, the advantages of CNTs (such as medication and quality delivery, photograph warm therapy) [19, 76] combined with imaging would make them excellent theranostic Nano-stages.

A wide range of animal diseases has been linked to CNTs without adequate surface functionalization. In addition to the limitations of each approach, one important test for the therapeutic use of CNTs is their long-term toxicity [102, 103]. CNTs without adequate surface functionalization have been known to cause poisoning *in vivo* [104–106]. Researchers have found that CNTs with well-planned surface coatings, such as PEGylation, are not toxic *in vitro* or *in vivo* [51, 79, 107–111] and are excreted from animals via renal or perhaps facial pathways [112–115]. In a number of studies, CNTs have been claimed to degrade when chemically oxidized [116, 117]. In any event, it is understandable that, while CNT-based biomedical imaging has enormous promise in basic biomedical research as well as *in vitro*/*ex vivo* clinical diagnosis, there will be significant challenges in using these nanomaterials in real-world clinical applications.

9.4 Conclusion

Several modalities of biomedical imaging have been outlined in this chapter. They offer versatile means of tracking, detecting, and imaging biological systems due to their intrinsic physical properties including resonance Raman scattering, NIR photoluminescence, and strong NIR optical absorption, along with various external organic dyes, radionucleotides, and magnetic labels. However, CNT-based biomedical imaging still has many challenges to overcome before it can be widely used for clinical imaging, but these nanomaterials hold great promise in the biomedical field, and may promise novel tools for molecular imaging in biological systems.

References

1. Ago, H., Petritsch, K., Shaffer, M.S., Windle, A.H., Friend, R.H., Composites of carbon nanotubes and conjugated polymers for photovoltaic devices. *Adv. Mater.*, 11, 1281–1285, 1999.
2. Ouyang, M., Huang, J.L., Lieber, C.M., Fundamental electronic properties and applications of single-walled carbon nanotubes. *Acc. Chem. Res.*, 35, 1018–1025, 2002.
3. Lee, N.S., Chung, D.S., Han, I.T., Kang, J.H., Choi, Y.S., Kim, H.Y., Park, S.H., Jin, Y.W., Yi, W.K., Yun, M.J., Jung, J.E., Lee, C.J., You, J.H., Jo, S.H., Lee, C.G., Kim, J.M., Application of carbon nanotubes to field emission displays. *Diam. Relat. Mater.*, 10, 265–270, 2001.
4. Rowell, M.W., Topinka, M.A., McGehee, M.D., Prall, H.-J., Dennler, G., Sariciftci, N.S., Hu, L., Gruner, G., Organic solar cells with carbon nanotube network electrodes. *Appl. Phys. Lett.*, 88, 233506-1–233506-3, 2006.
5. Besteman, K., Lee, J.O., Wiertz, F.G.M., Heering, H.A., Dekker, C., Enzyme-coated carbon nanotubes as single-molecule biosensors. *Nano Lett.*, 3, 727–730, 2003.
6. Wildoer, J.W.G., Venema, L.C., Rinzler, A.G., Smalley, R.E., Dekker, C., Electronic structure of atomically resolved carbon nanotubes. *Nature*, 391, 59–62, 1998.
7. Wu, H.-C., Chang, X., Liu, L., Zhao, F., Zhao, Y., Chemistry of carbon nanotubes in biomedical applications. *J. Mater. Chem.*, 20, 1036–1052, 2010.

8. De La Zerda, A., Zavaleta, C., Keren, S., Vaithilingam, S., Bodapati, S., Liu, Z., Levi, J., Smith, B.R., Ma, T.-J., Oralkan, O., Cheng, Z., Chen, X., Dai, H., Khuri-Yakub, B.T., Gambhir, S.S., Carbon nanotubes as photoacoustic molecular imaging agents in living mice. *Nat. Nanotechnol.*, 3, 557–562, 2008.
9. Liu, Z., Tabakman, S., Welsher, K., Dai, H., Carbon nanotubes in biology andmedicine: *In vitro* and *in vivo* detection, imaging and drug delivery. *Nano Res.*, 2, 85–120, 2009.
10. Wang, C., Ma, X., Ye, S., Cheng, L., Yang, K., Guo, L., Li, C., Li, Y., Liu, Z., Protamine functionalized single-walled carbon nanotubes for stem cell labeling and *in vivo* Raman/magnetic resonance/photoacoustic triple-modal imaging. *Adv. Funct. Mater.*, 22, 2363–2375, 2012.
11. Hong, S.Y., Tobias, G., Al-Jamal, K.T., Ballesteros, B., Ali-Boucetta, H., Lozano-Perez, S., Nellist, P.D., Sim, R.B., Finucane, C., Mather, S.J., Green, M.L.H., Kostarelos, K., Davis, B.G., Filled and glycosylated carbon nanotubes for *in vivo* radioemitter localization and imaging. *Nat. Mater.*, 9, 485–490, 2010.
12. Welsher, K., Liu, Z., Sherlock, S.P., Robinson, J.T., Chen, Z., Daranciang, D., Dai, H., A route to brightly fluorescent carbon nanotubes for near-infrared imaging in mice. *Nat. Nanotechnol.*, 4, 773–780, 2009.
13. Welsher, K., Sherlock, S.P., Dai, H., Deep-tissue anatomical imaging of mice using carbon nanotube fluorophores in the second near-infrared window. *Proc. Natl. Acad. Sci. U.S.A.*, 108, 8943–8948, 2011.
14. Hong, G., Shuo, D., Antaris, A.L., Chen, C., Zhang, B., Zhao, S., Atochin, D.N., Huang, P.L., Andreasson, K.I., Kuo, C.J. *et al.*, Through-skull fluorescence imaging of the brain in a new near-infrared window. *Nat. Photonics*, 8, 723–730, 2014.
15. Pan, J., Li, F., Choi, J.H., Single-walled carbon nanotubes as optical probes for bio-sensing and imaging. *J. Mater. Chem. B*, 5, 6511–6522, 2017.
16. Yudasaka, M., Yomogida, Y., Zhang, M., Tanaka, T., Nakahara, M., Kobayashi, N., Okamatsu-Ogura, Y., Machida, K., Ishihara, K., Saeki, K. *et al.*, Near-infrared photoluminescent carbon nanotubes for imaging of brown fat. *Sci. Rep.*, 7, 6272, 2017.
17. De La Zerda, A., Zavaleta, C., Keren, S., Vaithilingam, S., Bodapati, S., Liu, Z., Levi, J., Smith, B.R., Ma, T.-J., Oralkan, O. *et al.*, Carbon nanotubes as photoacoustic molecular imaging agents in living mice. *Nat. Nanotechnol.*, 3, 557–562, 2008.
18. De La Zerda, A., Liu, Z., Bodapati, S., Teed, R., Vaithilingam, S., Khuri-Yakub, B.T., Chen, X., Dai, H., Gambhir, S.S., Ultrahigh sensitivity carbon nanotube agents for photoacoustic molecular imaging in living mice. *Nano Lett.*, 10, 2168–2172, 2010.
19. Liu, Z., Tabakman, S., Sherlock, S., Li, X., Chen, Z., Jiang, K., Fan, S., Dai, H., Multiplexed five-color molecular imaging of cancer cells and tumor tissues

with carbon nanotube Raman tags in the near-infrared. *Nano Res.*, 3, 222–233, 2010.

20. Gaufrès, E., Marcet, S., Aymong, V., Tang, N.Y.-W., Favron, A., Thouin, F., Allard, C., Rioux, D., Cottenye, N., Verhaegen, M. *et al.*, Hyperspectral Raman imaging using bragg tunable filters of graphene and other low-dimensional materials: Hyperspectral Raman imaging using bragg tunable filters of graphene and other low-dimensional materials. *J. Raman Spectrosc.*, 49, 174–182, 2018.
21. Hong, S.Y., Tobias, G., Al-Jamal, K.T., Ballesteros, B., Ali-Boucetta, H., Lozano-Perez, S., Nellist, P.D., Sim, R.B., Finucane, C., Mather, S.J. *et al.*, Filled and glycosylated carbon nanotubes for *in vivo* radioemitter localization and imaging. *Nat. Mater.*, 9, 485–490, 2010.
22. Servant, A., Jacobs, I., Bussy, C., Fabbro, C., da Ros, T., Pach, E., Ballesteros, B., Prato, M., Nicolay, K., Kostarelos, K., Gadolinium-functionalised multi-walled carbon nanotubes as a t 1 contrast agent for MRI cell labelling and tracking. *Carbon*, 97, 126–133, 2016.
23. He, X., Gao, J., Gambhir, S.S., Cheng, Z., Near-infraredfluorescent nano-probes for cancer molecular imaging: Status and challenges. *Trends Mol. Med.*, 16, 574–583, 2010.
24. Chance, B., Near-infrared images using continuous, phase-modulated, and pulsed light with quantitation of blood and blood oxygenation, in: *Advances in Optical Biopsy and Optical Mammography*, Alfano, R.R. (Ed.), pp. 29–45, 1998.
25. Ishizawa, T., Fukushima, N., Shibahara, J., Masuda, K., Tamura, S., Aoki, T., Hasegawa, K., Beck, Y., Fukayama, M., Kokudo, N., Real-time identification of liver cancers by using indocyanine greenfluorescent imaging. *Cancer*, 115, 2491–2504, 2009.
26. Sekijima, M., Tojimbara, T., Sato, S., Nakamura, M., Kawase, T., Kai, K., Urashima, Y., Nakajima, I., Fuchinoue, S., Teraoka, S., An intraoperative-fluorescent imaging system in organ transplantation. *Transplant. Proc.*, 36, 2188–2190, 2004.
27. Morimoto, S., *In-vivo* imaging of tumors with protease activated near-infrared fluorescent probes. *Tanpakushitsu Kakusan Koso*, 52, 1774–1775, 2007.
28. Kim, S., Lim, Y.T., Soltesz, E.G., De Grand, A.M., Lee, J., Nakayama, A., Parker, J.A., Mihaljevic, T., Laurence, R.G., Dor, D.M., Cohn, L.H., Bawendi, M.G., Frangioni, J.V., Near-infrared fluorescent type II quantum dots for sentinel lymph node mapping. *Nat. Biotechnol.*, 22, 93–97, 2004.
29. Iizumi, Y., Okazaki, T., Ikehara, Y., Ogura, M., Fukata, S., Yudasaka, M., Immunoassay with single-walled carbon nanotubes as near-infrared fluorescent labels. *ACS Appl. Mater. Interfaces*, 5, 16, 7665–7670, 2013.
30. Lim, Y.T., Kim, S., Nakayama, A., Stott, N.E., Bawendi, M.G., Frangioni, J.V., Selection of quantum dot wavelengths for biomedical assays and imaging. *Mol. Imaging*, 2, 50–64, 2003.

31. Diao, S., Hong, G., Robinson, J.T., Jiao, L., Antaris, A.L., Wu, J.Z., Choi, C.L., Dai, H., Chirality enriched (12, 1) and (11, 3) single-walled carbon nanotubes for biological imaging. *J. Am. Chem. Soc.*, 134, 16971–16974, 2012.
32. Yi, H., Ghosh, D., Ham, M.-H., Qi, J., Barone, P.W., Strano, M.S., Belcher, A.M., M13 phage-functionalized single-walled carbon nanotubes as nanoprobes for second near-infrared window fluorescence imaging of targeted tumors. *Nano Lett.*, 12, 1176–1183, 2012.
33. Welsher, K., Sherlock, S.P., Dai, H., Deep-tissue anatomical imaging of mice using carbon nanotube fluorophores in the second near-infrared window. *Proc. Natl. Acad. Sci.*, 108, 8943–8948, 2011.
34. Robinson, J.T., Hong, G., Liang, Y., Zhang, B., Yaghi, O.K., Dai, H., *In vivo* fluorescence imaging in the second near-infrared window with long circulating carbon nanotubes capable of ultrahigh tumor uptake. *J. Am. Chem. Soc.*, 134, 10664–0669, 2012.
35. Kuzmany, H., Burger, B., Fally, M., Rinzler, A.G., Smalley, R.E., Effect of dimensionality in polymeric fullerenes and single-wall nanotubes. *Physica B*, 244, 186–191, 1998.
36. Jorio, A., Pimenta, M., Filho, A.S., Saito, R., Dresselhaus, G., Dresselhaus, M., Characterizing carbon nanotube samples with resonance Raman scattering. *New J. Phys.*, 5, 139.1–139.17, 2003.
37. Hong, G., Robinson, J.T., Cooke, J.P., Dai, H., Raaz, U., Xie, L., Huang, N.F., Lee, J.C., Multifunctional *in vivo* vascular imaging using near-infrared II fluorescence. *Nat. Med.*, 18, 1841–1846, 2012.
38. Welsher, K., Liu, Z., Sherlock, S.P., Robinson, J.T., Chen, Z., Daranciang, D., Dai, H., A route to brightlyfluorescent carbon nanotubes for near-infrared imaging in mice. *Nat. Nanotechnol.*, 4, 773–780, 2009.
39. Welsher, K., Sherlock, S.P., Dai, H., Deep-tissue anatomical imaging of mice using carbon nanotubefluorophores in the second near-infrared window. *Proc. Natl. Acad. Sci.*, 108, 8943–8948, 2011.
40. Robinson, J.T., Hong, G., Liang, Y., Zhang, B., Yaghi, O.K., Dai, H., *In vivo* fluorescence imaging in the second near-infrared window with long circulating carbon nanotubes capable of ultrahigh tumor uptake. *J. Am. Chem. Soc.*, 134, 10664–10669, 2012.
41. Ghosh, S., Bachilo, S.M., Weisman, R.B., Advanced sorting of single-walled carbon nanotubes by nonlinear density-gradient ultracentrifugation. *Nat. Nanotechnol.*, 5, 443–450, 2010.
42. Tu, X., Manohar, S., Jagota, A., Zheng, M., DNA sequence motifs for structure-specific recognition and separation of carbon nanotubes. *Nature*, 460, 250–253, 2009.
43. Liu, H., Nishide, D., Tanaka, T., Kataura, H., Large-scale single-chirality separation of single-wall carbon nanotubes by simple gel chromatography. *Nat. Commun.*, 2, 1–8, 2011.

44. Miyata, Y., Shiozawa, K., Asada, Y., Ohno, Y., Kitaura, R., Mizutani, T., Shinohara, H., Length-sorted semiconducting carbon nanotubes for high-mobility thinfilm transistors. *Nano Res.*, 4, 963–970, 2011.
45. Moshammer, K., Hennrich, F., Kappes, M.M., Selective suspension in aqueous sodium dodecyl sulfate according to electronic structure type allows simple separation of metallic from semiconducting single-walled carbon nanotubes. *Nano Res.*, 2, 599–606, 2009.
46. Diao, S., Hong, G., Robinson, J.T., Jiao, L., Antaris, A.L., Wu, J.Z., Choi, C.L., Dai, H., Chirality enriched (12,1) and (11,3) single-walled carbon nanotubes for biological imaging. *J. Am. Chem. Soc.*, 134, 16971–16974, 2012.
47. Hong, G., Tabakman, S.M., Welsher, K., Wang, H., Wang, X., Dai, H., Metal-enhanced fluorescence of carbon nanotubes. *J. Am. Chem. Soc.*, 132, 15920–15923, 2010.
48. Martin, R. and Falicov, L., Resonance Raman scattering, in: *Lightscattering in Solids I*, pp. 79–145, Springer, Berlin, Heidelberg, 1983.
49. Rao, A., Richter, E., Bandow, S., Chase, B., Eklund, P., Williams, K., Fang, S., Subbaswamy, K., Menon, M., Thess, A., Diameter-selective Raman scattering from vibrational modes in carbon nanotubes. *Science*, 275, 187–191, 1997.
50. Heller, D.A., Baik, S., Eurell, T.E., Strano, M.S., Single-walled carbon nanotube spectroscopy in live cells: Towards long-term labels and optical sensors. *Adv. Mater.*, 17, 2793–2799, 2005.
51. Liu, Z., Davis, C., Cai, W., He, L., Chen, X., Dai, H., Circulation and long-term fate of functionalized, biocompatible single-walled carbon nanotubes in mice probed by Raman spectroscopy. *Proc. Natl. Acad. Sci.*, 105, 1410–1415, 2008.
52. Lamprecht, C., Gierlinger, N., Heister, E., Unterauer, B., Plochberger, B., Brameshuber, M., Hinterdorfer, P., Hild, S., Ebner, A., Mapping the intracellular distribution of carbon nanotubes after targeted delivery to carcinoma cells using confocal Raman imaging as a label-free technique. *J. Phys. Condens. Matter*, 24, 164206–164216, 2012.
53. Heller, D., Kang, J.W., Lue, N., Dasari, R.R., Measuring uptake dynamics of multiple, identifiable carbon nanotube species via high-speed confocal Raman imaging of live cells. *223rd ECS Meeting*, May 12–17, 2013, ECS, 2013.
54. Zavaleta, C., De La Zerda, A., Liu, Z., Keren, S., Cheng, Z., Schipper, M., Chen, X., Dai, H., Gambhir, S.S., Noninvasive Raman spectroscopy in living mice for evaluation of tumortargeting with carbon nanotubes. *Nano Lett.*, 8, 2800–2805, 2008.
55. Smith, B.R., Zavaleta, C., Rosenberg, J., Tong, R., Ramunas, J., Liu, Z., Dai, H., Gambhir, S.S., High-resolution, serial intravital microscopic imaging of nanoparticle delivery and targeting in a small animal tumor model. *Nano Today*, 126–137, 2013.

56. Wagnieres, G.A., Star, W.M., Wilson, B.C., *In vivo* fluorescence spectroscopy and imaging for oncological applications. *Photochem. Photobiol.*, 68, 603–632, 1998.
57. Cao, Y.C., Jin, R., Mirkin, C.A., Nanoparticles with Raman spectroscopic fingerprints for DNA and RNA detection. *Science*, 297, 1536–1540, 2002.
58. Keren, S., Zavaleta, C., Cheng, Z., de La Zerda, A., Gheysens, O., Gambhir, S., Noninvasive molecular imaging of small living subjects using Raman spectroscopy. *Proc. Natl. Acad. Sci.*, 105, 5844–5849, 2008.
59. Liu, Z., Tabakman, S., Sherlock, S., Li, X., Chen, Z., Jiang, K., Fan, S., Dai, H., Multiplexed five-color molecular imaging of cancer cells and tumor tissues with carbon nanotube Raman tags in the near-infrared. *Nano Res.*, 3, 222–233, 2010.
60. Liu, Z., Li, X., Tabakman, S.M., Jiang, K., Fan, S., Dai, H., Multiplexed multicolor Raman imaging of live cells with isotopically modified single walled carbon nanotubes. *J. Am. Chem. Soc.*, 130, 13540–13541, 2008.
61. Wang, X., Wang, C., Cheng, L., Lee, S.-T., Liu, Z., Noble metal coated single-walled carbon nanotubes for applications in surface enhanced Raman scattering imaging and photothermal therapy. *J. Am. Chem. Soc.*, 134, 7414–7422, 2012.
62. Takagi, D., Homma, Y., Hibino, H., Suzuki, S., Kobayashi, Y., Single-walled carbon nanotube growth from highly activated metal nanoparticles. *Nano Lett.*, 6, 2642–2645, 2006.
63. Wang, D., Li, Z.-C., Chen, L., Templated synthesis of single-walled carbon nanotube and metal nanoparticle assemblies in solution. *J. Am. Chem. Soc.*, 128, 15078–15079, 2006.
64. Chen, Y.C., Young, R.J., Macpherson, J.V., Wilson, N.R., Silver-decorated carbon nanotube networks as SERS substrates. *J. Raman Spectrosc.*, 42, 1255–1262, 2011.
65. Chu, H., Wang, J., Ding, L., Yuan, D., Zhang, Y., Liu, J., Li, Y., Decoration of gold nanoparticles on surface-grown single-walled carbon nanotubes for detection of every nanotube by surface-enhanced Raman spectroscopy. *J. Am. Chem. Soc.*, 131, 14310–14316, 2009.
66. Chen, Z., Tabakman, S.M., Goodwin, A.P., Kattah, M.G., Daranciang, D., Wang, X., Zhang, G., Li, X., Liu, Z., Utz, P.J., Protein microarrays with carbon nanotubes as multicolor Raman labels. *Nat. Biotechnol.*, 26, 1285–1292, 2008.
67. Beqa, L., Fan, Z., Singh, A.K., Senapati, D., Ray, P.C., Gold nano-popcorn attached SWCNT hybrid nanomaterial for targeted diagnosis and photothermal therapy of human breast cancer cells. *ACS Appl. Mater. Interfaces*, 3, 3316–3324, 2011.
68. Beqa, L., Singh, A.K., Fan, Z., Senapati, D., Ray, P.C., Chemically attached gold nanoparticle–carbon nanotube hybrids for highly sensitive SERS substrate. *Chem. Phys. Lett.*, 512, 237–242, 2011.

69. Wang, X., Pang, Y., Ku, G., Xie, X., Stoica, G., Wang, L.V., Noninvasive laser-induced photoacoustic tomography for structural and functional *in vivo* imaging of the brain. *Nat. Biotechnol.*, 21, 803–806, 2003.
70. Ku, G. and Wang, L.V., Deeply penetrating photoacoustic tomography in biological tissues enhanced with an optical contrast agent. *Opt. Lett.*, 30, 507–509, 2005.
71. Hoelen, C., De Mul, F., Pongers, R., Dekker, A., Three-dimensional photoacoustic imaging of blood vessels in tissue. *Opt. Lett.*, 23, 648–650, 1998.
72. Yang, K., Hu, L., Ma, X., Ye, S., Cheng, L., Shi, X., Li, C., Li, Y., Liu, Z., Multimodal imaging guided photothermal therapy using functionalized graphene nanosheets anchored with magnetic nanoparticles. *Adv. Mater.*, 24, 1868–1872, 2012.
73. Ku, G., Zhou, M., Song, S., Huang, Q., Hazle, J., Li, C., Copper sulfide nanoparticles asa new class of photoacoustic contrast agent for deep tissue imaging at 1064nm. *ACS Nano*, 6, 7489–7496, 2012.
74. Agarwal, A., Huang, S., Donnell, M.O., Day, K., Day, M., Kotov, N., Ashkenazi, S., Targeted gold nanorod contrast agent for prostate cancer detection by photoacoustic imaging. *J. Appl. Phys.*, 102, 064701–064704, 2007.
75. Robinson, J.T., Welsher, K., Tabakman, S.M., Sherlock, S.P., Wang, H., Luong, R., Dai, H., High performance *in vivo* near-IR(N1μm) imaging and photothermal cancer therapy with carbon nanotubes. *Nano Res.*, 3, 779–793, 2010.
76. Liu, X., Tao, H., Yang, K., Zhang, S., Lee, S.-T., Liu, Z., Optimization of surface chemistry on single-walled carbon nanotubes for *in vivo* photothermal ablation of tumors. *Biomaterials*, 321, 44–151, 2011.
77. Moon, H.K., Lee, S.H., Choi, H.C., *In vivo* near-infrared mediated tumor destruction by photothermal effect of carbon nanotubes. *ACS Nano*, 3, 3707–3713, 2009.
78. Avti, P.K., Hu, S., Favazza, C., Mikos, A.G., Jansen, J.A., Shroyer, K.R., Wang, L.V., Sitharaman, B., Detection, mapping, and quantification of single walled carbon nanotubes in histological specimens with photoacoustic microscopy. *PloS One*, 7, e35064, 2012.
79. Wang, X., Pang, Y., Ku, G., Xie, X., Stoica, G., Wang, L.V., Noninvasive laser-induced photoacoustic tomography for structural and functional *in vivo* imaging of the brain. *Nat. Biotechnol.*, 21, 803–806, 2003.
80. Hoelen, C., De Mul, F., Pongers, R., Dekker, A., Three-dimensional photoacoustic imaging of blood vessels in tissue. *Opt. Lett.*, 23, 648–650, 1998.
81. Ku, G. and Wang, L.V., Deeply penetrating photoacoustic tomography in biological tissues enhanced with an optical contrast agent. *Opt. Lett.*, 30, 507–509, 2005.
82. Welsher, K., Liu, Z., Daranciang, D., Dai, H., Selective probing and imaging of cells with single walled carbon nanotubes as near-infraredfluorescent molecules. *Nano Lett.*, 8, 586–590, 2008.
83. Hoelen, C., De Mul, F., Pongers, R., Dekker, A., Three-dimensional photoacoustic imaging of blood vessels in tissue. *Opt. Lett.*, 23, 648–650, 1998.

84. Yang, K., Hu, L., Ma, X., Ye, S., Cheng, L., Shi, X., Li, C., Li, Y., Liu, Z., Multimodal imaging guided photothermal therapy using functionalized graphene nanosheets anchored with magnetic nanoparticles. *Adv. Mater.*, 24, 1868–1872, 2012.
85. Ku, G., Zhou, M., Song, S., Huang, Q., Hazle, J., Li, C., Copper sulfide nanoparticles asa new class of photoacoustic contrast agent for deep tissue imaging at 1064nm. *ACS Nano*, 6, 7489–7496, 2012.
86. Agarwal, A., Huang, S., O'Donnell, M., Day, K., Day, M., Kotov, N., Ashkenazi, S., Targeted gold nanorod contrast agent for prostate cancer detection by photoacoustic imaging. *J. Appl. Phys.*, 102, 064701–064704, 2007.
87. Robinson, J.T., Welsher, K., Tabakman, S.M., Sherlock, S.P., Wang, H., Luong, R., Dai, H., High performance *in vivo* near-IR(N1μm) imaging and photothermal cancer therapy with carbon nanotubes. *Nano Res.*, 3, 779–793, 2010.
88. Liu, X., Tao, H., Yang, K., Zhang, S., Lee, S.-T., Liu, Z., Optimization of surface chemistry on single-walled carbon nanotubes for *in vivo* photothermal ablation of tumors. *Biomaterials*, 32, 144–151, 2011.
89. Moon, H.K., Lee, S.H., Choi, H.C., *In vivo* near-infrared mediated tumor destruction by photo thermal effect of carbon nanotubes. *ACS Nano*, 3, 3707–3713, 2009.
90. Avti, P.K., Hu, S., Favazza, C., Mikos, A.G., Jansen, J.A., Shroyer, K.R., Wang, L.V., Sitharaman, B., Detection, mapping, and quantification of single-walled carbon nanotubes in histological specimens with photoacoustic microscopy. *PloS One*, 7, e35064, 2012.
91. Galanzha, E.I., Shashkov, E.V., Kelly, T., Kim, J.-W., Yang, L., Zharov, V.P., *In vivo* magnetic enrichment and multiplex photoacoustic detection of circulating tumour cells. *Nat. Nanotechnol.*, 4, 855–860, 2009.
92. de la Zerda, A., Bodapati, S., Teed, R., May, S.Y., Tabakman, S.M., Liu, Z., Khuri-Yakub, B.T., Chen, X., Dai, H., Gambhir, S.S., Family of enhanced photoacoustic imaging agents for high-sensitivity and multiplexing studies in living mice. *ACS Nano*, 6, 4694–4701, 2012.
93. Merbach, A.E. and Tóth, É., *The Chemistry of Contrast Agents in Medical Magnetic Resonance Imaging*, Wiley, Chichester (W. Sx.) etc, 2001.
94. Sitharaman, B., Kissell, K.R., Hartman, K.B., Tran, L.A., Baikalov, A., Rusakova, I., Sun, Y., Khant, H.A., Ludtke, S.J., Chiu, W., Super paramagnetic gadonanotubes are high performance MRI contrast agents. *Chem. Commun.*, 3915-3917, 2005.
95. Al Faraj, A., Cieslar, K., Lacroix, G., Gaillard, S., Canot-Soulas, E., Cremillieux, Y., *In vivo* imaging of carbon nanotube biodistribution using magnetic resonance imaging. *Nano Lett.*, 9, 1023–1027, 2009.
96. Al Faraj, A., Fauvelle, F., Luciani, N., Lacroix, G., Levy, M., Cremillieux, Y., CanetSoulas, E., *In vivo* biodistribution and biological impact of injected carbon nanotubes using magnetic resonance techniques. *Int. J. Nanomedicine*, 6, 351–361, 2011.

97. Cerpa, A., Koeber, M., Calle, D., Negri, V., Gavira, J.M., Hernanz, A., Briones, F., Cerdan, S., Ballesteros, P., Single-walled carbon nanotubes as anisotropic relaxation probes for magnetic resonance imaging. *MedChemComm*, 4, 669–672, 2013.
98. Doan, B.-T., Seguin, J., Breton, M., Le Beherec, R., Bessodes, M., Rodriguez-Manzo, J.A., Banhart, F., Beloeil, J.-C., Scherman, D., Richard, C., Functionalized single-walled carbon nanotubes containing traces of iron as new negative MRI contrast agents for *in vivo* imaging. *Contrast Media Mol. Imaging*, 7, 153–159, 2012.
99. Rivera, E.J., Sethi, R., Qu, F., Krishnamurthy, R., Muthupillai, R., Alford, M., Swanson, M.A., Eaton, S.S., Eaton, G.R., Wilson, L.J., Nitroxide radicals@US-tubes: new spin labels for biomedical applications. *Adv. Funct. Mater.*, 22, 3691–3698, 2012.
100. Sitharaman, B., Jacobson, B.D., Wadghiri, Y.Z., Bryant, H., Frank, J., The magnetic, relaxometric, and optical properties of gadolinium-catalyzed single-walled carbon nanotubes. *J. Appl. Phys.*, 113, 134308, 2013.
101. Tucker-Schwartz, J.M., Hong, T., Colvin, D.C., Xu, Y., Skala, M.C., Dual-modality photothermal optical coherence tomography and magnetic-resonance imaging of carbon nanotubes. *Opt. Lett.*, 37, 872–874, 2012.
102. Miyawaki, J., Yudasaka, M., Imai, H., Yorimitsu, H., Isobe, H., Nakamura, E., Iijima, S., Synthesis of ultrafine Gd2O3 nanoparticles inside single-wall carbon nanohorns. *J. Phys. Chem. B*, 110, 5179–5181, 2006.
103. Richard, C., Doan, B.-T., Beloeil, J.-C., Bessodes, M., Tóth, É., Scherman, D., Noncovalent functionalization of carbon nanotubes with amphiphilic Gd3+chelates: Toward powerful T1 and T2 MRI contrast agents. *Nano Lett.*, 8, 232–236, 2008.
104. Al Faraj, A., Cieslar, K., Lacroix, G., Gaillard, S., Canet-Soulas, E., Crémillieux, Y., *In vivo* imaging of carbon nanotube biodistribution using magnetic resonance imaging. *Nano Lett.*, 9, 1023–1027, 2009.
105. Choi, J.H., Nguyen, F.T., Barone, P.W., Heller, D.A., Moll, A.E., Patel, D., Boppart, S.A., Strano, M.S., Multimodal biomedical imaging with asymmetric single-walled carbon nanotube/iron oxide nanoparticle complexes. *Nano Lett.*, 7, 861–867, 2007.
106. Vittorio, O., Duce, S., Raffa, V., Cuschieri, A., Imaging and biomedical application of magnetic, in: *Carbon Nanotubes*, pp. 189–210, 2011.
107. Mackeyev, Y.A., Marks, J.W., Rosenblum, M.G., Wilson, L.J., Stable containment of radionuclides on the nanoscale by cut single-wall carbon nanotubes. *J. Phys. Chem. B*, 109, 5482–5484, 2005.
108. Monthioux, M., Filling single-wall carbon nanotube. *Carbon*, 40, 1809–1823, 2002.
109. Wu, H., Liu, G., Wang, X., Zhang, J., Chen, Y., Shi, J., Yang, H., Hu, H., Yang, S., Solvothermal synthesis of cobalt ferrite nanoparticles loaded on multiwalled carbon nanotubes for magnetic resonance imaging and drug delivery. *Acta Biomater.*, 7, 3496–3504, 2011.

110. Ananta, J.S., Matson, M.L., Tang, A.M., Mandal, T., Lin, S., Wong, K., Wong, S.T., Wilson, L.J., Single-walled carbon nanotube materials as T 2-weighted MRI contrast agents. *J. Phys. Chem. C*, 113, 19369–19372, 2009.
111. Gimble, J.M., Katz, A.J., Bunnell, B.A., Adipose-derived stem cells for regenerative medicine. *Circ. Res.*, 100, 1249–1260, 2007.
112. Caplan, A.I., Adult mesenchymal stem cells for tissue engineering versus regenerative medicine. *J. Cell. Physiol.*, 213, 341–347, 2007.
113. Vittorio, O., Duce, S.L., Pietrabissa, A., Cuschieri, A., Multiwall carbon nanotubes as MRI contrast agents for tracking stem cells. *Nanotechnology*, 22, 095706, 2011.
114. Wang, H., Wang, J., Deng, X., Sun, H., Shi, Z., Gu, Z., Liu, Y., Zhaoc, Y., Biodistribution of carbon single-wall carbon nanotubes in mice. *J. Nanosci. Nanotechnol.*, 4, 1019–1024, 2004.
115. Deng, X., Yang, S., Nie, H., Wang, H., Liu, Y., A generally adoptable radiotracing method for tracking carbon nanotubes in animals. *Nanotechnology*, 19, 075101, 2008.
116. Lacerda, L., Soundararajan, A., Singh, R., Pastorin, G., Al-Jamal, K.T., Turton, J., Frederik, P., Herrero, M.A., Li, S., Bao, A., Dynamic imaging of functionalized multi-walled carbon nanotube systemic circulation and urinary excretion. *Adv. Mater.*, 20, 225–230, 2008.
117. McDevitt, M.R., Chattopadhyay, D., Jaggi, J.S., Finn, R.D., Zanzonico, P.B., Villa, C., Rey, D., Mendenhall, J., Batt, C.A., Njardarson, J.T., PET imaging of soluble yttrium-86-labeled carbon nanotubes in mice. *Plos One*, 2, e907, 2007.

10

Functionalized Carbon Nanotubes for Artificial Bone Tissue Engineering

Sougata Ghosh[1] and Ebrahim Mostafavi[2,3]*

[1]Department of Microbiology, School of Science, RK University, Rajkot, Gujarat, India
[2]Stanford Cardiovascular Institute, Stanford University School of Medicine, Stanford, CA, USA
[3]Department of Medicine, Stanford University School of Medicine, Stanford, CA, USA

Abstract
Bone-related defects such as arthralgia, congenital malformation, osteoporosis, sports injuries, tumors, and traumatic accidents are often associated with insufficient self-repair and require additional medical intervention. Conventional orthopedic therapeutic approaches are occasionally challenged by non-integration, graft rejection, and implant failure. Advances in the area of nanotechnology have provided complementary and alternative approaches to ensure regeneration of the bone tissues by induction of osteogenesis followed by osseointegration. This chapter highlights the promises of carbon nanotubes (CNTs) which are considered as potential nanostructures for developing scaffolds for bone tissue engineering. The high surface area-to-volume ratio and superior electronic, mechanical, and thermal properties of CNTs provide an ideal surface for multifunctionalization of an array of biomolecules that include hydroxyapatite, polycaprolactone, polymethylmethacrylate, poly(lactide-co-glycolide), poly-L-lactic acid, polyvinyl alcohol, poly(ether ether ketone)-calcium polyphosphate cements, chitosan, collagen, bacterial cellulose, and silk fibroin. Further, strategies like solution-based processing, polymerization, melt-based processing, and grafting methods are used to fabricate CNTs based 2D/3D constructs that may combine cells, growth factors, and scaffolds as key elements in regenerative medicine. Similarly, CNTs based 3D-printing methods can result in desired microscale features or electrochemical properties to facilitate robust cell adhesion, proliferation, and differentiation. However, further

**Corresponding author*: ebimsv@stanford.edu; ebi.mostafavi@gmail.com

Jeenat Aslam, Chaudhery Mustansar Hussain and Ruby Aslam (eds.) Functionalized Carbon Nanotubes for Biomedical Applications, (225–256) © 2023 Scrivener Publishing LLC

investigation is required to understand the exact mechanisms behind osteoinductive effects of CNTs based nanocomposites and associated toxicity before it comes to practice. In view of the background, it is evident that CNTs are critical for developing regenerative medicine and bone tissue engineering that can significantly revolutionize future healthcare.

***Keywords*:** Carbon nanotubes, scaffolds, bone tissue regeneration, functionalization, osteogenic induction, osseointegration

10.1 Introduction

Bone defects affecting millions of people globally each year may be associated with traumatic accidents, sports injuries, arthralgia, genetic disorders, osteoporosis, and cancer. Bone tissue regeneration is challenged by critical-sized defects and deeper wounds which occasionally require medical interventions such as surgical procedures for grafting and implantation. The autografts, allografts, and xenografts are the most accepted treatment method for bone repair, where autograft facilitates efficient osteogenesis and osteoconduction without any immunogenicity [1, 2]. Several complications in this surgical process include chronic pain, bleeding, and injury to the nerves. Moreover, lack of donors and longer duration for the surgery are considered as major limitations [3, 4]. On the other hand, although allografts can induce osteogenesis and possess osteoconductive properties, graft rejection, contraction of transmissible diseases, and infections are some of the main drawbacks [5, 6]. Further, the mechanical properties of allografts are not identical to the load-bearing bone [7].

Alternative strategies for restoring, repairing, and replacing bone tissues using engineered biomaterials have recently gained more popularity. Nanotechnology is employed to develop scaffolds for delivery of cells, peptides, oligonucleotides, and drugs [8, 9]. Several biomaterials composed of metal, metal oxide, mesoporous silica, and polymeric nanostructures are used widely in tissue engineering [10, 11]. Carbon nanotubes (CNTs), which are rolled into single or multiple graphene sheets termed singled-walled CNTs (SWCNTs) and multi-walled CNTs (MWCNTs), respectively, are extensively explored for bone tissue engineering. The development in the area of CNTs based composite materials in the past two decades is depicted in Figure 10.1. These materials have exotic physico-chemical and opto-electronic properties that are notably distinct from the other allotropes of carbon, such as graphite, diamond, and fullerenes [12].

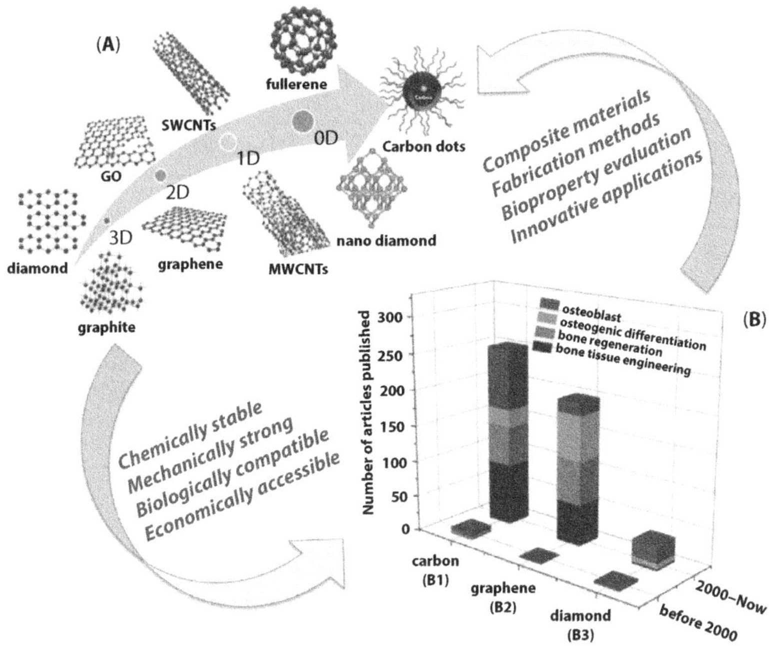

Figure 10.1 (A) Representative carbon-based materials: 3D graphite and diamond; 2D graphene and GO; 1D CNTs; 0D nano diamond, fullerene, and C-dots; (B) Number of articles published on the applications of carbon-based materials in BTE-related studies, search results obtained by Google Scholar on September 17, 2019. (B1) Blue column represents article number based on search results with the exact phrase "carbon" and "osteoblast" occurring in the title of a paper; the green column represents article number based on searching results with the exact phrase "carbon" and "osteogenic differentiation" occurring in the title of a paper; red column represents article number based on searching results with the exact phrase "carbon" and "bone regeneration" occurring in the title of a paper; black column represents article number based on searching results with the exact phrase "carbon" and "bone tissue engineering" occurring in the title of a paper. Columns of B2 (graphene), and B3 (diamond) are defined similarly as in the case for the column of B1 (carbon). Reprinted with permission from Peng Z, Zhao T, Zhou Y, Li S, Li J, Leblanc RM (2020) Bone tissue engineering via carbon-based nanomaterials. Adv. Healthcare Mater. 9(5), 1901495. Copyright © 2020 WILEY-VCH Verlag GmbH & Co. KGaA, Weinheim [16].

CNTs are mainly fabricated using arc discharge, chemical vapor deposition (CVD), and laser ablation methods followed by purification to remove the metal impurities from the catalysts used during the synthesis process. CNTs are significant materials for reinforcement, drug delivery [13], sensing, and bioimaging, hence considered as attractive theranostic agents [14, 15].

Table 10.1 CNTs based composite for bone tissue engineering.

CNTs	**Composite material**	**Functionalized ligand**	**Cells/models**	**References**
MWCNTs	hydroxyapatite	-	hBMSC	[22]
MWCNTs	hydroxyapatite	graphene oxide	MDCK cells	[23]
MWCNTs	hydroxyapatite	-	L929 mouse fibroblast cells	[24]
MWCNTs	hydroxyapatite	stainless steel surface	MG-63 human osteoblast cells	[25]
MWCNTs	hydroxyapatite	gellan, amylopectin	MG-63 human osteoblast cells	[26]
MWCNTs	poly(ε-caprolactone)	-	human adipose derived stem cells (hADSCs)	[27]
MWCNTs	poly(ε-caprolactone)	-	MG-63 human osteoblast cells	[29]
MWCNTs	poly(ε-caprolactone)	-	rat bone-marrow-derived stromal cells (BMSCs)	[30]
Amine-functionalized single-walled carbon nanotubes (aSWNTs)	poly(ε-caprolactone)	-	rat bone marrow-derived mesenchymal stem cells (rBMSCs)	[31]
MWCNTs	poly(ε-caprolactone)	nano hydroxyapatite	MG-63 human osteoblast cells	[32]
MWCNTs	polymethyl-methacrylate (PMMA)	chitosan	MG-63 human osteoblast cells	[33]
CNTs	polymethyl-methacrylate (PMMA)	hardystonite	MG-63 human osteoblast cells	[34]
CNTs	Poly(lactide-co-glycolide) (PLGA)	-	MC3T3-E1 osteoblasts	[35]

(Continued)

Table 10.1 CNTs based composite for bone tissue engineering. (*Continued*)

CNTs	Composite material	Functionalized ligand	Cells/models	References
MWCNTs	Poly(lactide-co-glycolide) (PLGA)	-	MC3T3-E1 osteoblasts	[36]
hydroxyl (OH), carboxylic acid (COOH) modified MWCNTs	Poly(lactide-co-glycolide) (PLGA)	-	MC3T3-E1 osteoblasts,	[37]
MWCNTs	Poly(L-lactic acid) (PLLA)	hydroxyapatite	periodontal ligament cells (PDLCs)	[39]
CNTs	Poly(L-lactic acid) (PLLA)	montmorillonite (MMT)	MG-63 cells	[40]
MWCNTs	Poly(L-lactic acid) (PLLA)	-	hFOB 1.19 cells	[41]
MWCNTs	polyvinyl alcohol (PVA)	biphasic calcium phosphate (BCP)	MC3T3-E1 subclone 14 preosteoblasts derived from newborn mouse calvaria	[42]
CNTs	polyvinyl alcohol (PVA)	-	Vero fibroblast-type cells, rat bone marrow-derived mesenchymal stem cells (rBMSCs)	[43]
MWCNTs	poly(ether ether ketone) (PEEK)	calcium polyphosphate	MC3T3-E1 osteoblast cells	[44]
MWCNTs	polymer polypropylene (PP)	hydoxyapatite nanorods (nHAs)	Human osteoblast cell line (Saos-2)	[45]

(*Continued*)

Table 10.1 CNTs based composite for bone tissue engineering. (*Continued*)

CNTs	Composite material	Functionalized ligand	Cells/models	References
SWCNTs	Chitosan	nHAP	human fetal osteoblasts (CRL-11372)	[46]
MWCNTs	Chitosan	HAP	MG-63 cells	[47]
MWCNTs	Collagen	HAP	bone marrow-derived MSCs (BMSCs)	[48]
Functionalized MWCNTs	Collagen	HAP, chitosan	MG-63 cells	[49]
Functionalized MWCNTs	Alginate	HAP	MG-63 cells	[50]
MWCNTs	bacterial cellulose (BC)	amphiphilic comb-like polymer (APCLP)	critical-sized defect in mouse calvaria	[51]
MWCNTs	Silk fibroin (SF)	nHAP, polyamide 66, dexamethasone	BMSCs	[52]

In view of this background, this chapter gives a detailed description of the recent advances on the CNTs based biomaterials for bone tissue engineering as listed in Table 10.1 [16]. Further, mechanisms behind the superior bone tissue repair and regeneration using CNTs based materials are also discussed at the molecular level. Eventually, the need for thorough toxicological studies, pharmacokinetics, and pharmacodynamic investigations is emphasized before the novel materials can enter the real-time therapeutic practice.

10.2 CNT-Based Scaffolds and Implants

Biomaterials used to fabricate scaffolds for bone tissue engineering should be rigid to bear external force. Rough surfaces of such materials can promote better cell adhesion and proliferation. Porous structures of the materials can help in better penetration of the body fluids and help in neovascularisation. Biodegradable materials are more preferred to avoid the need for surgical removal of the implant after healing the injury. Hence, CNTs with attractive physical and chemical properties are most preferred for the effective reinforcements of the scaffolds in combination with polymers such as collagen, gelatine, alginate, chitosan, and others [17]. Further, CNTs exhibit superior biocompatibility, which in combination with poly(L-lactide) displayed high direct-current conductivity, crystallization, plasticization of the polymer matrix, and growth inhibition of fibroblast cells [18]. Likewise, CNTs can also be encapsulated by poly(lactic acid) (PLA) nanofibers employing electrospinning [19]. Hence, this section elaborates on the CNTs based scaffolds and implants for bone tissue engineering.

10.2.1 Hydroxyapatite

Among several calcium phosphate bioceramics which are widely used for orthopedic and dental treatment, hydroxyapatite (HAP) nanoparticle $Ca_{10}(PO_4)_6(OH)_2$ is used as the main constituent as its predominant naturally in bones and teeth [20]. Nanostructured HAP is biocompatible and used in combination with CNTs to develop scaffolds with high tensile strength and toughness for load-bearing orthopedic implants [21]. A sol-gel method was used by Hooshmand *et al.* where initially $(NH_4)_2HPO_4$ solution was added to CNTs sol followed by drop-wise addition of $(Ca(NO_3)_2.4H_2O)$ [22]. Aging for 24 h resulted in the formation of the MWCNT/HAP nanocomposite. It is important to note that the HAP particles were homogenous with a spherical shape that formed clusters, while MWCNTs were dispersed within the clusters without any visible agglomeration. Low temperature for crystallization promoted the formation of more nucleation sites that prevented both dehydroxylation of HAP and oxidation of nanotubes. Further, SDS-adsorbed MWCNTs provided more negative charge on the surface and hence were completely covered with HAP crystal. Cellular attachment and efficient proliferation of the Human bone marrow stromal cells (hBMSC) cells on the MWCNT/HAP nanocomposite indicated biocompatibility of the scaffolds and the bone tissue engineering potential.

In a recent study, Jyoti *et al.* (2021) fabricated a composite by combining both graphene oxide-carbon nanotube (GCNTs) hybrid and HAP [23]. Initially, the MWCNTs were synthesized using the CVD method by reacting toluene and ferrocene. The resulting MWCNTs were further acid-functionalized (denoted as FCNTs) by reacting with 70% nitric acid. Graphene oxide (GO) was synthesized by a modified Hummers method, while the GCNTs hybrid was eventually synthesized by ultra-sonicating a mixture containing both GO and FCNTs in ethanol at a ratio of 1:1. The graphene flakes were rippled and crumpled structures of graphene flakes. Morphological features with elemental

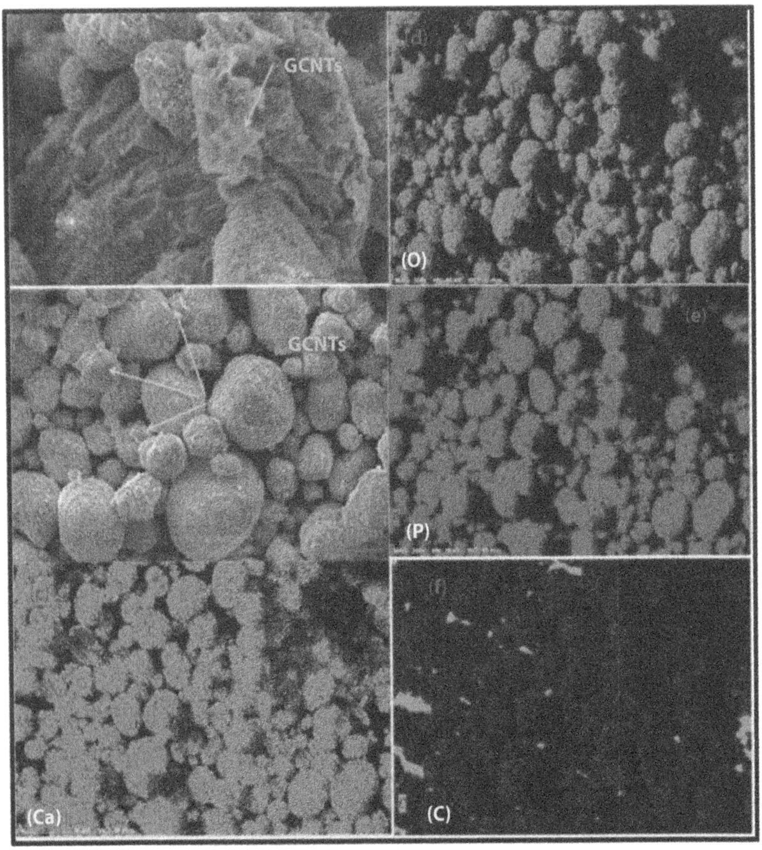

Figure 10.2 (a–b) SEM of GCNTs-HAP composites at different magnifications with the elemental mapping of HAP show the presence of (c) Calcium, (d) Oxygen, (e) Phosphorous, and (f) Carbon. Reprinted with permission from Jyoti J, Kiran A, Sandhu M, Kumar A, Singh BP, Kumar N (2021) Improved nanomechanical and *in-vitro* biocompatibility of graphene oxide-carbon nanotube hydroxyapatite hybrid composites by synergistic effect. Journal of the Mechanical Behavior of Biomedical Materials 117, 104376. Copyright © 2021 Published by Elsevier Ltd [23].

mapping of the nanocomposite exhibited strong O, P, and Ca signals attributed to the HAP component as evident from Figure 10.2. High nanoindentation hardness and elastic modulus of GCNTs-HAP indicated a highly reinforced nature that is ideal for mimicking bone tissue microenvironment. High cell viability in Madin-Darby Canine Kidney (MDCK) cells after treatment with GCNTs-HAP composites containing 2% of GCNTs nanofiller reflected the high biocompatibility apart from superior induction of cell proliferation. Hence, GCNTs can be a promising composite for promoting bone repair, regeneration, augmentation, and implantation.

Khalid and Suman (2017) reported HAP-reinforced MWCNTs with high biocompatibility. Initially, the MWCNTs were treated with 20% sulphuric acid and 20% nitric acid in a 3:1 ratio for one hour [24]. The washed and dried MWCNTs were then dispersed in Sodium Dodecyl Benzene Sulphonate (SDBS) followed by synthesis of HAP in the same solution. The HAP particles were densely oriented on the MWCNTs surface. The nanocomposite was nontoxic to L929 mouse fibroblast cell line.

Martinelli *et al.* (2018) deposited nanohydroxyapatite/superhydrophilic MWCNTs (nHAP/MWCNT) onto 316L stainless steel biomedical alloys by electrophoretic deposition using a voltage of -2V [25]. The outer and inner diameters of the MWCNTs were 60 nm and 50 nm, respectively. Figure 10.3 shows the morphology of the plate-like crystals of the electrodeposited nHAP on the stainless steel surface. The presence of nHAP in the composite was confirmed by the peaks at 961 cm^{-1}, 1030–1050 cm^{-1}, 420 cm^{-1}, 580 cm^{-1}, and 780 cm^{-1} in Fourier Transform Infrared (FTIR) spectra. Human osteoblast cells (MG-63) cultivated on nHAp/MWCNT showed upregulation of the expression of important genes such as alkaline phosphatase (ALP), osteopontin (OPN), and osteocalcin (OCN) that are associated with bone mineralization and maturation.

In an interesting study, Rajesh *et al.* (2016) reported novel tricomponent composite scaffolds composed of oxidized carbon nanotube (fMWCNT)–gellan–hydroxyapatite (HAP) and fMWCNT–amylopectin–HAP with high porosity of 82.82% and 91.76%, respectively [26]. The pore size of fMWCNT–gellan–HAP and fMWCNT–amylopectin–HAP were 49–51 μm and 55–85 μm, respectively, as evident from Figure 10.4. The HAP was uniformly dispersed on the polysaccharide-covered scaffolds. The biomaterial efficiently promoted the MG63 cell adhesion, proliferation, and mineralization.

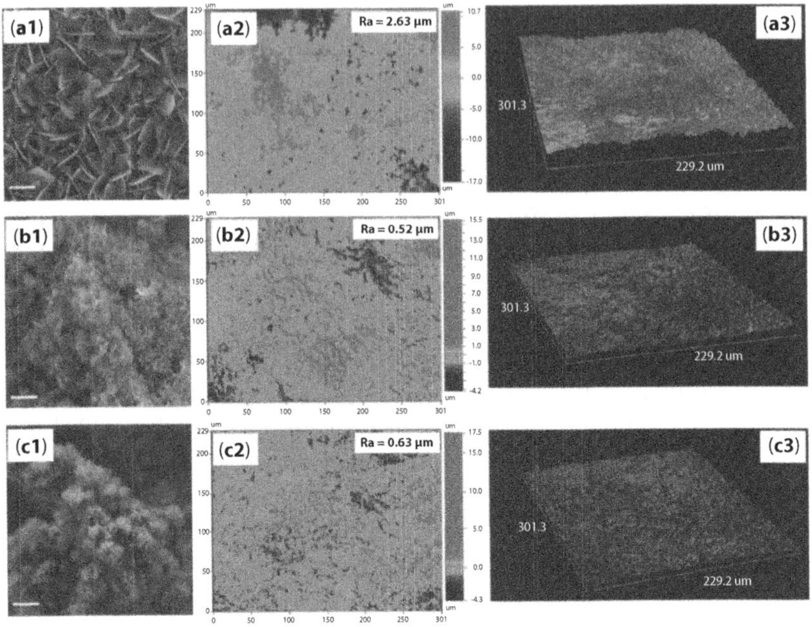

Figure 10.3 SEM and optical profilometry collected from deposited nHAp and nHAp/MWCNT films on stainless steel alloys. (a1) SEM illustrated plate-like crystals of nHAp electrodeposited onto 316L stainless steel alloy; (a2) optical images from profilometry collected at the top of nHAp thin films electrodeposited onto 316L stainless steel; (e3) 3D constructions extracted from the top of 316L stainless steel alloy; (b1) SEM illustrating needle-like crystals electrodeposited onto 316L stainless steel alloy; (b2) optical images from profilometry collected at the top of nHAp/MWCNT thin films electrodeposited onto 316L stainless steel; (c3) 3D constructions collected from the top of 316L stainless steel alloy illustrating nHAp/MWCNT 1% electrodeposited onto 316L stainless steel; (c1) SEM illustrated needle-like crystals electrodeposited onto 316L stainless steel alloy; (c2) optical images from profilometry collected at the top of nHAp/MWCNT thin films electrodeposited onto 316L stainless steel; (c3) 3D constructions extracted from the top of 316L stainless steel alloy illustrating nHAp/MWCNT 3% electrodeposited onto 316L stainless steel. SEM scale bars from Figure 10.3(a1,b1,c1) is 2 μm. Reprinted with permission from Martinelli NM, Ribeiro MJG, Ricci R, Marques MA, Lobo AO, Marciano FR (2018) *In vitro* osteogenesis stimulation via nano-hydroxyapatite/carbon nanotube thin films on biomedical stainless steel. Materials, 11, 1555 [25]. (Open access).

10.2.2 Polymers

Several polymers are incorporated with CNTs for developing scaffolds for bone tissue engineering. In this section, such polyester materials are discussed that are flexible, biodegradable, and biocompatible.

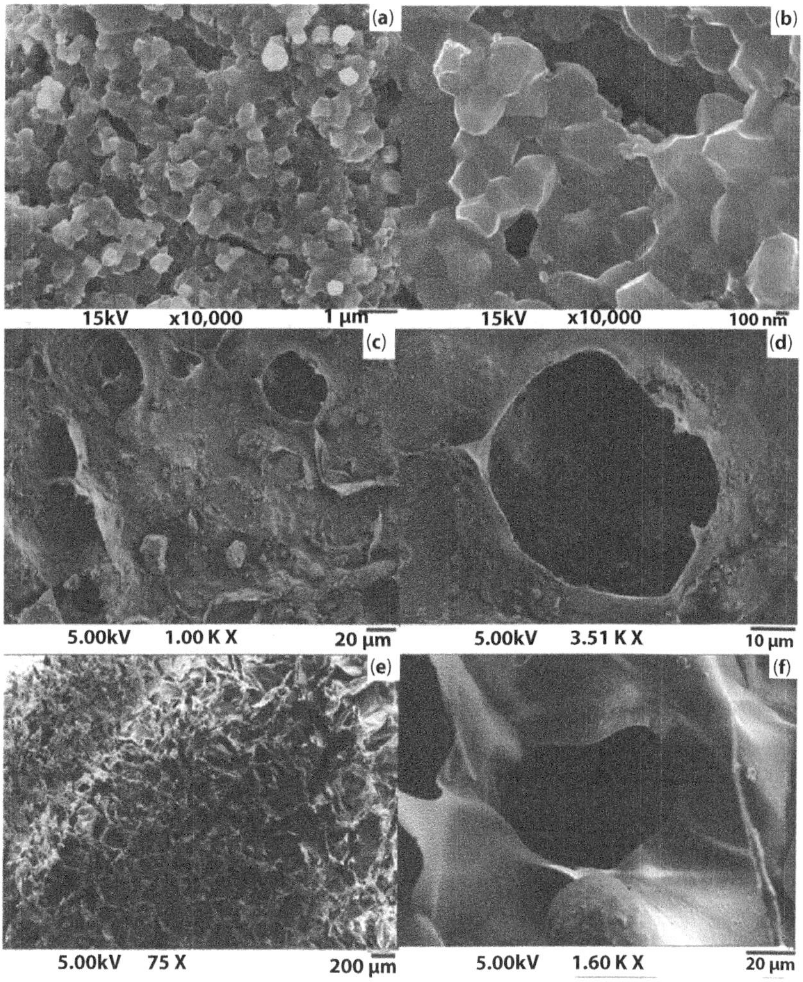

Figure 10.4 FE-SEM images of (a & b) HAP, and the (c & d) fMWCNT–gellan–HAP and (e & f) fMWCNT–amylopectin–HAP scaffolds. Reprinted with permission from Rajesh R, Ravichandran YD, Reddy MJK, Ryu SH, Shanmugharaj AM (2016) Development of functionalized multi-walled carbon nanotube-based polysaccharide-hydroxyapatite scaffolds for bone tissue engineering. RSC Adv. 6, 82385–82393. Copyright © 2016 The Royal Society of Chemistry [26].

10.2.2.1 Poly(ε-Caprolactone)

Huang *et al.* (2019) fabricated 3D printed scaffolds with the poly(ε-caprolactone) or (PCL) loaded with MWCNTs at different concentrations (0.25, 0.75, and 3 wt%) for bone tissue regeneration [27]. The pore size of the scaffolds was in a range from 366 μm and 397 μm. Contact angles in all of the scaffolds were higher than 90° indicating hydrophobic nature that was attributed to the intrinsic hydrophobicity of PCL. The cellular viability of human adipose-derived stem cells (hADSCs) was more than 80% in the PCL-based scaffolds and showed high cell density and cell bridging after 21 days, as shown in Figure 10.5.

Mostafavi *et al.* (2021) engineered composite filaments from a combination of PCL doped with zinc oxide nanoparticles (ZnO NPs) and hydroxyapatite microparticles for in situ bone printing. These scaffolds could adhere to wet bone tissue to prevent dislocation, and over time they showed osteoconductive and supported the osteodifferentiation of mesenchymal stem cells [28]. In another study, Mattioli-Belmonte *et al.*

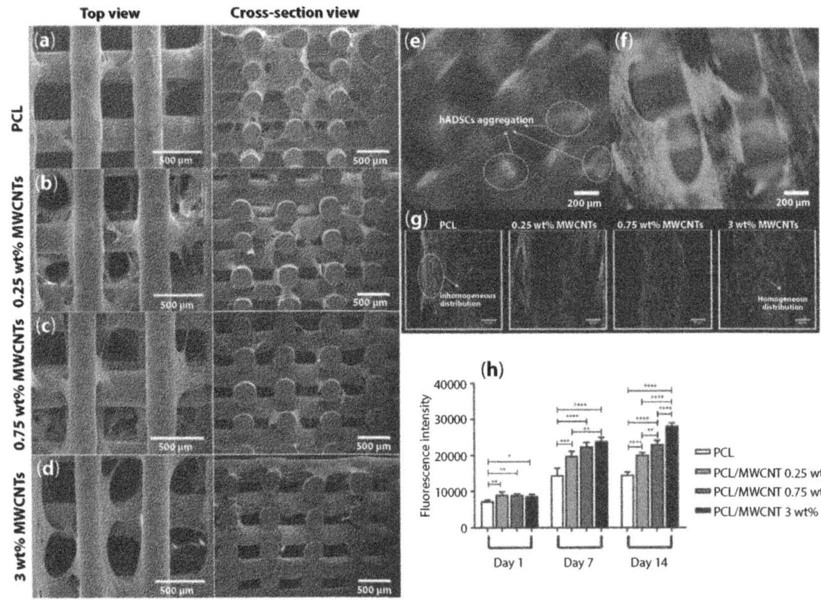

Figure 10.5 SEM of cell morphology at day14 showing the top-view and cross-section of the scaffolds (a–d); Fluorescent microscopy images of (e) PCL and (f) PCL/MWCNT 3 wt% and (g) confocal images of cell morphology on surfaces of all samples at day 14; (h) Alamar Blue results for all samples at days 1, 7 and 14. Reprinted with permission from Huang B, Vyas C, Roberts I, Poutrel QA, Chiang WH, Blaker JJ, Huang Z, Bártolo P (2019) Fabrication and characterization of 3D printed MWCNT composite porous scaffolds for bone regeneration. Mater. Sci. Eng. C 98, 266–278. Copyright © 2018 Published by Elsevier B.V [27].

(2012) microfabricated CNTs–PCL composite exhibited promising bone tissue regenerating effect by promoting proliferation in osteoblast-like cells (MG63) [29]. Interestingly the elastic modulus initially increased with increasing concentration of CNTs in PCL concentration until 12.5 mg/mL, beyond which it started to decrease. The elastic modulus of the nanocomposite varied between 10 and 75 MPa. The composites were more porous and irregular, which facilitated the evenly spreading of the osteoblasts. Similar composite scaffolds also facilitated the attachment and proliferation of the rat bone marrow-derived mesenchymal stem cells (rBMSCs) isolated from young (4–6 weeks old) Sprague Dawley rats. The cells were tightly anchored on the surface of the scaffolds and showed a fibroblast-like morphology, indicating that cells had successfully attached and were effectively spreading [30]. Similarly, amine-functionalized single-walled carbon nanotubes (aSWCNTs)-PCL composite-based electrospun scaffolds promoted the attachment, proliferation, differentiation of rBMSCs. It is important to note that superior mechanical property was obtained when 0.2 wt% aSWNT was incorporated in the scaffold. However, higher rBMSCs proliferation and differentiation was achieved when 0.5 wt% aSWNT was used [31].

One more study conducted by Yang *et al.* (2019) reported the osteoproductive potential of a three-phased multiwalled scaffold composed of MWCNTs, nHAP, and PCL [32]. The composite scaffold demonstrated smaller porosity and slow degradability in simulated body fluid (SBF). The higher proliferation of human osteosarcoma cell MG63 was observed on the scaffolds that efficiently differentiated into an osteogenic lineage that was confirmed by the high expression of ALP.

10.2.2.2 Polymethyl-Methacrylate

Among various bone cements, polymethyl-methacrylate (PMMA) is the most popularly used material that acts as a promising grouting agent in total joint replacement that promotes efficient load transfer between bones and implant. Composites prepared using PMMA, chitosan (CS) powder, and MWCNTs exhibited superior mechanical properties such as increased compressive and bending strengths [33]. The average size was 7.49 μm. The apatite-like deposition was observed on the surface after 28 days. The MG-63 cells could adhere efficiently to the composite surface and expressed high levels of ALP. Pahlevanzadeh *et al.* (2021) fabricated a novel bioactive polymethyl methacrylate-hardystonite (PMMA-HT)

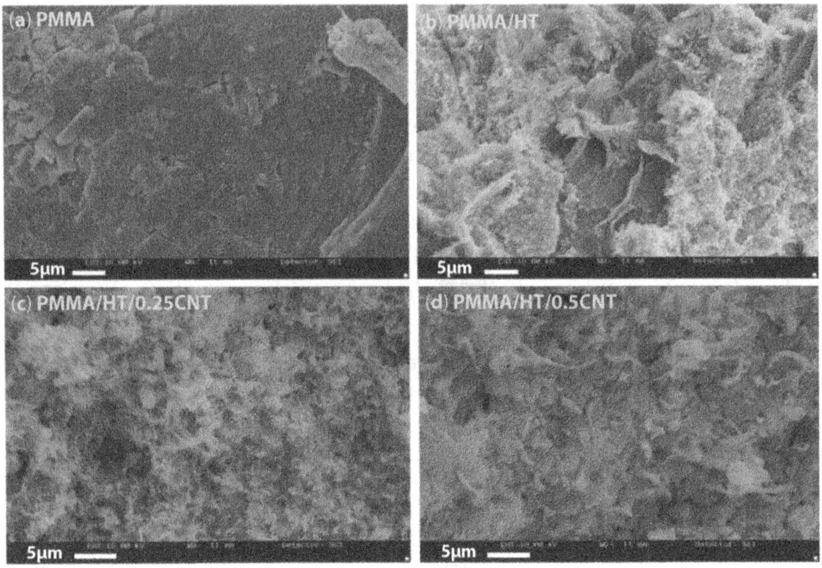

Figure 10.6 SEM images of fracture surfaces for (a) PMMA, (b) PMMA/HT, (c) PMMA/HT/0.25CNT, (d) PMMA/HT/0.5CNT. Reprinted with permission from Pahlevanzadeh F, Bakhsheshi-Rad HR, Kharaziha M, Kasiri-Asgarani M, Omidi M, Razzaghi M, Ismail AF, Sharif S, RamaKrishna S, Berto F (2021) CNT and rGO reinforced PMMA based bone cement for fixation of load-bearing implants: Mechanical property and biological response. J. Mech. Behav. Biomed. Mater. 116, 104320. Copyright © 2021 Elsevier Ltd [34].

porous bone cement (BC) reinforced with 0.25 and 0.5 wt% of CNTs, as evident from Figure 10.6 [34]. The incorporation of 0.5% CNTs enhanced the bending strength of PMMA/HT cements from 61.06 ± 0.68 to 82.07 ± 1.01 MPa. Even the tensile strength increased from 40.02 ± 0.71 MPa to 63.21 ± 1.70 MPa due to the presence of 0.5 wt% CNTs. Superior toughening, apatite formation, and cell interaction in CNTs supplemented composites was noticed when MG63 osteoblasts were grown on the bone cement composite indicating better promotion of implant fixation and load-bearing attributes.

10.2.2.3 Poly(Lactide-Co-Glycolide)

Poly(lactide-co-glycolide) (PLGA), a co-polymer of poly(lactic acid) (PLA), and poly(glycolic acid) (PGA) is considered as one of the most popular scaffold matrix polymers due to its biocompatibility, osteoconductivity, and suitable mechanical properties that are prerequisites for skeletal tissue repair and regeneration [35]. Porous scaffolds were fabricated by solvent casting/particulate leaching after dispersing the CNTs in PLGA solution. The cubic pore structure with 100 and 200 μm diameters was interconnected. Further, the incorporation of 1% CNTs enhanced the roughness of the composite from 7.17 ± 1.62 to 43.95 ± 10.74 nm. Similarly, the compressive modulus also increased by 3.6-fold with an average compressive modulus of 6.36 ± 0.55 MPa in 1% CNTs supplemented scaffolds. Not only the mechanical strength was increased, but also the surface roughness was enhanced, resulting in better adherence and cell division in MC3T3-E1 osteoblasts. The cell number increased around 2.9-fold per mm^2 on 1% CNT/PLGA films. Additionally, the same films enhanced the osteogenic differentiation associated with the highest level of ALP expression. Notably, 1.5-fold increase in the calcium deposit was observed after 28 days when the cells were grown on 1% CNT/PLGA scaffolds.

In another study, biodegradable PLGA/MWCNTs scaffolds were synthesized using a thermally induced phase separation (TIPS) technique [36]. The scaffolds exhibited porous morphology with interconnected cavities between 50 and 75 μm in size. These cavities were ideal for harboring cells of a specific type. Around 60% mass degradation was observed for the PLGA/MWCNTs 0.5%. It is important to note that the viability of MC3T3-E1 pre-osteoblast cell line reduced less than 30% in the presence of PLGA/MWCNTs composite with 0.1% of MWCNTs while 0.3 and 0.5% of MWCNTs in scaffolds showed high cytotoxicity.

Likewise, Mikael *et al.* (2014) developed PLGA based 3D porous scaffolds with pristine and modified (with hydroxyl (OH), carboxylic acid (COOH)) MWCNTs [37]. Mechanical properties such as compressive strength and modulus were significantly increased up to 35 MPa and 510.99 MPa, respectively, in the presence of 3% MWCNTs in the composite. The MC3T3-E1 cells exhibited superior viability, proliferation, and mineralization on the composite scaffolds. Inflammatory response after implantation was minimum with OH-modified MWCNTs. *In-vivo* studies were carried out by implanting the scaffolds subcutaneously in 28 male, retired breeder Sprague-Dawley rats of weight 450–500 g. After 4 weeks, the structural integrity of the implants started to decrease, which was attributed to the degradation of the polymer and invasion by host tissues which was more prominent by 8-12 weeks. A mild response and development of a fibrous capsule composed of a few lymphocytes, giant cells, polymorphonuclear (PMN) lymphocytes, and mainly macrophages were observed for the 3% OH-MWCNT supplemented scaffold.

10.2.2.4 Poly-L-Lactic Acid

Poly(L-lactic acid) (PLLA) is another polymer with superior biocompatibility that is used for designing scaffolds for tissue engineering [38]. In one such study, Mei *et al.* (2007) fabricated a composite membrane for guided tissue regeneration (GTR) by electrospinning a suspension consisting of PLLA, MWCNTs, and HAP [39]. The composite enhanced the adhesion and proliferation of periodontal ligament cells (PDLCs) by 30% and inhibited the adhesion and proliferation of gingival epithelial cells by 30%. On implantation of the composite membrane seeded PDLCs into the leg muscle pouches of immunodeficient mice. Interestingly, the composite promoted the formation of round to irregular bone with a higher calcium deposit. Further, the porous structure of the scaffolds facilitated the growth of the blood vessels that played a critical role in bone tissue regeneration.

Shuai *et al.* (2021) developed a self-assembled hybrid nanomaterial from acidified CNTs, aminopropyltriethoxysilane (KH550) [40]. The ammonium salt grafted CNTs successfully intercalated into the interlayers of large, flat, and layered montmorillonite (MMT). This was mainly attributed to the cation exchange reaction between the alkylammonium cations and the sodium cations of MMT. The MMT-CNT hybrid was then incorporated into PLLA that significantly increased the tensile strength and modulus by 113.04% and 111.46%, respectively. The composite scaffolds with MMT-CNT contents of 3%, 6%, 9%, and 12% were termed as PMC3, PMC6, PMC9, and PMC12, respectively. Likewise, the compressive strength and modulus were enhanced by 58.20% and 63.27%, respectively. Figure 10.7 shows that compared to PLLA, the MMT-CNT containing scaffolds are darker. The uniformly distributed pores in the scaffolds are advantageous as they can facilitate better cellular growth by allowing efficient transportation of nutrients, oxygen, and other essential molecules. Moreover, incorporation of the MMT-CNT in a varied amount ranging from 0% to 12% increased the water absorption rate of the scaffolds from 130.02% to 169.23%. Higher cell adhesion, growth, and proliferation of MG-63 cells were seen on the PLLA based scaffolds. Likewise, Vozzi *et al.*, 2013 reported microfabricated three-dimensional (3D) scaffolds composed of PLLA and MWCNTs that were biocompatible to the hFOB 1.19 cells. The cells exhibited better anchoring on the composite scaffold surface. However, the osteoblasts showed more spheroid morphology [41].

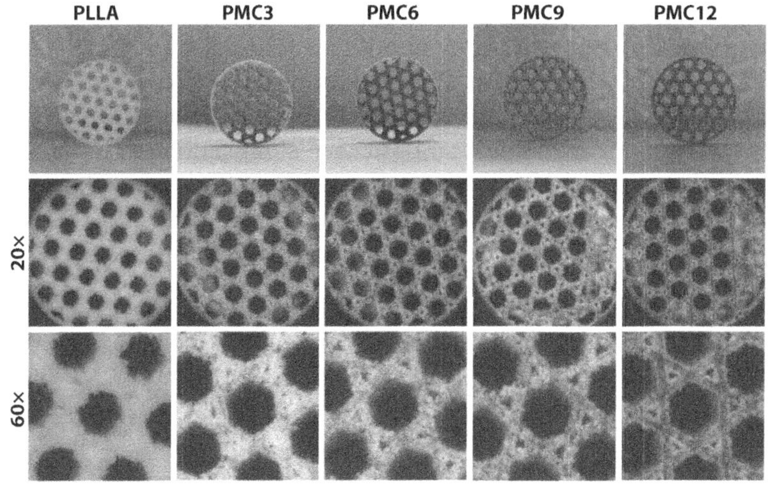

Figure 10.7 Digital and SEM images in different magnifications of the PLLA and PMC scaffolds. Reprinted with permission from Shuai C, Peng B, Liu M, Peng S, Feng P (2021) A self-assembled montmorillonite-carbon nanotube hybrid nano-reinforcement for poly-L-lactic acid bone scaffold. Mater. Today Adv. 11, 100158. Copyright © 2021 The Author(s). Published by Elsevier Ltd [40].

10.2.2.5 Polyvinyl Alcohol

As an integral part, the tissue engineering scaffolds have a supporting material that should be non-toxic, biocompatible, and stable. One such mechanically suitable supporting material with biotribological properties is polyvinyl alcohol (PVA) which is extensively studied for developing osteochondral scaffolds. Lan *et al.* (2019) developed an MWCNTs reinforced polyvinyl alcohol/biphasic calcium phosphate (PVA/BCP) scaffold employing a freeze-thawing and freeze-drying method [42]. The composite scaffold showed an enhanced compressive strength of 81 ± 6 kPa as compared to only PVA/BCP hydrogels (48 ± 2 kPa). The porous scaffolds exhibited high interconnectivity equivalent to 80 ± 0.6%. The viability of MC3T3-E1 subclone 14 preosteoblasts derived from newborn mouse calvaria was high on the composite PVA/BCP/CNT porous scaffolds. Further, the scaffold surface promoted high proliferation and spreading of the cells confirming the tissue regenerating potential.

Yet, in another study CNTs and carbon nanofibres (CNF) were produced by hot-filament chemical vapor deposition (HFCVD) for reinforcement of PVA-based hydrogel [43]. High biocompatibility was observed when the Vero fibroblast-type cells' were grown on the scaffolds. Interestingly, the composite induced significant osteogenic differentiation in the MSCs derived from rat bone marrow, as indicated by high alkaline phosphatase (ALP) activity and mineralization. Hence, it was speculated that the CNTs reinforced PVA hydrogels can be effectively used to treat osteochondral defects.

10.2.2.6 Others

Certain other biocompatible polymers are considered suitable for developing scaffolds for bone tissue engineering. Poly(ether ether ketone) (PEEK) is one of the most preferred polymers that can replace conventional surgical metal implants owing to its mechanical properties and stiffness identical to native bone tissues. In one such study, Cao *et al.* (2018) fabricated a bioactive composite made from MWCNTs, calcium polyphosphate, and PEEK [44]. Superior interfacial adhesion resulted in a uniform dispersion of the MWCNTs in the PEEK matrix. Interestingly, the MC3T3-E1 osteoblast cells exhibited high viability and better adhesion on the composite surface. Moreover, the composite promoted osteogenic differentiation of MC3T3-E1 cells that was confirmed by significantly high ALP activity. Likewise, Liao *et al.* (2013) reported another polymer polypropylene (PP) based biocomposites reinforced by MWCNTs and hydroxyapatite nanorods (nHAs) [45]. The MWCNTs were well dispersed in the polymeric matrix of the nanocomposites, as marked by the black arrows in Figure 10.8. Both the nHA and MWCNTs fillers might compete with each other for the nucleation of PP macromolecules. The MWCNTs promoted nucleation of PP in the composites containing 8% and 15% nHA, particularly the former with low nHA content. The incorporation of MWCNTs further enhanced the stiffness, tensile strength, and impact toughness of the PP/nHA nanocomposites. The human osteoblast cell line (Saos-2) was firmly attached to the biocomposite surface, and rapid proliferation led to covering the entire surface after 3 days. Interestingly, osteoblastic proliferation was found to be a function of the culturing time.

10.2.3 Biopolymers

Polymers such as chitosan, collagen, fibrin, hyaluronic acid, and alginate of biological origin called biopolymers are very significant due to their high

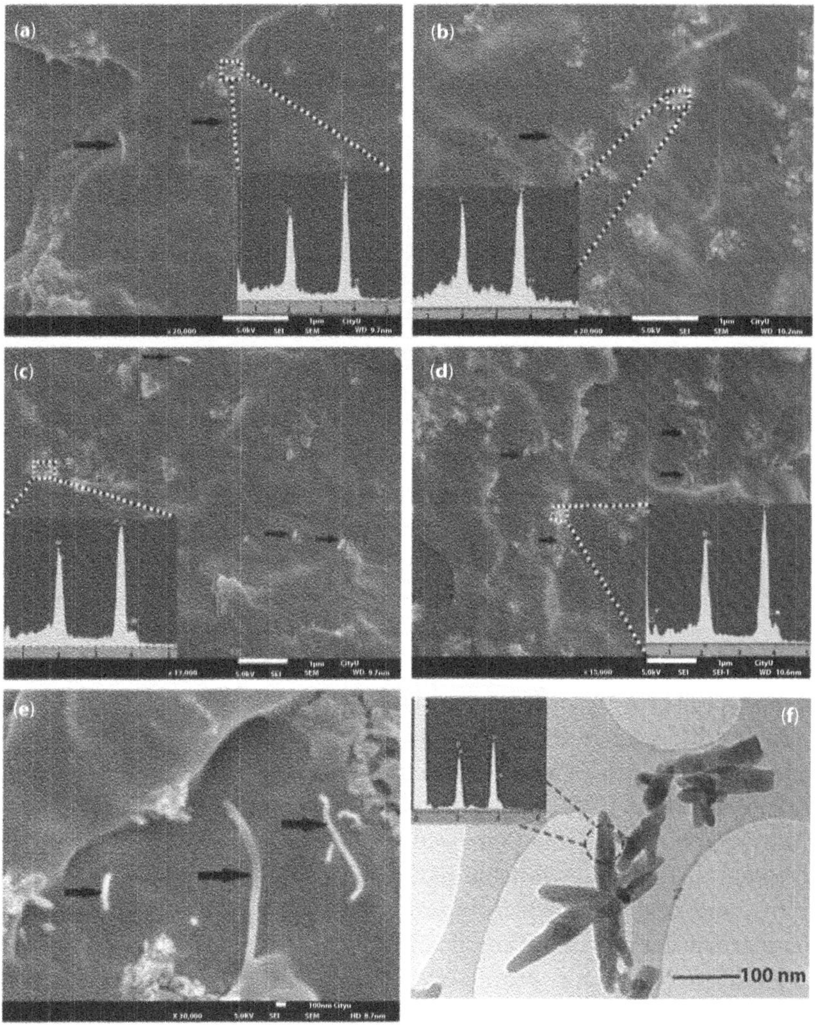

Figure 10.8 SEM morphologies for (a) PP/0.1%MWCNT–8%nHA, (b) PP/0.1%MWCNT–20%nHA, (c) PP/0.3%MWCNT–8%nHA, (d, e) PP/0.3%MWCNT–20% nHA nanocomposites. MWCNTs in these micrographs are denoted with black arrows. Dash square shows nHA fillers together with their EDX spectrum. (f) TEM micrograph of hydroxyapatite nanorods. Reprinted with permission from Liao CZ, Li K, Wong HM, Tong WY, Yeung KW, Tjong SC (2013) Novel polypropylene biocomposites reinforced with carbon nanotubes and hydroxyapatite nanorods for bone replacements. Mater. Sci. Eng. C Mater. Biol. Appl. 33, 1380–1388. Copyright © 2012 Elsevier B.V [45].

water content, good permeability for oxygen, nutrients, and other soluble signals. Several hydrogel scaffolds are designed using biopolymers as they can provide suitable mechanical properties like fibrous extracellular matrix identical to the native bone tissues.

10.2.3.1 Chitosan

Chitin is a biopolymer abundantly available in the exoskeletons of crustaceans and insects. Deacetylation of this chitin results in the formation of chitosan, which has promising therapeutic potential and is used in wound dressings and various tissue engineering applications. Being biodegradable porous chitosan sponges are used for fabrication of scaffolds for promoting the proliferation of osteoblasts. Im *et al.* (2012) reported 3D porous chitosan scaffolds with different concentrations of nHAP and SWCNTs employing a lyophilization procedure [46]. Initially, two types of SWCNTs were synthesized in an arc discharge with a magnetic field (B-SWCNT) and without a magnetic field (N-SWCNT) for reinforcement purposes to facilitate bone regeneration. Interestingly, the swelling ratios of all composite hydrogels exceeded 600%. The composite scaffolds supported the efficient attachment of the human fetal osteoblasts (CRL-11372), which exhibited long filopodia cell growth spreading onto the surfaces.

In another similar study, Venkatesan *et al.* (2011) employed a freeze-drying method to synthesize composite using chitosan and natural hydroxyapatite derived from *Thunnus obesus* bone (chitosan/HAP) and chitosan grafted with functionalized MWCNTs with HAP (f-MWCNT-g-chitosan/HAP) [47]. A notable reduction in the water uptake, retention ability, and degradation of composite scaffolds was observed. However, the thermal stability of the composite increased on incorporation of HAP and f-MWCNT. Equal porous structures with good interconnectivity were observed in Figure 10.9. Both HAP and f-MWCNT were homogeneously dispersed in the chitosan matrix with interconnected porosity of 70–200 μm(chitosan/HAP) and 46–200 μm(f-MWCNT-g-chitosan/HAP). Proliferation in MG-63 cells was doubled when cultured on the composite surface.

10.2.3.2 Collagen

Another naturally occurring biopolymer is collagen (Col) which shows low antigenicity, high biocompatibility, and provides the most favorable condition for cell attachment, proliferation, and differentiation for neurite, osteoblasts and fibroblasts. Jing *et al.* (2017) fabricated an innovative

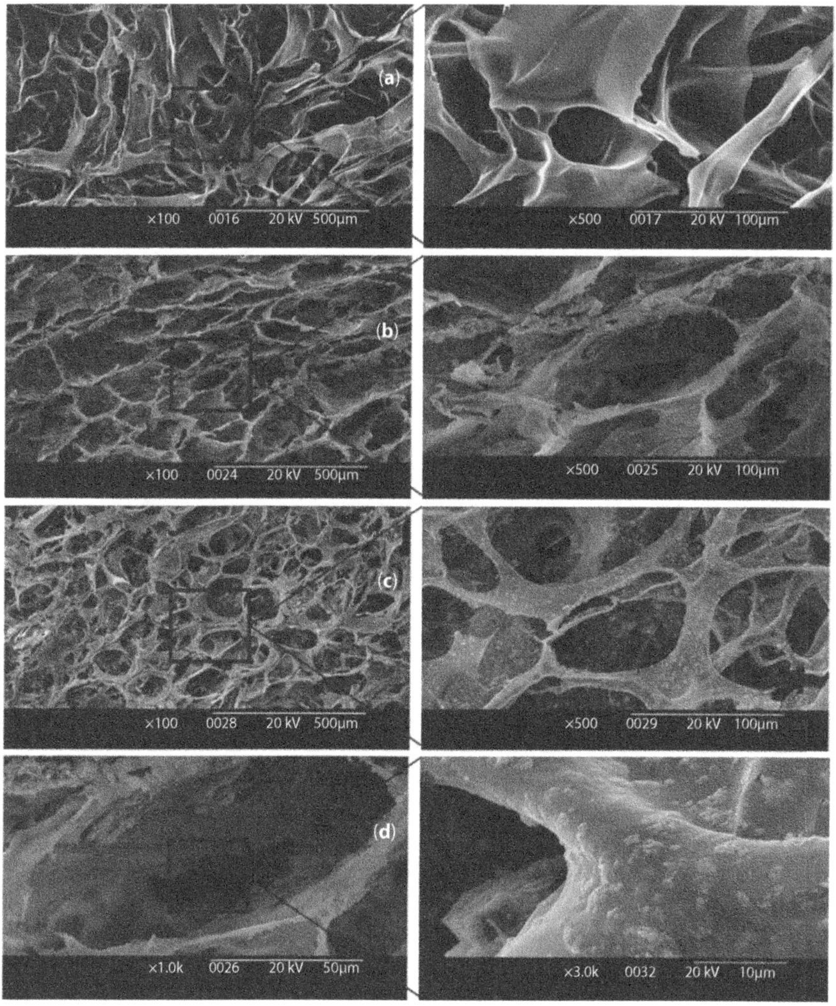

Figure 10.9 SEM images of (a) chitosan scaffold (b) chitosan/HAP scaffold (c) f-MWCNT-g-chitosan/HAP scaffold, and (d) dispersion of HAP particles in chitosan matrix. Reprinted with permission from Venkatesan J, Qian Z-J, Ryu B, Kumar NA, Kim SK (2011) Preparation and characterization of carbon nanotube-grafted-chitosan – Natural hydroxyapatite composite for bone tissue engineering. Carbohydr Polym 83, 569–577. Copyright © 2010 Elsevier Ltd [47].

porous scaffold by reinforcing MWCNTs into a Col-HAP composite [48]. This enhanced the stiffness by 10-fold. Further, this scaffold promoted the proliferation and spreading of the bone marrow-derived MSCs (BMSCs) that was confirmed by high levels of expression of bone sialoprotein (BSP) and osteocalcin (OCN). Dense connective tissues covered the defect site,

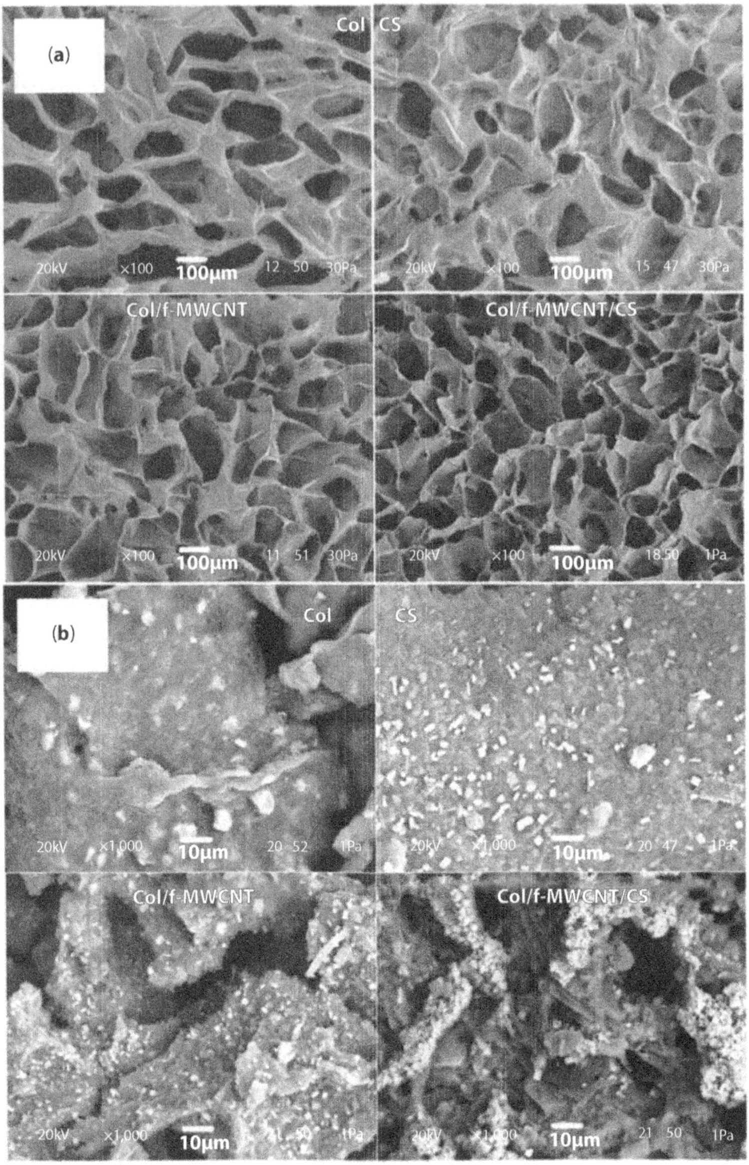

Figure 10.10 SEM micrographs of Col, CS, Col/f-MWCNT, and Col/f-MWCNT/CS scaffolds (a) before biomineralization at magnification ×50 and (b) after biomineralization at magnification ×1,000. Reprinted with permission from Türk S, Altınsoy I, GÇ, Ipek M, Özacar M, Bindal C (2018) 3D porous collagen/functionalized multiwalled carbon nanotube/chitosan/hydroxyapatite composite scaffolds for bone tissue engineering. Mater. Sci. Eng. C 92, 757–768. Copyright © 2018 Elsevier B.V [49].

followed by new bone formation that was triggered by the composite scaffold resulting in effective repair of the rat calvarial defects (8 mm diameter).

In another study, Türk *et al.* (2018) fabricated a novel collagen/functionalized MWCNTs/chitosan/hydroxyapatite composite scaffold represented as Col/f-MWCNT/CS/HA [49]. The composite was developed by freezing at up to −40°C at a 0.9°C/min rate followed by lyophilization (48 h, 0°C, and 200 mtorr). The compressive stress of the scaffold was in a range from 523 to 1112 kPa while swelling varied between 513.9 ± 27 and 481.05 ± 25%. The scaffolds were porous, which is advantageous for transporting the metabolites and nutrients for promoting cellular growth. The porosity ranged from 98 ± 0.15 to 95.7 ± 0.1%. A superior pore volume of 0.026 cc/g with interconnected porous microstructure was noticed, as evident from Figure 10.10. The pore size was in a range from 20 to 350 μm. The contact angle was between 87.8 and 76.7°. Further, the scaffolds showed a large surface area equivalent to 11.746 m^2/g. High viability of the MG-63 cells was noticed when grown on the composite scaffolds.

10.2.3.3 Others

Some unusual polymers are also explored for tissue engineering purposes. Seaweed-derived polysaccharide, alginate is mainly composed of 1,4-linked β-D-mannuronic acid and α-L-guluronic acid residues. This biomaterial can get cross-linked in the presence of divalent cations such as Ca^{2+}, resulting in gel formation. Alginate is considered as a significant biopolymer for drug delivery, wound dressings, and tissue engineering. In one such study, Rajesh and Ravichandran (2015) designed a tricomponent scaffold comprised of alginate, oxidized MWCNTs, and HAP using a freeze-drying method [50]. The HAP used in this scaffold was naturally derived from the chicken bones by thermal calcination at 800°C. The resulting fMWCNT–alginate–hydroxyapatite scaffold exhibited interconnected porosity with a pore size of 130–170 μm while the total porosity was 93.85%. The scaffold was biocompatible to MG-63 cells and did not show any toxicity. Instead, the composite scaffold promoted better cell proliferation, cell differentiation, and cell attachment in MG-63 cells.

In another study, Park *et al.* (2015) fabricated a 3D scaffold using bacterial cellulose (BC) incorporated with MWCNTs [51]. Initially, the bacteria *Gluconacetobacter xylinus*, was pre-cultured on a medium composed of 2.5% (w/w) mannitol, 0.5% (w/w) yeast extract, and 0.3% (w/w) bacto-peptone and eventually introduced in MWCNTs-dispersed culture medium at a ratio of 1:10. The reaction mixture was then incubated for 14 days at 28°C. The composite membrane composed of MWCNTs and BC was then harvested and purified by treatment with boiling 1 wt% sodium hydroxide for 2 h. Further, an amphiphilic comb-like polymer (APCLP) was adsorbed on the scaffold resulting in the formation of the MWCNT-BC-Syn with a core-shell structure, where MWCNTs were packed by BC nanofibril entanglements. The BC nanofibrils were 4.3 nm thick while around 3.9% of the MWCNTs surface was uncovered. Efficient regeneration of the bone tissue was noted when the scaffolds were implanted in the critical-sized defect in mouse calvaria.

Silk fibroin (SF) is another natural polymer extracted from silk fibrous protein. Its superior mechanical properties, tensile strength, and flexibility make it promising biomaterial tissue engineering. Moreover, its biodegradable nature, higher permeability, and slow-releasing potential make it ideal for biomedical applications. It is also advantageous for bone tissue regeneration as it can induce bone mineralization due to the predominance of carboxylic acid groups in its side chains. Yao *et al.* (2016) synthesized a nano-hydroxyapatite/polyamide 66 (nHAP/PA66) porous scaffold employing a phase inversion method and surface modified with which MWCNTs and SF using freeze-drying and cross-linking [52]. The resulting MWCNT/SF-nHA/PA66 scaffolds had an average diameter of ~500 μm while the porosity and compressive strength were 62.07% and 5.4 MPa, respectively. The water-absorbing rate of the scaffold was 49.11% after CNT/SF modification. In order to promote osteogenic differentiation in BMSCs, dexamethasone (DEX) was functionalized on the MWCNTs. The cumulative amount of DEX release was 97% in 0.5% SDS and 66% in PBS within 24h. Interestingly, large holes within the scaffolds were colonized by some of the cells, and DEX-loaded MWCNT/SF-nHA/PA66 scaffold induced more ALP secretion.

10.3 Intellectual Property Rights and Commercialization Aspects

Innovation in the area of functionalized CNTs has led to efficient commercially viable product development with potential marketability, as listed in Tables 10.2 and 10.3. In one such attempt, marine bioprocess waste, chitosan/hydroxyapatite (CNT-gCTS/HAp) complex scaffold was conjugated with CNTs to develop a bioactive material for bone tissue transplantation. The material showed promising thermal stability, interconnected pores, controlled biodegradability, and non-cytotoxicity [KR101273536B1] [53].

Table 10.2 Patents on CNTs based biomaterials for bone tissue engineering.

Patent number	Title	Publication date	Reference
KR101273536B1	Carbon nanotube-grafted-chitosan-hydroxyapatite composite for bone tissue engineering and a method of preparing the same	2013-06-14	[53]
KR101309416B1	Carbon nanotube-chitosan-silica nanohybrid for bone regeneration and method of preparation thereof	2013-09-23	[54]
KR101663350B1	CNT-gelatin-hydroxyapatite nanohybrids with fibrous multilayer core-shell structure for mimicking natural bone	2016-10-06	[55]
CN106178100A	Carbon Nanotubes/Chitosan complex microsphere surface forms the preparation method of orientation nano-apatite	2021-06-08	[56]
CN107537063B	A kind of complex stephanoporate bracket of carbon nanotubes and preparation method thereof	2019-10-11	[57]

(Continued)

Table 10.2 Patents on CNTs based biomaterials for bone tissue engineering. (*Continued*)

Patent number	Title	Publication date	Reference
KR20120052116A	Carbon nanotube-grafted-chitosan-hydroxyapatite composite for bone tissue engineering and a method of preparing the same	2012-05-23	[58]
US8614189B2	Carbon nanotube composite scaffolds for bone tissue engineering	2013-12-24	[59]

Table 10.3 Companies manufacturing scaffolds.

Name of company	Website	Ref.
LAYHER PTE LTD, Singapore	http://www.layher.sg/?gclid=Cj0KCQiAjc2QBhDgARIsAMc3SqSIKjWVgefEavamSBUTphH0i70QNju-tkz2JRGGChwKpvWvIcKI8YEaAu7OEALw_wcB	[60]
NORD, China	https://www.nordscaffolding.com/about-us/	[61]
YANCHENG LIANGGONG FORMWORK CO., LTD., China	https://lianggongform.en.made-in-china.com/?keyword=scaffolding%20supplier&gclid=Cj0KCQiAjc2QBhDgARIsAMc3SqQxcT35smk5wN0qcWMZnUlGyi5L7eLGkLv71JpckJzqotI98xjoGvIaAm6OEALw_wcB	[62]
Dura, China	https://www.durascf.com/about/	[63]

In another patent, a CNTs-chitosan-silica hybrid nanocomposite exhibited superior osteogenic potential, mechanical strength, and biocompatibility [KR101309416B1] [54]. CNTs-gelatin-hydroxyapatite based composite biomaterial was patented that showed a multilayer core-cell fiber bone mimetic structure [KR101663350B1] [55]. CNTs/chitosan complex microsphere were fabricated for an efficient orientation of nanoapatite that was speculated to be an efficient bone renovating biomaterial [CN106178100A] [56]. Chitosan, gelatin, hydroxyapatite, along with CNTs served as an excellent biomedical material for bone tissue engineering, clinical transplantation, and reparation [CN107537063B] [57]. Some scaffold manufacturing companies have started commercial synthesis of scaffolds for biomedical applications, as listed in Table 10.3.

10.4 Conclusion and Future Perspectives

Developing scaffolds for bone tissue engineering is challenging as the biomaterial should be able to provide adequate mechanical strength to the newly formed bone tissues. Hence, various nanomaterials are explored with superior reinforcement attributes. CNTs are considered as a viable solution to generate superior biomaterials that can be used for hard tissue engineering. However, the long-term effect of the CNTs on the target tissues should be thoroughly investigated. Similarly, it is necessary to exclusively study the biodistribution of the CNTs before they can be used in real-time therapy for bone tissue regeneration. Further, the physical properties such as length, thickness, and rigidity of the CNTs largely determine the toxic effects, and hence rational modification may help in developing nanostructures with minimum or no health hazards.

The large surface area of the CNTs can be rationally functionalized with the drugs and biomolecules. This may help trigger the cellular response in the site of injury, hence promoting more cell division, differentiation and mineralization. At the same time, functionalization with the contrast agents may help in bioimaging for effective monitoring of the healing process. Antimicrobial drugs or nanoparticles can be impregnated in the scaffolds to prevent the post-surgical risks of bacterial infections [64]. The drug release profiles should be carefully studied by selecting an array of chemical and biological polymers used for fabricating the scaffolds. However, thorough pharmacokinetics and pharmacodynamics studies should be carried out to ensure the safety of the novel CNTs based biomaterials that have a huge potential for revolutionizing regenerative medicine.

References

1. Trzeciak, T. *et al.*, Cells and nanomaterial-based tissue engineering techniques in the treatment of bone and cartilage injuries. *J. Nanosci. Nanotechnol.*, 16, 9, 8948–8952, 2016.
2. Mehranfar, S. *et al.*, The use of stromal vascular fraction (SVF), platelet-rich plasma (PRP) and stem cells in the treatment of osteoarthritis: An overview of clinical trials. *Artif. Cells Nanomed. Biotechnol.*, 47, 1, 882–890, 2019.
3. Spin-Neto, R. *et al.*, Clinical similarities and histological diversity comparing fresh frozen onlay bone blocks allografts and autografts in human maxillary reconstruction. *Clin. Implant Dentist. Relat. Res.*, 15, 4, 490–497, 2013.
4. Amorosa, L. *et al.*, Physiologic load-bearing characteristics of autografts, allografts, and polymer-based scaffolds in a critical sized segmental defect of long bone: An experimental study. *Int. J. Nanomedicine*, 8, 1637, 2013.
5. Roberts, T.T. and Rosenbaum, A.J., Bone grafts, bone substitutes and orthobiologics: The bridge between basic science and clinical advancements in fracture healing *Organogenesis*, 8, 4, 114–124, 2012.
6. Fernandez-Bances, I. *et al.*, Repair of long-bone pseudoarthrosis with autologous bone marrow mononuclear cells combined with allogenic bone graft. *Cytotherapy*, 15, 5, 571–577, 2013.
7. Pei, B. *et al.*, Applications of carbon nanotubes in bone tissue regeneration and engineering: Superiority, concerns, current advancements, and prospects. *Nanomaterials*, 9, 10, 1501, 2019.
8. Huang, B., Carbon nanotubes and their polymeric composites: The applications in tissue engineering. *Biomanuf. Rev.*, 5, 1, 1–26, 2020.
9. Mostafavi, E., Soltantabar, P., Webster, T.J., Nanotechnology and picotechnology: A new arena for translational medicine, in: *Biomaterials in Translational Medicine*, pp. 191–212, Elsevier, Netherlands, 2019. https://doi.org/10.1016/C2016-0-04209-2
10. Ghosh, S. and Webster, T.J., Metallic nanoscaffolds as osteogenic promoters: Advances, challenges and scope. *Metals*, 11, 9, 1356, 2021.
11. Mostafavi, E. *et al.*, Electroconductive nanobiomaterials for tissue engineering and regenerative medicine. *Bioelectricity*, 2, 2, 120–149, 2020.
12. Ghosh, S. and Webster, T.J., Mesoporous silica based nanostructures for bone tissue regeneration. *Front. Mater.*, 8, 213, 2021.
13. Zare, H. *et al.*, Carbon nanotubes: Smart drug/gene delivery carriers. *Int. J. Nanomedicine*, 16, 1681–1706, 2021.
14. Haniu, H. *et al.*, Basic potential of carbon nanotubes in tissue engineering applications. *J. Nanomater.*, 2012, 4, 4, 2012. https://doi.org/10.1155/2012/343747
15. Medina-Cruz, D. *et al.*, Green nanotechnology-based drug delivery systems for osteogenic disorders. *Expert Opin. Drug Deliv.*, 17, 3, 341–356, 2020.
16. Peng, Z. *et al.*, Bone tissue engineering via carbon-based nanomaterials. *Adv. Healthc. Mater.*, 9, 5, 1901495, 2020.

17. Abarrategi, A. *et al.*, Multiwall carbon nanotube scaffolds for tissue engineering purposes. *Biomaterials*, 29, 1, 94–102, 2008.
18. Zhang, D. *et al.*, Poly (L-lactide) (PLLA)/multiwalled carbon nanotube (MWCNT) composite: Characterization and biocompatibility evaluation. *J. Phys. Chem. B*, 110, 26, 12910–12915, 2006.
19. McCullen, S.D. *et al.*, Characterization of electrospun nanocomposite scaffolds and biocompatibility with adipose-derived human mesenchymal stem cells. *Int. J. Nanomedicine*, 2, 2, 253, 2007.
20. Truong, L.B. *et al.*, Advances in 3D-printed surface-modified Ca-Si bioceramic structures and their potential for bone tumor therapy. *Mater. (Basel)*, 14, 14, 3844, 2021. https://doi.org/10.3390/ma14143844
21. Perkins, B.L. and Naderi, N., Suppl-3, M7: Carbon nanostructures in bone tissue engineering. *Open Orthop. J.*, 10, 877, 2016.
22. Hooshmand, T. *et al.*, Development of sol-gel-derived multi-wall carbon nanotube/hydroxyapatite nanocomposite powders for bone substitution. *J. Compos. Mater.*, 48, 4, 483–489, 2014.
23. Jyoti, J. *et al.*, Improved nanomechanical and *in-vitro* biocompatibility of graphene oxide-carbon nanotube hydroxyapatite hybrid composites by synergistic effect. *J. Mech. Behav. Biomed. Mater.*, 117, 104376, 2021.
24. Khalid, P. and Suman, V., Carbon nanotube-hydroxyapatite composite for bone tissue engineering and their interaction with mouse fibroblast L929 *in vitro*. *J. Bionanoscience*, 11, 3, 233–240, 2017.
25. Martinelli, N.M. *et al.*, *In vitro* osteogenesis stimulation via nanohydroxyapatite/carbon nanotube thin films on biomedical stainless steel. *Materials*, 11, 9, 1555, 2018.
26. Rajesh, R. *et al.*, Development of functionalized multi-walled carbon nanotube-based polysaccharide-hydroxyapatite scaffolds for bone tissue engineering. *RSC Adv.*, 6, 85, 82385–82393, 2016.
27. Huang, B. *et al.*, Fabrication and characterisation of 3D printed MWCNT composite porous scaffolds for bone regeneration. *Mater. Sci. Eng. C*, 98, 266–278, 2019.
28. Mostafavi, A. *et al.*, *In situ* printing of scaffolds for reconstruction of bone defects. *Acta Biomater.*, 127, 313–326, 2021.
29. Mattioli-Belmonte, M. *et al.*, Tuning polycaprolactone-carbon nanotube composites for bone tissue engineering scaffolds. *Mater. Sci. Eng. C*, 32, 2, 152–159, 2012.
30. Pan, L. *et al.*, Multiwall carbon nanotubes/polycaprolactone composites for bone tissue engineering application. *Colloids Surf. B Biointerfaces*, 93, 226–234, 2012.
31. Tohidlou, H. *et al.*, Amine-functionalized single-walled carbon nanotube/polycaprolactone electrospun scaffold for bone tissue engineering: *in vitro* study. *Fiber. Polym.*, 20, 9, 1869–1882, 2019.
32. Yang, H., Li, J., Liao, Q., Guo, H., Chen, H., Zhu, Y., Cai, M., Lv., H., *In vitro* evaluation of a novel multiwalled carbon nanotube/nanohydroxyapatite/

polycaprolactone composite for bone tissue engineering. *J. Mater. Res.*, 34, 4, 532–544, 2019.

33. Eil Bakhtiari, S.S., Karbasi, S., Tabrizi, S.A.H., Ebrahimi-Kahrizsangi, R., Salehi, H., Evaluation of the effects of chitosan/multiwalled carbon nanotubes composite on physical, mechanical and biological properties of polymethyl methacrylate-based bone cements. *Mater. Technol.*, 35, 5, 267–280, 2020.
34. Pahlevanzadeh, F., Bakhsheshi-Rad, H.R., Kharaziha, M., Kasiri-Asgarani, M., Omidi, M., Razzaghi, M., Ismail, A.F., Sharif, S., Krishna, S.R., Berto, F., CNT and rGO reinforced PMMA based bone cement for fixation of load bearing implants: Mechanical property and biological response. *J. Mech. Behav. Biomed. Mater.*, 116, 104320, 2021. https://doi.org/10.1016/j.jmbbm.2021.104320
35. Cheng, Q., Rutledge, K., Jabbarzadeh, E., Carbon nanotube-poly (lactide-co-glycolide) composite scaffolds for bone tissue engineering applications. *Ann. Biomed. Eng.*, 41, 5, 904–916, 2013.
36. Diaz, E. *et al.*, Hydrolytic degradation and cytotoxicity of poly (lactic-co-glycolic acid)/multiwalled carbon nanotubes for bone regeneration. *J. Appl. Polym. Sci.*, 137, 10, 48439, 2020.
37. Mikael, P.E., Amini, A.R., Basu, J., Arellano-Jimenez, M.J., Laurencin, C.T., Sanders, M.M., Carter, C.B., Nukavarapu, S.P., Functionalized carbon nanotube reinforced scaffolds for bone regenerative engineering: Fabrication, *in vitro* and *in vivo* evaluation. *Biomed. Mater.*, 9, 3, p.035001, 2014. https://doi.org/10.1088/1748-6041/9/3/035001
38. Farahani, A., Zarei-Hanzaki, A., Abedi, H.R., Tayebi, L., Mostafavi, E., Polylactic acid piezo-biopolymers: Chemistry, structural evolution, fabrication methods, and tissue engineering applications. *J. Funct. Biomater.*, 12, 4, p71, 2021. https://doi.org/10.1088/1748-6041/9/3/035001
39. Mei, F. *et al.*, Improved biological characteristics of poly (L-lactic acid) electrospun membrane by incorporation of multiwalled carbon nanotubes/hydroxyapatite nanoparticles. *Biomacromolecules*, 8, 12, 3729–3735, 2007.
40. Shuai, C. *et al.*, A self-assembled montmorillonite-carbon nanotube hybrid nanoreinforcement for poly-L-lactic acid bone scaffold. *Mater. Today Adv.*, 11, 100158, 2021.
41. Vozzi, G., Corallo, C., Daraio, C., Pressure-activated microsyringe composite scaffold of poly (L-lactic acid) and carbon nanotubes for bone tissue engineering. *J. Appl. Polym. Sci.*, 129, 2, 528–536, 2013.
42. Lan, W. *et al.*, Carbon nanotube reinforced polyvinyl alcohol/biphasic calcium phosphate scaffold for bone tissue engineering. *RSC Adv.*, 9, 67, 38998–39010, 2019.
43. Rodrigues, A.A. *et al.*, Polyvinyl alcohol associated with carbon nanotube scaffolds for osteogenic differentiation of rat bone mesenchymal stem cells. *Carbon*, 50, 2, 450–459, 2012.

44. Cao, J. *et al.*, Bioactive poly (etheretherketone) composite containing calcium polyphosphate and multi-walled carbon nanotubes for bone repair: Mechanical property and *in vitro* biocompatibility. *J. Bioact. Compat. Polym.*, 33, 5, 543–557, 2018.
45. Liao, C.Z. *et al.*, Novel polypropylene biocomposites reinforced with carbon nanotubes and hydroxyapatite nanorods for bone replacements. *Mater. Sci. Eng. C*, 33, 3, 1380–1388, 2013.
46. Im, O. *et al.*, Biomimetic three-dimensional nanocrystalline hydroxyapatite and magnetically synthesized single-walled carbon nanotube chitosan nanocomposite for bone regeneration. *Int. J. Nanomedicine*, 7, 2087, 2012.
47. Venkatesan, J., Qian, Z.-J., Ryu, B.M., Kumar, N.A., Kim, S.-K., Preparation and characterization of carbon nanotube-grafted-chitosanâ-natural hydroxyapatite composite for bone tissue engineering. *Carbohydr. Polym.*, 83, 2, 569–577, 2011.
48. Jing, Z., Wu, Y., Su, W., Tian, M., Jiang, W., Cao, L., Zhao, L., Zhao, Z., Carbon nanotube reinforced collagen/hydroxyapatite scaffolds improve bone tissue formation *in vitro* and *in vivo*. *Ann. Biomed. Eng.*, 45, 9, 2075–2087, 2017.
49. Türk, S. *et al.*, 3D porous collagen/functionalized multiwalled carbon nanotube/chitosan/hydroxyapatite composite scaffolds for bone tissue engineering. *Mater. Sci. Eng. C*, 92, 757–768, 2018.
50. Rajesh, R. and Ravichandran, Y.D., Development of a new carbon nanotube-alginate-hydroxyapatite tricomponent composite scaffold for application in bone tissue engineering. *Int. J. Nanomedicine*, 10, Suppl 1, 7, 2015.
51. Park, S. *et al.*, *In situ* hybridization of carbon nanotubes with bacterial cellulose for three-dimensional hybrid bioscaffolds. *Biomaterials*, 58, 93–102, 2015.
52. Yao, M.-Z. *et al.*, Fabrication and characterization of drug-loaded nanohydroxyapatite/polyamide 66 scaffolds modified with carbon nanotubes and silk fibroin. *Int. J. Nanomedicine*, 11, 6181, 2016.
53. 중-지, 김.벤., Carbon nanotube-grafted-chitosan- hydroxyapatite composite for bone tissue engineering and a method of preparing the same. South Korea Patent KR101273536B1, 2013.
54. 김해원김정주신원상, Carbonnanotube-chitosan-silica nanohybrid for bone regeneration and method of preparation thereof. South Korea Patent KR101309416B1, 2013.
55. 신원상김해원, CNT-gelatin-hydroxyapatite nanohybrids with fibrous multilayer core-shell structure for mimicking natural bone. South Korea Patent KR101663350B1, 2016.
56. 黄棣杜晶晶胡银春牛璐璐魏延连小洁陈维毅, Carbon nanotubes/chitosan complex microsphere surface forms the preparation method of orientation nano-apatite. China Patent CN106178100A, 2021.
57. 宋克东蒋思宇卢延国李文芳李丽颖刘天庆, A kind of complex stephanoporate bracket of carbon nanotubes and preparation method thereof. China Patent CN107537063B, 2019.

58. K. Se-Kwon, R. Bomi J. Benkatasan, C. Zhong-Ji, Carbon nanotube-grafted-chitosan- hydroxyapatite composite for bone tissue engineering and a method of preparing the same. South Korea Patent KR20120052116A, 2013.
59. S.G. Kumbar, C.T. Laurencin, S.P. Nukavarapu, Carbon nanotube composite scaffolds for bone tissue engineering. US Patent US8614189B2, 2013.
60. LAYHER PTE LTD, Singapore. Website: http://www.layher.sg/?gclid=Cj0K-CQiAjc2QBhDgARIsAMc3SqSIKjWVgefEavamSBUTphH0i70QNju-tkz-2JRGGChwKpvWvIcKI8YEaAu7OEALw_wcB. Accessed on 24th February, 2022.
61. NORD, China. Website: https://www.nordscaffolding.com/about-us/ Accessed on 24th February, 2022.
62. YANCHENG LIANGGONG FORMWORK CO., LTD., China Website: https://lianggongform.en.made-in-china.com/?keyword=scaffolding%20supplier&gclid=Cj0KCQiAjc2QBhDgARIsAMc3SqQxcT35smk5wN0qcW-MZnUlGyi5L7eLGkLv71JpckJzqotI98xjoGvIaAm6OEALw_wcB Accessed on 24th February, 2022.
63. Dura, China. Website: https://www.durascf.com/about/ Accessed on 24th February, 2022.
64. Ibrahim, D.M., Sani, E.S., Soliman, A.M., Zandi, N., Mostafavi, E., Youssef, A.M., Allam, N.K., Annabi, N., Bioactive and elastic nanocomposites with antimicrobial properties for bone tissue regeneration. *ACS Appl. Bio Mater.*, 3, 5, 3313–3325, 2020.

11

Application of Functionalized Carbon Nanotubes in Biomimetic/ Bioinspired Systems

Mohammad Mobin[1]*, Ruby Aslam[1], Saman Zehra[1], Jeenat Aslam[2] and Shahidul Islam bhat[1]

[1]Corrosion Research Laboratory, Department of Applied Chemistry, Faculty of Engineering and Technology, Aligarh Muslim University, Aligarh, India
[2]Department of Chemistry, College of Science, Taibah University, Yanbu, Al-Madina, Saudi Arabia

Abstract

Over the past billion years, biological systems have developed various ways to achieve optimal bulk and structure unification. The discovery of biological properties has led to an increased interest in studying these materials through modern characterization techniques. Biomimetics is a field of research that allows scientists to mimic the biological properties of organisms for the development of new materials and processes. The complexity of biological materials and their surfaces can be distinguished from one another through their morphological and chemical properties. In nature, hierarchical structures with dimensions of features ranging from the macroscale to the nanoscale are commonly formed to provide properties of interest. Scientists and engineers are now working on developing new engineered materials that can mimic the properties of biological systems. For the past two decades, the development of carbon nanotubes (CNTs)-reinforced composites has been increasing at a fast pace. This paper presents a detailed study on the current breakthroughs in the manufacturing of bio-inspired hybrid materials. Various bio-inspired functionalities of CNT-based composites, including bioadhesives, injectability, antibacterial properties, and degradability applicable to advanced drug delivery systems and medical devices will be discussed. The paper also reviews the future potential of these materials.

**Corresponding author*: drmmobin@hotmail.com

Jeenat Aslam, Chaudhery Mustansar Hussain and Ruby Aslam (eds.) Functionalized Carbon Nanotubes for Biomedical Applications, (257–280) © 2023 Scrivener Publishing LLC

***Keywords*:** Bioactive scaffolds, composite materials, CNTs, bio-mimicking

11.1 Introduction

The structure and function of biological materials have been continuously refined over millions of years. Through the study of live organisms, researchers can then apply these principles and technologies to manmade materials to emulate biological functions and performance is in a central place of biomimetics [1]. The term "biomimetic" was coined in the 1960s and is derived from the Greek words "bios" (nature) and "mimesis" (imitate) (imitation) [2]. Due to the increasing number of people who get injured and sick every year, the demand for medical devices that can help treat and cure these conditions has become very high. In the US, over 8 million surgeries per year are performed annually for these patients. Each year, over 100,000 people die while waiting for an organ transplant. The global market for bio-based medical devices was valued at USD 35.5 billion dollars in 2020. It is expected to reach USD 47.5 billion dollars by 2025 at a compound annual growth rate of 6.0% [3].

The complexity of biological systems has been optimized through the years to enable them to adapt to changes in their environment [4, 5]. For instance, Nacre and bone have incredibly tough and resilient characteristics. They can endure different kinds of impacts much beyond their constituent parts [6]. The feet of a gecko may produce enough adhesion to sustain many times its body weight, allowing it to climb walls and move quickly across ceilings [7]. Despite growing in a muddy area, Lotus leaves have great water repellence and appear to never get dirty [8]. Peacock feathers and Morpho butterfly wings have vivid and iridescent hues attributable to components that evolved independently of pigments in other species yet are akin to pigments in other species. There is a rising effort to discover techniques to synthesize composites with these extraordinary features mirroring those observed in nature, in addition to research activities committed to further understanding the causes underlying these interesting structural properties in nature. Understanding the design and properties of nature is also important to developing materials that are like those found in nature. Various techniques such as chemical vapor deposition, lithography, and self-assembly can be used to create advanced materials that mimic the properties of nature [9, 10].

11.2 Naturally Occurring Materials

11.2.1 Nacre and Bone

The nacre, which is a typical shell or bone structure, is often used to demonstrate how natural composite materials can achieve exceptional properties. Its laminated composite structure is made up of aligned ceramic platelets and a small fraction of polymer. The nacre has a "brick and mortar" microstructure [11]. The mortar is made up of various proteins and chitin, and it is approximately 20 to 50 nm thick. The bricks are made up of 0.5 μm thick and 10–20μm wide aragonite ($CaCO_3$) [12]. The former provides strength while the latter provides ductility. The main components of mortar's toughening mechanism include crack bridging, aragonite-plate sliding, and energy dissipation [13]. The properties of these materials have led to the development of composite materials that have improved fracture toughness and strength–toughness balance.

11.2.2 Petal Effect and Gecko Feet

There are two types of natural surfaces that have ultrahigh adhesion. The first is the petal effect, which allows plants i.e., roses to keep their petals in a spherical shape even when turned upside down (Figure 11.1) [14].

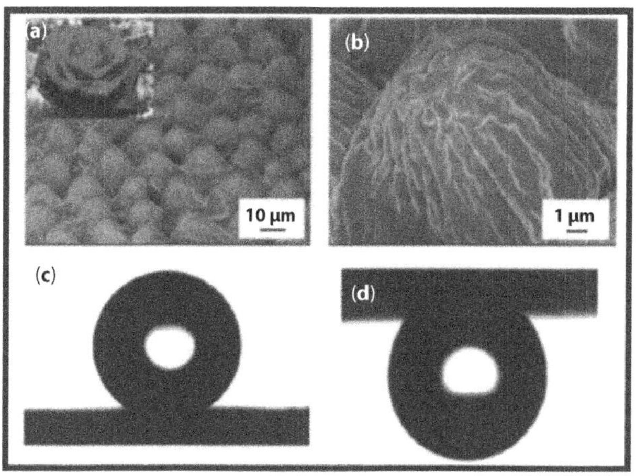

Figure 11.1 SEM images of the surface of a red rose petal (a, b). Shape of a water droplet on the petal's surface (c) and when it is turned upside down (d). Republished from Ref. [14] with permission from the American Chemical Society.

The second one is animals like the gecko lizards can cling to a variety of surfaces, such as moist, large, small, smooth, or rough ones. These creatures' adhesive force can hold far more than their own weight [15]. A recent study revealed that the strong adhesion of the petals and feet of the geckos can be attributed to their periodic arrays of micro-and nanostructures. Red rose petals with nanofolds on the top are made with micropapillae.

The gecko feet are made up of hundreds of tiny spatula-like structures having a diameter of 200-500 nm that are arranged into a series of well-aligned keratinous hairs which are known as setae and are 30–130 μm in length and 5 μm in diameter. These spatulas bend when they encounter any surface and interact with the surface through van der Waals forces which allows the gecko to climb up walls or across the ceiling.

11.2.3 Lotus Effect

The leaves of plants have a unique wettability or water repellence that prevents water droplets from adhering to their surfaces. This phenomenon, which is known as the lotus effect, can cause the leaves to bounce off and leave behind dirt particles. This property is known as the "lotus effect" or "self-cleaning" property [16]. Studies suggest that the layer of

Figure 11.2 Images of superhydrophobic lotus leaf (inset image is a water droplet on the surface of a leaf) (a). Low magnification SEM image of the lotus leaf surface structure (b). High-resolution SEM image of a single papilla, consisting of cilium-like nanostructures (c). Republished from Ref. [20] with permission from the American Institute of Physics.

epicuticular wax on top of randomly scattered micro-papillae contributes to the non-wetting properties of the papillae (Figure 11.2) and low surface energy epicuticular wax covering on these papillae are responsible for the non-wetting property [17]. The surface area of the liquid-solid contact rises with roughness, resulting in a larger energy barrier. Additionally, the rough surface retains air beneath the water droplet, forcing it to sit partially on an air layer. All these characteristics combine to create a superhydrophobic surface (contact angle > 150° and sliding angle 5°) [18, 19].

11.2.4 Structural Colors, Antireflection, and Light Collection

A photonic structure is a type of optical structure that affects the flow of light. It can produce various optical effects such as antireflection and structural color and light collection. Many birds, insects, fish have structures with optical effects that allow them to change their appearance through the viewing angle. These structural colors can be used to attract mates or scare away predators. For instance, some insects and birds, such as the butterfly wings and the beetle scales, have structures with optical effects that allow them to change their appearance through the viewing angle [21]. Antireflective coatings allow insects to capture more photons, which helps them see better in low-light situations. Antireflection can also be used as a form of camouflage [22]. Brittle stars use photonic elements formed of calcite to capture light, allowing them to detect shadows and flee predators into dark crevices. The appeal of these structural colors has prompted a spate of attempts to duplicate them. If successfully manufactured, such structures might be used as sensing Bragg reflectors, high-resolution optical filters, and a replacement for pigment-based coatings [23].

11.3 Bioinspired Functionalized CNTs Material

One-dimensional CNTs are made by a rolling sheet of graphene, which is an advanced material that can be used for various applications. They can be categorized into two types based on the number of graphene sheets used: single-walled CNTs (SWCNTs) and multi-walled CNTs (MWCNTs). MWCNTs also contain double-walled CNTs. These materials are used in the production of structural components for aerospace and automotive applications due to their outstanding strength and flexibility. CNT reinforced fibers are stiffer than steel and resistant to external wear & tear. They have high electrical conductivity which makes them ideal for use in conductive polymer composites. The global carbon nanotube market was valued at $15.3 billion by 2017 and is expected to reach $130.2 billion by 2030, growing at a CAGR of 16.3% from 2021 to 2030. The increasing demand for CNTs in various applications such as medical devices and aerospace and defense and coatings is driving the market [24].

Since the beginning of the 21st century [25], CNTs are also expected to play a major role in the biomedical industry such as drug delivery, cancer therapy, tissue engineering, body implants, medical devices, and dental filling materials due to their various advantages, such as their small size, chemical stability, large surface area to mass ratio, and electronic polyaromatic structure. They are expected to be better drug carriers than the currently available drug carriers such as nanoparticles, polymers, niosomes, microparticles, and liposomes [25, 26]. They can be used as a drug delivery system by directly penetrating cells or by conjugating with various therapeutic molecules [25]. The medicine might be administered to the surface of the CNTs or put inside them. Both the medication and the CNT carrier enter the cells by endocytosis or insertion and diffusion in two different drug delivery conjugates using CNT systems [27, 28]. Surface attachment is ineffective when compared to internalization. In the former methodology, the conjugated drug-releasing pharmacological agents are degraded inside the cells, whereas in the latter method, the drug can be degraded in the physiological fluids before being internalized by the cells [29, 30]. CNTs were first binded to antineoplastic and antibacterial drugs for cancer and infection treatment, respectively, employing these cutting-edge medication delivery strategies. Although these properties are linked to highly desirable mechanical, electrical, and chemical properties for medical applications, CNTs' biological applications are limited by their initial hydrophobic surface, which is prone to severe aggregation when exposed to aqueous solutions and has been found to be harmful to human health. Researchers, manufacturers, and those involved in the creation of medicinal infusions are also in danger [27]. To solve this issue in biomedical applications, carbon nanotube functionalization is essential. A chemical functionalization is a common approach for increasing CNT dispersibility, decreasing toxicity, and imparting desirable features in biological contexts [31]. After surface modification by targeting ligands and drug molecules, the resulting functionalized CNTs (f-CNTs) are less toxic [32] and immunogenic [33, 34], and show significant superiority as drug delivery devices capable of high drug payload and specific distribution into [35] biological systems [36, 37].

Functionalized CNT-based composites can replicate extracellular matrix (ECM) components and boost cellular connections and osteogenic potential by linking with different proteins and other macromolecules. The use of osteogenic drugs or proteins is sometimes required for bone remodeling. By acting as a delivery channel for pharmaceuticals, growth hormones, proteins, and nucleic acids, carbon nanotubes with diverse functional groups can greatly influence the behavior and differentiation of osteoblasts and osteoclasts in bone tissue [38]. Due to 1) a mechanism of greatly increased fluid flow, CNT membranes offer an interesting chance to replicate natural protein channels, 2) the possibility to use 'gatekeeper' chemistry at the pores' entry, 3) the ability to perform biochemical reactions on gatekeeper molecules, as well as 4) the ability to chemically functionalize each side of the membrane individually. Biomimetic nanomembranes are a subclass of synthetic nanomembranes that share some features with natural biological membranes, either partially or totally.

The focus of this article is on recent advances in bioinspired hybrid materials made from functionalized carbon nanotubes. The following section will go through a few of the biomimetic materials research that has been published on the CNTs. Finally, many bioinspired hybrid material fabrication applications, current challenges, and future development are discussed.

By pouring a mixture of gases at 750°C over SiO_2/Si wafers, Qu *et al.* were able to build vertically aligned multi-walled carbon nanotube (VAMWNT) arrays [39]. The sheer force of the dry adhesives produced is about ten times that of a foot gecko. Due to the alignment of the carbon nanotube's curly segments, strong binding was observed along the shear direction and facile lifting in the normal direction was also exhibited. Due to the higher packing densities and the better structure of conjugated carbon, a new type of single-walled carbon nanotube known as VA-SWNT could be produced. This new material exhibits strong van der Waals forces and a vertically aligned structure [40]. These nanotubes had a macroscopic normal adhesion force of up to 29 N cm^{-2}. Due to the unusual thermal and electric properties of VA-SWNT, these dry adhesives exhibit reversible semiconducting behavior under load and good thermal resistance. The Ge *et al.* [41] were able to create a synthetic gecko tape. The tape was made with aligned CNT films and exhibited a high level of normal adhesion force of up to 29 N cm^{-2}. The tape was made with VA-SWNT, which has mechanical and electric properties that are different from those of other commercially available nanotube films. These dry adhesives exhibit reversible semiconducting behavior under load and good thermal resistance.

Majumder *et al.* used carbon nanotube-based membranes with "gatekeeper" chemistry to replicate the skin's natural pores [42]. Carbon nanotubes were implanted into chelate of Ca^{2+} ions with sodium alginate (SA) and polyacrylic acid (PAA) by Wei *et al.* [43]. The prepared

Ca-PAA-SA-CNTs hydrogel was further used to monitor human movements such as finger bending, knee bending, and breathing [43]. The produced hydrogel had excellent rheological features, such as stretchability, self-healing, and 3D printing. The self-healing property of Ca-PAA-SA-CNTs hydrogels was shown in Figure 11.3a. The resistive and capacitive

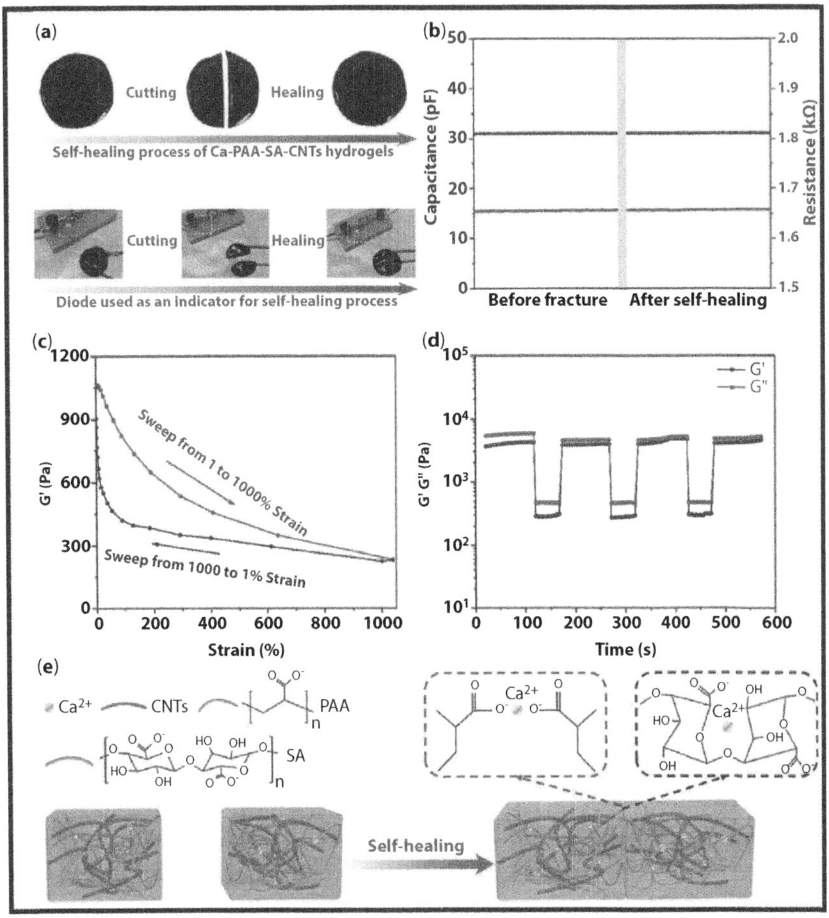

Figure 11.3 (a) Cutting and self-healing processes of a circular Ca-PAA-SA-CNTs hydrogel and the corresponding self-healing processes with a diode used as an indicator. (b) Resistance and capacitance of Ca-PAA-SA-CNTs hydrogels before fracture and after self-healing. (c) Self-healing efficiency of Ca-PAA-SA-CNTs hydrogels quantitatively evaluated by calculating the recovery rate of G′ at strain sweeps from 1 to 1000% and back to 1% at a frequency of 1 Hz. (d) G′ and G″ of Ca-PAA-SA-CNTs hydrogels subjected to a strain sweep test with alternate small oscillation forces (1% strain) and large oscillation forces (500% strain). (e) Schematic illustration of the self-healing process. Republished from Ref. [43] with permission from American Chemical Society.

signals of the Ca-PAA-SA-CNTs hydrogels were characterized using an LCR meter (Figure 11.3b). Meanwhile, the rheometer's strain sweep mode was utilized to quantify the hydrogel's self-healing efficiency, which was obtained by calculating the recovery of storage modulus (G′). The hydrogel's self-healing efficiency was assessed to be 86.6% during the process of altering the strain from 1% (G′ = 1051 Pa) to 1000% (G′ = 218 Pa) and then back to 1% (G′ = 910 Pa), as shown in Figure 11.3c. The hydrogel that was created was viscoelastic (Figure 11.3d). The improvement of the self-healing abilities of Ca-PAA-SA-CNTs hydrogels could be achieved by the chelation of Ca^{2+} and carboxyl group of PAA. The formation of an egg-box structure between Ca^{2+} and SA further improves the self-healing performances, as shown in Figure 11.3e.

A study conducted by X Li *et al.* [44] discovered that multi-walled carbon nanotubes (MWNTs) and graphite (GP) compacts could stimulate the formation of bone in the dorsal musculature of ddy mice. The results indicated that the presence of MWNTs could stimulate the activation of inducible cells in soft tissues, which could then form inductive bone. Kuang *et al.* [45] constructed a peptide recognition element (PRE) (Figure 11.4) that may spontaneously coat SWNTs and give target selectivity. This method can be used to make SWNT-FET devices that can be tuned to detect traces of certain targets with PREs.

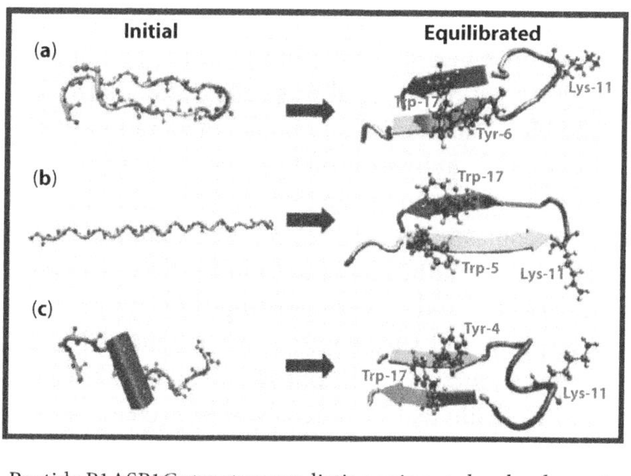

Figure 11.4 Peptide P1ASP1C structure prediction using molecular dynamics simulations. Starting three different initial conformations a, b, and c (left panels) and equilibrated structures (right panels). Interaction between Trp17-Phe18 from one side and Tyr4-Trp6-Tyr5 from the other side stabilizes the sheet structure, while the bulkiness of Asn9- Asn10-Lys11-Thr12 destabilizes the helix structure. Republished from Ref. [45] with permission from American Chemical Society.

In another study, the researchers [46] were able to produce high-quality chitosan/carbon nanotubes/hydroxyapatite (CS/CNTs/Hap) composite materials with a high elastic modulus (509.9-1089.1 MPa) and high compressive strength (33.2-105.5 MPa). *In vitro* cell line tests on preosteoblast, MC3T3-E1 cells revealed that the composite materials can effectively stimulate osteoblast proliferation onto the surface of CS/CNTs/Hap. The authors proposed that the composites generated are suitable for bone tissue engineering, according to the research. A biomimetic scaffold made from chitosan, hydroxyapatite, and MWCNTs (CTS/MWCNT/nHAp) was created by mixing these components into an ice segregation-induced self-assembly approach [47]. They tested their scaffold's efficiency by introducing multiple Mesenchymal Stem Cells (MSCs) derived from the periosteum. Because of their structural organization, surface properties, and cell survival, CTS/MWCNT/nHAp scaffolds can be used as a suitable material for bone tissue engineering. The hydroxyapatite component was then combined with carbon nanotubes and chitosan to improve its biological properties. The resulting biomimetic scaffold is then coated with an alpha-beta titanium alloy i.e., Ti-6Al-4V alloy to improve the biological and mechanical qualities of orthopedic material [48].

Another work [49] used glycerol phosphate to crosslink chitosan-grafted carbon nanotubes (MWCNTs and SWCNTs) and chitosan-grafted hydroxyapatite complexes. Using hydrogel nanofillers with a faster sol/gel transition increased the mechanical characteristics of the final material. Injectable composites of cross-linked carbon nanotubes and hydroxyapatite in thermosensitive gels showed promise, with good bioactivity, longer drug release, enhanced mechanical properties, and a faster gelation time. To improve the biocompatibility of a bone regeneration implant, a nano-structured CNTs-chitosan hybrid solution was created and electrophoretically deposited onto a titanium plate [50]. The effects of the CNTs-chitosan hybrid on the osteoblastic MC3T3-E1 cell line were investigated, and the results revealed that cell adhesion was stimulated. Furthermore, the nano-topographical property of the hybrid coating aided in the massive amount of protein adsorption and sustained release. Gholizadeh *et al.* [51] used MWCNTs, chitosan, and b-Glycerophosphate composites to make CNTs-chitosan composite scaffolds. The water absorption characteristics, porosity, and electrical conductivity of the composites were improved by increasing f-MWCNT to 1 w/v%. The compressive and tensile strength of the scaffold was also increased by adding up to 0.5% carbon nanotubes. The cell line results indicate that the composite obtained between 0.1 and 0.5 w/v% was appropriate for bone tissue engineering application because the functionalized MWCNTs make the nanotubes more biocompatible and soluble in solvents, and the polymeric structure of the composites has not shown significant cytotoxicity, unlike other structured composites. In another work [52], the chitosan-MWCNTs scaffold was employed to investigate bone repair and regeneration. Using a water-soluble tetrazolium salt test and double staining techniques, the cytotoxic, apoptotic, and necrotic effects of MWCNTs-chitosan scaffold on chondrocyte cell lines were examined. The results suggest that the scaffold material had no effect on the viability of chondrocyte cell lines, implying that CNTs-chitosan composites are not toxic to cells and that the double staining method can distinguish between necrotic and apoptotic effects of chitosan-MWCNT nanocomposite on different cell lines. The scaffold's mechanical capabilities are demonstrated by the stress-strain graph, and the inclusion of CNTs increased the elongation strength. Highly porous collagen composites functionalized with CNTs, chitosan, and HAp (Col/f-MWCNT/CS/HA) scaffolds were created using the freeze-drying method [53]. The Ca/P ratio of the reported Col/f-MWCNT/CS/HA composite was 1.52, which is like the Ca/P ratio of natural bone (1.6). The -OH, $-NH_2$, and -C=O groups of collagens have formed hydrogen connections with the $-NH_2$ and -OH groups of chitosan, as well as the C=O and -OH groups of f-MWCNTs (Figure 11.5). To biomineralize the composites, a biomimetic technique using simulated body fluid was applied. The elastic modulus and compressive stress of the composite were measured before (523 and 37 kPa) and after mineralization (523 and 1112 and 57 kPa), respectively, to ensure structural unity during bone regeneration.

The swelling (513.9-481.05%), porosity (98-95.7%), and contact angle (87.8°-76.7°) of Col/f-MWCNT/CS/HA were measured and compared to collagen, chitosan, and Col/f-MWCNT before and after biomineralization. Figure 11.6 shows how HAs were evenly distributed over the whole scaffold, with only minor aggregation.

Sharmeen *et al.* [54] presented the results of their study on the properties of MWCNTs/gelation-chitosan composite materials functionalized with polyethylene glycol. They found that these materials have better thermal and mechanical properties, stiffness, wetness, and microfibrillar within pore walls which make them be used in the development of bone tissue engineering. Their improved mechanical and thermal properties can be utilized in porous gelatin/chitosan composite materials. An antibiotic drug, ciprofloxacin, was incorporated into a nanocomposite matrix. The antimicrobial susceptibility test and dissolution test were performed to evaluate the effects of MWCNTs on the drug's release. The dissolution test (Figure 11.7) revealed that the drug's release was initially sharp. However, it steadily decreased after being subjected to the antimicrobial susceptibility

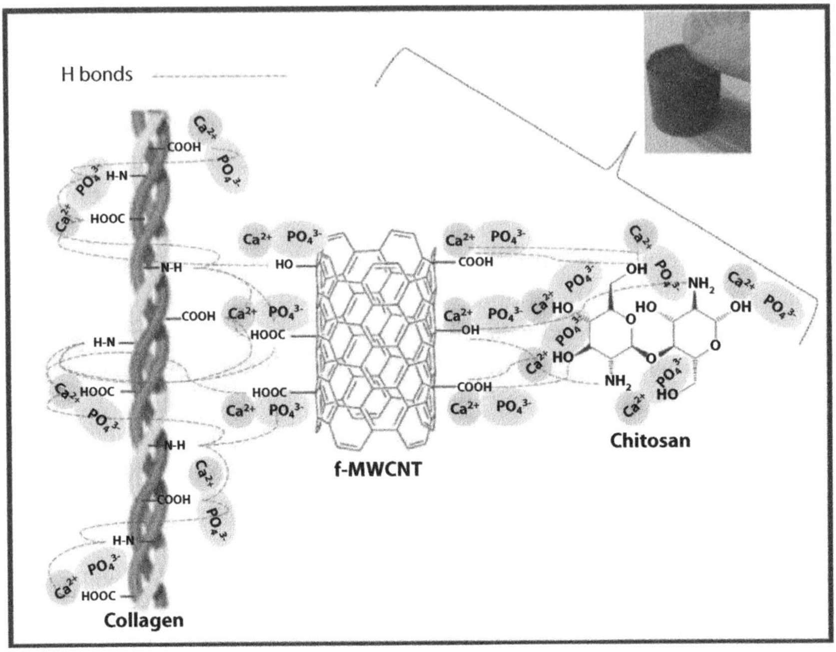

Figure 11.5 Biomineralization of collagen functionalized with CNTs, chitosan, and HAp composites using simulated body fluid. Republished from Ref. [53] with permission from Elsevier.

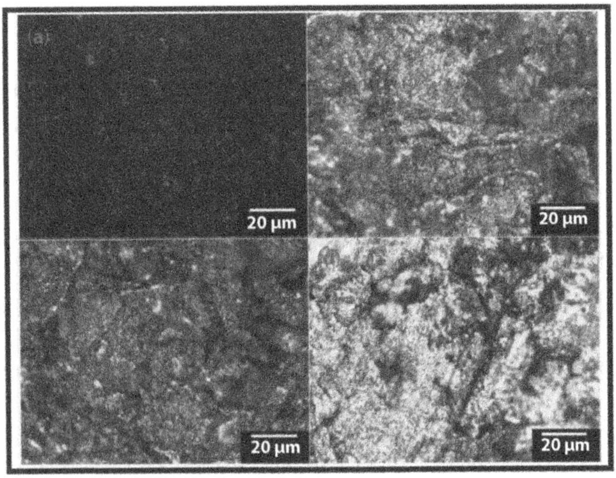

Figure 11.6 Optical microscopy images of fabricated 3D (a) Col, (b) CS, (c) Col/f-MWCNT and (d) Col/f-MWCNT/CS scaffolds. Republished from Ref. [53] with permission from Elsevier.

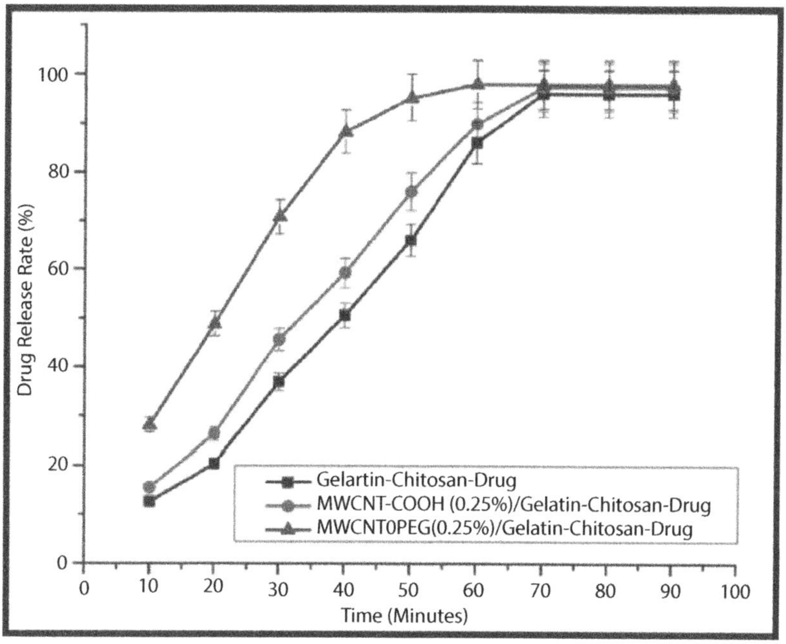

Figure 11.7 The release rate (%) of ciprofloxacin from different composite films at pH 7.4. Republished from Ref. [54] with permission from Elsevier.

test showing a sustained release. Brine shrimp lethality test revealed that the MWCNTs/gelatin-chitosan nanocomposites did not show any cytotoxicity.

The combination of CNTs and sodium hyaluronate can help in the healing of tooth sockets faster in rats [55]. The researchers also found no effect of the HY-CNTs on the cardiovascular function of the rats on the direct ingestion even after 7 days. A second study revealed that the titanium surfaces coated with the sodium hyaluronate (HY)-functionalized CNTs bio-composites exhibited increased mRNA expression of various type I and III collagen, bone morphogenetic proteins 2 and 4, and osteocalcin, as well as mineralized bone nodule deposition [56, 57]. To repair bone in rat tibiae, CNTs were combined with hyaluronate (HY) bio composites. The bone defect was generated with a 1.6 mm diameter drill on the 7th and 14th days, and histological and morphometric examinations were performed. The ensuing histomorphometry demonstrated an enhanced percent with highly structured and denser bone trabeculae. Furthermore, increased levels of VEGF, collagen I (Col I), bone morphogenetic protein-2 (BMP-2), and osteocalcin (OCN) were seen in the Tibiae of an animal treated with HY-CNTs [58].

The 3D poly(lactic acid) multiwalled carbon nanotube (PLA/MWCNTs) nanocomposites scaffolds were created using a pressure-activated microsyringe microfabrication technique that allows for the direct fabrication of selected microstructures [59]. The intrinsic mechanical characteristics of the scaffolds can be varied by changing the ratio of CNTs to PLA, ranging from 60 to 170 MPa. The built scaffold demonstrated higher stiffness, porosity (60-75%), and cell survival (>75%) than a pure 3D microfabricated PLA scaffold on human fetal osteoblasts (hFOBs).

Poly(d, L lactic acid) was added to vertically aligned-CNTs and hydroxyapatite in another study [60]. The *in vivo* investigation revealed that the scaffold induced bone remodeling. It was created using a poly(d, l-lactic acid)/vertically aligned CNTs/ hydroxyapatite bio-based scaffold by electrodeposition and immersion in the simulated physiological fluid. Human chondrocyte adhesion is aided by the produced material, which also reduces type I Collagen mRNA levels [61]. A 3D printed structure reinforced with polylactic acid/CNT filaments is created via melt extrusion [62]. After a 24-hour incubation period, the newly created composite filaments possessed mechanical properties and porosity comparable to pure polymer scaffolds, and biocompatibility was discovered on human MSCs (Mesenchymal Stem Cells).

Photoacoustic (PA) differentiation of bone marrow-derived marrow stromal cells to osteoblasts was described by Sitharaman *et al.* [63] utilizing SWCNT-Poly(Lactic-Co-Glycolic Acid) (PLGA) films. Quantitative (alkaline phosphatase, calcium, and osteopontin) and qualitative (alizarin red stain) assays were used to assess cell differentiation. The osteo-differentiation of bone marrow-derived marrow stromal cells was improved by PA stimulation and SWCNTs in PLGA films, according to the study. Gupta *et al.* [64] used SWCNT and PLGA composites to explore the interactions between human bone marrow mesenchymal stem cells (BMSCs) and MC3T3-E1 cells in terms of cell proliferation, growth, mineralization, gene expression, and extracellular matrix synthesis. The findings revealed that the uniform distribution of SWCNTs on the PLGA matrix had no effect on the degradation rate. The biocompatibility of the composites was further demonstrated by the MC3T3-E1 and hBMSCs cells' normal and non-stressed morphologies. Gupta *et al.* [65] confirmed *in vivo* biocompatibility of SWCNT/PLGA composite in a subcutaneous implant rat model. The consistent distribution of composite matches human trabecular bone in appearance, according to the research. The composites' potential was evaluated over a 12-week period, and the results revealed that PLGA degrades slowly. The composite material's slow degradation rate may be advantageous for long-term investigations since the lower percentage of SWCNTs in the composite can considerably reduce implant toxicity. f-SWCNTs on the PLGA matrix had no effect on the degradation rate. The biocompatibility of the composites was further demonstrated by the MC3T3-E1 and human bone marrow-derived mesenchymal stem cells (hBMSCs) normal and non-stressed morphologies. Pan *et al.* [66] employed a solvent evaporation method to create MWNTs-PCL scaffolds. These composite materials' mechanical properties, such as tensile strength, swelling ratio, and degradation percentages, were discovered. Cell growth and differentiation were facilitated by the composites' increased hydrophilicity and protein adsorption. As a result, the article demonstrates that PLGA-based carbonaceous composites can aid bone regeneration. The PCL-HA-CNT scaffold [67] was implanted in rat subcutaneous tissue for four weeks and generated soft fibrous tissues with neo-blood vessels on the scaffold's three-dimensional holes without causing discomfort.

A three-phased MWCNTs/nanohydroxyapatite/polycaprolactone composite [68] scaffold (mwCNT/nHA/PCL) exhibited excellent interfacial force and great miscibility and the scaffold has minimal porosity and delayed degradation, as well as big, crystalline hydroxyapatite in SBF solution. Furthermore, the scaffold enhances MG63 cell proliferation and differentiation.

Later, the polymeric matrix of polycaprolactone-hydroxyapatite (PCL-HA) scaffolds loaded with CNTs were 3D printed to stimulate the proliferation of MG-63 cells [69]. The scaffold made up of 2% carbon nanotubes has outstanding mechanical and electrical conductivity. The scaffold's compressive strength (4 MPa) was also compatible with trabecular bone. Flores-Cedillo *et al.* [70] used three different MWCNTs/PCL composites for bone regeneration: randomly dispersed MWCNTs/PCL, aligned-MWCNTs/PCL, and BGP-modified with aligned-MWCNTs/PCL. The mechanical and physicochemical properties of these composites were fascinating; for example, β-glycerol phosphate (BGP) modified MWCNTs/PCL had a tensile strength that was comparable to cancellous bone replacement (10-20 MPa). MWCNTs (0.3 wt%) did not produce toxicity in human dental pulp stem cells (HDPSCs), according to this study. Furthermore, BGP modified MWCNTs/PCL composites showed a greater proliferation of HDPSCs for up to 21 days.

11.4 Challenges and Solutions in Using CNTs

1. Due to their hydrophobic nature, CNTs are not ideal for the development of bio-compatible composite materials. Their poor dispersibility is a major impediment to the creation of such products. The strong non-polar covalent interactions and hydrophobic surfaces of sp^2 hybrid carbon atoms in CNTs are responsible for their hydrophobicity. To improve their biocompatibility and aqueous solubility, continuous efforts should be made to functionalize biomolecules with CNTs.
2. Due to the unknown toxicity of carbon nanotubes, their usage as tissue engineering nanomaterials is of concern. Several studies have shown that their interaction with proteins, nucleic acids, and osteoblast-like human cells can increase cell affinity for various tissue engineering applications. Due to this, CNTs can cause inflammation on the surface of scaffolds and implants. This could result in loosening or migration of the desired implant location as well as the release of metal ions into the surrounding tissues. After CNTs are removed from implants and scaffolds, cross barriers can then distribute and become toxicological agents in the physiological environment. Functionalization can greatly boost bundled CNT dispersion and biocompatibility in an aqueous environment. New strategies and technologies are needed to overcome these issues. The surface of carbon nanotubes can be adjusted to prevent their negative effects either by adjusting the surface of carbon nanotubes or changing the manufacturing and purification procedures.

11.5 Conclusion and Perspectives

The development of bio-inspired interfacial materials that mimic the natural functions of nature will have a significant impact on the energy and environment. Until now, developing a possible bioactive scaffold that resembles the extracellular matrix of genuine bone has been difficult. In the medical field, nanomaterial research and development have gained a lot of attention and demand. CNT-based composite materials outperform conventional biomaterials in terms of mechanical, electrical, and chemical properties, making them a versatile platform in the biomedical field. CNT-based composites have become popular in tissue engineering. The enhanced physicochemical properties of carbon nanotubes have improved cellular interactions, adhesion, proliferation, and osteogenesis differentiation, resulting in more bone replacement and growth. Because of their large surface area, biocompatibility, and stimulation, CNTs' activity as nanocarriers for drug administration and cellular transportation for bone disorders and deformities has been expanded. Even though CNTs have made significant advances in the field of bone regeneration and repair, there are still some hurdles that must be overcome through the collaborative efforts of researchers from around the world, from the laboratory to clinical application. Despite significant progress, the state of the art of CNT-based materials has not yet reached the maturity required for industrial applications in terms of scalability, stability, and reliability, which are complicated by their operating environments and the lack of simple strategies to exquisitely control local structural texture and chemical composition at multiple length scales.

Suggested Readings

- Bio-Inspired Nanotechnology from Surface Analysis to Applications 2014, Eds. Marc R. Knecht, Tiffany R. Walsh, 9781461494461, Springer 2014.
- Functional Properties of Bio-inspired Surfaces Characterization and Technological Applications 2009, Eds. Eduardo A. Favret, Néstor O. Fuentes, World Scientific, 9789812837028.
- https://www.marketsandmarkets.com/Market-Reports/biomaterials-393.html?gclid=Cj0KCQjw0PWRBhDKARIsAPKHFGi58CY12cNV0DDc-wK1c_iY4tt45QaU0xlbz-TZXEbB2OP4I-7clS50aAkM-EALw_wcB
- https://www.marketsandmarkets.com/Market-Reports/carbon-nanotubes-139.html
- https://www.alliedmarketresearch.com/carbon-nanotube-market
- https://www.fortunebusinessinsights.com/industry-reports/medical-devices-market-100085
- Functional Properties of Bio-inspired Surfaces Characterization and Technological Applications 2009.

References

1. Ritchie, R.O., The conflicts between strength and toughness. *Nat. Mater.*, 13, 435, 2014.
2. Liu, K.S., Yao, X., Jiang, L., Recent developments in bio-inspired special wettability. *Chem. Soc. Rev.*, 39, 3240, 2010.
3. https://www.marketsandmarkets.com/Market-Reports/biomaterials-393.html?gclid=Cj0KCQjw0PWRBhDKARIsAPKHFGi58CY12cNV0D-Dc-wK1c_iY4tt45QaU0xlbzTZXEbB2OP4I-7clS50aAkM-EALw_wcB
4. Yang, W., Chen, I.H., Gludovatz, B., Zimmermann, E.A., Ritchie, R.O., Meyers., M.A., Natural flexible dermal armor. *Adv. Mater.*, 25, 31, 2013.
5. Meyers, M.A., Chen, P.Y., Lin, A.Y.M., Seki, Y., Biological materials: Structure and mechanical properties. *Prog. Mater. Sci.*, 53, 1, 2008.
6. Song, F., Soh, A.K., Bai, Y.L., Structural and mechanical properties of the organic matrix layers of nacre. *Biomaterials*, 24, 3623, 2003.
7. Autumn, K., Liang, Y.A., Hsieh, S.T., Zesch, W., Chan, W.P., Kenny, T.W., Fearing, R., Full, R.J., Adhesive force of a single gecko foot-hair. *Nature*, 405, 681, 2000.
8. Feng, L., Li, S.H., Li, Y.S., Li, H.J., Zhang, L.J., Zhai, J., Song, Y.L., Liu, B., Jiang, Q.L., Zhu, D.B., Super-hydrophobic surfaces: From natural to artificial. *Adv. Mater.*, 14, 1857, 2002.
9. Ge, L., Sethi, S., Ci, L., Ajayan, P.M., Dhinojwala, A., Carbon nanotube-based synthetic gecko tapes. *Proc. Natl. Acad. Sci. U.S.A.*, 104, 10792, 2007.
10. Wang, J.R., Qiao, J.L., Wang, J.F., Zhu, Y., Jiang, L., Bioinspired hierarchical alumina–graphene oxide–poly (vinyl alcohol) artificial nacre with optimized strength and toughness. *ACS Appl. Mater. Interfaces*, 7, 9281, 2015.
11. Wegst, U.G.K., Bai, H., Saiz, E., Tomsia, A.P., Ritchie, R.O., Bioinspired structural materials. *Nat. Mater.*, 14, 23, 2015.
12. Meyers, M.A., Chen, P.Y., Lin, A.Y.M., Seki, Y., Biological materials: Structure and mechanical properties. *Prog. Mater. Sci.*, 53, 1, 2008.
13. Evans, A.G., Suo, Z., Wang, R.Z., Aksay, I.A., He, M.Y., Hutchinson, J.W., Model for the robust mechanical behavior of nacre. *J. Mater. Res.*, 16, 2475, 2001.
14. Feng, L., Zhang, Y.A., Xi, J.M., Zhu, Y., Wang, N., Xia, F., Jiang, L., Petal effect: A superhydrophobic state with high adhesive force. *Langmuir*, 24, 4114, 2008.
15. Tian, S.B., Li, L., Sun, W.N., Xia, X.X., Han, D., Li, J.J., Gu, C.Z., Robust adhesion of flower-like few-layer graphene nanoclusters. *Sci. Rep.*, 2, 1–7, 2012.
16. Hansen, W.R. and Autumn, K., Evidence for self-cleaning in gecko setae. *Proc. Natl. Acad. Sci. U.S.A.*, 102, 385, 2005.
17. Jin, X., Yang, S., Li, Z., Liu, K.S., Jiang, L., Preparation of nano-silica with radial wrinkle structures for self-cleaning and superhydrophobic coatings. *Sci. China Chem.*, 55, 2327, 2012.

18. Wang, S. and Jiang, L., Definition of super hydrophobic states. *Adv. Mater.*, 19, 3423, 2007.
19. Lafuma, A. and Quere, D., Superhydrophobic states. *Nat. Mater.*, 2, 457, 2003.
20. Liu, K.S., Zhang, M.L., Zhai, J., Wang, J., Jiang, L., Bioinspired construction of Mg–Li alloys surfaces with stable super hydrophobicity and improved corrosion resistance. *Appl. Phys. Lett.*, 92, 183103, 2008.
21. Parker, A.R. and Townley, H.E., Biomimetics of photonic nanostructures. *Nat. Nanotechnol.*, 2, 347, 2007.
22. Zhang, T.J., Ma, Y.R., Qi, L.M., Bioinspired colloidal materials with special optical, mechanical, and cell-mimetic functions. *J. Mater. Chem. B*, 1, 251, 2013.
23. Kim, S., Shin, J., Yoo, S., Kim, S., Jeon, B., Moon, C., Jang, J.E., *International Conference on Nano-Bio Sensing, Imaging, and Spectroscopy 95230F*, 2015.
24. https://www.marketsandmarkets.com/Market-Reports/carbon-nanotubes-139.html
25. He, H., Pham-Huy, L.A., Dramou, P., Xiao, D., Zuo, P., Pham-Huy, C., Carbon nanotubes: Applications in pharmacy and medicine. *BioMed. Res. Int.*, 2013 2013, 578290, 2013.
26. Lamberti, M., Zappavigna, S., Sannolo, N., Porto, S., Caraglia, M., Advantages and risks of nanotechnologies in cancer patients and occupationally exposed workers. *Expert Opin. Drug Deliv.*, 11, 1087–1101, 2014.
27. Madani, S.Y., Mandel, A., Seifalian, A.M., A concise review of carbon nanotube's toxicology. *Nano Rev.*, 3, 4, 2013.
28. Hirlekar, R., Yamagar., M., Garse., H., Carbon nanotubes and its applications: A review. *Asian J. Pharm. Clin. Res.*, 2, 17, 2009.
29. Patole, S.P., Alegaonkar, P.S., Lee, H.C., Yoo, J.B., Optimization of water-assisted chemical vapor deposition parameters for super growth of carbon nanotubes. *Carbon*, 46, 1987, 2008.
30. Lamberti, M., Pedata, P., Sannolo, N., Porto, S., Rosa, A.D., Caraglia, M., Carbon nanotubes: Properties, biomedical applications, advantages and risks in patients and occupationally exposed workers. *Int. J. Immunopathol. Pharmacol.*, 28, 4–13, 2015.
31. Battigelli, A., Menard-Moyon., C., Da Ros, T., Prato, M., Bianco, A., Endowing carbon nanotubes with biological and biomedical properties by chemical modifications. *Adv. Drug Deliv. Rev.*, 65, 1899, 2013.
32. Hong, E.J., Choi, D.G., Shim, M.S., Targeted and effective photodynamic therapy for cancer using functionalized nanomaterials. *Acta Pharm. Sin. B*, 6, 297, 2016.
33. Gaillard, C., Duval., M., Dumortier., H., Bianco, A., Carbon nanotubecoupled cell adhesion peptides are non-immunogenic: A promising step toward new biomedical devices. *J. Pept. Sci.*, 17, 139, 2011.

34. Medepalli, K., Alphenaar, B., Raj., A., Sethu., P., Evaluation of the direct and indirect response of blood leukocytes to carbon nanotubes (CNTs). *Nanomed. Nanotechnol.*, 7, 983, 2011.
35. Dimiev, A.M., Khannanov, A., Vakhitov, I., Kiiamov, A., Shukhina, K., Tour, J.M., Revisiting the mechanism of oxidative unzipping of multiwall carbon nanotubes to graphene nanoribbons. *ACS Nano*, 12, 3985, 2018.
36. Liu, Z., Chen, K., Davis, C., Sherlock, S., Cao, Q., Chen, X., Drug delivery with carbon nanotubes for *in vivo* cancer treatment. *Cancer Res.*, 68, 6652, 2008.
37. Xiang, C., Zhang, Y., Guo, W., Liang, X.-J., Biomimetic carbon nanotubes for neurological disease therapeutics as inherent medication. *Acta Pharm. Sin. B*, 10, 239, 2020.
38. Cirillo, G., Hampel, S., Spizzirri, U.G., Parisi, O.I., Picci, N., Iemma, F., Carbon nanotubes hybrid hydrogels in drug delivery: A perspective review. *Biomed. Res. Int.*, 2014, 825017, 2014.
39. Qu, L.T., Dai, L.M., Stone, M., Xia, Z.H., Wang, Z.L., Carbon nanotube arrays with strong shear binding-on and easy normal lifting-off. *Science*, 322, 238, 2008.
40. Qu, L. and Dai, L., Gecko-foot-mimetic aligned single-walled carbon nanotube dry adhesives with unique electrical and thermal properties. *Adv. Mater.*, 19, 3844, 2007.
41. Ge, L., Sethi, S., Ci, L., Ajayan, P.M., Dhinojwala, A., Carbon nanotube-based synthetic gecko tapes. *Proc. Natl. Acad. Sci. U.S.A.*, 104, 10792, 2007.
42. Majumder, M., Stinchcomb, A., Hinds, B.J., Towards mimicking natural protein channels with aligned carbon nanotube membranes for active drug delivery. *Life Sci.*, 86, 563, 2010.
43. Wei, J., Xie, J., Zhang, P., Zou, Z., Ping, H., Wang, W., Xie, H., Shen, J.Z., Lei, L., Fu, Z., Bioinspired 3D printable, self-healable, and stretchable hydrogels with multiple conductivities for skin-like wearable strain sensors. *ACS Appl. Mater. Interfaces*, 13, 2952, 2021.
44. Li, X., Liu, H., Niu, X., Yu, B., Fan, Y., Feng, Q., Cui, F., Watari, F., The use of carbon nanotubes to induce osteogenic differentiation of human adipose-derived MSCs *in vitro* and ectopic bone formation *in vivo*. *Biomaterials*, 33, 4818, 2012.
45. Kuang, Z., Kim, S.N., Crookes-Goodson, W.J., Farmer, B.L., Naik, R.R., Biomimetic chemosensor: Designing peptide recognition elements for surface functionalization of carbon nanotube field effect transistors. *ACS Nano*, 4, 452, 2010.
46. Chen, L., Hu, J., Shen, X., Tong, H., Synthesis and characterization of chitosanemultiwalled carbon nanotubes/hydroxyapatite nanocomposites for bone tissue engineering. *J. Mater. Sci. Mater. Med.*, 24, 1843, 2013.
47. Fonseca-García, A., Garcia-Carvajal, Z.Y., Martínez-Lopez, V., Ruvalcaba, E., Landa-Solís, C., Solis, L., Ibarra, C., Gutierrez, M.C., Terrones, M., Sanchez, I.C., del Monte, F., Velasquillo, M.C., Luna-Barcenas, G., Effect of

doping in carbon nanotubes on the viability of biomimetic chitosan-carbon nanotubes-hydroxyapatite scaffolds. *J. Biomed. Mater. Res. A*, 102, 3341, 2014.
48. Gopi, D., Nithiya, S., Shinyjoy, E., Rajeswari, D., Kavitha, L., Carbon nanotubes/ carboxymethyl chitosan/mineralized hydroxyapatite composite coating on Ti-6Al-4V alloy for improved mechanical and biological properties. *Ind. Eng. Chem. Res.*, 53, 7660, 2014.
49. Yasmeen, S., Lo, M.K., Bajracharya, S., Roldo, M., Injectable scaffolds for bone regeneration. *Langmuir*, 30, 12977, 2014.
50. Patel, K.D., Kim, T.-H., Lee, E.-J., Han, C.-M., Lee, J.-Y., Singh, R.K., Kim, H.-W., Nanostructured biointerfacing of metals with carbon nanotube/chitosan hybrids by electrodeposition for cell stimulation and therapeutics delivery. *ACS Appl. Mater. Interfaces*, 6, 20214, 2014.
51. Gholizadeh, S., Moztarzadeh, F., Haghighipour, N., Ghazizadeh, L., Baghbani, F., Shokrgozar, M.A., Allahyari, Z., Preparation and characterization of novel functionalized multiwalled carbon nanotubes/chitosan/b-glycerophosphate scaffolds for bone tissue engineering. *Int. J. Biol. Macromol.*, 97, 365, 2017.
52. Ilbasmis-Tamer, S., Ciftci, H., Turk, M., Degim, T., Tamer, U., Multiwalled carbon nanotube-chitosan scaffold: Cytotoxic, apoptotic, and necrotic effects on chondrocyte cell lines. *Curr. Pharm. Biotechnol.*, 18, 327, 2017.
53. Türk, S., Altınsoy, I., ÇelebiEfe, G., Ipek, M., Ozacar, M., Bindal, C., 3D porous collagen/functionalized multiwalled carbon nanotube/chitosan/hydroxyapatite composite scaffolds for bone tissue engineering. *Mater. Sci. Eng. C*, 92, 75776, 2018.
54. Sharmeen, S., Rahman, A.F.M.M., Lubna, M.M., Salem, K.S., Islam, R., Khan, M.A., Polyethylene glycol functionalized carbon nanotubes/gelatin-chitosan nanocomposite: An approach for significant drug release. *Bioact. Mater.*, 3, 236, 2018.
55. Joviano-Santos, J.V., Sa, M.A., de Maria, M.L.A., Almeida, T.C.S., Geraldo, V., Oliveira, S., Ladeira, L.O., Ferreira, A.J., Evaluation of cardiovascular toxicity of carbon nanotubes functionalized with sodium hyaluronate in oral regenerative medicine. *Braz. J. Med. Biol. Res. Rev. Bras. Pesqui. Medicas E Biol.*, 47, 560, 2014.
56. Sa, M.A., Ribeiro, H.J., Valverde, T.M., Sousa, B.R., Martins-Júnior, P.A., Mendes, R.M., Ladeira, L.O., Resende, R.R., Kitten, G.T., Ferreira, A.J., Single-walled carbon nanotubes functionalized with sodium hyaluronate enhance bone mineralization. *Braz. J. Med. Biol. Res.*, 49, 4888, 2015.
57. Sa, M.A., Andrade, V.B., Mendes, R.M., Caliari, M.V., Ladeira, L.O., Silva, E.E., Silva, G.B., Correa-Júnior, J.D., Ferreira, A.J., Carbon nanotubes functionalized with sodium hyaluronate restore bone repair in diabetic rat sockets. *Oral. Dis.*, 19, 484, 2013.
58. Andrade, V.B., Sa, M.A., Mendes, R.M., Martins-Júnior, P.A., Silva, G.A.B., Sousa, B.R., Caliari, M.V., Avila, E.S., Ladeira, L.O., Resende, R.R., Ferreira, A.J., Enhancement of bone healing by local administration of carbon

nanotubes functionalized with sodium hyaluronate in rat Tibiae. *Cells Tissues Organs*, 204, 137, 2017.

59. Vozzi, G., Corallo, C., Daraio, C., Pressure-activated microsyringe composite scaffold of poly (L-lactic acid) and carbon nanotubes for bone tissue engineering. *J. Appl. Polym. Sci.*, 129, 528, 2013.
60. Siqueira, I.A.W.B., Corat, M.A.F., das Cavalcanti, B.N., Neto, W.A.R., Martin, A.A., Bretas, R.E.S., Marciano, F.R., Lobo, A.O., *In vitro* and *in vivo* studies of novel poly(d, l-lactic acid), superhydrophilic carbon nanotubes, and nanohydroxyapatite scaffolds for bone regeneration. *ACS Appl. Mater. Interfaces*, 7, 9385, 2015.
61. Stocco, T.D., Antonioli, E., Elias, C., de, M.V., Rodrigues, B.V.M., de Brito Siqueira, I.A.W., Ferretti, M., Marciano, F.R., Lobo, A.O., Cell viability of porous poly(d, l-lactic acid)/vertically aligned carbon nanotubes/nanohydroxyapatite scaffolds for osteochondral tissue engineering. *Materials*, 12, 849, 2019.
62. Kim, H.-B., Patel, D.K., Seo, Y.-R., Lim, K.-T., 3D-Printed scaffolds with reinforced poly (lactic acid)/carbon nanotube filaments based on melt extrusion. *J. Biosyst. Eng.*, 44, 120, 2019.
63. Sitharaman, B., Avti, P.K., Schaefer, K., Talukdar, Y., Longtin, J.P., A novel nanoparticle-enhanced photoacoustic stimulus for bone tissue engineering. *Tissue Eng. A*, 17, 1851, 2011.
64. Gupta, A., Woods, M.D., Illingworth, K.D., Niemeier, R., Schafer, I., Cady, C., Filip, P., El-Amin, S.F., Single walled carbon nanotube composites for bone tissue engineering. *J. Orthop. Res.*, 31, 1374, 2013.
65. Gupta, A., Liberati, T.A., Verhulst, S.J., Main, B.J., Roberts, M.H., Potty, A.G.R., Pylawka, T.K., El-Amin III, S.F., Biocompatibility of single-walled carbon nanotube composites for bone regeneration. *Bone Joint Res.*, 4, 70, 2015.
66. Pan, L., Pei, X., He, R., Wan, Q., Wang, J., Multiwall carbon nanotubes/polycaprolactone composites for bone tissue engineering application. *Colloids Surf. B Biointerfaces*, 93, 226, 2012.
67. Dorj, B., Won, J.-E., Kim, J.-H., Choi, S.-J., Shin, U.S., Kim, H.-W., Robocasting nanocomposite scaffolds of poly(caprolactone)/hydroxyapatite incorporating modified carbon nanotubes for hard tissue reconstruction. *J. Biomed. Mater. Res. A*, 101A, 1670, 2013.
68. Yang, H., Li, J., Liao, Q., Guo, H., Chen, H., Zhu, Y., Cai, M., Lv, M., *In vitro* evaluation of a novel multiwalled carbon nanotube/nanohydroxyapatite/polycaprolactone composite for bone tissue engineering. *J. Mater. Res.*, 34, 532, 2019.
69. Goncalves, E.M., Oliveira, F.J., Silva, R.F., Neto, M.A., Fernandes, M.H., Amaral, M., Vallet-Regı, M., Vila, M., Three-dimensional printed PCL-hydroxyapatite scaffolds filled with CNTs for bone cell growth stimulation. *J. Biomed. Mater. Res. B Appl. Biomater.*, 104B, 1210, 2016.

70. Flores-Cedillo, M.L., Alvarado-Estrada, K.N., Pozos-Guillen, A.J., Murguía-Ibarra, J.S., Vidal, M.A., Cervantes-Uc, J.M., Rosales-Ibanez, R., Cauich-Rodríguez, J.V., Multiwall carbon nanotubes/polycaprolactone scaffolds seeded with human dental pulp stem cells for bone tissue regeneration. *J. Mater. Sci. Mater. Med.*, 27, 35, 2016.

12

Functionalized Carbon Nanotubes: Applications in Tissue Engineering

Ajahar Khan*, **Khalid A. Alamry**† **and Raed H. Althomali**

Department of Chemistry, Faculty of Science, King Abdulaziz University, Jeddah, Saudi Arabia

Abstract

Tissue engineering is an arrangement of bioactive molecules, cells, and scaffolds for the reconstruction of injured tissues. Functionalized carbon nanotubes (CNTs) have emerged in the present time as pioneering components for the fabrication of the next-generation scaffolds to reconstruct injured tissue. Polymer-functionalized CNTs could exhibit various properties, including higher compatibility and ability of complexation with polymer matrix, improved dispersibility in different solvents, and responsivity to environmental stimuli. Furthermore, CNTs have broadly been investigated for their potential in tissue engineering because of their exceptional mechanical, physicochemical, and thermal features as well as suitable biocompatibility. CNTs based scaffolds have acquired significant recognition because of their brilliant attributes, which offer a synthetic but feasible microenvironment for chondrogenic differentiation, and cell proliferation. In this chapter, we aim to give an overview of the structural, physicochemical characteristics along with interactions and biodegradation of CNTs-based polymeric scaffolds with the biomolecules and outline their impacts on the extracellular matrix in the regeneration of neural, cardiac, bone, and cartilage tissue engineering. This chapter also provides a current state of information accessible investigating the utilization of the polymeric composites comprising CNTs and the development of the 3D scaffolds as well as the directions of future research and challenges.

Keywords: Tissue engineering, Functionalized carbon nanotubes, Cartilage regeneration, Polymeric scaffolds

**Corresponding author*: arkhan.029@gmail.com

†*Corresponding author*: kaalamri@kau.edu.sa

Jeenat Aslam, Chaudhery Mustansar Hussain and Ruby Aslam (eds.) Functionalized Carbon Nanotubes for Biomedical Applications, (281–318) © 2023 Scrivener Publishing LLC

12.1 Introduction

Tissue engineering is an interdisciplinary area of science focusing on bioengineering, life science, pharmaceutical and materials science, and speculations of medicine that attempt to replace, reconstruct, and design damaged or diseased tissues/organs with biologic alternatives that can maintain and restore the regular function of the original organ [1–4]. Considerable benefits in the fields of tissues or organ transplantation, along with advancement in engineering and materials science, have aided in the ongoing advancement of regenerative medicine and tissue engineering. Particularly in the applications of tissue engineering, the evolution of nanomaterials is one of the promising research interests [5]. Furthermore, nanotechnology is emerging as a unique area of research that utilizes the excellent chemical and physical features of materials developed in the nano range, which are not accessible in bulk solids, therefore, it invoked great significance among worldwide scientific communities [6, 7]. Nanotechnology demands utilizing materials which acquire at least one physical measurement in 1-100 nm range to the fabricated systems, devices, and structures that have novel characteristics. The combination of tissue engineering and nanotechnology offers a broad platform for nano-materials in healthcare applications [8]. In the biomedical field, great interest and demands in the research of bone tissue engineering have been generated because of the various types of bone disorders witnessed at present time. Worldwide around ~20 million people are suffered from various diseases and accidents every year that involve bone disorders [9]. Furthermore, in recent times bone disorders such as rheumatic diseases, osteoporosis, infection, back pain, fracture, scoliosis, and tumors are the main concern because of the obesity, scarceness of regular physical exercise, and growing population [10]. Since the advancement of tissue engineering increasing there is a great demand for evaluating tools and improved monitoring of engineered tissues along with the development of newly hybrid biomaterials to assist tissue growth. Therefore, the preparation of promising bioactive material is a great requirement for tissue engineering that can mimic natural organs.

Since the finding of carbon nanotubes (CNTs) by Iijima [11], nanotechnology-based on carbon has received great attention in comparison to the other nanoscale materials and appeared like a considerable material for several biomedical applications. They also possess biomechanical properties [12], good functionalization [13], high porosity [14, 15], followed by excellent tensile strength and thermal conductivity due to their nanostructure and strength of the bonds between carbon atoms. CNTs have also acquired significant accomplishments in genetic engineering and might act as promising diagnostic tools and biosensors in biomedical research [16, 17]. The combination of polymer matrix with CNTs shows a variety of physicochemical and structural reinforcement properties, such as biocompatibility and flexibility, reduction of thrombosis, the introduction of angiogenesis, improvement of strength, and alteration of gene expression toward tissue engineering [18]. Typically, the length of the CNTs is in the range of micrometers and the diameters are in the nanometers range. It has been reported that the excellent structural morphology and aspect ratio (diameter and length) of CNTs affect their characteristics [19]. The theoretical surface area of single walled carbon nanotubes (SWNTs) (1, 315 m^2/g) produces it as unique material for the applications of biomedical and nanorobotics. The inter distance amid each layer of multi-walled carbon nanotubes (MWNTs) is about 0.36 nm and the theoretical surface area of MWNTs is 100 m^2/g [20, 21]. To replicate the morphology of fibrous proteins in the extracellular matrix, CNTs can be embedded within a hydrogel. According to previous reports CNTs had been used in hydrogel to improve the stiffness of the bulk gel [22, 23]. CNTs based hydrogels have been widely demonstrated for a wide range of applications in the human body, including brain circuit stimulation, skin growth, osteoblastic cell proliferation, and differentiation [12]. On the other side, CNTs exhibit such characteristics that make them a promising candidate for the future progress of tissue developments [24]. The important characteristic feature of CNTs in tissue regeneration is intrinsic morphology along with their anisotropic electrical and mechanical features. As anisotropic materials CNTs mimic the anisotropic alignment of cardiomyocytes present in the cardiac tissue by promoting the electroactive cardiac cells aligning in one direction. Moreover, the inherent CNT anisotropy might be united with synthetic approaches that permit the development of aligned morphologies to further gain the benefit of this feature. Therefore, several techniques including chemical vapor deposition [25], dielectrophoresis [26], or electrospinning [27], can be employed to design CNTs aligned cardiac constructs and encourage the improvement in the cellular alignment of cardiomyocytes. However, the utilization of CNTs incorporated scaffolds has not been reported so far for the application of promoting cell migration. Therefore, to evaluate the probability of utilizing CNT-based scaffolds as the synthetic extracellular matrix, their physical, mechanical, and biological properties need to be studied. In this review, we have overviewed investigations involving CNTs which are significant to the field of tissue engineering relative to their utilization on their own or in composition with suitable polymers.

12.2 Structural, Physical, and Chemical Properties

CNTs are cylindrical carbon tubes assigning nanometer diameters and exhibit remarkable structural and electronic properties determine by length, diameter, and chirality. The carbon atoms in the backbone of CNTs present in a sequence of condensed benzene rings through a very large aspect ratio. As allotropes of carbon like graphene, CNTs exhibit the covalently bonded carbon atoms via sp2 hybridizations with one another. According to the cylindrical tubes numbers, CNTs might be comprised of a single cylinder-usually known as single-walled-CNT (SWNT), double-walled (two concentric cylinders), or multiple-concentric cylinders generally called multi-walled CNTs (MWNTs) [20]. SWNTs possessed hollow-long structures with one-atom-thick walls, whereas MWNTs concentrically surrounded a central CNT and possessed multi-layers of graphitic carbon tubes. In general, the diameter of SWCNTs centered in the range of 1–2 nm, whereas for MWNTs centered around 10-100 nm having length differing from 50 nm to 1 cm [28, 29]. The structural morphology of SWCNTs involves sidewalls and tips. Three separate tubular structures can be identified via the orientation of rolling up a graphene film: chiral, armchair, and zigzag, that is designated with the chiral indices (m, n) [30]. The variations among diameter and chiral angle lead to different properties along with the different structures. The chiral and zigzag CNTs possess similar properties to semiconductors while armchair CNTs share analogous electrical properties to metals [31, 32]. SWNTs and MWNTs exhibited excellent mechanical strength, displaying tensile strength around 11-200 GPa and Young's Modulus around 0.27–1.34 TPa [33–36]. CNTs also presented excellent thermal conductivity (5000 Wm^{-1} K) and superb electrical conductivity (104 S/cm^2) [18, 37, 38]. The electrical properties and conductive structure of CNTs might be concluded through studying the electrical behavior of graphene [39, 40].

CNTs are linked through powerful nonpolar covalent bonds and their hydrophobic behavior on the surface produces them immiscible in the majority of organic solvents and water. Moreover, the high aspect ratio and Van de Waals forces of CNTs usually result in agglomeration [41]. Hence, the uniform dispersion of CNTs is a major challenge related to the CNTs utilization, deteriorating the mechanical, chemical, and electrical properties. To avoid this limitation, functionalizations of CNTs have been widely investigated. To functionalized CNT, two frequently used strategies are based on the covalent (chemical bond) and non-covalent (physical adsorption) functionalizations [42, 43]. The non-covalent approach is comparatively easy and can be carried out under mild conditions based on physical bonding because of diverse adsorption forces, i.e. π-π interactions, hydrogen bonding, electrostatic force, and Van der Waals force [42–45]. The non-covalent functionalization techniques can be achieved in two ways: The first one is describes wrapping polymer chains or surfactants on the CNTs sidewalls and another approach can be accomplished via π-π stacking interactions between π-electrons of graphitic sheets on the CNTs surface and aromatic rings of grafted materials [32, 46]. The grafted polymer and surfactants can act as a dispersing agent to improve the solubility of CNTs in an aqueous medium [47, 48]. Noncovalent technique functionalized CNTs that possess high affinity towards biomolecules like enzymes, proteins, RNA, DNA, and peptides producing them attractive target delivery carriers to deliver biomolecules toward organs or cells [48–51]. On the other side, the covalent functionalization technique is focused on the preparation of covalent linkage among the main backbone of CNTs and functional groups. The defective carbon atoms at the end or on the sidewalls of both SWNTs and MWNTs might be oxidized through powerful oxidants (e.g. H_2SO_4/HNO_3), generating carboxylated fractions or -COOH groups [32, 49]. To further enhance the activity of CNTs, permitting the linking of hydrophilic moieties, the -COOH are usually changed into acid chloride followed by amidation or esterification reaction [52–54]. In a study, Mooney *et al.* report the influence of the carboxylated-SWNT on human mesenchymal stem cells (hMSC) proliferation, viability, and differentiation [55]. It has been investigated that the COOH-functionalized-SWNT was easily dispersed in hMSC media and was least toxic to the cells. In Figure 12.1 (A) the green channel displayed that biotinylated COOH-functionalized-SWNT were uptake through the hMSC and for the first 24 h primarily occupied a cytoplasmic location. A considerable amount of the labeled CNTs had assumed a nuclear location (Figure 12.1 Ac) after 6 days in culture. The hMSC did not show any evidence of green fluorescence that was not exposed to CNT (Figure 12.1 Ad). Furthermore, several molecules, such as biomolecules, metals, and synthetic polymers are found to be grafted to the carboxylated CNTs surface [56]. You *et al.* [57] reported that various types of aqueous soluble ionic polymer backbones, for example anionic (e.g. poly(acrylic acid), cationic (e.g. poly(2-(dimethylamino) ethylmethacrylate), and zwitterionic polymers (e.g. poly(3-(N-(3-methacrylamidopropyl)-N, N-dimethyl) ammoniopropane sulfonate)), were straightforwardly grafted to the MWCNTs surface through surface reversible addition-fragmentation chain transfer polymerization, assigning improve solubility. Redondo-Gomez *et al.* [58] studied the modification of the MWNT surface using cholic acid through a chemical approach. The functionalized-MWNTs displayed facial bioamphiphilic nature, enhancing the stability of dispersion in water, low-polarity and organic polar solvents.

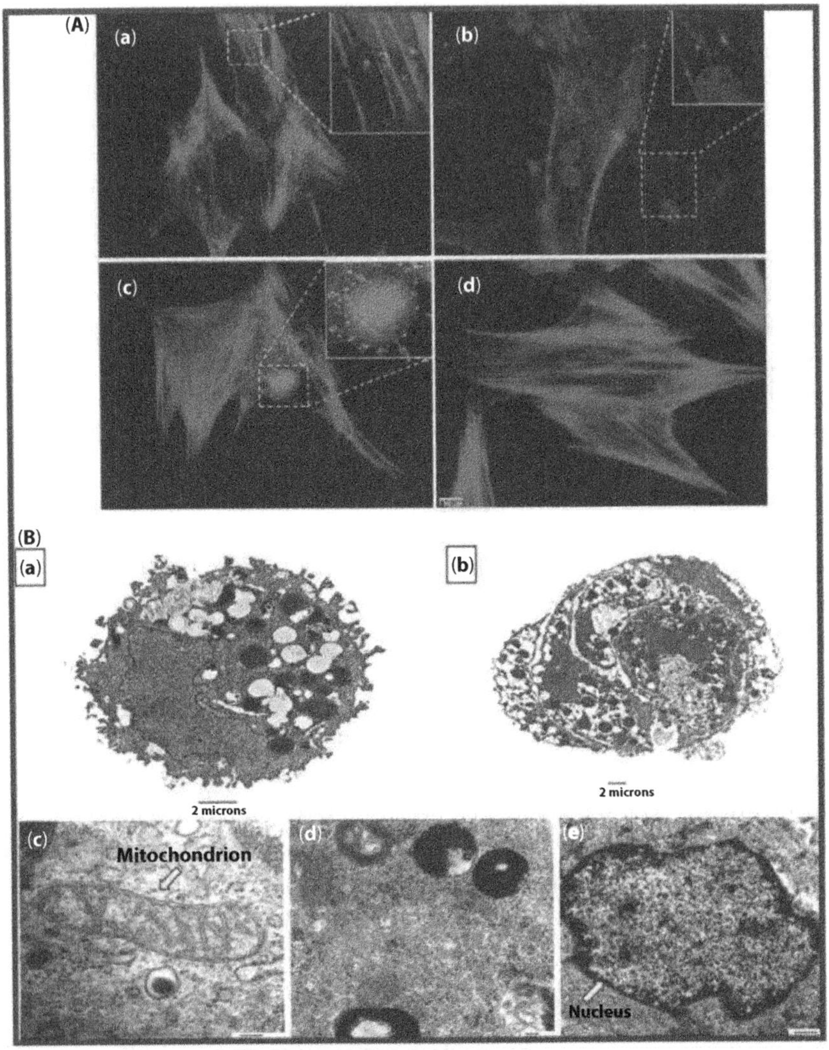

Figure 12.1 (A) COOH-functionalized SWNT uptake via cell i.e. biotinylated CNT in the cell after (a) 24 h, (b) 48 h, and (c) 6 days and (d) hMSC alone (fluorescent images scale bar 130 μm). (B) (a) hMSC seeded on alginate and (b) human mesenchymal stem cells uncovered to carboxylated-SWNT and seeded on alginate for 24 h, (c) mitochondria, (d) cytoplasm and ribosomes on the endoplasmic reticulum, (e) and cell nucleus. Reprinted with permission from Ref. [55]. Copyright © 2008 American Chemical Society.

The covalent functionalization technique stabilizes the functionality and processability, being a robust and highly stable approach in comparison to the non-covalent functionalization technique [59].

12.3 Interactions and Biodegradation of CNTs with Biomolecule

To enhance the biocompatibility of nanotubes, bioconjugation of CNTs is an appropriate approach to avoid possible health problems [60–62] followed by allowing the CNTs assembly into highly ordered structural form. To provide superb electronic and mechanical characteristics of CNTs to the microscale structures is of significant importance, which might be employed as active components [63]. As already studied for spherical polystyrene nanoparticles [64], CNTs can readily interact with a wide range of biomolecules (for instance proteins or lipids), creating a nanoparticles-biointerface in biological environments [65]. This interface consists of thermodynamic and kinetics exchanges, and dynamic physico-chemical interactions between the surfaces of biological components and the surface of the nano-materials, such as proteins and phospholipids [66]. In nanoparticles-cell interactions the liquid/solid interface is a significant task, which is generally not focused after entering the nanoparticles into a biological fluid (for example blood serum) and proteins compete for the surfaces of nanoparticles and lead to the construction of a protein corona [64]. Due to the definite physico-chemical characteristics of the nanoparticles, various progression can occur at this interface [66]. The gastro-intestinal tract (through digestion), blood vessels (through intravenous injection), the lungs (via inhalation), and the skin are the most possible way via which CNTs can penetrate into the human body [67], however, the pulmonary system (i.e. inhalation) is the main vital (human) exposure pathway for CNTs. Furthermore, CNTs can directly deposit onto the pulmonary surfactant layer positioned at the air/liquid interface of a lining layer of liquid casing the cells at the luminal side of the lung wall. This surfactant layer comprises mainly phospholipids (85-90%), with different proteins [68]. Furthermore, when deposited onto highly oriented pyrolytic graphite (HOPG) surfaces, adenine-functionalized-SWNTs combined into well-ordered structures like stripe [69]. In a similar approach, oxidized SWNTs were functionalized with derivatives of uracyl via amidation, emerging in toroidal structures (diameter ranges ~50 nm in), when deposited onto the surfaces of HOPG [70]. Sonication is a potent technique for dispersing bundled SWNTs in the aqueous solution of SWNTs and ribonucleic acid (RNA), double stranded deoxyribonucleic acid (dsDNA), and single stranded deoxyribonucleic acid (ssDNA) [71–73]. In the obtained dispersion, ssDNA helically wraps around the SWNT through π-π stacking of the aromatic nucleotide bases of the ssDNA. Furthermore, separating the surplus deoxyribonucleic acid (DNA) will neither influence the SWNTs dispersion nor be responsible for flocculation, demonstrating that the interaction of DNA-SWNT is comparatively strong. In addition, a report investigated that the phospholipids sited in pulmonary surfactant are capable to combine nonspecifically on the MWNTs backbone, [29] eventually stimulating the adsorption of blood plasma proteins [68, 74]. This prototype of adsorbed proteins is hypothesized to considerably affect the combination among biological systems and MWNTs as well as its effects. To understand this highly complex and dynamic interface further investigation has been accomplished. The hexagonal arrangement of carbon atoms of CNTs produced them to be highly biopersistent. SWNTs have been demonstrated to be biodegradable by natural enzymatic catalysis [75]. Allen *et al.* studied that, COOH-SWCNTs can be degraded by the use of the horseradish peroxidase (oxidative compounds) and hydrogen peroxide (the latter at low concentrations) at 48 °C in a cell-free environment after 16 weeks [76]. In a further investigation, it was studied that in a similar environment carboxylated SWNTs as well as pristine-SWNT both are biodegradable [76]. Although, whether peroxidase intermediates that are released within biofluids or human cells are involved in CNTs' biodegradation is not presently understood. First of all, Kagan *et al.* investigated that reactive radical intermediates and hypochlorite generated from human neutrophils (the enzyme myeloperoxidase) are competent for SWNTs degradation *in vitro* (neutrophils and macrophages (lower degree)) [77]. The CNTs' biodegradation inferred an innovative side of controlling the CNTs' interaction with biomolecules. The comprehension of the bio/nano interface is very essential and needs further, in detail study, as such interaction may conclude to a huge degree whether a nanomaterial is biocompatible, bio-available, or might probably draw unfavorable influences on human fitness. To accomplish this understanding, new concepts, ideas, and analytical techniques are required.

12.4 Bio-Security of CNT-Based Scaffolds Toward *In Vivo* Analyses

To enhance the mechanical strengths of polymers employed in bone regeneration, it has been proposed to introduce CNTs. Though, the concern arises related to the bio-security of the CNTs *in vivo*. CNTs are not considered biodegradable materials like some synthetic polymers or collagen. In contrast, as a scaffold material CNTs will migrate to other body organs and remain in bone. Moreover, the biodegradability of the used scaffold material in tissue engineering is not a key parameter. Correspondingly, for materials that consist of inadequate biodegradability, there is a chance of inflammatory reactions that can occur from the degradation or remnants' of by-products [78, 79]. Therefore, low degradability may essentially be a benefit for materials-based scaffolds in tissue engineering [80]. The strength and flexibility of the CNTs can positively contribute to the excellence of the engineered tissue. According to the literature survey, only a few studies of *in vivo* investigations have been carried out to evaluate the biocompatibility and propriety of scaffolds fabricated using CNT [81, 82]. Bhattacharya *et al.* studied the influence of a layer-by-layer assembled CNT-composite *in vivo* on bone tissue of rats and osteoblasts *in vitro* [83]. The fabrication of composite was carried out using polyethyleneimine and SWNTs on polycaprolactone films [83]. *In vitro* analyses suggested that the cell differentiation of osteoblasts and matrix mineralization were enhanced on the CNT-based composites in comparison to the pristine titanium implants. Further, *In vivo* analyses suggested that the fabricated CNT-based composites stimulated bone repair and bone formation without inflammation or rejection signs after implanting in critical-sized rat calvarial defects. Lewitus *et al.* reported a unique strategy for producing CNT-fibers (CNF) composed with the agarose (polysaccharide) [84]. The obtained results suggested that the obtained CNT-fibers were not only non-toxic but also biologically modifiable and biocompatible. Furthermore, agarose/CNT hybrid material is revealed to assist response and cell attachment both *in vivo* and *in vitro* and found promising material for applications of biointerfacing with the nervous system and neural tissue engineering. Zande *et al.* reported a magnetic resonance imaging method to analyze the CNTs' distribution pattern following release from a degrading poly(lactic-co-glycolic acid) scaffold *in vivo* [85]. Poly(lactic-co-glycolic acid) scaffold with incorporated gadolinium-labeled SWNTs subcutaneously was received by five rats. The obtained results suggested that magnetic resonance imaging is a brilliant approach to analyze the biological fate of gadolinium-labeled single-walled CNTs. Kaisai *et al.* studied the influence of carbon nanohorns on the regeneration of bone and their potential application for guided tissue engineering [86]. They evaluated a new material based on polytetrafluoroethylene membranes with carbon nanohorns for hard-tissue reconstruction in a rat calvarial model by implanting specimens and their possible application for guided bone regeneration. After implanting of two weeks, they noticed that the formation of fresh bone was enhanced under the polytetrafluoroethylene membrane with carbon nanohorns in comparison to the pristine polytetrafluoroethylene. It was found that the growth of new bone begins from the edge of the presented bone and propagates to the center of the defect. However, after implanting of 8 weeks, no major variation was observed in the amount of freshly prepared bone noticed amid the membranes and the emergence of macrophages was reduced. Therefore they suggested the biocompatibility of the carbon nanohorns that promote bone regeneration in the early stage *in vivo* with a small inflammatory response and found that the formation of new bone was observed to be linked with the occurrence of macrophages. Saito *et al.* demonstrated the bone-tissue compatibility of CNTs by implanting MWNTs in mouse skulls [87]. The pathologic investigation suggested that MWNTs had no effect on bone even in CNTs contact with it and did not produce powerful inflammatory reactions, indicating adequate bone-tissue compatibility. To determine the effect on the curing of bone, MWNTs were then implanted in tibial defects shaped in mice. Histopathology analyses showed that MWNTs were noticed in the bone matrix as well as in bone marrow. In another report, they implanted a 3D web based on recombinant human bone morphogenetic protein 2 (rhBMP-2)/carbon fiber porous composition into the dorsal muscle pouches of mice [80]. The insertion of web leads to the formation of fresh ectopic bone and the values of the bone mineral content and bone mineral density were considerably more than those procured a collagen sheet. The interface between the bone matrix and carbon fibers suggests that the fibers are integrated directly into the bone matrix, demonstrating higher bone-tissue compatibility. It was also revealed that the rhBMP-2/carbon-fiber web composite restore a critical-size bone injury within a short stretch of time. Abarrategi *et al.* studied the effect of MWNT/chitosan scaffolds after implanting with BMP-2 for a period of three weeks into the subcutaneous pockets of rats [88]. This research group noticed ectopic bone formation, considerable degradation of the scaffold, and the non-appearance of chronic inflammation throughout the complete period of implantation. For the first time, Facca *et al.* reported *in vivo* analysis of plasma-sprayed CNT-reinforced hydroxyapatite coating on Ti implants incorporated in rodents' bone [89]. It was found that the growth of bone was normal around hydroxyapatite-CNT coated implants along with no cytotoxicity or unfavorable consequence of CNT addition on cells and tissues of bone. The addition of CNT convinces higher osseointegration as compared to hydroxyapatite. It was also observed that the elastic modulus of the fresh bone was corresponding with the composites' modulus, which revealed that the hydroxyapatite and CNTs' occurrence improved the mechanical compatibility of the implant with the bone and lowered the probability of implant failure or fracture. The construction of CNT-based 3D scaffolds having potential for tissue regeneration tested *in vivo* is summarized in Table 12.1.

Table 12.1 Summary of the reviewed CNT-scaffolds tested *in vivo* for biomedicine applications.

S. no.	Material	Fabrication method	Application	*In vivo* implant	Ref.
1	Poly (d, l-lactic acid)/ graphene/MWNT-oxides	Oxygen plasma etching	Bone regeneration	Tibia of rats	[90]
2	CNT- membrane	Electrochemically	Transdermal nicotine delivery	Skin of hairless guinea pig	[91]
3	MWNT/collagen	Freeze-casting	Bone regeneration	Ectopic bone (mouse)	[87]
4	Polycaprolactone fumarate-CNT	Photocrosslinking	Nerve regeneration applications	PC-12 cell	[92]
5	CNT/Agarose Fibers	Electrospinning	Neural Tissue Engineering	Rat cerebral cortex	[84]
6	CNT-hydroxyapatite scaffold	Electrophoresis	Bone tissue engineering	Bone mesenchymal stem cell lines	[93]
7	poly(3, 4-ethylenedioxythiophene) (PEDOT)-CNT/sucrose scaffold	Vapor phase polymerization	Tissue engineering	C8-D1A astrocyte	[94]

(*Continued*)

Table 12.1 Summary of the reviewed CNT-scaffolds tested *in vivo* for biomedicine applications. (*Continued*)

S. no.	Material	Fabrication method	Application	*In vivo* implant	Ref.
8	CNT-Alginate gel	Gel formation/ freeze-casting	Tissue engineering	Subcutaneous (rat)	[95]
9	Poly(3-hydroxybutyrate)/ MWNT	Solvent casting and particulate leaching	Tissue engineering	Subcutaneous	[96]
10	Poly(propylene fumarate)/ SWNT	Particulate leaching	Tissue engineering	Femur condyle and subcutaneous (rabbit)	[97]
11	MWNT/collagen	Coating	Tissue engineering	Bone	[82]
12	CNT-fibers	-	Restoration of Myocardial Conduction	Myocardial	[98]

12.5 CNTs Towards the Bone Compatibility

The compatibility of bone-tissue of CNTs and their effects on bone regeneration are significant concerns, although the CNTs' consequences on bone have not been described. In a bone implantation testing report, it was observed that the CNT/alumina composite exhibited better compatibility of bone tissue and was directly associated with new bone. The *in vitro* cell attachment analysis was carried out for fibroblasts, chondrocytes, osteoblasts, and smooth muscle cells where the composite of CNT/alumina displayed cell attachment and proven bone tissue compatibility and good biocompatibility [99]. The discussed *in vivo* investigations suggested quite improved bone compatibility of carbon nanotubes, some *in vitro* experiments have displayed cytotoxic consequences for bone cells. Because of their good affinity with CNTs, nutrients in the culturing medium requisite for the growth of cells can combine with CNTs, lowering the cell proliferation rate [55]. It has been studied that CNTs may also bind to the receptors at the cellular level and on the cell membrane block the pores, or even perforate the whole cell membrane or physically rupture [100]. This behavior can also permit the gene or drug delivery into cells by means of CNTs (such as impalefection). After internalizing into the cells, the fabrication of reactive O_2 species might be stimulated because of the increased surface reactivity and the high surface-to-volume ratio of CNTs. High concentrations of reactive oxygen species are generally linked with the spoil intracellular components through uncontrolled oxidation and can consequence in cell apoptosis [100–102]. Liu *et al.* studied the reserve proliferation, mineralization, and osteogenic differentiation of human mesenchymal stem cells on COOH-CNTs [103]. In preference to the reactive oxygen species mechanism, they revealed that the articulation of adipocyte differentiation specific genes and osteoblast specific genes was significantly decreased in the occurrence of COOH-CNTs. The microscopic investigation suggested the connections of CNTs with proteins positioned in the cytoplasm or on the cell membrane. Therefore it was revealed that the toxicity arises from modulation during the Smad-dependent bone morphogenetic protein signaling approach [103]. The obtained data was in a good accord with the research of Mu *et al.* who reported the repression of bone morphogenetic protein signaling using COOH-SWNTs [104]. SWNTs are considered to be promising materials for controlling intracellular signaling cascades and regulating genetic regulation. Furthermore, *in vitro* experiments showed that carbon nanotubes have also revealed a dose-dependent influence toward gene expression in human fibroblasts [105]. In order to consider the adverse effects, several techniques have been carried out to reduce the cytotoxic issues in the last two decades, for instance by surface functionalization, purification, shape engineering, and compositing with suitable biocompatible materials [106, 107]. The metal impurities (such as catalysts) might be remaining as by product through the fabrication of CNTs. Such metal impurities can propel more cytotoxicity to bone cells [102]. The appearance of toxicity can possibly reduce via eliminating these impurities using acid oxidation and supporting the growth of cells [108]. Nimmagadda *et al.* studied the overall physical and chemical influence of different SWNT fabrications on *in vitro* cell metabolic activity and viability [109]. They described the 3T3 cells (mouse fibroblasts) behaviors of on 3 different SWNTs modified with glucosamine [109]. The obtained results showed that metabolic activity and cell viability 3T3 cells of were influenced by the preparation method and concentration of CNTs. Purification and functionalization of CNT surface to reduce hydrophobicity or to incorporate functional groups considerably mitigated the cytotoxicity [109]. Poly(ethylene glycol)ylation (PEGylation) is one of the significant approaches to improving bone biocompatibility. Sonicating of the COOH-CNTs in the solution of dimethylformamide and treating with oxalyl chloride, along with the addition PEG under mechanical agitation at elevated temperature is a common method for fabricating hydrophilic PEGylated CNTs [110]. Nayak *et al.* fabricated a thin film based on PEGylated MWNTs and determined their capability to effects human mesenchymal stem cells' morphology, differentiation, and proliferation into osteoblasts [111]. They reported that these functionalized nanotubes did not possess cytotoxicity. In addition, the viability of human mesenchymal stem cells was increased and the osteogenic differentiation of human mesenchymal stem cells was stimulated, with simultaneous bone matrix mineralization. It is significantly noted that the assay being employed to evaluate the CNTs' cytotoxicity is also serious as other approaches may generate different readings and results [112]. Therefore it is essential to indicate the assay utilized when one investigates the cytotoxicity effects of CNTs.

12.6 Applications of Functionalized CNTs in Tissue Engineering

12.6.1 Functionalized CNTs for Cardiac Tissue Engineering

Among the other carbon-nanomaterial based composites, CNTs attributed promising features that produce them as potential candidates for the fabrication of future cardiac implants [113]. Numerous types of research investigations to encapsulate the carbon nanomaterials into cardiac tissue regenerations have targeted the utilization of graphene derivatives, gold nanoparticles, and CNTs based materials [114–117]. The incorporation of CNTs has permitted overcoming several drawbacks, offering patches with improved mechanical and conductive properties [113]. CNTs accelerated the maturation of cardiomyocytes and enhanced the functionality of cardiac tissue in comparison to other conductive carbon-based nanomaterials, for instance, reduced graphene oxide (GO) [118, 119]. Moreover, CNTs bound to or encapsulated into polymeric moieties act more similar to particulate materials rather than nanometric materials, accordingly no interaction with the cellular cytoskeleton, with no inflammation, cytotoxicity or any significant adverse effect when investigated *in vitro* or *in vivo* for the application of cardiac tissue engineering [120]. CNTs offered electrical signals to glutaraldehyde cross-linked gelatin hydrogels [121]. The 3-dimonsional scaffolds stimulated electrical activity and cardiomyocyte contractile *in vitro*. The results appeared to be *in vivo* analysis of the patch implanted into rat infarcted hearts, noticing its fruitful biohybridization among the host myocardium and the successive cardiac tissue reconstruction [121]. The native myocardium extracellular matrix comprises a wide range of aligned nanofibrous-like and electrically conductive proteins, mostly laminin and collagen that facilitate the propagation of electrical signals alongside the cardiac cells to ultimately assist the synchronous contraction of the cardiac tissue. The fabrication of nanofibrous matrices with oriented morphologies to mimic biologically was found to promote the construction of aligned engineered cardiac tissue with mature phenotype and improved performance [122]. This type of orientation can lead to the aligned association of cardiac cells, gap junctions, and sarcomeres, to facilitate the electric proliferation and succeeding synchronized contraction. Additionally, microelectrodes using aligned-CNT at lower voltages can permit pacing in close association with cells, accordingly regulating the electrical pulses amid tissue and cells constructs followed by controlling their function and behavior. Such type of morphology, which attains close resemblance to the nanofibrous microstructure of the native extracellular matrix, allows localized stimulation along with guiding cardiac cells. Furthermore, Ahadian *et al.* [123] aligned CNTs in gelatin methacrylate through dielectrophoresis. In this finding, mouse embryoid bodies were cultured in the microwells developed on the scaffolds of aligned CNT-hydrogel. The utilization of CNT-hydrogel on mouse embryoid bodies upon electrical stimulation confirmed a better cardiac differentiation, in comparison to randomly-dispersed or nonconductive CNT hydrogel controls [123]. Wu *et al.* developed a 3D hybrid scaffold using an aligned

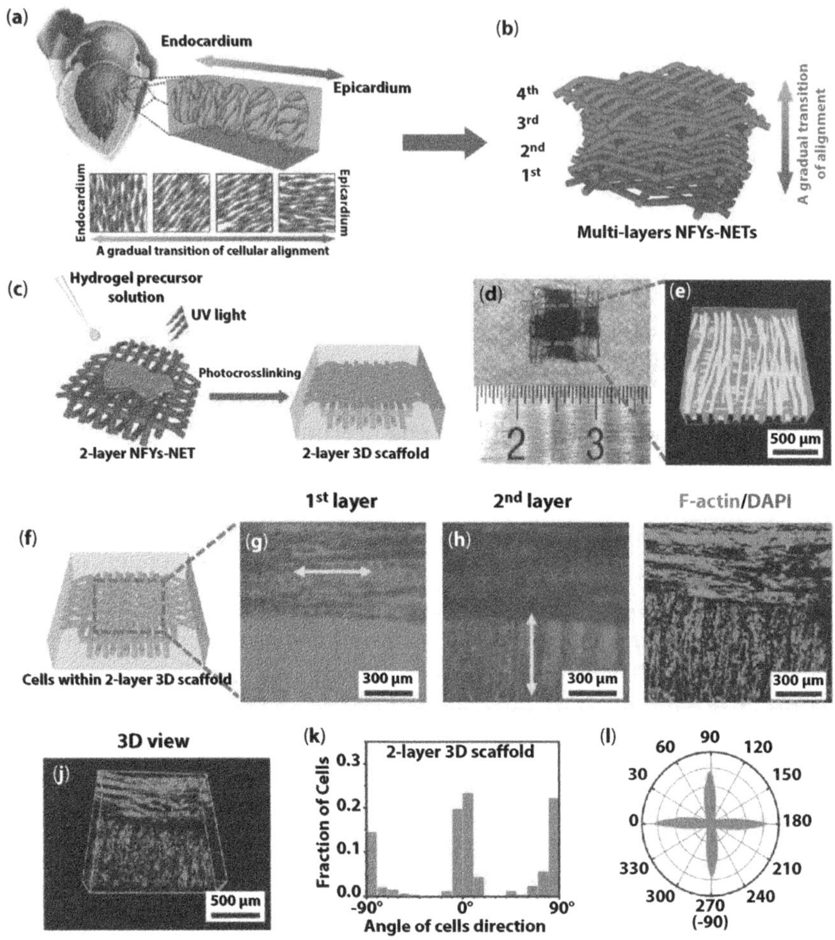

Figure 12.2 Fabrication of 2-layer 3D aligned conductive nanofiber yarns network (NFYs-NET)/Gel hybrid scaffolds and cardiomyocytes cultivation. (a) Schematic representation of myocardium representing a steady transition of aligned cell layers from endocardium toward epicardium. (b) Schematic representation of NFYs-NETs assembled multiple layers with the steady transition of orientation. (c) The possible approach of 2-layer 3D scaffold construction through incorporating two layers of NFYs-NET with orthogonal adaptation after photo-cross-linking in methacrylated gelatin hydrogel shell. The gross image (d) and 3D representation of confocal image (e) of 2-layer 3D scaffolds, where hydrogel was stained with green and aligned conductive nanofiber yarns network was stained with red, respectively. (f) The representation of cells culturing on two aligned conductive nanofiber yarns network layers in shell of hydrogel. Fluorescent images of cardiomyocytes through staining F-actin (green) on the scheme layer with horizontal (g) and vertical direction (h) in the fabricated scaffold. The top view (i) and 3D view (j) of confocal images of cells through staining F-actin/DAPI. The quantitative distribution investigation of cellular orientation verified that cells were aligned on each layer while were perpendicular on different layers (k, l). Reprinted with permission from Ref. [124]. Copyright© 2017 American Chemical Society.

conductive nanofiber yarns network (combination: CNT, silk fibroin, and polycaprolactoned) via weaving in a hydrogel shell for mimicking the native cardiac tissue structure. The fabricated nanofiber yarns network was demonstrated a huge probability for constructing 3D cardiac anisotropy towards cardiac tissue engineering [124]. The incubation of cardiomyocytes in the nanofiber yarns network scaffolds verified their improved biocompatibility and aptitude to assist cellular elongation and alignment, improving cardiomyocytes' function and maturation (Figure 12.2).

12.6.2 Functionalized CNTs for Neuronal Tissue Regeneration

Nerve damage/injury can happen through virus infection, hypoglycemia, anoxia, diabetes, and external trauma [18]. However nerve axons are competent to regenerate, the size of the damage is restricted to 5 mm and adjustment is necessarily required to surgically fix the injured nerve ends. Hence, numerous types of nerve therapies have been investigated such as nerve autograft, coaptation, nerve conduit, and nerve allograft [125]. An approach which is utilized to fix the injured central nervous system cannot be left conveying the axons regrowth, the regeneration of neurons, and the plastic recasting of neuronal circuitry which are emerged from stem cells [126–128]. Although this is a very complicated matter, the growth of axons requires overcoming inhibited and unfavorable environments and needs target recognition, the rebuild of functional synapses, and suitable axonal spatial organization. Stem cells are able to differentiate into a specific neuronal lineage and need to be alive [18]. In this view, the artificial implants based on synthetic material, for instance, 3D scaffolds are needed to offer a bioactive and biocompatible atmosphere, permitting the maintenance, differentiation, and attachment of nerve cells and ultimately involving the construction of functional neuronal assembly [127, 129, 130]. Additionally, the surface offered for the regeneration of the nerve requires having suitable dimensions to limit any nerve compression and damage and to permit the discharge of neurotrophic factors and diffusion using the proximal site of axon damage [131]. To permit *in vivo* manipulation flexibility is essential, requiring proper elastic modulus and ultimate tensile strength (0.58 and 1.4 MPa respectively for decellularized nerve tissue) [132–134]. In due course, nanotopography possesses huge significance as the neurons' contact with their growing substitutes arises at the nanoscale level for example neural cell adhesion molecules-NCAM, integrins, and N-cadherin which is very delicate to the substrates nanoatmosphere [135, 136]. Several kinds of research have been studied the utilization of CNTs directly as a platform for neuronal tissue regeneration and outcomes suggested that carbon nanotubes are competent to assist neurons attachment, ease the generation of more elaborate and longer neurite as well as promote cell differentiation [137–139]. The combination of CNTs with surprising mechanical properties and high electrical conductivity provides a route for non-conductive materials towards neuronal tissue regeneration along with minimizing the toxicity behavior of CNTs. Due to these properties; several reports have demonstrated the composition of biocompatible polymers with CNTs to fabricate 2D electrical conductive membranes or meshes for neural tissue regeneration. Shao *et al.* [140] studied a unique approach to fabricating a membrane that comprises PDDA (poly(dimethyldiallylammonium chloride) and CNTs through a layer-by-layer congregation. Fascinatingly, outcomes suggested that the negatively charged CNTs and positively and negatively charged PDDA can show more homogeneous CNTs dispersion in the polymer backbones. The prepared substrates generated an effective regulatory signal for neural stem cells with respect to adhesion, outgrowth, differentiation, proliferation, and electrophysiological advancement of neural stem cells (NSCs) derived neurons. Furthermore, they suggested a plausible method that the integrin-mediated contacts among NSCs and substrates initiated focal adhesion kinases (FAK), leading to activation of a progression of signaling for example a wingless-related integration site (Wnt) signaling pathway which controls NSCs proliferation, mitogen-activated protein kinase (MAPK) signaling route which controls differentiation of neural stem cells, and a phosphatidylinositol 3-kinase and protein kinase B (PI3K-AKT) signaling pathway to controls the cell survival [140]. Vicentini *et al.* [141] reported 3 types of organic-functionalized carbon nanostructures namely, reduced-graphene oxide (GO), carbon nanohorns, and *p*-methoxyphenyl functionalized-MWNTs, employed as nanofillers for the fabrication of well-dispersed and homogeneous composites of poly(l-lactic acid), a biodegradable and biocompatible U.S. food and drug administration (FDA)-approved polymer. A 2D film was fabricated by all these composites through electrospinning for neural tissue regeneration. The MWNT containing material possessed higher elongation break and Young's modulus and but reduced electrical resistance. These fabricated scaffolds demonstrated improved biocompatibility and assist the proliferation of SH-SY5Y (human neuronal precursor cell line).

12.6.3 Functionalized CNT for Cartilage Tissue Engineering

Biocompatible and eco-friendly polymers are broadly utilized in the construction of scaffolds for cartilage tissue regeneration. Biopolymers have good rates of degradation but possess lower mechanical strengths [142].

In this regard, synthetic polymers with high mechanical stability such as polyglycolic acid (PGA), poly (lactide-co-glycolide) (PLGA), polylactic acid (PLA), and polycaprolactone (PCL), have been evaluated for regenerative medicine and tissue engineering. A composition of biocompatible materials such as CNT and graphene oxide (GO) with a suitable polymeric matrix can advance the biomechanical characteristics of the polymeric matrix. In a report, PLGA and COOH-MWNTs based nanocomposite film was found to be enhanced the biocompatibility of osteoblasts and proved to be a suitable implant substrate for tissue engineering [143]. This report confirmed the cellular reactions of osteoblasts grown on the fabricated film. The experimental outcomes suggested that interaction with the nanocomposite films modulated osteogenic differentiation in osteoblasts, stimulated DNA synthesis and S phase accumulation, and improved cell proliferation [143]. According to the literature, chondrocytes was cultured in composite mixtures of agarose (2%) and two preparations of nanotubes, i.e. covalent polyethylene glycol (PEG)-functionalized SWNTs, and SWNT-COOH [144]. The carboxyl functionalized SWCNT displayed better viability than a control scaffold, whereas the scaffold based on SWCNT-PEG displayed reserved proliferation. Fascinatingly, a larger amount of SWCNT-PEG regardless of the lowest cell viability accomplished more advanced extracellular matrix (ECM) production. Karbasi *et al.* fabricated Poly(3-hydroxybutyrate) (PHB)/chitosan scaffolds comprising MWNTs through electrospinning for cartilage tissue engineering [145]. The obtained results verified that mechanical strengths are approximately similar to characteristics of cartilage. Therefore, PHB-chitosan/MWNTs electrospun scaffold possessed appropriate mechanical and structural behaviors which can be employed for cartilage tissue regeneration [145]. In a study, Zadehnajar *et al.* prepared PCL/gelatin based scaffold with 1 wt.% of MWNTs using the electrospinning technique [146]. It was found that the bioactivity and hydrophilicity of the PCL/gelatin/MWNTs were increased due to MWNTs. The mechanical stability of the PCL-gel/MWNTs scaffold was increased to 2.92 MPa in comparison to the PCL/geletin because of the presence of 1 wt% of MWNTs. The weight loss of the PCL/geletin scaffold after 14 days was more in comparison to the PCL-gel/MWNTs scaffold. The PCL/geletin scaffold was completely degraded within 32 days due to the fragile interaction amid the polymers. The rate of degradation of PCL/geletin/MWNTs scaffold was reduced because of the ionic bonds among the carboxyl group of MWNTs and amine group of geletin which enhanced the chemical strengthen of its structure. According to the methylthiazolyldiphenyl-tetrazolium bromide (MTT) investigation, both electrospun scaffolds were cytocompatible for up to seven days [146]. Furthermore, the poly 3-hydroxybutyrate-chitosan-1 wt% MWNTs functionalized by COOH (MWNTs)/silk nano-micro scaffolds were developed via electrospinning for cartilage tissue engineering [147]. Chitosan, MWNTs, and silk promoted the attachment and development of chondrocytes. The carboxyl groups in MWNTs simulated body fluid, facilitating the deposition of phosphorous and calcium that can behave similarly to ceramic reinforcement and improved bioactivity of the scaffold. Additionally, scaffold comprising 1 wt% MWNTs attributed enhanced tensile strength (46.68 ± 2.20 MPa) that might be appropriate for cartilage tissue engineering. The calcium concentrations in the scaffold comprising MWNTs were improved from 2.20 wt% to 3.73 wt% after 28 days, due to adsorption of phosphate with the amine group of chitosan and depositions of Ca^{2+} with carboxyl groups of MWNTs. Besides the MWNTs degrade in about 56 days because of the ionic bond among carboxyl and amine groups in poly 3-hydroxybutyrate and MWNTs, it can properly serve for long-term cartilage tissue engineering [147].

12.6.4 CNT for Bone Tissue Regeneration

Bone is a type of connective tissue which is a vascular and extremely specialized form and found to be very promising towards maintaining the skeleton shape. During movement defending the soft tissues in cranial, transferring the forces of muscular contraction, pelvic and thoracic cavities, regulating the ECM, and maintaining ions, blood pH, and blood production [148]. Bone has the ability to regenerate and heal itself in the case of partial fracture, or damage. However, bone is unable to heal itself in pathological fractures, primary tumor resection or traumatic bone loss, when injuries surpass a critical size (5 mm) [149]. Therefore, various types of therapies were evaluated such as xenografts, allografts, and autografts but these approaches display different drawbacks for example donor site morbidity and limited supply for autografts, risk of diseases transmission and rejection for allografts, and poor clinical outcome and risk of immunogenicity for xenografts [150–152]. As a result, the utilization of synthetic bone 3D scaffolds, that afford the essential reinforcement for cell attachment, differentiation, and proliferation are requisite. Literature survey reports that CNTs block displayed strong alkaline phosphatase (ALP) action with the occurrence of recombinant human bone morphogenetic proteins (BMP-2), representing better osteogenesis activities. Tanaka *et al.* [153] studied the capability of the 3D porous CNT scaffold for bone regenerative medicine and comparison of CNTs was also carried out with hydroxyapatite *in vivo* and *in vitro*. The experimental outcomes suggested that CNTs displayed improved absorption and release capability of protein. *In vivo* analyses indicated that porous morphology of CNT had better osteoconduction, greater proliferation of cell and higher bone generation upon incorporation with recombinant human BMP-2 [153]. Tanaka *et al.* [154] also reported the preparation CNTs based 3D block structure and demonstrated the efficacy of CNTs' as scaffold materials for bone tissue engineering and is a comparison with PET-reinforced collagen. They found that the prepared scaffold displayed considerable variations with the femoral bone of rat with analogous compressive strength of 62.1 and 61.86 Mpa respectively. It was also found that CNT scaffolds reported prior cell adhesion than PET-reinforced collagen scaffolds. Khalid *et al.* [155] produced MWNT/hydroxyapatite scaffolds using different loadings of CNTs (1, 3, and 5 wt%). Investigations through human osteoblast sarcoma cell lines indicated the dose-dependent cytotoxicity of the fabricated scaffolds and the viability of cells reduced with the rise in content of CNT. Oyefusi *et al.* [156] produced promising bone implant materials by successfully grafting hydroxyapatite onto graphene nanosheets and CNTs. These materials were further examined in terms of cell differentiation and proliferation rate of the temperature-sensitive human fetal osteoblastic cell lines. The obtained outcomes verified that both CNTs-hydroxyapatite and graphene-hydroxyapatite felicitated cell differentiation and growth. In another report, Gupta *et al.* [157] developed composite microspheres based on polylactic-co-glycolic acid and SWNT for bone tissue regeneration. The considered MC3T3-E1 cells (MC3T3-E1 subclone 4, precursor osteoblasts) were investigated in terms of gene expression, proliferation, and adhesion. They reported that MC3T3-E1 cells adhered, exhibited normal, and grew, the addition of SWNT and non-stressed morphology strengthened gene expression and rate of cell proliferation in comparison to pure polylactic-co-glycolic acid scaffolds [157]. Facca *et al.* report *in vivo* analyses of implants based on plasma-sprayed-CNT-reinforced-hydroxyapatite (HA) coating on Ti incorporated in rodents' bone [89]. They noticed normal bone growth in the vicinity of HA-CNT implants and found that the presence of CNT stimulates larger osseointegration in comparison to HA. For this during one month Ti rods or beads with 3 different coatings: (i) uncoated Ti (group 1), (ii) HA coating (group 2), and (iii) HA/4 wt % CNT (group 3) coating were implanted in the external condyle of rodents' bone (Figure 12.3(I-a)). They found that after one month, for all groups the cortical defect was restored because of the development of the neocortical bone Figure 12.3(II). As shown in Figure 12.3(III) with toluidine blue coloration, the gray layer (HA-CNT coating) was observed to be firmly connected

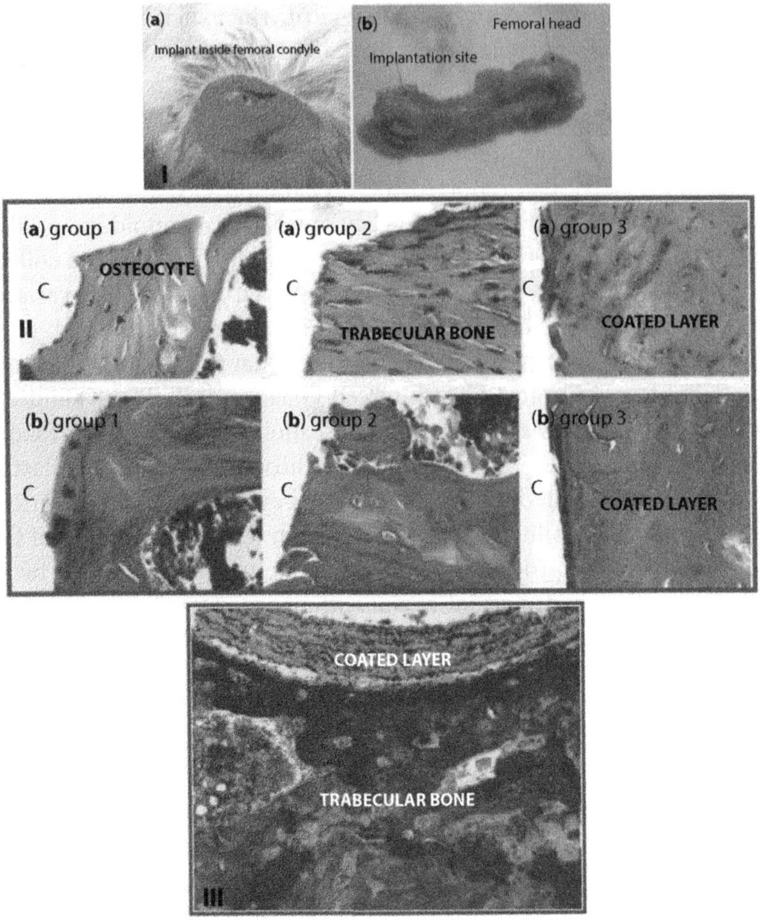

Figure 12.3 (I)(a) A rod incorporated inside the femoral condyle (through a bony defect). (b) femoral bone of rat got from group 3: HA-CNT-coated Ti implant. (II) Experimental outcomes with rod-shaped implants for rat bones (histological data 40×). (a) Mallory coloration views and (b) hematoxylin-eosine coloration view thick trabecular, normal, and hematopoietic marrowbones, without any inflammatory tissues and reactions. A thicker for group 3 (HA-CNT coating) and attached to newly grown bone tissues closed to the cavity (c) caused by the retrieved implant. (III)n for group 3 implant with toluidine blue coloration(histological image (5×)) displaying HA-CNT coating firmly connected to freshly developed bone tissues and near to the coating normal neobonealong with normal cell morphology. Reprinted with permission from Ref. [89]. Copyright © 2011 American Chemical Society.

to the freshly developed bone tissues. According to this study *in vivo* analysis verified that for the HA composite CNTs as nanoreinforcement can be employed for applications of orthopedic without any adverse effect [89].

12.7 Future Perspectives and Challenges

The fabrication of new effective nanomaterials and their utilization in tissue engineering is really important for the regrowth or repair of damaged or injured tissue. To meet this demand several researchers try to construct new biomaterials along with the existing nanotechnology, through different compositions of various nanomaterials. The construction of scaffold materials proficient in generating preferred tissue and cellular behavior is greatly favored in the area of tissue regeneration. Several reports have suggested that due to their exceptional properties, CNTs are potential candidates in fabricating such materials with finely tunable and improved physico-chemical and mechanical characteristics. A variety of multifunctional hybrids with desired structures have been demonstrated to be efficient in improving the adhesion, directing the differentiation of bone-related cells, and promoting proliferation, along with offering appropriate sites for the deposition of minerals. Furthermore, pretty desirable *in vivo* bone biocompatibility has also been noticed after implanting these materials inside animals. Despite having too many reports, progress in this field is still in its beginning and a lot of research is required to completely investigate the limitations and potential of CNTs based materials in tissue engineering. The directions of future research along with the challenges that need to be addressed are as follows:

(i) Advanced structural morphologies along with the 3D porous structure that replicates the structure of the original bone are needed. These sophisticated morphologies not only produce mechanical hold against the *in vivo* actions but also promote the nutrients' transportation.
(ii) The mechanical properties of CNTs and their derivatives required further improvement i.e. up to the natural bone level in order to lower the conflict at the inorganic-organic interface.
(iii) The early stages of cell culture *in vitro* is the center for most research investigations, hence detailed *in vivo* investigations should be focused on, followed by the polymeric materials' degradation along with the discharge of CNTs.

(iv) The long-term biocompatibility of CNTs based nanomaterials need to be addressed cautiously. Prior to any clinical trials, the entire knowledge of the biocompatibility of these nanomaterials should be focused on. Moreover, more efficient approaches to reducing their cytotoxicity need to be addressed. So that adverse effects of these materials could be minimized according to the precautionary principle during the fabrication, examining, and clinical applications. There is still a need to take a big step in the stability, utilization, and biosafety of these nanomaterials.

12.8 Conclusion

In this chapter, we summarized the basic characteristics, structural, physical, and chemical properties, bone compatibility along with interactions and biodegradation of different types of CNTs based nanomaterials, and their typical applications in tissue engineering. Carbon nanotubes with excellent physiochemical characteristics such as electrical, thermal, mechanical, and structural properties have been extensively employed in electronic, photovoltaics, and energy storage applications and nowadays for the application of biomedical fields. Recently, CNTs are considered the most favorable candidates with immense potential for the successful engineering of fabricating to stimulate the adhesion, differentiation, and proliferation of injured tissue. CNTs based scaffolds permit the enhancement of the electrical activity as well as the direction of the nanotubes promoting tissue differentiation and growth. However, the main concern when employing pristine CNT constructs is the matter of toxicity. Numerous studies report that to minimize the toxicity and enhance the dispersity in the aqueous phase, CNTs can be modified with different functional groups (for instance -COOH and -OH and) using different approaches. Furthermore, the combination of CNTs along with the polymeric scaffolds for the fabrication of constructs for tissue engineering has been verified to improve viability, enhance mechanical integrity, improve electrophysiological performance along with better cell adhesion, and uniformity. For tissue engineering applications, CNTs have been primarily employed as reinforcement materials in the polymeric framework, revamping electrical and mechanical features and permitting the development of scaffolds for cardiac, bone, cartilage, and neural tissue engineering. The composition of polymer matrix with CNTs can enhance the nerve cell responses by accelerating the conductivity of the scaffold related to the nerve. It can also promote the biological responses, elastic strength, and conductivity of the scaffolds related to the cardiac tissue. Additionally, CNTs are employed to reinforce the biological and mechanical features for bone tissue engineering. In the future, the development of nanotechnologies that are rationally designed and related to the current advancement in CNTs based nanostructures will solve most of the problems encountered in current tissue engineering applications.

Important Websites

https://www.prescouter.com/2017/03/applications-carbon-nanotubes/
https://nanoten.com/NTSite/
https://www.nanowerk.com/nanotechnology/introduction/introduction_to_nanotechnology_22.php
https://www.nanocyl.com/expertise-center/health-safety-environment/

References

1. Shi, J., Votruba, A.R., Farokhzad, O.C., Langer, R., Nanotechnology in drug delivery and tissue engineering: From discovery to applications. *Nano Lett.*, 10, 3223–3230, 2010.
2. Asl, M.A., Karbasi, S., Beigi-Boroujeni, S., Zamanlui Benisi, S., Saeed, M., Evaluation of the effects of starch on polyhydroxybutyrate electrospun scaffolds for bone tissue engineering applications. *Int. J. Biol. Macromol.*, 191, 500–513, 2021.
3. Li, Y., Liu, Y., Li, R., Bai, H., Zhu, Z., Zhu, L., Zhu, C., Che, Z., Liu, H., Wang, J., Huang, L., Collagen-based biomaterials for bone tissue engineering. *Mater. Des.*, 210, 110049, 2021.
4. Guo, J.L., Kim, Y.S., Mikos, A.G., Biomacromolecules for tissue engineering: Emerging biomimetic strategies. *Biomacromolecules*, 20, 2904–2912, 2019.
5. Zheng, X., Zhang, P., Fu, Z., Meng, S., Dai, L., Yang, H., Applications of nanomaterials in tissue engineering. *RSC Adv.*, 11, 19041–19058, 2021.
6. Capek, I., *Nanocomposite structures and dispersions science and nanotechnology - fundamental principles and colloidal particles*, pp. 1–69, Elsevier, Amsterdam, 2006.
7. Kumar, S., Nehra, M., Kedia, D., Dilbaghi, N., Tankeshwar, K., Kim, K.-H., Nanodiamonds: Emerging face of future nanotechnology. *Carbon N. Y.*, 143, 678–699, 2019.

8. Peng, H., Liu, X., Wang, R., Jia, F., Emerging nanostructured materials for musculoskeletal tissue engineering. *J. Mater. Chem. B Mater. Biol. Med.*, 2, 6435–6461, 2014.
9. Arumugam, S., Ramamoorthy, P., Chakkarapani, L.D., Biodegradable dendrimer functionalized carbon nanotube-hybrids for biomedical applications. *J. Polym. Res.*, 26, 182, 2019.
10. De Witte, T.M., Fratila-Apachitei, L.E., Zadpoor, A.A., Peppas, N.A., Bone tissue engineering via growth factor delivery: From scaffolds to complex matrices. *Regen. Biomater.*, 5, 197–211, 2018.
11. Iijima, S., Helical microtubules of graphitic carbon. *Nature*, 354, 56–58, 1991.
12. Resende, R., Tonelli, F.M.P., Santos, A.K., Gomes, K.N., Lorençon, E., Ladeira, L.O., Guatimosim, S., Carbon nanotube interaction with extracellular matrix proteins producing scaffolds for tissue engineering. *Int. J. Nanomedicine*, 7, 4511, 2012.
13. Sun, Y.-P., Fu, K., Lin, Y., Huang, W., Functionalized carbon nanotubes: Properties and applications. *Acc. Chem. Res.*, 35, 1096–1104, 2002.
14. Lalwani, G., Gopalan, A., D'Agati, M., Srinivas Sankaran, J., Judex, S., Qin, Y.-X., Sitharaman, B., Porous three-dimensional carbon nanotube scaffolds for tissue engineering. *J. Biomed. Mater. Res. Part A*, 103, 3212–3225, 2015.
15. Esteves, I.A.A.C., Cruz, F.J.A.L., Müller, E.A., Agnihotri, S., Mota, J.P.B., Determination of the surface area and porosity of carbon nanotube bundles from a Langmuirian analysis of sub- and supercritical adsorption data. *Carbon N. Y.*, 47, 948–956, 2009.
16. Deng, P., Xu, Z., Li, J., Simultaneous determination of ascorbic acid and rutin in pharmaceutical preparations with electrochemical method based on multi-walled carbon nanotubes–chitosan composite film modified electrode. *J. Pharm. Biomed. Anal.*, 76, 234–242, 2013.
17. Shi Kam, N.W., O'Connell, M., Wisdom, J.A., Dai, H., Carbon nanotubes as multifunctional biological transporters and near-infrared agents for selective cancer cell destruction. *Proc. Natl. Acad. Sci.*, 102, 11600–11605, 2005.
18. Hopley, E.L., Salmasi, S., Kalaskar, D.M., Seifalian, A.M., Carbon nanotubes leading the way forward in new generation 3D tissue engineering. *Biotechnol. Adv.*, 32, 1000–1014, 2014.
19. Kim, H., Wang, M., Lee, S.K., Kang, J., Nam, J.-D., Ci, L., Suhr, J., Tensile properties of millimeter-long multi-walled carbon nanotubes. *Sci. Rep.*, 7, 9512, 2017.
20. Dorri Moghadam, A., Omrani, E., Menezes, P.L., Rohatgi, P.K., Mechanical and tribological properties of self-lubricating metal matrix nanocomposites reinforced by carbon nanotubes (CNTs) and graphene – A review. *Compos. Part B Eng.*, 77, 402–420, 2015.
21. Birch, M.E., Ruda-Eberenz, T.A., Chai, M., Andrews, R., Hatfield, R.L., Properties that Influence the Specific Surface Areas of Carbon Nanotubes and Nanofibers. *Ann. Occup. Hyg.*, 57, 1148–1166, 2013.

22. Bhattacharyya, S., Guillot, S., Dabboue, H., Tranchant, J.-F., Salvetat, J.-P., Carbon nanotubes as structural nanofibers for hyaluronic acid hydrogel scaffolds. *Biomacromolecules*, 9, 505–509, 2008.
23. Liu, C., Zhang, J., He, J., Hu, G., Gelation in carbon nanotube/polymer composites. *Polym. (Guildf)*, 44, 7529–7532, 2003.
24. Martinelli, V., Bosi, S., Peña, B., Baj, G., Long, C.S., Sbaizero, O., Giacca, M., Prato, M., Mestroni, L., 3D Carbon-nanotube-based composites for cardiac tissue engineering. *ACS Appl. Bio Mater.*, 1, 1530–1537, 2018.
25. Shin, S.R., Shin, C., Memic, A., Shadmehr, S., Miscuglio, M., Jung, H.Y., Jung, S.M., Bae, H., Khademhosseini, A., Tang, X.S., Dokmeci, M.R., Aligned carbon nanotube-based flexible gel substrates for engineering biohybrid tissue actuators. *Adv. Funct. Mater.*, 25, 4486–4495, 2015.
26. Ramón-Azcón, J., Ahadian, S., Estili, M., Liang, X., Ostrovidov, S., Kaji, H., Shiku, H., Ramalingam, M., Nakajima, K., Sakka, Y., Khademhosseini, A., Matsue, T., Dielectrophoretically aligned carbon nanotubes to control electrical and mechanical properties of hydrogels to fabricate contractile muscle myofibers. *Adv. Mater.*, 25, 4028–4034, 2013.
27. Kharaziha, M., Shin, S.R., Nikkhah, M., Topkaya, S.N., Masoumi, N., Annabi, N., Dokmeci, M.R., Khademhosseini, A., Tough and flexible CNT–polymeric hybrid scaffolds for engineering cardiac constructs. *Biomaterials*, 35, 7346–7354, 2014.
28. Gajendiran, M., Choi, J., Kim, S.-J., Kim, K., Shin, H., Koo, H.J., Kim, K., Conductive biomaterials for tissue engineering applications. *J. Ind. Eng. Chem.*, 51, 12–26, 2017.
29. Ku, S.H., Lee, M., Park, C.B., Carbon-based nanomaterials for tissue engineering. *Adv. Healthc. Mater.*, 2, 244–260, 2013.
30. Balasubramanian, K. and Burghard, M., Chemically functionalized carbon nanotubes. *Small*, 1, 180–192, 2005.
31. Sajid, M.I., Jamshaid, U., Jamshaid, T., Zafar, N., Fessi, H., Elaissari, A., Carbon nanotubes from synthesis to *in vivo* biomedical applications. *Int. J. Pharm.*, 501, 278–299, 2016.
32. He, H., Pham-Huy, L.A., Dramou, P., Xiao, D., Zuo, P., Pham-Huy, C., Carbon nanotubes: Applications in Pharmacy and medicine. *BioMed. Res. Int.*, 2013, 1–12, 2013.
33. Iijima, S., Brabec, C., Maiti, A., Bernholc, J., Structural flexibility of carbon nanotubes. *J. Chem. Phys.*, 104, 2089–2092, 1996.
34. Demczyk, B., Wang, Y., Cumings, J., Hetman, M., Han, W., Zettl, A., Ritchie, R., Direct mechanical measurement of the tensile strength and elastic modulus of multiwalled carbon nanotubes. *Mater. Sci. Eng. A*, 334, 173–178, 2002.
35. Yu, M.F., Files, B.S., Arepalli, S., Ruoff, R.S., Tensile loading of ropes of single wall carbon nanotubes and their mechanical properties. *Phys. Rev. Lett.*, 84, 5552–5555, 2000.

36. Thostenson, E.T., Ren, Z., Chou, T.-W., Advances in the science and technology of carbon nanotubes and their composites: A review. *Compos. Sci. Technol.*, 61, 1899–1912, 2001.
37. Nardecchia, S., Carriazo, D., Ferrer, M.L., Gutiérrez, M.C., del Monte, F., Three dimensional macroporous architectures and aerogels built of carbon nanotubes and/or graphene: Synthesis and applications. *Chem. Soc. Rev.*, 42, 794–830, 2013.
38. Orlita, M., Faugeras, C., Plochocka, P., Neugebauer, P., Martinez, G., Maude, D.K., Barra, A.-L., Sprinkle, M., Berger, C., de Heer, W.A., Potemski, M., Approaching the dirac point in high-mobility multilayer epitaxial graphene. *Phys. Rev. Lett.*, 101, 267601, 2008.
39. Dresselhaus, M.S. and Avouris, P., *Carbon nanotubes synthesis, structure, properties, and applications*, M.S. Dresselhaus, G. Dresselhaus, P. Avouris, (Eds.), pp. 1–9, Springer Berlin, Heidelberg, 2001.
40. Avouris, P., Appenzeller, J., Martel, R., Wind, S.J., Carbon nanotube electronics. *Proc. IEEE*, 9, 1772–1784, 2003.
41. Newman, P., Minett, A., Ellis-Behnke, R., Zreiqat, H., Carbon nanotubes: Their potential and pitfalls for bone tissue regeneration and engineering. *Nanomed. Nanotechnol. Biol. Med.*, 9, 1139–1158, 2013.
42. Meng, S., Liu, Z., Shen, L., Guo, Z., Chou, L.L., Zhong, W., Du, Q., Ge, J., The effect of a layer-by-layer chitosan–heparin coating on the endothelialization and coagulation properties of a coronary stent system. *Biomaterials*, 30, 2276–2283, 2009.
43. Merum, S., Veluru, J.B., Seeram, R., Functionalized carbon nanotubes in bioworld: Applications, limitations and future directions. *Mater. Sci. Eng. B*, 223, 43–63, 2017.
44. Gupta, S., Murthy, C.N., Prabha, C.R., Recent advances in carbon nanotube based electrochemical biosensors. *Int. J. Biol. Macromol.*, 108, 687–703, 2018.
45. Oliveira, S.F., Bisker, G., Bakh, N.A., Gibbs, S.L., Landry, M.P., Strano, M.S., Protein functionalized carbon nanomaterials for biomedical applications. *Carbon N. Y.*, 95, 767–779, 2015.
46. Ajayan, P.M. and Tour, J.M., Nanotube composites. *Nature*, 447, 1066–1068, 2007.
47. Kaur, P., Shin, M.-S., Park, J.-S., Verma, G., Sekhon, S.S., Supramolecular modification of Carbon Nanofibers with Poly(diallyl dimethylammonium) chloride and Triton X-100 for electrochemical application. *Int. J. Hydrogen Energy*, 43, 6575–6585, 2018.
48. Behnam, B., Shier, W.T., Nia, A.H., Abnous, K., Ramezani, M., Non-covalent functionalization of single-walled carbon nanotubes with modified polyethyleneimines for efficient gene delivery. *Int. J. Pharm.*, 454, 204–215, 2013.
49. Zhang, Y., Bai, Y., Yan, B., Functionalized carbon nanotubes for potential medicinal applications. *Drug Discovery Today*, 15, 428–435, 2010.
50. Singh, R., Pantarotto, D., McCarthy, D., Chaloin, O., Hoebeke, J., Partidos, C.D., Briand, J.-P., Prato, M., Bianco, A., Kostarelos, K., Binding and

condensation of plasmid dna onto functionalized carbon nanotubes: Toward the construction of nanotube-based gene delivery vectors. *J. Am. Chem. Soc.*, 127, 4388–4396, 2005.

51. Pantarotto, D., Singh, R., McCarthy, D., Erhardt, M., Briand, J.-P., Prato, M., Kostarelos, K., Bianco, A., Functionalized carbon nanotubes for plasmid DNA gene delivery. *Angew. Chem. Int. Ed.*, 43, 5242–5246, 2004.
52. Meng, L., Fu, C., Lu, Q., Advanced technology for functionalization of carbon nanotubes. *Prog. Nat. Sci.*, 19, 801–810, 2009.
53. Zhu, W., Minami, N., Kazaoui, S., Kim, Y., Fluorescent chromophore functionalized single-wall carbon nanotubes with minimal alteration to their characteristic one-dimensional electronic states. *J. Mater. Chem.*, 13, 2196–2201, 2003.
54. Kim, W.-J., Nair, N., Lee, C.Y., Strano, M.S., Covalent Functionalization of single-walled carbon nanotubes alters their densities allowing electronic and other types of separation. *J. Phys. Chem. C*, 112, 7326–7331, 2008.
55. Mooney, E., Dockery, P., Greiser, U., Murphy, M., Barron, V., Carbon nanotubes and mesenchymal stem cells: Biocompatibility, proliferation and differentiation. *Nano Lett.*, 8, 2137–2143, 2008.
56. Prabhavathi, G., Yamuna, R., Jafer, A.C., Covalent functionalization and solubilization of multi-walled carbon nanotubes by using zinc and copper complexes of meso-tetra(4-aminophenyl) porphyrin. *J. Organomet. Chem.*, 861, 219–229, 2018.
57. You, Y.-Z., Hong, C.-Y., Pan, C.-Y., Directly growing ionic polymers on multi-walled carbon nanotubes via surface RAFT polymerization. *Nanotechnology*, 17, 2350–2354, 2006.
58. Redondo-Gómez, C., Orozco, F., Michael Noeske, P.-L., Soto-Tellini, V., Corrales-Ureña, Y.R., Vega-Baudrit, J., Cholic acid covalently bound to multi-walled carbon nanotubes: Improvements on dispersion stability. *Mater. Chem. Phys.*, 200, 331–341, 2017.
59. Nosek, M., Sainio, J., Joensuu, P.M., 2,2′-bipyridine-functionalized single-walled carbon nanotubes: The formation of transition metal complexes and their charge transfer effects. *Carbon N. Y.*, 129, 175–182, 2018.
60. Firme, C.P. and Bandaru, P.R., Toxicity issues in the application of carbon nanotubes to biological systems. *Nanomed. Nanotechnol. Biol. Med.*, 6, 245–256, 2010.
61. Negri, V., Pacheco-Torres, J., Calle, D., López-Larrubia, P., *Surface-modified nanobiomaterials for electrochemical and biomedicine applications*, pp. 177–217, Springer, Cham, 2020.
62. Saito, N., Haniu, H., Usui, Y., Aoki, K., Hara, K., Takanashi, S., Shimizu, M., Narita, N., Okamoto, M., Kobayashi, S., Nomura, H., Kato, H., Nishimura, N., Taruta, S., Endo, M., Safe clinical use of carbon nanotubes as innovative biomaterials. *Chem. Rev.*, 114, 6040–6079, 2014.
63. Yan, Y., Chan-Park, M.B., Zhang, Q., Advances in carbon-nanotube assembly. *Small*, 3, 24–42, 2007.

64. Cedervall, T., Lynch, I., Lindman, S., Berggard, T., Thulin, E., Nilsson, H., Dawson, K.A., Linse, S., Understanding the nanoparticle-protein corona using methods to quantify exchange rates and affinities of proteins for nanoparticles. *Proc. Natl. Acad. Sci.*, 104, 2050–2055, 2007.
65. Monopoli, M.P., Walczyk, D., Campbell, A., Elia, G., Lynch, I., Baldelli Bombelli, F., Dawson, K.A., Physical–chemical aspects of protein corona: Relevance to *in vitro* and *in vivo* biological impacts of nanoparticles. *J. Am. Chem. Soc.*, 133, 2525–2534, 2011.
66. Nel, A.E., Mädler, L., Velegol, D., Xia, T., Hoek, E.M.V., Somasundaran, P., Klaessig, F., Castranova, V., Thompson, M., Understanding biophysicochemical interactions at the nano–bio interface. *Nat. Mater.*, 8, 543–557, 2009.
67. Oberdörster, G., Oberdörster, E., Oberdörster, J., Nanotoxicology: An emerging discipline evolving from studies of ultrafine particles. *Environ. Health Perspect.*, 113, 823–839, 2005.
68. Gasser, M., Rothen-Rutishauser, B., Krug, H.F., Gehr, P., Nelle, M., Yan, B., Wick, P., The adsorption of biomolecules to multi-walled carbon nanotubes is influenced by both pulmonary surfactant lipids and surface chemistry. *J. Nanobiotechnol.*, 8, 31, 2010.
69. Singh, P., Kumar, J., Toma, F.M., Raya, J., Prato, M., Fabre, B., Verma, S., Bianco, A., Synthesis and characterization of nucleobase–carbon nanotube hybrids. *J. Am. Chem. Soc.*, 131, 13555–13562, 2009.
70. Singh, P., Toma, F.M., Kumar, J., Venkatesh, V., Raya, J., Prato, M., Verma, S., Bianco, A., Carbon nanotube-nucleobase hybrids: Nanorings from uracil-modified single-walled carbon nanotubes. *Chem. - Eur. J.*, 17, 6772–6780, 2011.
71. Xu, Y., Pehrsson, P.E., Chen, L., Zhang, R., Zhao, W., Double-stranded dna single-walled carbon nanotube hybrids for optical hydrogen peroxide and glucose sensing. *J. Phys. Chem. C*, 111, 8638–8643, 2007.
72. Zheng, M., Jagota, A., Semke, E.D., Diner, B.A., Mclean, R.S., Lustig, S.R., Richardson, R.E., Tassi, N.G., DNA-assisted dispersion and separation of carbon nanotubes. *Nat. Mater.*, 2, 338–342, 2003.
73. Vogel, S.R., Kappes, M.M., Hennrich, F., Richert, C., An unexpected new optimum in the structure space of dna solubilizing single-walled carbon nanotubes. *Chem. - Eur. J.*, 13, 1815–1820, 2007.
74. Schleh, C., Rothen-Rutishauser, B., Kreyling, W.G., The influence of pulmonary surfactant on nanoparticulate drug delivery systems. *Eur. J. Pharm. Biopharm.*, 77, 350–352, 2011.
75. Allen, B.L., Kichambare, P.D., Gou, P., Vlasova, I.I., Kapralov, A.A., Konduru, N., Kagan, V.E., Star, A., Biodegradation of single-walled carbon nanotubes through enzymatic catalysis. *Nano Lett.*, 8, 3899–3903, 2008.
76. Allen, B.L., Kotchey, G.P., Chen, Y., Yanamala, N.V.K., Klein-Seetharaman, J., Kagan, V.E., Star, A., Mechanistic investigations of horseradish peroxidase-catalyzed degradation of single-walled carbon nanotubes. *J. Am. Chem. Soc.*, 131, 17194–17205, 2009.

77. Kagan, V.E., Konduru, N.V., Feng, W., Allen, B.L., Conroy, J., Volkov, Y., Vlasova, I.I., Belikova, N.A., Yanamala, N., Kapralov, A., Tyurina, Y.Y., Shi, J., Kisin, E.R., Murray, A.R., Franks, J., Stolz, D., Gou, P., Klein-Seetharaman, J., Fadeel, B., Star, A., Shvedova, A.A., Carbon nanotubes degraded by neutrophil myeloperoxidase induce less pulmonary inflammation. *Nat. Nanotechnol.*, 5, 354–359, 2010.
78. Khang, D., Sato, M., Price, R.L., Ribbe, A.E., Webster, T.J., Selective adhesion and mineral deposition by osteoblasts on carbon nanofiber patterns. *Int. J. Nanomed.*, 1, 65–72, 2006.
79. Price, R.L., Waid, M.C., Haberstroh, K.M., Webster, T.J., Selective bone cell adhesion on formulations containing carbon nanofibers. *Biomaterials*, 24, 1877–1887, 2003.
80. Aoki, K., Usui, Y., Narita, N., Ogiwara, N., Iashigaki, N., Nakamura, K., Kato, H., Sano, K., Ogiwara, N., Kametani, K., Kim, C., Taruta, S., Kim, Y.A., Endo, M., Saito, N., A thin carbon-fiber web as a scaffold for bone-tissue regeneration. *Small*, 5, 1540–1546, 2009.
81. Haniu, H., Saito, N., Matsuda, Y., Tsukahara, T., Usui, Y., Narita, N., Hara, K., Aoki, K., Shimizu, M., Ogihara, N., Takanashi, S., Okamoto, M., Kobayashi, S., Ishigaki, N., Nakamura, K., Kato, H., Basic potential of carbon nanotubes in tissue engineering applications. *J. Nanomater.*, 2012, 1–10, 2012.
82. Hirata, E., Uo, M., Takita, H., Akasaka, T., Watari, F., Yokoyama, A., Multiwalled carbon nanotube-coating of 3D collagen scaffolds for bone tissue engineering. *Carbon N. Y.*, 49, 3284–3291, 2011.
83. Bhattacharya, M., Wutticharoenmongkol-Thitiwongsawet, P., Hamamoto, D.T., Lee, D., Cui, T., Prasad, H.S., Ahmad, M., Bone formation on carbon nanotube composite. *J. Biomed. Mater. Res. Part A*, 96A, 75–82, 2011.
84. Lewitus, D.Y., Landers, J., Branch, J.R., Smith, K.L., Callegari, G., Kohn, J., Neimark, A.V., Biohybrid carbon nanotube/agarose fibers for neural tissue engineering. *Adv. Funct. Mater.*, 21, 2624–2632, 2011.
85. van der Zande, M., Sitharaman, B., Walboomers, X.F., Tran, L., Ananta, J.S., Veltien, A., Wilson, L.J., Álava, J.I., Heerschap, A., Mikos, A.G., Jansen, J.A., *In vivo* magnetic resonance imaging of the distribution pattern of gadonanotubes released from a degrading poly(lactic-co-glycolic acid) scaffold. *Tissue Eng. Part C Methods*, 17, 19–26, 2011.
86. Kasai, T., Matsumura, S., Iizuka, T., Shiba, K., Kanamori, T., Yudasaka, M., Iijima, S., Yokoyama, A., Carbon nanohorns accelerate bone regeneration in rat calvarial bone defect. *Nanotechnology*, 22, 065102, 2011.
87. Usui, Y., Aoki, K., Narita, N., Murakami, N., Nakamura, I., Nakamura, K., Ishigaki, N., Yamazaki, H., Horiuchi, H., Kato, H., Taruta, S., Kim, Y.A., Endo, M., Saito, N., Carbon nanotubes with high bone-tissue compatibility and bone-formation acceleration effects. *Small*, 4, 240–246, 2008.
88. Abarrategi, A., Gutiérrez, M.C., Moreno-Vicente, C., Hortigüela, M.J., Ramos, V., López-Lacomba, J.L., Ferrer, M.L., del Monte, F., Multiwall carbon

nanotube scaffolds for tissue engineering purposes. *Biomaterials*, 29, 94–102, 2008.
89. Facca, S., Lahiri, D., Fioretti, F., Messadeq, N., Mainard, D., Benkirane-Jessel, N., Agarwal, A., *In vivo* osseointegration of nano-designed composite coatings on titanium implants. *ACS Nano*, 5, 4790–4799, 2011.
90. Silva, E., Vasconcellos, L.M.R., de Rodrigues, B.V.M., dos Santos, D.M., Campana-Filho, S.P., Marciano, F.R., Webster, T.J., Lobo, A.O., PDLLA honeycomb-like scaffolds with a high loading of superhydrophilic graphene/multi-walled carbon nanotubes promote osteoblast *in vitro* functions and guided *in vivo* bone regeneration. *Mater. Sci. Eng. C*, 73, 31–39, 2017.
91. Gulati, G.K., Berger, L.R., Hinds, B.J., A preclinical evaluation of a programmable CNT membrane device for transdermal nicotine delivery in hairless Guinea pigs. *J. Control. Release*, 293, 135–143, 2019.
92. Zhou, Z., Liu, X., Wu, W., Park, S., Miller II, A.L., Terzic, A., Lu, L., Effective nerve cell modulation by electrical stimulation of carbon nanotube embedded conductive polymeric scaffolds. *Biomater. Sci.*, 6, 2375–2385, 2018.
93. Liu, L., Yang, B., Wang, L.-Q., Huang, J.P., Chen, W.-Y., Ban, Q., Zhang, Y., You, R., Yin, L., Guan, Y.-Q., Biomimetic bone tissue engineering hydrogel scaffolds constructed using ordered CNTs and HA induce the proliferation and differentiation of BMSCs. *J. Mater. Chem. B*, 8, 558–567, 2020.
94. Dominguez-Alfaro, A., Alegret, N., Arnaiz, B., González-Domínguez, J.M., Martin-Pacheco, A., Cossío, U., Porcarelli, L., Bosi, S., Vázquez, E., Mecerreyes, D., Prato, M., Tailored methodology based on vapor phase polymerization to manufacture PEDOT/CNT scaffolds for tissue engineering. *ACS Biomater. Sci. Eng.*, 6, 1269–1278, 2020.
95. Kawaguchi, M., Fukushima, T., Hayakawa, T., Nakashima, N., Inoue, Y., Takeda, S., Okamura, K., Taniguchi, K., Preparation of carbon nanotube-alginate nanocomposite gel for tissue engineering. *Dent. Mater. J.*, 25, 719–725, 2006.
96. Misra, S.K., Ansari, T.I., Valappil, S.P., Mohn, D., Philip, S.E., Stark, W.J., Roy, I., Knowles, J.C., Salih, V., Boccaccini, A.R., Poly(3-hydroxybutyrate) multifunctional composite scaffolds for tissue engineering applications. *Biomaterials*, 31, 2806–2815, 2010.
97. Shi, X., Sitharaman, B., Pham, Q.P., Liang, F., Wu, K., Edward Billups, W., Wilson, L.J., Mikos, A.G., Fabrication of porous ultra-short single-walled carbon nanotube nanocomposite scaffolds for bone tissue engineering. *Biomaterials*, 28, 4078–4090, 2007.
98. McCauley, M.D., Vitale, F., Yan, J.S., Young, Pasquali, M., *In Vivo* Restoration of myocardial conduction with carbon nanotube fibers. *Circ. Arrhythm. Electrophysiol.*, 12, 007256, 2019.
99. Ogihara, N., Usui, Y., Aoki, K., Shimizu, M., Narita, N., Hara, K., Nakamura, K., Ishigaki, N., Takanashi, S., Okamoto, M., Kato, H., Haniu, H., Ogiwara, N., Nakayama, N., Taruta, S., Saito, N., Biocompatibility and bone

tissue compatibility of alumina ceramics reinforced with carbon nanotubes. *Nanomedicine*, 7, 981–993, 2012.
100. Han, Z.J., Levchenko, I., Kumar, S., Yajadda, M.M.A., Yick, S., Seo, D.H., Martin, P.J., Peel, S., Kuncic, Z., Ostrikov, K., Plasma nanofabrication and nanomaterials safety. *J. Phys. D. Appl. Phys.*, 44, 174019, 2011.
101. Kong, M.G., Keidar, M., Ostrikov, K., Plasmas meet nanoparticles—Where synergies can advance the frontier of medicine. *J. Phys. D. Appl. Phys.*, 44, 174018, 2011.
102. Fisher, C., Rider, A.E., Jun Han, Z., Kumar, S., Levchenko, I., Ostrikov, K., Applications and nanotoxicity of carbon nanotubes and graphene in biomedicine. *J. Nanomater.*, 2012, 1–19, 2012.
103. Liu, D., Yi, C., Zhang, D., Zhang, J., Yang, M., Inhibition of proliferation and differentiation of mesenchymal stem cells by carboxylated carbon nanotubes. *ACS Nano*, 4, 2185–2195, 2010.
104. Mu, Q., Du, G., Chen, T., Zhang, B., Yan, B., Suppression of human bone morphogenetic protein signaling by carboxylated single-walled carbon nanotubes. *ACS Nano*, 3, 1139–1144, 2009.
105. Ding, L., Stilwell, J., Zhang, T., Elboudwarej, O., Jiang, H., Selegue, J.P., Cooke, P.A., Gray, J.W., Chen, F.F., Molecular Characterization of the cytotoxic mechanism of multiwall carbon nanotubes and nano-onions on human skin fibroblast. *Nano Lett.*, 5, 2448–2464, 2005.
106. Kam, N.W.S. and Dai, H., Carbon nanotubes as intracellular protein transporters: Generality and biological functionality. *J. Am. Chem. Soc.*, 127, 6021–6026, 2005.
107. Chen, R.J., Zhang, Y., Wang, D., Dai, H., Noncovalent sidewall functionalization of single-walled carbon nanotubes for protein immobilization. *J. Am. Chem. Soc.*, 123, 3838–3839, 2001.
108. Ng, R., Zang, R., Yang, K.K., Liu, N., Yang, S.T., Three-dimensional fibrous scaffolds with microstructures and nanotextures for tissue engineering. *RSC Adv.*, 2, 10110, 2012.
109. Nimmagadda, A., Thurston, K., Nollert, M.U., McFetridge, P.S., Chemical modification of SWNT alters *in vitro* cell-SWNT interactions. *J. Biomed. Mater. Res. Part A*, 76A, 614–625, 2006.
110. Zhao, B., Hu, H., Yu, A., Perea, D., Haddon, R.C., Synthesis and characterization of water soluble single-walled carbon nanotube graft copolymers. *J. Am. Chem. Soc.*, 127, 8197–8203, 2005.
111. Nayak, T.R., Jian, L., Phua, L.C., Ho, H.K., Ren, Y., Pastorin, G., Thin films of functionalized multiwalled carbon nanotubes as suitable scaffold materials for stem cells proliferation and bone formation. *ACS Nano*, 4, 7717–7725, 2010.
112. Stout, D.A. and Webster, T.J., Carbon nanotubes for stem cell control. *Mater. Today*, 15, 312–318, 2012.
113. Huang, B., Carbon nanotubes and their polymeric composites: The applications in tissue engineering. *Biomanuf. Rev.*, 5, 3, 2020.

114. Roshanbinfar, K., Mohammadi, Z., Sheikh-Mahdi Mesgar, A., Dehghan, M.M., Oommen, O.P., Hilborn, J., Engel, F.B., Carbon nanotube doped pericardial matrix derived electroconductive biohybrid hydrogel for cardiac tissue engineering. *Biomater. Sci.*, 7, 3906–3917, 2019.
115. Alagarsamy, K.N., Mathan, S., Yan, W., Rafieerad, A., Sekaran, S., Manego, H., Dhingra, S., Carbon nanomaterials for cardiovascular theranostics: Promises and challenges. *Bioact. Mater.*, 6, 2261–2280, 2021.
116. Amin, R., Sink, E., Narayan, S.P., Abdel-Hafiz, M., Mestroni, L., Peña, B., Nanomaterials for cardiac tissue engineering. *Molecules*, 25, 5189, 2020.
117. Dozois, M.D., Bahlmann, L.C., Zilberman, Y., Tang, X., Carbon nanomaterial-enhanced scaffolds for the creation of cardiac tissue constructs: A new frontier in cardiac tissue engineering. *Carbon N. Y.*, 120, 338–349, 2017.
118. Lee, J., Manoharan, V., Cheung, L., Lee, S., Cha, Shin, S.R., Nanoparticle-based hybrid scaffolds for deciphering the role of multimodal cues in cardiac tissue engineering. *ACS Nano*, 13, 12525–12539, 2019.
119. Ashtari, K., Nazari, H., Ko, H., Tebon, P., Akhshik, M., Akbari, M., Khademhosseini, A., Electrically conductive nanomaterials for cardiac tissue engineering. *Adv. Drug Deliv. Rev.*, 144, 162–179, 2019.
120. Menezes, B.R.C., de Rodrigues, K.F., da Silva Fonseca, B.C., Ribas, R.G., do Amaral Montanheiro, T.L., Thim, G.P., Recent advances in the use of carbon nanotubes as smart biomaterials. *J. Mater. Chem. B*, 7, 1343–1360, 2019.
121. Zhou, J., Chen, J., Sun, H., Qiu, X., Mou, Y., Liu, Z., Zhao, Y., Li, X., Han, Y., Duan, C., Tang, R., Wang, C., Zhong, W., Liu, J., Luo, Y., Xing, M., Wang, C., Engineering the heart: Evaluation of conductive nanomaterials for improving implant integration and cardiac function. *Sci. Rep.*, 4, 3733, 2015.
122. Zhao, G., Zhang, X., Li, B., Huang, G., Xu, F., Zhang, X., Solvent-free fabrication of carbon nanotube/silk fibroin electrospun matrices for enhancing cardiomyocyte functionalities. *ACS Biomater. Sci. Eng.*, 6, 1630–1640, 2020.
123. Ahadian, S., Yamada, S., Ramón-Azcón, J., Estili, M., Liang, X., Nakajima, K., Shiku, H., Khademhosseini, A., Matsue, T., Hybrid hydrogel-aligned carbon nanotube scaffolds to enhance cardiac differentiation of embryoid bodies. *Acta Biomater.*, 31, 134–143, 2016.
124. Wu, Y., Wang, L., Guo, B., Ma, P.X., Interwoven aligned conductive nanofiber yarn/hydrogel composite scaffolds for engineered 3D cardiac anisotropy. *ACS Nano*, 11, 5646–5659, 2017.
125. Sedaghati, T., Yang, S.Y., Mosahebi, A., Alavijeh, M.S., Seifalian, A.M., Nerve regeneration with aid of nanotechnology and cellular engineering. *Biotechnol. Appl. Biochem.*, 58, 288–300, 2011.
126. Rossignol, S., Plasticity of connections underlying locomotor recovery after central and/or peripheral lesions in the adult mammals. *Philos. Trans. R. Soc. B Biol. Sci.*, 361, 1647–1671, 2006.
127. Fabbro, A., Prato, M., Ballerini, L., Carbon nanotubes in neuroregeneration and repair. *Adv. Drug Deliv. Rev.*, 65, 2034–2044, 2013.

128. Farin, A., Liu, C.Y., Langmoen, I.A., Apuzzo, M.L.J., Biological restoration of central nervous system architecture and function. *Neurosurgery*, 65, 831–859, 2009.
129. Geller, H.M. and Fawcett, J.W., Building a bridge: Engineering spinal cord repair. *Exp. Neurol.*, 174, 125–136, 2002.
130. Subramanian, A., Krishnan, U.M., Sethuraman, S., Development of biomaterial scaffold for nerve tissue engineering: Biomaterial mediated neural regeneration. *J. Biomed. Sci.*, 16, 108, 2009.
131. Mackinnon, S.E. and Dellon, A.L., Clinical Nerve Reconstruction with a bioabsorbable polyglycolic acid tube. *Plast. Reconstr. Surg.*, 85, 419–424, 1990.
132. Hudson, T.W., Evans, G.R.D., Schmidt, C.E., Engineering strategies for peripheral nerve repair. *Orthop. Clin. North Am.*, 31, 485–497, 2000.
133. Arslantunali, D., Budak, G., Hasirci, V., Multiwalled CNT-pHEMA composite conduit for peripheral nerve repair. *J. Biomed. Mater. Res. Part A*, 102, 828–841, 2014.
134. Borschel, G.H., Kia, K.F., Kuzon, W.M., Dennis, R.G., Mechanical properties of acellular peripheral nerve. *J. Surg. Res.*, 114, 133–139, 2003.
135. Myers, J.P., Santiago-Medina, M., Gomez, T.M., Regulation of axonal outgrowth and pathfinding by integrin-ECM interactions. *Dev. Neurobiol.*, 71, 901–923, 2011.
136. Geiger, B., Spatz, J.P., Bershadsky, A.D., Environmental sensing through focal adhesions. *Nat. Rev. Mol. Cell Biol.*, 10, 21–33, 2009.
137. Galvan-Garcia, P., Keefer, E.W., Yang, F., Zhang, M., Fang, S., Zakhidov, A.A., Baughman, R.H., Romero, M.I., Robust cell migration and neuronal growth on pristine carbon nanotube sheets and yarns. *J. Biomater. Sci. Polym. Ed.*, 18, 1245–1261, 2007.
138. Hu, H., Ni, Y., Montana, V., Haddon, R.C., Parpura, V., Chemically functionalized carbon nanotubes as substrates for neuronal growth. *Nano Lett.*, 4, 507–511, 2004.
139. Jan, E. and Kotov, N.A., Successful differentiation of mouse neural stem cells on layer-by-layer assembled single-walled carbon nanotube composite. *Nano Lett.*, 7, 1123–1128, 2007.
140. Shao, H., Li, T., Zhu, R., Xu, X., Yu, J., Chen, S., Song, L., Ramakrishna, S., Lei, Z., Ruan, Y., He, L., Carbon nanotube multilayered nanocomposites as multifunctional substrates for actuating neuronal differentiation and functions of neural stem cells. *Biomaterials*, 175, 93–109, 2018.
141. Vicentini, N., Gatti, T., Salerno, M., Hernandez Gomez, Y.S., Bellon, M., Gallio, S., Marega, C., Filippini, F., Menna, E., Effect of different functionalized carbon nanostructures as fillers on the physical properties of biocompatible poly(l-lactic acid) composites. *Mater. Chem. Phys.*, 214, 265–276, 2018.
142. Smith, L.A. and Ma, P.X., Nano-fibrous scaffolds for tissue engineering. *Colloids Surf. B Biointerfaces*, 39, 125–131, 2004.

143. Jiang, Y.-Q., Ouyang, Z., Zhang, H.-G., Wang, Y.-F., Kang, N., Lin, C.-L., Hou, Z.-Q., Ye, S.-F., A molecular basis for enhanced biocompatibility of osteoblasts on poly(lactic-coglycolic acid)-multi-wall carbon nanotubes nanocomposite. *Curr. Nanosci.*, 8, 587–597, 2012.
144. Szymański, T., Mieloch, A.A., Richter, M., Trzeciak, T., Florek, E., Rybka, J.D., Giersig, M., Utilization of carbon nanotubes in manufacturing of 3D cartilage and bone scaffolds. *Mater. (Basel)*, 13, 4039, 2020.
145. Karbasi, S. and Alizadeh, Z.M., Effects of multi-wall carbon nanotubes on structural and mechanical properties of poly(3-hydroxybutyrate)/chitosan electrospun scaffolds for cartilage tissue engineering. *Bull. Mater. Sci.*, 40, 1247–1253, 2017.
146. Zadehnajar, P., Akbari, B., Karbasi, S., Mirmusavi, M.H., Preparation and characterization of poly ε -caprolactone-gelatin/multi-walled carbon nanotubes electrospun scaffolds for cartilage tissue engineering applications. *Int. J. Polym. Mater. Polym. Biomater.*, 69, 326–337, 2020.
147. Mirmusavi, M.H., Zadehnajar, P., Semnani, D., Karbasi, S., Fekrat, F., Heidari, F., Evaluation of physical, mechanical and biological properties of poly 3-hydroxybutyrate-chitosan-multiwalled carbon nanotube/silk nano-micro composite scaffold for cartilage tissue engineering applications. *Int. J. Biol. Macromol.*, 132, 822–835, 2019.
148. Dorati, R., Colonna, C., Genta, I., Conti, B., *Active implants and scaffolds for tissue regeneration*, M. Zilberman, (Ed.), pp. 259–285, Springer, Berlin, Heidelberg, 2011.
149. Lee, J., Farag, M.M., Park, E.K., Lim, J., Yun, H., A simultaneous process of 3D magnesium phosphate scaffold fabrication and bioactive substance loading for hard tissue regeneration. *Mater. Sci. Eng. C*, 36, 252–260, 2014.
150. Denry, I. and Kuhn, L.T., Design and characterization of calcium phosphate ceramic scaffolds for bone tissue engineering. *Dent. Mater.*, 32, 43–53, 2016.
151. Sanz, M. and Vignoletti, F., Key aspects on the use of bone substitutes for bone regeneration of edentulous ridges. *Dent. Mater.*, 31, 640–647, 2015.
152. Oryan, A., Alidadi, S., Moshiri, A., Maffulli, N., Bone regenerative medicine: Classic options, novel strategies, and future directions. *J. Orthop. Surg. Res.*, 9, 18, 2014.
153. Tanaka, M., Sato, Y., Zhang, M., Haniu, H., Okamoto, M., Aoki, K., Takizawa, T., Yoshida, K., Sobajima, A., Kamanaka, T., Kato, H., Saito, N., *In vitro* and *in vivo* evaluation of a three-dimensional porous multi-walled carbon nanotube scaffold for bone regeneration. *Nanomaterials*, 7, 46, 2017.
154. Tanaka, M., Sato, Y., Haniu, H., Nomura, H., Kobayashi, S., Takanashi, S., Okamoto, M., Takizawa, T., Aoki, K., Usui, Y., Oishi, A., Kato, H., Saito, N., A three-dimensional block structure consisting exclusively of carbon nanotubes serving as bone regeneration scaffold and as bone defect filler. *PLoS One*, 12, e0172601, 2017.

155. Khalid, P., Hussain, M., Rekha, P., Arun, A., Carbon nanotube-reinforced hydroxyapatite composite and their interaction with human osteoblast *in vitro*. *Hum. Exp. Toxicol.*, 34, 548–556, 2015.
156. Oyefusi, A., Olanipekun, O., Neelgund, G.M., Peterson, D., Stone, J.M., Williams, E., Carson, L., Regisford, G., Oki, A., Hydroxyapatite grafted carbon nanotubes and graphene nanosheets: Promising bone implant materials. *Spectrochim. Acta Part A Mol. Biomol. Spectrosc.*, 132, 410–416, 2014.
157. Gupta, A., Main, B.J., Taylor, B.L., Gupta, M., Whitworth, C.A., Cady, C., Freeman, J.W., El-Amin, S.F., *In vitro* evaluation of three-dimensional single-walled carbon nanotube composites for bone tissue engineering. *J. Biomed. Mater. Res. Part A*, 102, 4118–4126, 2014.

13

Functionalized Carbon Nanotubes for Cell Tracking

Sagar Salave, Dhwani Rana, Jyotsna Vitore and Aakanchha Jain*

National Institute of Pharmaceutical Education and Research (NIPER)-Ahmedabad, Palaj, Opp Air Force Station, Gandhinagar, Gujarat, India

Abstract

Carbon nanotubes owing to their attributes of high surface area, unique needle-like structure, and capability of circumventing the physical and biochemical blockage of the blood-brain barrier are extensively explored for theragnostic applications. Combining the surface of carbon nanotubes with desired functional groups imparts unique properties to carbon nanotubes and enhances their applicability. Functionalized carbon nanotubes of anticipated strength and conductivity properties provide us with options to explore. Functionalized carbon nanotubes-based biosensors are extensively used for diagnostic purposes and one wide application of functionalized carbon nanotubes is in cell tracking. We intend to comprehensively explore the application of functionalized carbon nanotubes for cell tracking and advances made in this regard.

***Keywords*:** Carbon nanotubes, cell tracking, functionalization, surface coating, cell-specific tracking

Abbreviations

CLSM	Confocal laser scanning microscopy
CNTs	Carbon nanotubes
FITC	Fluorescein isothiocyanate
GNT	Gadonanotube
HSPC	Hematopoietic stem/progenitor cells
LDH	Layered double hydroxide
MRI	Magnetic resonance imaging
PLPEG	Phospholipid-polyethylene glycol

**Corresponding author*: jainaakanchha83@gmail.com; aakanchha.jain@niperahm.ac.in

Jeenat Aslam, Chaudhery Mustansar Hussain and Ruby Aslam (eds.) Functionalized Carbon Nanotubes for Biomedical Applications, (319–338) © 2023 Scrivener Publishing LLC

SPIONs Super magnetic iron oxide nanoparticles
SW Single-walled

13.1 Introduction

In all living organisms, cells are the structural and fundamental unit. Governing the morpho-dynamic behavior of cells remains the most crucial part of understanding the physiological processes and underlying cellular behavior in healthy as well as diseased conditions [1]. This presiding is at times carried out by imaging and analysis, typically involving tracking as well as quantification of multiple cells in phase-contrast microscopy or differential interference contrast microscopy, intravital microscopy, time-lapse fluorescence, and computational algorithms [2, 3]. Individual cells vary substantially in their behavior and temporal evolution. These typical attributes help remarkably in not only identification but also understanding cell to cell differences and the prevailing heterogeneity amongst them [4]. Thus, cell tracking usually refers to the spatial-temporal studies of living cells' motility [5]. These two aspects are the chief image processing stages in cell tracking. Cell segmentation, the spatial aspect, refers to the method of dividing the image into biologically significant segments and generating a new image whereas cell association, the temporal aspect, implies to the method of identification and linkage of segmented cells for obtaining cell trajectories from frame to frame in the image sequence. Briefly, segmenting individual cells and connecting cells over time is the usual cell

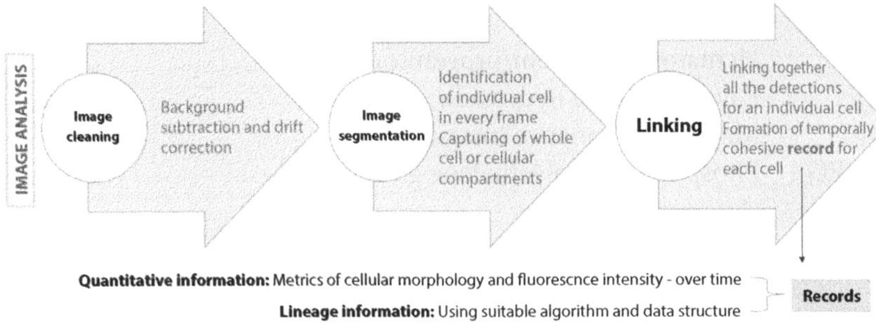

Figure 13.1 Cell tracking basics.

tracking in practice. Newer trends are evolving for improving robustness by employing the use of models and probabilistic estimation methods [2]. The collected dynamic data can be integrated with end-point genomic assays to discover and get a better understanding of cellular behavior [4]. Figure 13.1 represents the generalized concept of cell tracking and associated features.

The usefulness of cell tracking lies in accurate capturing of the cellular events and their dynamics, cell behaviors, cellular motions, and quantification of migration, morphological changes, and rebuilding of cell lineages [1, 6, 7].

Some therapies like adoptive immunotherapy as well as cell-based therapy have been evaluated as new therapeutics in several clinical practices and diseases [5, 8–10]. Various cell-based therapeutic strategies require the determination of location, distribution, and viability by cells with their biological fortune relating to cellular activation as well as differentiation. Cell tracking methods serve to elucidate the new biological mechanism [11, 12]. There are two foremost doctrines of cell tracking [13]; i) Direct labelling and ii) Indirect labelling. Direct labelling is simple which does not require the genetic modification of the cell [14]. But the limitation of direct labelling system is the tendency of label to get diluted upon cell division resulting into a reduced number of labels in the individual cell. Additionally, the label gets disseminated asymmetrically to the progeny cells all through the cell division.

Indirect labelling method uses introduction of a reporter gene into the cell followed by its consequent translation into enzymes, receptors, bioluminescent or fluorescent protein [15]. Based on the stability of the reporter gene, labelled cells could be observed throughout the complete life span of cells. Upon passing of this reporter gene onto the progeny cells, expanding cell population imaging is possible which gives the cell viability information.

13.2 Carbon Nanotubes

Carbon nanotubes (CNTs) have long been referred to as versatile system due to their remarkable features in the domains of biological science, targeted drug delivery, and tissue engineering [16–18]. Figure 13.2 depicts the role of CNTs as a versatile platform for cell tracking. To date, a large variety of applications of CNTs has been explored to explain the mechanism of CNTs in various research areas [19]. Cell tracking stands as the most significant approach of CNTs for measuring various metabolic activity and live status of cells. CNTs pave the way for non-invasive cell tracking in the recent era [20]. It owns the characteristics features including high surface

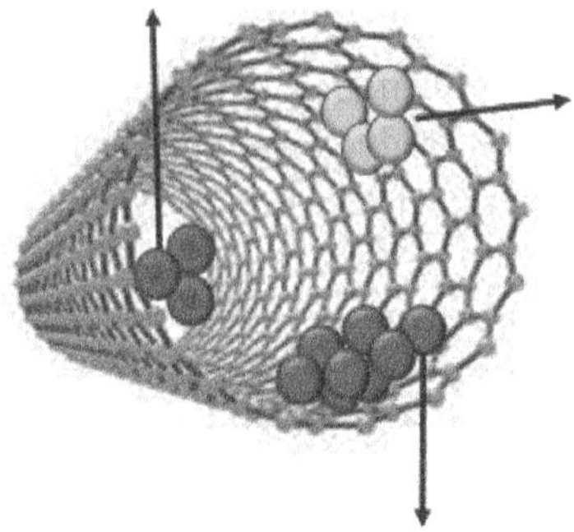

Figure 13.2 Schematic representation of CNTs as a versatile platform for cell tracking.

area, unique – needle-like structure, ability to carry targeted moiety through the cell membrane. Single-walled (SW) and multiple walled CNTs are extensively explored as a biosensor in the medical field [21]. In current time, CNT-based bioimaging is widely investigated. SWCNTs exhibiting quasi-one dimensions possess a narrow band almost the 1eV that allows emission of fluorescence through NIR I region ranging from 700-900 nm whilst NIR-II region ranging from 1100-1400 nm. Due to the stronger absorbance at the NIR region CNTs play a crucial role as the photoacoustic imaging contrast agent. In addition, it is used in cell labelling grounded on Raman imaging, and magnetic resonance imaging (MRI) [22].

CNTs are known for their potential to uniquely equip and transfer the drug molecules across physiological membranes. Functionalized nanotubes, resulting from post-synthesis surface modification of nanotubes, are excellent carriers, efficient delivery systems, and have theragnostics applications [23]. Apart from these applications, CNT hybrids are explored for tracking administered nanocarriers for assessing in real-time their spatial distribution as well as hence measuring their biodistribution profile [24].

For the potential application of CNTs essential and wide functionalization to make them capable to tune their properties.

Several strategies have been encountered to functionalize the CNTs as represented in Figure 13.3. These strategies are followed under two main categories [25]; i) Mechanical technique, and, ii) Chemical technique. The mechanical technique involves high shear missing, high impact mixing, ultracentrifugation, and sonication whereas the chemical method involves alteration of the surface energy of CNTs. Chemical modification aims for surface chemistry manipulations of CNTs by the method of adsorption

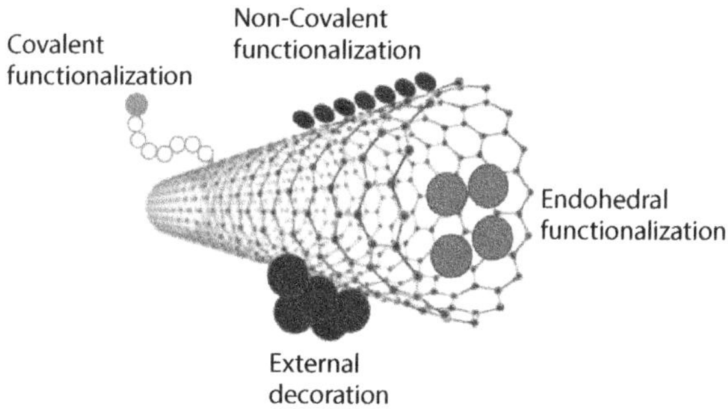

Figure 13.3 Commonly used strategies for CNTs functionalization.

involving non-covalently binding and covalent method involving functionalization of CNTs by covalent bonding [26]. Functionalized forms of CNTs may have different electrical, mechanical, and optical characteristics compared to the original form [27].

Lamm *et al.* studied the merits of fluorescence microscopy, as a method having powerful non-invasive way of penetrating nanoparticles in the physiological domains. They investigated induction of fluorescence energy transfer system amongst a molecule of lysophospholipid molecule and a SWCNT upon cellular uptake. They also described that coating nanotube with RNA and suspending fullerenes with phenolic acids facilitates their translocation through cellular membranes and shuttling amongst cellular organelles. Further, employing the use of molecular simulations could facilitate and guide such experimentations [28].

Moreover, owing to the exceptional attributes of CNTs, it has opened new avenues in the treatment of antimicrobial disorders [29, 30]. Jin-WooKim demonstrated the anti-microbial effects of CNTs shielded with bacterial filters. It was observed that bacteria bind the CNTs, as they possess high binding affinity and further get physically destroyed. The results of this study revealed that CNTs have the superior capability of NIR-responsive bunching PT contrast agents, that could be applied for the diagnosis of pathogenic organisms as well as antimicrobial agents at a single bacterium level [31].

Cellular mechanisms including cell internalization and uptake can be understood from the CNTs based cell tracking [24]. Numerous researchers have demonstrated the mechanism of cell permeability via cell membrane through passive transport or endocytosis. In addition, a reduction in cellular uptake has been observed with the help of fluorescently labelled macromolecules decorated on CNTs [32, 33].

There are reports in the literature where two new CNT-based contrast agents were firstly synthesized by Wilson and co-workers and evaluated their performance in 2005. This material is used as a magnetic contrast agent, which is referred to as Gadonanotube (GNT). It has the length between 20-80 nm. GNTs exhibited the greatest T1relaxivity (r1) known, with values of up to 170 mM^{-1}per s^{-1} per Gd3+ ion at 40 °C and 1.5 T. GNTs are amphiphilic in nature when disperse in surfactants which induced biocompatibility and hence it easily penetrated through local membrane. The synthesized GNTS consisting Gd3+-ion clusters or Gd3+ was further evaluated on mesenchymal stem cells MSCs. The results suggested that CNTs based contrast agent was taken up by mesenchymal stem cells (MSCs) in absence of any transfection agent and toxicity was not found on the either of used CNTs addition to this, Gd3+n-ion clusters containing ultra-short SWCNTs in the US-tubes are linear Superparamagnetic molecular magnets with MRI efficacies of 40 to 90 times larger than any $Gd3^{+}$ based contrast agent in existing medical custom. Studies show that GNTs may denote a substantial revolution in design of contrast agent for high-field imaging [34].

Various ongoing research suggests the use of ferromagnetic metals such as Fe, Ni, Co for the synthesis of super magnetic iron oxide nanoparticles (SPIONs) straight in the apertures of silica nanotubes (SNTs). These silica nanoparticles have characteristic features of having good magnetic property. The integration also showed promising results in improvising the MRI contrast agent. In another work, Bai *et al.* [35] developed SPIONs straightway in the apertures of SNTs. In another report, amino group functionalized CNTs (f-CNTs) were detected in two different cell types i) multidrug-resistant cells (K562A) and, ii) the parent cells (K562S) at a different incubation temperature viz. 4°C rather than 37°C. This result suggests that fluorescent-tagged CNTs can aid the permeation through the cell membrane by transverse endocytosis [36].

Similarly, Kostarelos *et al.* elucidated the capability of fluorescently labelled CNTs to interact with cells and govern the important variables involved in these interactions. For this, they developed several f-CNTs such as ammonium-functionalized CNT, fluorescein isothiocyanate (FITC) f-CNT, CNT bi-functionalized with ammonium groups and FITC, acetamid of-CNT, CNT bi-functionalized with methotrexate, and FITC, shortened CNT bi-functionalized with amphotericin Band FITC, and shortened CNT bi-functionalized with ammonium groups and FITC using an amide linkage. Confocal laser scanning microscopy (Hematopoietic stem/progenitor cells) epifluorescence microscopy was employed to track and image the fluorescently labelled CNTs and/or inherently luminescent f-CNTs. Interactions amongst fluorescently labelled CNTs and an extensive diversity of viable cells were then examined and it was found out that cellular internalization of fluorescently labelled CNTs occurs in the former ones. It was settled that fluorescently labelled CNTs hold a capability to be engulfed by prokaryotic as well as mammalian cells and to intracellularly traffic between various cell-based barriers through energy-independent processes [37].

13.2.1 Cellular Interaction of CNTs

Cell trafficking of CNTs has immense implications in medical science and the crux of it lies in the communication of CNTs and cell membranes, controlled by the regulations of molecular cellular biology as well as the fundamental attributes of the nanotubes [11, 12, 38].

CNTs are endowed with super magnetic properties by the coating of magnetic nanoparticles on the surface [39, 40]. Owning to their magnetic properties, they are mostly having high reverse relaxivity [41]. Giuseppe Lamanna appraised the role of nanoparticle- CNT (NP/CNT) hybrids functioning like contrast agent for MRI along with their connections with cells. This multimodal complex opened the doors for new platform strategies in drug delivery as well as in theragnostic application. Although, there are various methods used for the fabrication of CNTs with NP resulting into extremely anisotropic ferromagnetic nanostructures. Ligand exchange or chemo-selective ligation method is preferred to use. The study results suggested that with this technique, well-organized, stable, and effective linking of magnetic iron oxide NPs to MWCNTs were obtained. The fabricated hybrid showed efficient internalization in tumor cells without indicating toxicity effects. It was observed that NP/CNTs hybrid could label cells and show magnetic mobility even at the single-cell level with high resolution of MRI. Biodistribution analysis also highlights the application of this tool in cell tracking [36].

13.3 Cellular Tracking via CNT

Carbon nanotube having the size in the range of (~100 to 300 nm) has successfully been used for tracking the molecular motion of the living cells over time from millisecond to hours. In order to follow motorized movement in cells by fluorescence microscopy needs to fulfil certain criteria including (i) stable, non-bleaching fluorescent probes, (ii) effective targeting of probes to selective compounds, and (iii) high signal-to-noise ratio (S/N) in imaging. CNT-based tracking is invasive in nature and has been used in the observation of kinesin-1 motor protein in COS -7 cells. Cells are dynamic organizations and molecular force generation is a dynamic process at a supramolecular scale. CNTs can observe the directed kinesin-driven transport on microtubules in addition to that it can also observe the fluctuations due to moving kinesin bound to the microtubules. The microtubules track is embedded in the viscoelastic actin cytoskeleton, which turns fluctuation and generates stress by cytoplasmic myosin [42].

Gadolinium f-MWCNTs were developed to work like contrast agent intending MRI labelling as well as tracking [38]. Diethylenetriaminepentaacetic acid (DTPA) aiding tight attachment of gadolinium atom on the surface of nanotube was done. Figure 13.4 depicts development of DTPA modified MWCT for development of contrast agent. Developed contrast agents were known to be constant for about two weeks in water as well as mouse serum. Three-fold higher r1 relaxivity was observed compared to a clinically approved contract agent. The contrast efficiency investigated at 7T was obtained 6.6 $mM^{-1}s^{-1}$ in human umbilical vein endothelial cells [38].

In one learning, Singh *et al.* established the biopharmaceutical parameters of f-CNTs following *i.v.* administration. They functionalized water-soluble, SWCNT with a chelating molecule diethylenetriaminepentaacetic and labelled with indium7 (^{111}In) for the purpose of imaging. The developed f-SWCNT was administered intravenously and was radioactively traced employing the use of gamma scintigraphy. This ensured that f-SWCNT was not retained in liver or spleen; organs related to reticuloendothelial system and were quickly cleared from the blood stream via the route of renal excretion. Also, urine excretion studies exploiting both f-SWCNT and f-MWCNT and electron microscopy analysis of the same discovered that both the nanotubes were expelled as intact nanotubes. This quick blood clearance along with a half-life of 3 hours indicated the clinical potential of CNTs [43]. Further, Wang *et al.* fabricated radio-labelled iron oxide

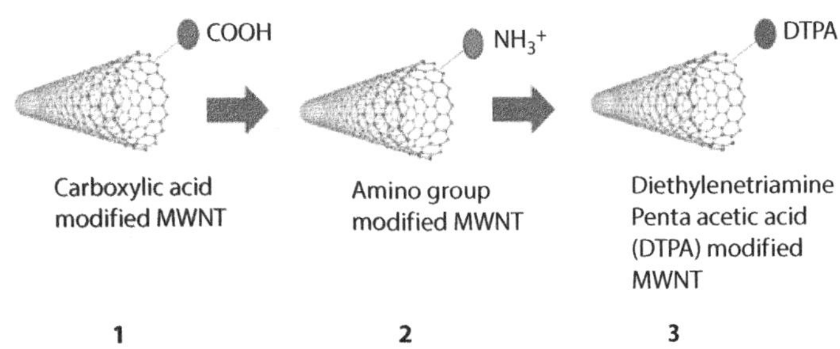

Figure 13.4 Schematic representation of DTPA modified MWNT for MRI cell labelling and tracking.

decorated MWCNT as dual magnetic resonance and single-photon emission computed tomography imaging materials. Using *in situ* generation hybrids of different amounts of Fe_2O_3 were synthesized and it was known from physicochemical characterizations that magnetic properties to hybrids were imparted due to the occurrence of SPION. These hybrids were then radio-labelled with technetium-99m via functionalized bisphosphonate and the quantitative biodistribution in mice was carried out using SPECT/CT imaging and γ-scintigraphy. In this examination, the existence of SPION was recognized by Perls stain whereas the presence of MWCNT was recognized using Neutral Red stain. Further, the histological examination revealed no abnormalities in mice. Transmission electron microscope images of tissues of spleen and liver discovered co-localization of SPION and MWCNT within the same intracellular vesicles, and hence indicated the *in vivo* solidity of the hybrids following intravenous injection. This study divulged the potential of SPION-MWCNT hybrids as dual MRI and SPECT contrast agents for *in vivo* use [44].

Lamanna *et al.* assessed the potential of SPION coated MWCNTs (NP/CNT) hybrids as a contrast agent for MRI and interactions of the same with cells. Nanotubes covered with NP allowed conversing magnetic properties to CNTs. The capability of the hybrids to magnetically govern and influence cells was also inspected. It was found that NP/CNTs could be deployed using a remote magnetic field with improved contrast in MRI. The developed system showed internalization in the neoplastic cells without demonstrating significant cytotoxicity. The labelled cells displayed magnetic mobility and hence could be magnetically manipulated perceived at a single cell level through high-resolution MRI [36].

T.S.T. Balamurugan *et al.* prepared the Prussian blue microcubes functionalized graphenated nanotubes modified electrodes to track the *in vivo* H_2O_2 generation in mammalian cells. The synthesis process was straightforward and the fabricated sensor presented detection limit at nanomolar range, reproducibility, and durability [45].

Besides cell tracking, CNTs possess potential applications in therapeutic modalities. Emma Mooney *et al.* have established a study demonstrating the effect of CNT on hMSC renewal, cellular ultra-structure, nuclear location metabolic activity, and differentiation. Screening based on conductance value among the various CNTs such as SWCNT, MWCNT, COOHf-SWCNT, and OH f-MWCNT was done to recognize the optimal kind of CNT for usage with hMSC. It has been stated that COOHf-SWCNT showed minimum toxicity to the cells. In this research, the f-CNT was distributed in hMSC media for the evaluation of all parameters. COOH functionalized and OHf-SWCNT showed a harmful outcome on the cells. However, at low concentration (0.00128 mg/mL) viability of cells and metabolic rate was enhanced compared with control. At high concentration (0.032 mg/mL) both OH functionalized and COOHf-CNTs found detrimental effect. The cells were microscopically analyzed and COOH-f-WCNT was biotinylated with 1 × 10^3 mol NHS biotin for analysis. The study revealed that SWCNT travelled through the cell wall to a nucleus 24 hours later. Authors have explained that an innovative electrophysiological milieu for electrically stimulating HMSC encourage differentiation towards a cardiomyocyte linkage [46].

13.3.1 Effect of the Surface Coating of CNTs in Single-Particle Tracking

Gao *et al.* evaluated different SWCNT external modifications for single-particle tracking applications in complex biomedical samples. The study involved screening various coatings in order to encapsulate the CNTs having luminescent property and displaying nominal cell related toxicity and negligible non-specific interactions with the cell membrane. Amongst all the screened coatings, phospholipid-polyethylene glycol (PLPEG) and pluronic F108-coated SWCNTs exhibited minimal acute cellular cytotoxicity of 1–4 days at 1 µg/mL and least cell-based interaction when compared to some other extensively utilized biodegradable surface coatings. Also, PLPEG-coated SWCNTs displayed brighter luminescence than F108-coated CNTs and could be viewed at video rate for some time at the single tube level while diffusing in a composite aqueous network and hence concluded to be the optimum coating for single nanotube tracking application in biomedical applications [47].

13.4 3D Tracking Using CNTs

Reuel *et al.* established 3D tracking of fluorescent SWCNT following the use of an orbital tracking microscope within live HeLa cells. The tracking system had engrained spatial resolutions of 7-15 nm and a temporal resolution of 32 ms for visualization of chitosan-wrapped SWCNTs. The study involved the determination of the viscosity regime (above 250 cP) and the rotational diffusion coefficient could be used for estimation of length. The obtained tracking data were used for spatially mapping of corral volumes, determining active transport velocity, and calculating local viscosities within the cell. Within the cell, the coral volume was found to be 0.27- 1.32 μm^3, active transport velocity was found to be 455 nm/s and local viscosity was 54-179 cP. The SWCNTs could also be used as sensors in living cells, changing the fluorescence signal by about 4-13% in order to permit disjunction of the sensor signal from fluctuations owing to rotation of the SWCNT while gauging with a time resolution of 32 ms [22].

13.4.1 Detection of Single Protein Molecules Through CNTs

Imaging and tracking of single-molecule coupled with florescent probes are known as single-molecule tracking. SMT eases the investigators to straightway witness molecular behavior and interaction in the living cells. Several molecular interactions and reactions occur in the living cells. SMT allows the researcher to observe distinct molecules at work in living cells precisely [48].

In 2017, MIT chemical engineers established arrays of CNT sensors that could detect single protein molecules being secreted from cells. These chemically modified CNTs could be consumed for protein production by distinct cells. This kind of fabrication can be employed for tracking even minor quantities of protein like tracking viral infection, nursing cells' production of effective proteins, or even learning contamination of edible stuffs. Upon functionalization of these sensitive molecular sensing platforms-CNTs, one could observe the stochastic fluctuations of single molecules attaching to them [49]. In this regard, Landry *et al.* showed label-free recognition of specific proteins from bacteria-*E. coli* and yeast-*P. pastoris* immobilized in a microfluidic chamber, evaluating efflux of protein from individual organism in real-time. The array was invented leveraging the use of non-covalent conjugation of an aptamer-anchor polynucleotide sequence to near-infrared emissive SWCNT, employing an adjustable chemical spacer that could optimize sensor response. From various cell lines, proteins like unlabeled RAP1 GTPase and HIV integrase were specifically distinguished using large near-infrared fluorescent turn-on responses [50].

Moreover, the current research area also focuses the H2s monitoring in the biomedical domains due to the generation of a third endogenous gasotransmitter. Muhammad Asif *et al.* have demonstrated a simple strategy with the help of a sensitive electrochemical sensor and 2Dnanosheet-shaped layered double hydroxide (LDH) covered with CNTs nanohybrid (CNTs@LDH). The synthesis was carried out by the coprecipitation method. The results showed that NTs@CuMn-LDH nanohybrid possess noteworthy sensing attributes for H2S determination. The lower detection limit was 0.3 nM (S/N = 3), with greater specificity, durability, and reproducibility.

This structural integration strategy was a new platform to detect the abiotic H2s efflux formed by sulfate-reducing bacteria upon use in real-time assay [51].

13.4.2 Stem Cell Labeling and Tracking Through CNTs

Uludag *et al.* presented an extremely effective and harmless mCNT-mediated stem cells labelling techniques. Fluorescent and magnetic FITC-mCNTs were synthesized and magnetic-field mediated hematopoietic stem/progenitor cells (HSPC) uptake of FITC-mCNT was observed. The developed system was sent for fluorescence-activated cell sorter analysis for determining the uptake efficiencies. The results were suggestive of efficient internalization of FITC-mCNT by HSPC in a time and concentration-dependent manner. Compared to various nanoparticular uptake using innate or modified SPION, FITC-mCNT labelling or uptake was found more effective for the purpose of labelling stem cells. Moreover, the developed CNT showed no adverse effect on macrophages, no noteworthy outcome on chondrogenic, osteogenic, or adipogenic differentiation of hMSC, and was found non-cytotoxic [52].

Vittorio *et al.* inspected the role of MWCNTs with low metal impurities (2.57% iron) as MRI contrast agents for tracking stem cells. Cellular viability and proliferation was not seen to be affected by the developed MWCNTs, suggesting that such attributed exhibited by MWCNTs could aid in *in vivo* purpose of stem cell tracking/imaging as well as in MWCNT-mediated targeted electro-chemotherapeutics [53].

13.4.3 Labelling and Tracking of Human Pancreatic Cells Through CNTs

In one study, Syed *et al.* compared the SPION ferumoxide (Endorem) and MWCNTs for islet cell labeling as well as tracking. They incubated INS-1 E cells and human pancreatic islets obtained from twelve non-diabetic cadaveric organ donors with 50 μg/ml Endorem or 15 μg/ml MWCNTs and studied after 7 or 14 days for assessing cell survival, function, beta-cell morphology, ultrastructure, and *in-vitro* and *in-vivo* MRI. Well-maintained morphology, as well as ultrastructure, was observed for both INS-1 E and human islets at the time of incubation in light and electron microscopy examination. The presence of MWCNTs was observed within beta cells whereas alpha cells were devoid of it when seen under electron microsocpe. Effective transplantation of MWCNT labelled human islets into subcutis of rats localized in the anticipated site was carried out employing the use of magnetic field while tracking was performed using MRI. The findings were suggestive that MWCNTs could be a potential labelling multifarious and could be used along with human islets for experimental as well as transplantation studies [54]. This study highlighted the possible mechanism of CNT for cell internalization. The very first-time MWCNTs were reported for entering human beta cells through direct membrane penetration and endosomal pathways. This study highlights the potential of MWCNTs for islets grafting tools in tracing after engraftment and post-transplant monitoring.

13.4.4 CNT as Macrophage Carrying Microdevices

Cell tracking techniques employing the use of MRI and positron emission tomography require ionizing radiation which is not the case with Raman imaging. In fact, Raman imaging is referred to be extremely material-dependent, could be excited by NIR permitting deep tissue penetration at the centimeter scale. CNTs on the other hand display strong resonance-enhanced Raman modes possessing an exclusive Raman spectrum. Wang *et al.* developed an adaptable system to fabricate the microdevices and prepare microdevices-live cells complexes. This technique was further extended for the production of CNT built microdevices for tracking single macrophages using Raman scattering. The fabricated microdevices consisted of polyelectrolyte trilayer, CNTs, and PLGA. The generated Raman signals by the developed CNTs system in microdevices were perceived at a centimeter scale with the sample-lens distance under NIR excitation, suggesting its *in-vivo* potential. The fabricated microdevices were taken up by the principal macrophages while not affecting the cell viability as well as phagocytic capability of the cells. These single macrophages having microdevices could be spotted by Raman scattering and hold potential for non-invasive multiplex tracking of single macrophages *in vivo* [55].

13.4.4.1 Intracellular Fluctuations and CNT

A quantifiable investigation focusing on molecular motions in cells was carried out by Fakhri *et al.*, wherein the authors employed extremely steady near-infrared luminescence of SWCNTs targeted to kinesin-1 motor proteins in COS-7 cells. The authors stated that scheme of active random "stirring" institutes a mean of transportation, which differ from thermal diffusion as well as directed motor activities. The observed high-frequency action was thermally driven. Non-equilibrium dynamics were observed higher than 100 milliseconds. Apart from this author also observed robust random dynamics obliged by myosin that lead to increased nonspecific transport [56].

13.4.5 Limitations of CNTs

CNTs are explored versions of carbon allotropes found in the same category of fullerenes. Various MWCNTs are regarded to cause cell-cycle arrest, perturbations in multiple cellular pathways, increase in apoptosis/necrosis, and stimulation of genes concerned in cell-cycle regulation, cellular metabolism, transport, and reaction to stress. There are reports regarding SWCNT for topical skin application [57]. However, exposure to this material on the skin surface can lead to skin-associated issues including carbon fiber dermatitis, hyperkeratosis, etc. There are diverse studies reported with toxicity evaluation of CNTs. For instance, SWCNT has been studied on cell culture of immortalized human epidermal keratinocytes. It revealed that on the exposure up to 18 hours on cell culture, respective parameters such as oxidative stress as well as cellular toxicity were observed. Despite these, multiple reports are found in the literature for the mixture of Ferric compound with CNTs causing iron associated overload cell damage and provoking the other clinical condition such as heart diseases, cancers, hormonal abnormalities, skin diseases, immune system dysfunctions, liver diseases and diabetes [58]. The iron catalyst compounds inside the SWCNT material are sheathed in a carbon-based material. However, the nano-sized particles have been reported more toxic because of their large surface area.

13.5 Concluding Remarks and Future Perspective

Functionalization and designating of CNTs have been proved as versatile and highly efficient strategy for enhancing the CNT application in the biomedical field. Surface modification is novel emerging approach to improve the theragnostic potential of CNTs as it influences biocompatibility, targeting and organ distribution. The applicability of f-CNTs in the cell tracking is gaining researchers' interest. However, cell tracking is the sophisticated and invasive tool to monitor the particles and cellular components at cellular and molecular level. Novel approaches have been listed in this chapter with the desired fabrication specification of CNTs in terms of cell tracking application. Numerous microscopic and fluorescent techniques are used in reported literature to continuously monitor the cells internal environment and cellular uptake mechanism. The cell specific application of f-CNTs is illustrated with their examples and results. Many researchers have claimed positive results for the cell viability and internalization of CNTs. Besides that, it is important to consider the cellular toxicity that can causes due to the various functional groups and carbon containing skeleton of CNTs. In addition, due to its nanoscale size range, after internalization into the cells, the CNTs interact with the living organism and might cause toxicity. Hence, the understanding and exploring the toxicity events and its exact mechanism followed by administration helps to fabricate and select the safe and efficient materials in the development of CNTs. In future, there is scope for exploration of different functionalized material for the cell tracking application by invasive way.

Important Links

https://dspace.mit.edu/handle/1721.1/73371
https://scholarship.rice.edu/bitstream/handle/1911/90940/RevisedBIOMATERIALS.pdf?sequence=4&isAllowed=y
https://tigerprints.clemson.edu/cgi/viewcontent.cgi?referer=https://www.google.com/&httpsredir=1&article=1279&context=all_dissertations

Acknowledgment

The authors are thankful to former Director, NIPER-A for providing necessary literature search facilities.

References

1. Emami, N., Sedaei, Z., Ferdousi, R., Computerized cell tracking: Current methods, tools and challenges. *Vis. Inform.*, 5, 1, 1–13, 2021.
2. Meijering, E., Dzyubachyk, O., Smal, I., Van Cappellen, W.A., Tracking in cell and developmental biology. *Semin. Cell Dev. Biol.*, 20, 8, 894–902, 2009.
3. Ulman, V., Maška, M., Magnusson, K.E.G., Ronneberger, O., Haubold, C., Harder, N., Matula, P., Matula, P., Svoboda, D., Radojevic, M., Smal, I., Rohr, K., Jaldén, J., Blau, H.M., Dzyubachyk, O., Lelieveldt, B., Xiao, P., Li, Y., Cho, S.Y., Dufour, A.C., Olivo-Marin, J.C., Reyes-Aldasoro, C.C., Solis-Lemus, J.A., Bensch, R., Brox, T., Stegmaier, J., Mikut, R., Wolf, S., Hamprecht, F.A., Esteves, T., Quelhas, P., Demirel, Ö., Malmström, L., Jug, F., Tomancak, P., Meijering, E., Muñoz-Barrutia, A., Kozubek, M., Ortiz-De-Solorzano, C.,

An objective comparison of cell tracking algorithms. *Nat. Methods*, 14, 12, 1141, 2017.

4. Moen, E., Borba, E., Miller, G., Schwartz, M., Bannon, D., Koe, N., Camplisson, I., Kyme, D., Pavelchek, C., Price, T., Kudo, T., Pao, E., Graf, W., Van Valen, D., Accurate cell tracking and lineage construction in live-cell imaging experiments with deep learning. *bioRxiv*, 803205, 2019.
5. Svensson, C.M., Medyukhina, A., Belyaev, I., Al-Zaben, N., Figge, M.T., Untangling cell tracks: Quantifying cell migration by time lapse image data analysis. *Cytom. Part A*, 93, 3, 357–370, 2018.
6. Majumdar, R., Steen, K., Coulombe, P.A., Parent, C.A., Non-canonical processes that shape the cell migration landscape. *Curr. Opin. Cell Biol.*, 57, 123–124, 2019.
7. Shrier, D., Tompsett, M., Shrier, L., Adult mother–daughter relationships: A review of the theoretical and research literature. *Guilford Press. J. Am. Acad. Psychoanal. Dyn. Psychiatry*, 32, 1, 91–115, 2004.
8. Cheng, M., Chen, Y., Xiao, W., Sun, R., Tian, Z., NK cell-based immunotherapy for malignant diseases. *Cell. Mol. Immunol.*, 10, 3, 230, 2013.
9. Zhao, L. and Cao, Y.J., Engineered T cell therapy for cancer in the clinic. *Front. Immunol.*, 10, 2250, 2019.
10. Xie, G., Dong, H., Liang, Y., Ham, J.D., Rizwan, R., Chen, J., CAR-NK cells: A promising cellular immunotherapy for cancer. *EBioMedicine*, 59, 102975, 2020.
11. Kircher, M.F., Gambhir, S.S., Grimm, J., Noninvasive cell-tracking methods. *Nat. Rev. Clin. Oncol.*, 8, 11, 677–688, 2011.
12. Srinivas, M., Boehm-Sturm, P., Figdor, C.G., de Vries, I.J., Hoehn, M., Labeling cells for *in vivo* tracking using 19F MRI. *Biomaterials*, 33, 34, 8830–8840, 2012.
13. Hong, H., Yang, Y., Zhang, Y., Cai, W., Non-invasive cell tracking in cancer and cancer therapy. *Curr. Top. Med. Chem.*, 10, 12, 1237, 2010.
14. Kim, M.H., Lee, Y.J., Kang, J.H., Stem cell monitoring with a direct or indirect labeling method. *Nucl. Med. Mol. Imaging (2010)*, 50, 4, 275, 2016.
15. Weissleder, R., Moore, A., Mahmood, U., Bhorade, R., Benveniste, H., Chiocca, E.A., Basilion, J.P., *In vivo* magnetic resonance imaging of transgene expression. *Nat. Med.*, 6, 3, 351–354, 2000.
16. Zare, H., Ahmadi, S., Ghasemi, A., Ghanbari, M., Rabiee, N., Bagherzadeh, M., Karimi, M., Webster, T.J., Hamblin, M.R., Mostafavi, E., Carbon nanotubes: Smart drug/gene delivery carriers. *Int. J. Nanomedicine*, 16, 1681, 2021.
17. Simon, J., Flahaut, E., Golzio, M., Overview of carbon nanotubes for biomedical applications. *Mater. (Basel)*, 12, 4, 624, 2019.
18. Dubey, R., Dutta, D., Sarkar, A., Chattopadhyay, P., Functionalized carbon nanotubes: Synthesis, properties and applications in water purification, drug delivery, and material and biomedical sciences. *Nanoscale Adv.*, 3, 20, 5722–5744, 2021.

19. Zare, H., Ahmadi, S., Ghasemi, A., Ghanbari, M., Rabiee, N., Bagherzadeh, M., Karimi, M., Webster, T.J., Hamblin, M.R., Mostafavi, E., Carbon nanotubes: Smart drug/gene delivery carriers. *Int. J. Nanomed.*, 16, 1681–1706, 2021.
20. Hong, H., Yang, Y., Zhang, Y., Cai, W., Non-invasive cell tracking in cancer and cancer therapy. *Curr. Top. Med. Chem.*, 10, 12, 1237–1248, 2010.
21. Sireesha, M., Jagadeesh Babu, V., Kranthi Kiran, A.S., Ramakrishna, S., A review on carbon nanotubes in biosensor devices and their applications in medicine. *Nanocomposites*, 4, 2, 36–57, 2018.
22. Reuel, N.F., Dupont, A., Thouvenin, O., Lamb, D.C., Strano, M.S., Three-dimensional tracking of carbon nanotubes within living cells. *ACS Nano*, 6, 6, 5420–5428, 2012.
23. Cui, H.F., Vashist, S.K., Al-Rubeaan, K., Luong, J.H.T., Sheu, F.S., Interfacing carbon nanotubes with living mammalian cells and cytotoxicity issues. *Chem. Res. Toxicol.*, 23, 7, 1131–1147, 2010.
24. Costa, P.M., Bourgognon, M., Wang, J.T.W., Al-Jamal, K.T., Functionalised carbon nanotubes: From intracellular uptake and cell-related toxicity to systemic brain delivery. *J. Control. Release*, 241, 200–219, 2016.
25. Mallakpour, S. and Soltanian, S., Surface functionalization of carbon nanotubes: Fabrication and applications. *RCS Adv.*, 8, 109916–109935, 2016.
26. Meng, L., Fu, C., Lu, Q., Advanced technology for functionalization of carbon nanotubes. *Prog. Nat. Sci.*, 19, 7, 801–810, 2009.
27. Norizan, M.N., Moklis, M.H., Ngah Demon, S.Z., Halim, N.A., Samsuri, A., Mohamad, I.S., Knight, V.F., Abdullah, N., Carbon nanotubes: Functionalisation and their application in chemical sensors. *RSC Adv.*, 10, 71, 43704–43732, 2020.
28. Lamm, M.H. and Ke, P.C., Cell trafficking of carbon nanotubes based on fluorescence detection. *Methods Mol. Biol.*, 625, 135–151, 2010.
29. Seo, Y., Hwang, J., Kim, J., Jeong, Y., Hwang, M.P., Choi, J., Antibacterial activity and cytotoxicity of multi-walled carbon nanotubes decorated with silver nanoparticles. *Int. J. Nanomed.*, 9, 1, 4621–4629, 2014.
30. Liu, D., Mao, Y., Ding, L., Carbon nanotubes as antimicrobial agents for water disinfection and pathogen control. *J. Water Health*, 16, 2, 171–180, 2018.
31. Kim, J.W., Shashkov, E.V., Galanzha, E.I., Kotagiri, N., Zharov, V.P., Photothermal antimicrobial nanotherapy and nanodiagnostics with self-assembling carbon nanotube clusters. *Lasers Surg. Med.*, 39, 7, 622–634, 2007.
32. Kam, N.W.S. and Dai, H., Carbon nanotubes as intracellular protein transporters: Generality and biological functionality. *J. Am. Chem. Soc.*, 127, 16, 6021–6026, 2005.
33. Xiao, H., Yang, L., Zou, H., Yang, L., Le, X.C., Analysis of oxidized multi-walled carbon nanotubes in single K562 cells by capillary electrophoresis with laser-induced fluorescence. *Anal. Bioanal. Chem.*, 387, 1, 119–126, 2006.

34. Sitharaman, B., Kissell, K.R., Hartman, K.B., Tran, L.A., Baikalov, A., Rusakova, I., Sun, Y., Khant, H.A., Ludtke, S.J., Chiu, W., Laus, S., Tóth, E., Helm, L., Merbach, A.E., Wilson, L.J., Superparamagnetic gadonanotubes are high-performance MRI contrast agents. *Chem. Commun. (Camb)*, 32, 3915–7, 2005.
35. Issa, B., Obaidat, I.M., Albiss, B.A., Haik, Y., Magnetic nanoparticles: Surface effects and properties related to biomedicine applications. *Int. J. Mol. Sci.*, 14, 14, 2013.
36. Lamanna, G., Garofalo, A., Popa, G., Wilhelm, C., Bégin-Colin, S., Felder-Flesch, D., Bianco, A., Gazeau, F., Ménard-Moyon, C., Endowing carbon nanotubes with superparamagnetic properties: Applications for cell labeling, MRI cell tracking and magnetic manipulations. *Nanoscale*, 5, 10, 4412–4421, 2013.
37. Kostarelos, K., Lacerda, L., Pastorin, G., Wu, W., Wieckowski, S., Luangsivilay, J., Godefroy, S., Pantarotto, D., Briand, J.P., Muller, S., Prato, M., Bianco, A., Cellular uptake of functionalized carbon nanotubes is independent of functional group and cell type. *Nat. Nanotechnol.*, 2, 2, 108–113, 2007.
38. Servant, A., Jacobs, I., Bussy, C., Fabbro, C., Ros, T., Pach, E., Ballesteros, B., Prato, M., Nicolay, K., Kostarelos, K., Gadolinium-functionalised multi-walled carbon nanotubes as a T1 contrast agent for MRI cell labelling and tracking. *Carbon*, 97, 126–133, 2016.
39. Zhou, H. and Wang, X., Decoration of magnetic nanoparticles on functionalized multi-wall carbon nanotubes by *in situ* method. *Adv. Mater. Res.*, 601, 37–40, 2013.
40. Masotti, A. and Caporali, A., Preparation of magnetic carbon nanotubes (Mag-CNTs) for biomedical and biotechnological applications. *Int. J. Mol. Sci.*, 14, 12, 24619, 2013.
41. Tomczyk, M.M., Boncel, S., Herman, A., Krawczyk, T., Jakóbik-Kolon, A., Pawlyta, M., Krzywiecki, M., Chrobak, A., Minoshima, M., Sugihara, F., Kikuchi, K., Kuźnik, N., Oxygen functional groups on MWCNT surface as critical factor boosting T2 relaxation rate of water protons: Towards improved CNT-based contrast agents. *Int. J. Nanomed.*, 15, 7433–7450, 2020.
42. Yokomizo, H., Possingham, H.P., Hulme, P.E., Grice, A.C., Buckley, Y.M., Olofson, P., Grijalva, E., Kerr, D., Hogle, I., Thornton, W., Fakhri, N., Wessel, A.D., Willms, C., Pasquali, M., Klopfenstein, D.R., MacKintosh, F.C., Schmidt, C.F., We thank the personnel of the invasive spartina project, including high-resolution mapping of intracellular fluctuations using carbon nanotubes. *Biol. Invasions*, 127, 839–849, 2006.
43. Singh, R., Pantarotto, D., Lacerda, L., Pastorin, G., Klumpp, C., Prato, M., Bianco, A., Kostarelos, K., Tissue biodistribution and blood clearance rates of intravenously administered carbon nanotube radiotracers. *Proc. Natl. Acad. Sci.*, 103, 9, 3357–6662, 2006.
44. Wang, J.T.W., Cabana, L., Bourgognon, M., Kafa, H., Protti, A., Venner, K., Shah, A.M., Sosabowski, J.K., Mather, S.J., Roig, A., Ke, X., Van Tendeloo,

G., De Rosales, R.T.M., Tobias, G., Al-Jamal, K.T., Magnetically decorated multiwalled carbon nanotubes as dual MRI and spect contrast agents. *Adv. Funct. Mater.*, 24, 13, 1880–1894, 2014.

45. Balamurugan, T.S.T., Mani, V., Hsieh, C.C., Huang, S.T., Peng, T.K., Lin, H.Y., Real-time tracking and quantification of endogenous hydrogen peroxide production in living cells using graphenated carbon nanotubes supported prussian blue cubes. *Sens. Actuators B Chem.*, 257, 220–227, 2018.
46. Mooney, E., Dockery, P., Greiser, U., Murphy, M., Barron, V., Carbon Nanotubes and mesenchymal stem cells: Biocompatibility, proliferation and differentiation. *Nano Lett.*, 8, 8, 2137–2143, 2008.
47. Gao, Z., Danné, N., Godin, A.G., Lounis, B., Cognet, L., Evaluation of different single-walled carbon nanotube surface coatings for single-particle tracking applications in biological environments. *Nanomaterials*, 7, 11, 393, 2017.
48. Kusumi, A., Tsunoyama, T.A., Hirosawa, K.M., Kasai, R.S., Fujiwara, T.K., Tracking single molecules at work in living cells. *Nat. Chem. Biol.*, 10, 7, 524–532, 2014.
49. New sensors can detect single protein molecules, MIT News | Massachusetts Institute of Technology, Massachusetts, US, https://news.mit.edu/2017/new-sensors-detect-single-protein-molecules-0123 (accessed Dec 10, 2021).
50. Landry, M.P., Ando, H., Chen, A.Y., Cao, J., Kottadiel, V.I., Chio, L., Yang, D., Dong, J., Lu, T.K., Strano, M.S., Single-molecule detection of protein efflux from microorganisms using fluorescent single-walled carbon nanotube sensor arrays. *Nat. Nanotechnol.*, 12, 4, 368–377, 2017.
51. Li, D.W., Qu, L.L., Hu, K., Long, Y.T., Tian, H., Monitoring of endogenous hydrogen sulfide in living cells using surface-enhanced raman scattering. *Angew. Chem. - Int. Ed.*, 54, 43, 12758–12761, 2015.
52. Gul-Uludag, H., Lu, W., Xu, P., Xing, J., Chen, J., Carbon nanotube-mediated labelling platforms for stem cells. *Carbon Nanotub. - Growth Appl.*, 2011.
53. Vittorio, O., Duce, S.L., Pietrabissa, A., Cuschieri, A., Multiwall carbon nanotubes as MRI contrast agents for tracking stem cells. *Nanotechnology*, 22, 9, 095706, 2011.
54. Syed, F., Riggio, C., Masini, M., Bugliani, M., Battaglia, V., Novelli, M., Suleiman, M., Vittorio, O., Boggi, U., Filipponi, F., Marselli, L., Bartolozzi, C., Masiello, P., Raffa, V., Marchetti, P., Labeling and tracking of human pancreatic islets using carbon nanotubes. *J. Biomed. Nanotechnol.*, 11, 4, 730–738, 2015.
55. Abstract: Carbon Nanotube-Based Microdevices for Tracking Single Macrophages By Raman Scattering (2016 Annual Meeting), https://aiche.confex.com/aiche/2016/webprogram/Paper471251.html (accessed Dec 10, 2021).
56. Fakhri, N., Wessel, A.D., Willms, C., Pasquali, M., Klopfenstein, D.R., MacKintosh, F.C., Schmidt, C.F., High-resolution mapping of intracellular fluctuations using carbon nanotubes. *Science*, 344, 6187, 1031–1035, 2014.

57. Ding, L., Stilwell, J., Zhang, T., Elboudwarej, O., Jiang, H., Selegue, J.P., Cooke, P.A., Gray, J.W., Chen, F.F., Molecular characterization of the cytotoxic mechanism of multiwall carbon nanotubes and nano-onions on human skin fibroblast. *Nano Lett.*, 5, 12, 2448–2464, 2005.
58. Shvedova, A., Castranova, V., Kisin, E., Schwegler-Berry, D., Murray, A., Gandelsman, V., Baron, P., Exposure to carbon nanotube material: Assessment of nanotube cytotoxicity using human keratinocyte cells. *J. Toxicol. Environ. Health A*, 66, 20, 1909–1926, 2011, https://doi.org/10.1080/713853956.

14

Functionalized Carbon Nanotubes for Treatment of Various Diseases

Ajahar Khan*, Khalid A. Alamry and Raed H. Althomali

Department of Chemistry, Faculty of Science, King Abdulaziz University, Jeddah, Saudi Arabia

Abstract

In present times functionalized carbon nanotubes (CNTs) have received more attention from therapeutic formulation scientists due to their unique intrinsic physical and chemical characteristics. The aspect ratio is notion to be accountable for their brilliant cell penetration ability, semi/anisotropic conductivity, as well as their axis, is best for integration with muscular and nervous tissue, and ability to penetrate biological membranes. These unique features of functionalized-CNTs enable them for a wide range of biomedical applications, such as infectious diseases, cancer remediation, diagnosis, delivery of drugs, and central nervous system disorders applications. Their adaptable physicochemical properties facilitate the introduction of various pharmaceutically significant. CNTs can be easily functionalized with different functional groups to allow customizing of biological recognition. Herein, we summarized widely the characteristics of CNTs, such as their mechanisms performance, and huge potential in biomedicine. This work also describes several functionalization approaches utilized for CNTs that originated new opportunities for researchers to enhance the potential of delivered therapeutics. Consequently, the previous reports have surveyed the current progress to emphasize the current application status in pharmaceuticals. The potential biomedical applications of CNTs with their biosafety/toxicology description and the way to conquer the cytotoxicity are also reviewed with future perspectives.

***Keywords*:** Diagnosis, functionalized-CNTs, cancer treatment, gene delivery, biomedical applications

**Corresponding author*: arkhan.029@gmail.com

Jeenat Aslam, Chaudhery Mustansar Hussain and Ruby Aslam (eds.) Functionalized Carbon Nanotubes for Biomedical Applications, (339–376) © 2023 Scrivener Publishing LLC

14.1 Introduction

The promising field of nanomaterials bridges the physical sciences with biological sciences using chemical techniques in emergent novel platforms and devices for disease diagnosis and remediation by understanding biological systems [1]. Nanomaterials comprise sizes in the range of 1 to 100 nanometers, similar to various biological macromolecules for instance deoxyribonucleic acid (DNA) plasmids, antibodies, and enzymes. Therefore nanomaterials demonstrate attractive physical characteristics, different from both the molecular and bulk scales, exhibiting innovative spaces for the research and application of the biomedical-related fields. There are numerous types of nanomaterials already studied or under research that could be considered for biomedical applications [2–4]. With the increasing demands of potential nanomaterials, carbon nanotubes (CNTs) are reviewed as the most promising candidate in the nanotechnology field and the construction of CNT-based materials is estimated to produce huge societal interests [5]. CNTs have concerned the interest of researchers worldwide; their incredible strength, wonderful physico-chemical features, and miniature size make them particularly useful for different functions. Additionally, owing to unique surface properties like thermal, electrical and optical properties, CNTs have attained considerable interest as an imaging and diagnosis agent [6, 7] in the field of biomedical and became a most wonderful nanomaterial in recent drug development [8].

CNTs were accidentally discovered as a by-product of fullerene by Sumio Iijima in 1991 at the time of study surfaces of graphite electrodes during electric arc discharge [9]. This study along with the structural analysis of nanotube offered a novel course to the research of carbon, which harmonized the interest for and actions within organized research of fullerene. Based on different arrangements in graphene cylinders, two types of CNTs are found i.e., single-walled carbon nanotubes (SWNTs) and multi-walled carbon nanotubes (MWNTs) classified based on the numbers of carbon layers present. A single, seamless and uniform graphene tube consisting of a single layer of carbon atoms is SWNTs and in comparison to SWNTs, MWNTs are formed by multiple, single concentric graphene tubes attaining several layers of carbon atoms and have much larger diameters [10]. While MWNTs attain less affluent and striking optical features than SWNTs, their performance in living environments can be diverse from that of SWNTs because of their bigger sizes, which can provide dissimilar substrates for different causes, including delivery of bigger biomolecules such as DNA plasmids into cells [11, 12]. CNTs' solubility is a great challenge in executing several research formulations but covalent functionalization of

CNTs can resolve the problem of solubility in a better way. Furthermore, functionalization with triethylene glycol, polyethylene glycol, and with proteins/oligonucleotides biomaterials could also enhance the solubility of CNTs [13]. A broad variety of bioactive materials can be conjugated with the CNTs and provide new routes in pharmaceutical applications. The construction of the bioelectronic device has allowed better chances concerning the immobilization of proteins and DNA over the CNTs surface. The CNTs also exhibit superb anti-oxidant activity against photo and thermal degradation of polymers [14].

The excellent electrical conductivity allows CNTs to be used as optical and electronic biosensors for the detection of microorganisms and bioactive molecules such as cells, DNA, and proteins. CNTs exhibit brilliant features of high optical absorption in the Raman scattering [15], photoluminescence [16], infrared region [17], echogenic, and photoacoustic features which are beneficial in tracking and visualization of nanotubes in the living system. In addition to the above-raised characteristics, the healthcare applications of CNTs in vaccine delivery, neural regeneration [18], gene and drug delivery, muscle and bone regeneration [19], and tissue engineering [20], are key achievements of researchers. CNTs can be utilized for the cause of diagnosis as a therapeutic imaging agent, biomolecular detector, nanotweezers, and biosensor [21]. The diminished size of particles of CNTs promotes the surface area contact then redoubles the uptake through the cellular membrane. Hydrophobic nature and size in nano range help CNTs in improved cellular internalization, manifesting nanotubes influence as significant carriers towards brain delivery [22]. Straightforward functionalized-CNTs produce them vital carriers for specific delivery of anticancer genes and drugs with lower adverse effects [23]. In biology and medicine, CNTs have been widely used as nanocarriers to deliver various proteins, peptides, small interfering ribonucleic acids (siRNAs), and plasmid DNAs [24]. Furthermore, hybridization and composition with many metallic nanoparticles, various biomaterials, and polymers produce them alter the properties of CNTs which improves its application in biology and medicine [25].

This chapter plans to briefly emphasize the current state of investigations and recent advances into the promising biological consequences of CNTs, with a look at how their biocompatibility is suitable for disease treatment. Furthermore, the significance of CNT composites their toxicities studies for *in vivo* and *in vitro* cellular uptake has been discussed. Sidewise an insight into how to enhance the suitability, comparability, reliability, and validity, of currently used examination techniques for establishing the possible biological effects of CNTs has also been reviewed.

14.2 CNTs: Basic Structure, and Synthesis Methods

14.2.1 Structure and Synthesis of CNTs

As previously discussed, during fullerene synthesis CNTs were accidentally discovered as a by-product [9]. CNTs are graphene's covered sheets comprising 2 end caps analogous to semi-C60 [9] and classified into two types i.e., SWNTs are single graphene cylinders, flexible, smaller in diameter, and allow imaging assist while MWNTs are a complex nesting of graphene cylinders with large surface area and, highly proficient endohedral filling [26, 27]. The CNTs possess a diameter in the range of 0.4-100 nm, while can attain a length up to diverse micrometers. In order to synthesize CNTs, three main methods i.e., laser-ablation, arc-discharge, and catalytic growth (chemical vapor deposition) are established and proposed over the last two decades. CNTs can be synthesized in 2 forms i.e. as a powder (substrate-free) and on the substrate. The substrate comprises vertically aligned CNTs, which are commonly used for electronic and sensors systems.

14.2.2 Arc Discharge Technique

The first time the CNTs were synthesized through the arc discharge technique [9]. In the arc discharge technique carbon is vaporized in the inert atmosphere (argon or helium) from a graphite cathode heated through applied current. Herein, the function of higher pressure (50-700 mbar) and high-current-voltage (50-100 amps) creates highly-temperature discharge among two electrodes positioned 1 mm distant, which leads to evaporation of carbon rods followed by depositing on the rod-shaped cathode tubes. Several metal catalysts for example Pt, Co, Fe, Rh, or Ni, are used for SWCNTs synthesis. The properties of CNTs may be adjusted by introducing catalysts in the cathode and by changing the current intensity [26].

14.2.3 Laser Ablation Technique

In this technique, a continuous laser of strong power is applied to vaporize targeted graphite within a furnace at a temperature of 1200°C with helium or argon gas at a pressure of 500 torr. The CNTs synthesized by the laser ablation technique exhibit high purity, controlled size with lesser structural defects, chirality, and around 70% yield [28]. The laser ablation technique can be employed for the synthesis of both SWNTs and MWNTs. By this method CNTs of fine crystallinity can be developed, though, fullerenes, amorphous carbon, and catalysts are left as the final product.

14.2.4 Catalytic Chemical Vapor Deposition Technique

This technique is focused on the catalytic decomposition of a gas comprising carbon on a heterogeneous or homogeneous catalyst. In catalytic chemical vapor deposition technique, the hydrocarbon molecules (such as acetylene, benzene, ethane, and toluene, etc.) are dissociated into the reactive carbon compound around 500–1000°C in a quartz tube positioned in the furnace. Nucleation of the catalyst through chemical etching by thermal annealing or ammonia is accomplished after the catalyst deposition on the substrate. The carbon resource persists in the gas phase in the reaction chamber and is converted to the atomic level, that diffuses for the metal catalyst on which the development of CNT arises [29]. The CNTs synthesized by catalytic chemical vapor deposition technique can be lengthening in millimeters range. The temperature required in this technique is lesser than that desired for the laser ablation and arc discharge techniques; however, these CNTs can display various limitations. In a report, Araga *et al.* studied the synthesis of MWNTs by using plasma-enhanced chemical vapor deposition method at relatively low temperature (450°C) straight on the charcoal surface derived from coconut-shell [30]. Herein, minerals including potassium, sodium, magnesium, and calcium found in the biomass behave like catalysts, and the low temperature terminate the catalytic behavior of Fe. The MWNTs synthesized, which were characterized by Raman spectroscopy, X-ray spectroscopy, elemental detection, and scanning electron microscopy exhibited smaller diameters (25–30 nm) [30].

14.3 Functionalization of CNTs

Even surfaces and inherent hydrophobicity with no hanging bonds present, pure CNTs are incompatible and chemically inert toward nearly all inorganic and organic solvents and exceptionally tricky to disperse in aqueous phases. These limitations have produced a major drawback for CNTs applications in biology and medicine. Therefore, it is needed to carry out further modification on the surface of CNTs that not only enhance their miscibility/dispersibility but also make them able to interact with biological systems [31]. Nowadays, scholars have already studied that in both types of CNTs carbon atoms reveal chemical reactivity for several reagents and therefore may be believed as newly hybrid types of macromolecular carbon form. After functionalizations, the biocompatibility and miscibility of CNTs in water might be improved. In this regard, two main approaches i.e. covalent and non-covalent functionalizations can be carried to improve their water solubility [32].

14.3.1 Covalent Functionalization

Covalent functionalizations could be designated as 'defect functionalizations' as barely faulty C-atoms at the end or on the sidewall of CNTs might be oxidized using good oxidants to produce carboxylated fractions or carboxylic acid groups, which can be modified chemically through esterification or amidation [33]. The covalent modifications on the sidewall carbon atoms are accomplished by rehybridization or converting the configurations of sp2 carbon into sp3, hence, producing covalency among the attacking moieties [34]. This can be achieved with the radicals' addition cycloaddition and halogenation of azomethine ylides. The cycloaddition of azomethine ylides (Prato reaction) is the most frequently used pathway. The surface modifications of defective carbon atoms are carried out through creating defects by oxidation to get -COOH moieties. These -COOH functionalities might be used to graft different specific groups including antibodies or protein and peptides via chemical reactions such as esterification or amidation. Several biological molecules [35], metals [36], and polymers [33] can be employed as the backbone of carboxylated CNTs. In covalent modifications of CNTs addition reactions were also adopted, which were carried out from those established for fullerenes or conventional for graphite surfaces [37, 38]. Besides of single group functionalization of CNTs, the plan of multi-functionalization (i.e. double or triple covalent modification) is probably significant, particularly for the delivery of drugs. The double covalent functionalization might be accomplished either through amidation/esterification and 1,3-dipolar cycloaddition; double arylation; double amidation/esterification; amidation and cyclopropanation; arylation and amidation; double 1,3-dipolar cycloaddition; 1,3-dipolar cycloaddition and arylation etc. Triple covalent functionalizations can be achieved through concurrent functionalizations with diverse aryl diazonium salts [39]. The above-discussed functionalizations, however, may be responsible for the loss of optical and electrical features, which are beneficial for recognizing CNTs in biological environments [18, 40]. In addition, a literature survey reported that the covalently surface functionalization associated with biological screening is a potential strategy to lower the CNTs' toxicity [41].

14.3.2 Non-Covalent Functionalization

Apart from covalent functionalization, noncovalent functionalization of CNTs permits preserving intrinsic optical and electrical properties.

These modifications can be done through adsorption/wrapping of surfactants, polymers, and biopolymers on the tubular surface through π-π stacking and van der Waals interactions. Two strategies have been carried for non-covalent functionalizations of carbon nanotubes. The first one the CNT sidewall is wrapped by polymer chains [42] and the other is through π-π stacking interaction among π-electrons of graphite sheets and aromatic rings of the loaded materials on the surface of CNTs [43]. This non-covalent functionalization will disrupt the van der Waals forces that ground SWNTs to be combined in the form of bundles [44] and may also enhance their water solubility. Non-covalent functionalizations of CNTs are also reported for the applications of medical. CNTs showed a good affinity for RNA and single-stranded DNA and with the polynucleotide molecules develop an electrostatic complex, which may not only convey water miscibility but also be responsible for genes delivery within cells [11, 45]. A number of intrinsic physical characteristics of CNTs, such as Raman scattering and near-infrared fluorescence can be reserved and the conjugated electronic morphology of CNTs will not be hindered [46, 47]. Polymers such as poly (meta phenylene vinylene), tetraalkylammonium, polyethylene glycol (PEG), poly (vinylpyrrolidone), and poly(oxyethylene), etc., when immersed in a solution containing polymer are able to wrap around the SWNTs [42]. The CNTs wrapped with polymer allow added support in the stabilization of CNTs dispersion. PEG is mainly employed for anchoring a broad variety of molecules (biomolecules, drugs, etc.). Due to its good solubility and biocompatibility under different physiological surroundings, PEG is widely employed for different causes in pharmaceutical applications [48]. PEGylated CNTs are proficient to reduce the uptake using the reticuloendothelial substrate, in mice block the binding of non-specific serum proteins [49] and extend the blood circulating time [50]. In addition, poly(vinylpyrrolidone) used for anchoring drug molecules simplifies the process of bioimaging [18, 51]. The surfactants like Triton X-100, sodium dodecyl sulfate (SDS), and sodium dodecylbenzene sulfonate (SDBS), etc., adsorb onto the CNTs backbone using π-interactions to enhance their permeability, solubility, and dispersibility using biological membranes [52]. However advancement in the CNTs solubility can be lower (0.1 to 0.9 mg/mL), yet it is still suitable for biomedical application [52]. The CNTs solubilization in biomaterials (nucleic acids, proteins, and peptides) is a doubtlessly much appropriate technique to combine with biological substrates. It has been reported that a straightforward sonication with single-stranded DNA is the simplest route to solubilize CNTs. The complexes based on DNA-CNTs displayed good solubility and exceptional stability. Amphiphilic peptides (KV6-$CONH_2$, A6RGD, RGD,

EAK8, EAK12, and A6R etc.) exhibited a proficient approach to disperse CNT [53, 54]. M. Arnold *et al.* studied the noncovalent functionalization and dispersion of CNTs in water via peptide amphiphiles. The aqueous dispersibility of peptide amphiphiles-CNTs is because of the disclosure of hydrophilic heads to water whereas the hydrophobic tails are still attached to CNTs surface. The peptide amphiphiles-CNTs showed improved dispersibility in the aqueous phase along with offering bio-functionality to CNTs [55].

14.4 Toxicity/Bio-Safety Profile of Carbon Nanotubes

The clinical trials of CNTs and their nanocomposites for biomedical applications are only possible when their security is fully confident and their toxicology in living systems is understood [56]. However, the literature survey based on toxicity reports on CNTs and biological behaviors toward biological systems suggested that the conflict in toxicity reports can be ascribed to various experimental variable conditions such as integrity of CNTs, cells, doses, and functionalizations of CNTs [57]. The toxicological behavior of CNT-based nanocomposites is mostly concentrated on different parameters such as impurities of metal, type and length of CNTs, and its functionalizations. The metal catalyst impurities originating during the growing process of CNTs are the main source of CNTs cytotoxicity [58]. In the current era of biomedicine, CNTs present a new promising platform whereupon various problems linked with traditional techniques of the therapeutic field are conveyed simultaneously [59]. Focusing on all the challenges researchers primarily harped about the goodness of CNTs based nanomaterials and insisted that no evident toxicity was noticed via CNTs [60]. Apart from this, few researchers affirmed that CNTs generated toxic behavior, mainly during pulmonary and intravenous incorporation of nanomaterials [61]. There was a discrepancy noticed in several reports, which has depressed its wide applications in the area of biomedical. It was absolutely apparent that a unique approach must be investigated for the fabrication technique of CNTs and the characterization requires to be carried out through valid approaches and models. Although the difficulties faced, the surface modification of CNTs using diethylentri amine pentaacetic dianhydride [23], functionalization with PEG [62], surface coating with vitamin E [63], manipulation with genetic materials [64], surface coating with immunoglobulins [65], conjugation with PEG and arginine-glycine-aspartate (RGD) peptides [23], and adhering to blood proteins [66] are the few techniques amid

Table 14.1 Toxicity analyses of CNTs.

S. no.	Carbon nanotubes	Dose	Exposure time	Model	Toxicity	Ref
1.	CNTs-OH-PAA@ PEGMA	50 μL	12 or 24h	Cell lines HepG2	Cell viability can stay around 100% for 24 h. Displayed low good biocompatibility and cytotoxicity	[72]
2.	Pure CNTs	25 mg	4 weeks	Guinea pigs	CNTs does not promote any abnormalities of measurable inflammation or pulmonary function	[73]
3.	SWNT and MWNT	Large amount of CNT	24, 48, 72 and 96 h	Albino rabbits	No signs of health risk associated to allergic risk and skin irritation	[74]
4.	HiPco™ SWNT,	0.1 or 0.5 mg	Intratracheal instillation	7 or 90 days	Mortality was noticed with the over dose Stimulated dose-dependent epithelioid granulomas	[75]
5.	Pure laser SWNT	1 or 5 mg	Intratracheal instillation	From 24 h and up to 3 months	Within 24 h post-instillation the high dose produce mortality. Pulmonary inflammation with non dose dependent granulomas	[76]

(*Continued*)

Table 14.1 Toxicity analyses of CNTs. (*Continued*)

S. no.	Carbon nanotubes	Dose	Exposure time	Model	Toxicity	Ref
6.	HiPco™ and SWNT	Aerosol	Human volunteers	30 min and 11–16 h	CNT in the range of 0.7 to 53 μg/m^3 (HiPco material formed observable big clumps on the filter) Visible contamination (from 0.2 to 6 mg)	[77]
7.	SWNT	1.0. 40 and 200 mg	Mice	3 months	No noticeable hepatic harm is found in the histopathological analyses amid all the groups. There was no change in spleen	[78]
8.	Hat-stacked carbon nanofibers	-	Rats	1–4 weeks	Regular practice of inflammation for external bodies, with no strict inflammatory action. No inhibition of wound healing. No acute toxicity in the subcutaneous tissue.	[79]
9.	Functionalized -MWNTs	1 to 25 μg/mL	Liver cancer cells (HepG2)	24 h	Reduced the growth of cancer cell at a very low concentration.	[80]

various others that showed achievement. To describe the CNTs in their purest form are particularly difficult to atomize in aqueous solutions and organic solvents. Hence, the toxicity of pure CNTs is more than modified CNTs, which hampers biological reactions. A suitable functionalization technique is required to prepare the CNTs dispersion in solvents through interacting with other molecules to the nanotubes' surface [67]. In this respect, the covalent or noncovalent functionalization techniques can also minimize the toxicity associated with non-functionalized CNTs. Most of the toxicological reports are based on extraordinarily high doses and *in vitro* investigations with cell lines; therefore it is needed to be very careful to extrapolate these outcomes to living beings [68]. In recent times, advancement has been based on the elimination, biodistribution, and translocation of CNTs after precise introduction, producing the initial step for secure biomedical uses of CNTs [69]. It has been established [41, 70] that the biocompatibility and toxicity of CNTs can be monitored intentionally by material and chemical functionalizations [71]. A summary of a few previously reported studies describing the CNTs toxicity is displayed in Table 14.1.

14.5 Investigating the Promising Biomedical Effects of Functionalized CNTs

CNTs exhibit exclusive properties which provide them for a wide range of applications in the field of biomedical science. Due to hydrophobic characteristics, CNTs are appropriate for several healthcare applications [81]. Although the capability to modify their surface (coatings, chemistry, and charge) and morphological (stiffness and aspect ratio of length and width) properties, the promising effects of CNTs are presently not completely understood [82]. Despite the definite application will signify the specific exposure to the human being [83], as previously reported, it is accepted that to humans the primary way of CNT exposure is through inhalation, and accordingly the lung is considered as the main organ of interest with respect to their consideration to endorse any unfavorable health issues or, alternatively, add to presented lung disease states [84]. The CNTs possess the ability to interact with the receptors of the cell surface which permits their internalization because of the occurrence of various functional sites. These receptor-mediated approaches permit specific cell loading that could assist the drop in the dose of a drug [85].

14.5.1 Functionalized CNTs-Based Remediation of Infectious Diseases

Current breakouts of the rigorous acute respiratory syndrome (SARS), avian, and swine flu have shown that infectious diseases have emerged as serious global concerns related to the public health issue. Few infectious diseases, for instance, acquired immune deficiency syndrome (AIDS), have emerged as tedious, and no efficient remedies have been obtainable till now. The healthcare exercises of nanotechnology have been considered as a potential candidate in order to provide the fast diagnosis and efficient treatment of contagious diseases. Kang *et al.* [86, 87] reported that pure SWNTs attributed an antimicrobial potential in a size-dependent approach, suggesting that CNTs based materials can be assigned as potential aspirants for antimicrobial therapeutics. The surface modification of CNTs may create a position for the interaction of bioactive moieties, along with protecting the secondary structure [88] and, therefore, extract definite anti-epitope antibodies [89]. This type of coupling facilitated the recognition of antibodies to the conjugates. CNTs can play a vital role in the therapy of viral disease via producing high-sensitive detecting systems. As previously reported a coordinated biosensor [90] developed by SWNTs and gold nanoparticles and has been investigated for sensing the nanomolar range of human immunodeficiency virus-1 protease (HIV-1 PR), an aspartic protease accountable for maturation and assembly of virion [91]. The recognition of highly-delicate detection of this protease was encouraging to rapid fabrication of effectual HIV-1 PR inhibitors. On the other hand diagnosis of viral infection is the electrical sensing of hepatitis C virus RNA [92]. Distinctive electronic characteristics and a large surface-to-volume ratio produced CNTs a potential candidate for developing bio-detectors with the high-sensitivity, which were significantly required in the viral ailment diagnosis and the fabrication of fresh anti-viral drugs. Therefore, it is expected that CNTs can considerably offer to the nursing of infectious diseases in the future.

14.5.2 Functionalized CNTs for the Treatment of Central Nervous System Disorders (CNS)

CNS disorders comprise brain tumors and neurodegenerative ailments. Due to the constrained anatomical contact (blood-brain barrier) and the complicated and unique environment of the CNS, it is further complicated to treat and diagnose CNS disorders in comparison to the other ailments. However, nanotechnology is the potential to modernize the rank quo in this area. Due to the nano dimensions and attainable exterior or external functionalizations, nanomaterials are capable to cross the blood-brain hurdle through different selected mechanisms and, therefore, they might work as valuable carriers for delivery towards selecting brain. Several types of nanomaterials have been successfully used as appropriate delivery devices for treating brain tumors and neurodegenerative defects [93, 94]. As a favorable therapeutic candidate, CNTs had been employed in neurosciences [95]. The outcomes of such investigations have revealed that both pure and chemically modified CNTs have a constructive influence on neuronal growth, while cytotoxicity was also demonstrated by them. It has been suggested that the charge on the CNT can be managed to regulate the outgrowth of neurite [96], and it has been investigated that functionalization of CNTs with brain derived neurotrophic factor or nerve growth aspect can stimulate neurons growth on the CNTs based scaffold [97]. Keefer *et al.* [98] confirmed that CNTs coated stainless steel and conventional tungsten wire electrodes can improve electrical stimulation as well as neurons recording in culture (monkeys and rats). The advantage could be assigned to the ability of CNTs to increase charge transfer and reduce the impedance of electrodes. Electrodes coated with CNT are suggested to enhance current and to support the construction of long-lasting brain-machine interfacing systems; which will assist the treatment and diagnosis of CNS disorders. The literature review also indicated that the fabrications of a few chemotherapeutic agents based on nanoparticles (e.g. doxorubicin (DOX) [99] possessed the probability for the systematized brain tumors' chemotherapy with high efficiency than the reported agents. MWNTs functionalized with siRNA and DNA were addressed to be internalized through brain microglia (migratory monocytes resultant macrophage in the brain that is supposed to be an efficient site for the treatment of cancer) *in vivo* and *in vitro* without involving cytokine and proliferative modifications [100, 101]. These outcomes suggested that MWNTs can perform as a secure nano-carrier delivery source for immune therapies of brain cancers (e.g. gliomas). However, this research is still in its beginning stage, and further investigation is required.

14.5.3 Functionalized CNTs for Gene Delivery

As a nanocarrier, the utilization of functionalized-CNTs for gene therapies is an emerging field of research interest that is presently under examination. In comparison to drugs based on small molecules, biomolecules including microRNA (miRNA), small interfering RNA (siRNA), and double stranded DNA (dsDNA), etc., are unable to go through the cellular membranes and simply deteriorate through nucleases [11, 102, 103]. However, various nonviral and viral vectors are present for the delivery of such nucleic acids, although controlling the immunogenicity of these biomolecules has grown as a major concern. In this regard, functionalized-CNTs look like capable delivery devices as functionalized-CNTs do not stimulate immune response inherently [89]. Recently it has also been found that CNTs can deliver bulky cargo loads of biomolecules to targeted positions. Under physiological conditions, the ammonium-functionalized CNTs are cationic in nature and also reported to translocate DNA and proficiently complex [104] within the cells [11, 105] and capable to deliver genes. Pantarotto *et al.* reported that ammonium-functionalized CNTs possess the ability to combine with plasmid DNA by electrostatic interactions [11]. The association with mammalian cells allows the ammonium-functionalized CNTs uptake into the cells through penetrating the cell membranes. The CNTs having reduced cytotoxicity and ammonium-functionalized CNTs coupled plasmid DNA is delivered to cells proficiently; the levels of gene expression are up to 10 times high than those accomplished with pristine DNA. This report permits consideration of this advanced technique as promising for gene delivery. It was also concluded that a unique permutation feature assignable to soluble CNTs evidence the probability of these structures as candidates of superior delivery systems for various therapeutic applications. The potential of CNTs for gene therapeutics has been further investigated for applications of gene silencing [106, 107]. Complexes of siRNA strands and SWNTs functionalized with a hydrocarbon tail were adopted for target killing of cancer cells. In addition advancements of this system are still essential to authenticate the *in vivo* use. Behnam *et al.* studied that functionalized SWNTs associated with polyethyleneimines accomplished enhanced transfection efficiency in comparison to underivatized polyethyleneimines [108]. A nearly 19-fold increase is being noticed in the functionalized-SWNTs with the smallest polyethyleneimines. Additionally, polyethyleneimines-functionalized SWNTs were found competent gene delivery vectors *in vivo* along with tail vein injection in mice with the prevalent expression happening with the vector polyethyleneimines-functionalized-SWNTs using polyethyleneglycol as a linker. Pan *et al.* reported the fabrication of polyamidoamine dendrimer-modified MWNTs for the delivery of genes into human cancer cells [109]. The polyamidoamine dendrimer modified MWNTs totally conjugated with fluorescein isothiocyanate (FITC)-labeled antisense c-myc oligonucleotides were incubated with liver cancer cell line liver hepatocellular carcinoma (HepG2) cells, breast cancer cell line (MDA-MB-435 cells), and MCF-7 cells (human breast), and revealed to penetrate into tumor cells within 15 min through laser confocal microscopy. This was observed that fabricated CNTs composites inhibited the dose-dependent means and cell growth in time followed by down-controlled the expression of the C-Myc protein and C-Myc gene. This report also favors that polyamidoamine dendrimer modified MWNTs

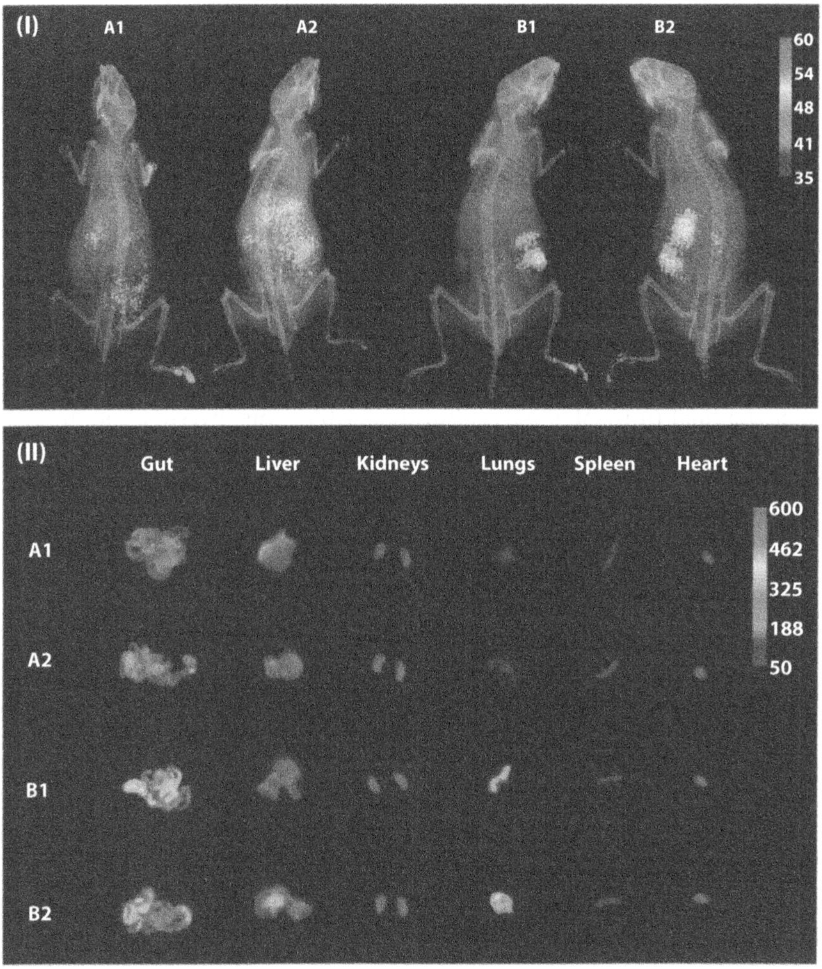

Figure 14.1 (I) The fluorescently-labeled images for *in vivo* biodistribution (A) rCNTs and (B) fCNTs injected in mice. Images taken after the injection of 24 h of (A1) rCNTs, (A2) short rigid (s-rCNTs), (B1) fCNTs and (B2) short flexible (s-fCNTs). (II) The fluorescently labeled images for *ex vivo* organ biodistribution after 24 h of injection (A) rCNTs and (B) fCNTs. Fluorescent images overlaid with X-ray images regarding to the aggregation of (A1) rCNTs, (A2) s-rCNTs, (B1) fCNTs and (B2) s-fCNTs in the gut, liver, kidneys, lungs, spleen and heart. "Reprinted with permission from [110]. Copyright (2017) American Chemical Society."

composites have possibilities in applications for instance drug or gene delivery for molecular imaging and cancer treatment. Cifuentes-Rius *et al.* demonstrated rigid r-CNTs and flexible f-CNTs as promising delivery vectors because of their large aspect ratio and ability to translocate via plasma membranes as nanoneedle [110]. By paying close attention to the physicochemical characteristics of CNTs, r-CNTs and f-CNTs were coated using polyallylamine (ppAA) through plasma enhanced chemical vapor deposition. The ppAA-coated f-CNTs have been confirmed to bind larger amounts of plasmid-green fluorescent protein (pGFP) than r-CNTs, their tangled nature complicates the appropriate discharge of the loaded DNA and the internalization of the material into the cell. Fluorescence imaging of *ex vivo* and *in vivo* investigation can present a semi-quantitative determination of the CNTs response upon intravenous administration and permits observing of the similar matter at various time intervals. As shown in Figure 14.1 (I) after 24 hours of injection one mouse was imaged for each CNT group. The mice were culled and the main organs (lungs, gut, liver, kidneys, heart, and spleen) were promptly harvested and imaged, after the final time interval. Within the 24-h of this investigation, no side effects were evident in the mice injected with fluorescently labeled CNTs. Particularly, they suggested that that by selecting the r-CNTs over f-CNTs, accumulation might be eliminated, which is essential for the fruitful utilization of CNTs in gene and drug delivery [110].

14.5.4 Implication of Functionalized CNTs in Cancer Diagnosis and Treatment

For cancer diagnosis and treatment, CNTs might be employed as a drug carrier as they possess the property of attaining the specific site followed by assembling in the cancer cells and forming the interaction with targeting moieties [111]. As a 1-dimensional structure CNTs might be employed for loading receptors that may associate through transducer and analyte as a result it can be used as biosensors which are more responsive to the chemical and electrical changes within the abrupt surrounding. As CNTs exhibit superb chemical stability, inertness, and mechanical strength [112] therefore they could be used as a reliable scaffold for tissue regeneration because it provides a gap to grow the cells and could be modified for stimulating orientation and growth of cells in an appropriate manner [113]. Shikori *et al.* prepared nanocomposite based on CNT and chitosan functionalized bioactive glass and employed for bone tissue regeneration with a sluggish biodegradability up to 30 days and compression strength of 5.95 ± 0.5 MPa [114]. The fabricated nanocomposite scaffolds were tested on MG63 osteoblast cell line, chitosan/bioactive glass/CNT nanocomposite in 80:20:2 wt% ratio found to be the finest ratio related to the cellular strength and viability. Therefore, describing that the horizons for using CNT nanocomposite scaffold could be improved by means of straightforward functionalization with definite biocompatible polymers [115]. The CNTs as a 3-dimensional network or in different arrangements can be employed for the immobilization of uniform protein. This can be defined through the investigation carried out via Song *et al.*, where 3-dimensional CNT pillars were developed over silicon plate against which an oral squamous cell carcinoma and cytokeratin-19 antibody marker was immobilized. The developed system was sufficiently able to estimate cytokeratin-19 antibody even at nanogram level (0.5 ng/mL). Therefore noticeably suggests the significance of CNTs in constructing systems for detecting different biomarkers demonstrating the symptoms and traces of infection [116]. Moreover, CNTs are also being employed to improve the electric and thermoelectric transport features of several polymers, because of an important interaction among the polymer and CNTs [117].

It has been reported that CNT displays sensitivity regarding K-channels as per dose, this produces CNT as a feasible therapy for anticancer [118]. The CNTs applications are elaborated toward particular cancer therapy. Toward brain cancer, glioblastomas astrocytoma, and anaplastic are highly invasive and eminent, for such kind of tumor, polyethylene glycol along with SWNTs and fluorescein-labeled integrin monoclonal antibody has been employed [119]. An analysis carried out via Taghdisi *et al.* suggested that how functionalized-CNT can be used for targeting blood cancer [120]. This report suggested that when SWNTs loaded daunorubicin were functionalized through sgc8c aptamer an active targeting to the leukemic cells was feasible, therefore one more times signifying the limitless probabilities toward functionalization of CNT for disease targeting [120]. The CNTs in breast cancer have also been employed, which is found as responsible for cancer-causing death among women. Researchers employ the reality of over-expression of human epidermal growth factor receptor 2 (HER2)/neureceptors, which has proliferated as targeted drug delivery. The modification of SWNT as a carrier toward small interfering (siRNA)-murine double minute 2 (MDM2) (siRNA-MDM2) and siRNA complexes was done for assessment in B-Cap-37 cells. Furthermore, the DOX loading in SWNTs coated with fluorescein and monoclonal antibody is utilized for colon cancer, which presented that in cancer cells it had specificity towards serum albumin receptors [121]. The primary cause of liver cancer includes hepatocellular carcinoma which is alcoholic cirrhosis, metabolic disorder, Hepatitis B, and Hepatitis C virus. CNTs conjugated dendrimer with antisense avian myelocytomatosis virus oncogene cellular homolog oligonucleotide displayed proficient drug delivery and targeting effect in the tumor cells [109]. This delivery of nucleic acids (anionic molecules) was facilitated owing to the functionalization of CNT with amine terminated dendrimers as they produced the necessary +ve charge which allowed proficient (~95%) entrapment of antisense avian virus oncogene. These studies describe the usefulness of surface functionalization of CNT which can utilize to obey the preferred effect [109]. Furthermore, MWNTs have also been employed in the treatment and diagnosis of cancer. Figure 14.2 displays the procedure of cancer treatment using MWNTs. In a study, dendrimer modified MWNTs were used for delivering doxorubicin drugs at lower pH [122]. It has also been reported that MWNTs are appropriate carriers for genes, peptides, and proteins as such macromolecules get simply deteriorated through enzymes found inside cells or on the cells surfaces. Guo *et al.* fabricated carrier cationic MWNTs-NH_3^+ employed to deliver the apoptotic siRNA against polo-like kinase (siPLK1) in calu6 tumor xenografts by direct intertumoral injections [123]. On the other side, it has been established by Wang *et al.* that narrow MWNTs of 9.2 nm (average diameter) had increased the affinity of non-reticular endothelial tissues, in comparison to broader MWNTs (average diameter 39.5 nm). This report also suggested that a larger aspect ratio of narrow MWNTs can be useful in the biomedical field because of the higher tissue accumulation [124].

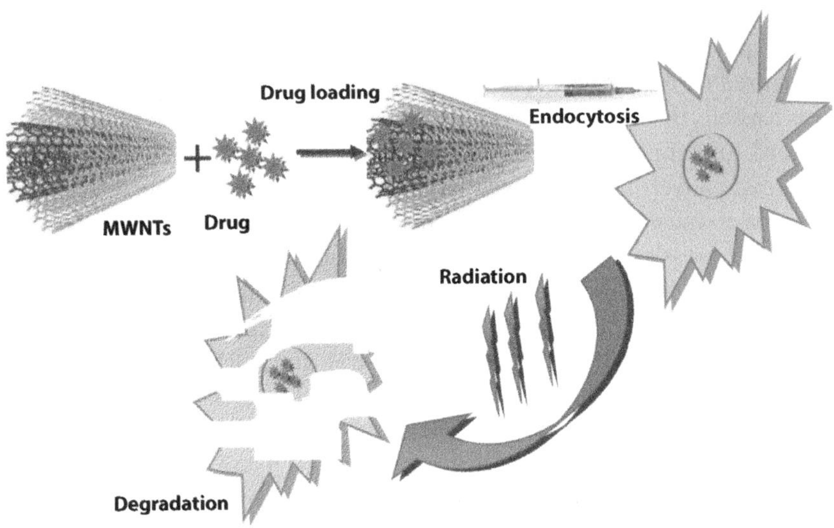

Figure 14.2 Possible graphical illustration of cancer treatment through MWNTs.

14.5.5 Functionalized CNTs for Drug Targeting and Release

For designing a broad variety of drug delivery scaffolds CNTs have been employed for the treatment of several ailments. Definite exclusive properties of CNTs (adequate chemical stability, large surface area, high mechanical strength, and drug loading capacity) produce them brilliant nanocarriers for the delivery of different drugs [125]. Anticancer drug delivery scaffolds based on CNTs are frequent which implies specific targeting acquired by functionalization [126]. It can be calibrated sensitively to the microenvironment of the tumor which results in toxicity reduction along with reduce side effects. The prevention of cell proliferation and multi-drug resistance of tumor cells and the cell cycle can be possibly reduced in the affected cancer cells through the employment of a CNT-based scaffold [127]. The fabrication of peptide-functionalized SWNTs has revealed enhanced tumor-targeting and a high anti-tumor effect [85]. For immunization CNTs also behaves like a carrier against antigens [13]. Substances like acetylcholine are difficult to be transferred to the brain by means of normal procedure can be straightforwardly transported with the help of SWNT [128]. Although, covalent functionalization of SWNT with ester and amide bonds permits slow discharge of drug for a satisfactory period of the interval which sources the raise in the solubility of organic and aqueous phase [129]. MWNTs can be employed for association with neural tissue cells and are helpful for constructing competent drug and gene delivery sources [130]. The ability of SWNT towards the drugs' inclusion is larger. The functionalized CNTs encapsulated with drug penetrate the cells via crossing the cell membrane through passive diffusion or by endocytosis and gets internalized within the nucleus and the cell organelles. Exocytosis and enzymatic degradation are the two main approaches that help CNTs to get rid of the cell [131]. The unique cocoon-like MWNTs and polyethylene glycol-based nanoparticles assign a large probability to be employed as nano-biomaterials and can be loaded simply with curcumin (a natural anticancer drug) [132]. Currently, in a recent report, the cisplatin (anticancer drug) has been investigated to have antileishmanial activity. The report suggested the enhanced antileishmanial activity and less cytotoxic nature of cisplatin bonded MWNTs at an especially lower amount highlighting the huge probability of CNTs as drug nanocarriers [133]. In another report, CNT buckypapers have been found to be constructive for transdermal drugs delivery. The *in vitro* analysis established the ability of CNT buckypapers in regulating the release of drugs [134]. Furthermore, CNTs can be covalently conjugated with small drug molecules for *in vitro* delivery. Drug cargoes and fluorescent dyes were concurrently associated to 1,3-dipolar cycloaddition modified CNTs through amide bonds for the anti-cancer drug delivery [135] or an antifungal drug [136] into cells. In an inspiring investigation, Moon *et al.* analyzed the *in-vivo* and *in-vitro* photothermal efficiency of polyethyleneglycol-conjugated SWNTs. The

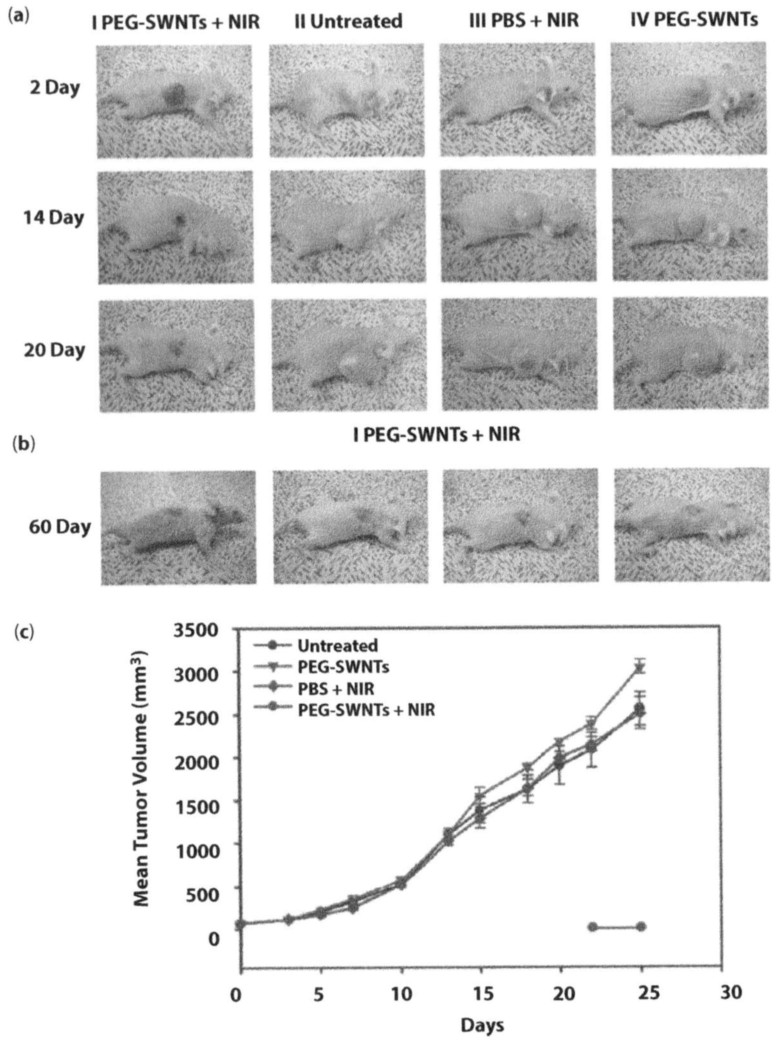

Figure 14.3 Photothermal effects of polyethyleneglycol (PEG)-SWNTs for destruction of tumor. (a) the mice treated I, PEG-SWNTs+NIR; II, untreated; III, PBS+NIR; IV, PEG+SWNTs. (b) Four mice treated with I, PEG-SWNTs+NIR after 60 days of treatments. (c) Time-dependent tumor growth curves of KB tumor cell xenografts. After sample treatments volumes of tumor were measured three times a week. "Reprinted with permission from [137]. Copyright (2009) American Chemical Society."

Table 14.2 Advanced and current drug/gene delivery systems based on CNTs.

S. no.	CNTs material	Gene/drug	Application	Ref
1.	Polymer-doxorubicin-CNT	Doxorubicin	Targeted drug delivery for cancer	[139]
2.	Functionalized-CNT	Dacarbazine	Drug delivery system for the anticancer	[140]
3.	CNT-hydrogel	Rhodamine B and doxorubicin	Simultaneous monitoring and programmed release of dual drugs	[141]
4.	Functionalized-SWNT	Doxorubicin	Pathway to assemble a higher density of doxorubicin drug on the surfaces of functionalized-CNTs and its delivery	[142]
5.	Functionalized CNTs nanocarrier	Cladribine	Anti-cancer treatment	[143]
6.	Non-covalent functionalized SWNT-polyethyleneimines	Plasmid DNA	Competent gene delivery vectors *in vivo* as well as tail vein injection in mice	[108]
7.	Functionalized-MWNT	mRNA and C2C12 cells	The up-regulation in mRNA level in caspase 3/7 due to the presence of CNTs	[144]

(Continued)

Table 14.2 Advanced and current drug/gene delivery systems based on CNTs. (*Continued*)

S. no.	CNTs material	Gene/drug	Application	Ref
8.	Functionalized-CNTs into hydrophobic drug crystals	Sulfamethoxazole and griseofulvin	Enhance aqueous solubility	[145]
9.	pH-responsive-functionalized-SWNT-polyethylenimine-betaine	doxorubicin and siRNA	Permutation of chemotherapy and gene therapy	[146]
10.	CNT-(Fe)/hydroxyapatite	Doxorubicin	Drug transport in high external magnetic fields (magnetic targeted delivery system)	[147]
11.	Chitosan-CNT based thermosensitive hydrogels	Methotrexate	Effectively controls growth of tumor cell	[148]
12.	Boron nitride/CNTs	Efavirenz	Anti-HIV drug delivery	[149]
13.	Functionalized SWNTs	Flutamide	Enhanced drug adsorption on the functionalized-CNT	[150]
14.	MWNT/lentinan/gemcitabine composite	Gemcitabine	Chemo-photothermal synergistic cancer treatment	[151]
15.	Oxidized MWNTs	Metformin	*In vitro* biological activity and inhibited cellular uptake of CNT	[152]

(*Continued*)

Table 14.2 Advanced and current drug/gene delivery systems based on CNTs. (*Continued*)

S. no.	CNTs material	Gene/drug	Application	Ref
16.	Cu/CNTs	Cu nanoparticles	Without any toxicity enhanced antimicrobial potential against biofilms	[153]
17.	Covalent-functionalized CNTs based drug carrier	Doxorubicin	Increased encapsulation efficiency	[154]
18.	Starch/MWNT/glucose	Zolpidem	For drug delivery	[155]
19.	Functionalized-MWNTs with nanoneedle structure	-	Gene delivery	[156]
20.	Functionalized-MWNTs/ cyclodextrins/polyethylenimine	Cidofovir	Antiviral drug delivery system	[157]

fabricated system determined length in the range of 50–300 nm and diameter in the range of 2–5 nm. On the basis of the *in vitro* outcomes, further *in vivo* photo-thermal analysis was performed with nude mice bearing human epidermoid mouth carcinoma KB tumor cells. As demonstrated in Figure 14.3a, the mice medicated with polyethyleneglycol-conjugated SWNTs and near-infrared light (NIR) irradiation (I) represented fully demolition of tumors within twenty days. When treated with (I) a clear examination of black round marks was noticed on the mice skins (Figure 14.3a, I, 2 Day, and 14 Day). Figure 14.3 exhibits the destruction of tumor volume with control, polyethyleneglycol-SWNT+NIR, phosphate buffer saline (PBS)+NIR (PBS+NIR), and polyethyleneglycol-SWNT after 2, 14, and 20 days. Greater tumor destruction is shown in the case of polyethyleneglycol-SWNT+NIR in comparison to the untreated group [137]. From these investigations, it has been suggested that the functionalization of both SWNTs and MWNTs was effectively accomplished to deliver the payload to the targeting place. It was also noticed that CNTs in the functionalized form displayed enhanced antiproliferative effect and *in-vitro* drug release than the free drugs. The functionalization of polyethyleneglycol is also valuable to lower the toxicity problems that emerged due to the employment of CNTs [138]. Table 14.2 displayed the advanced and current applications of CNT as a drug/gene carrier, however, there are many further gene and drug delivery applications studied for CNTs.

14.6 Future Prospective

CNTs open up an innovative era for different sensible purposes and are functional in fulfilling the fundamental physics at the non-metric level. CNTs acknowledged the various challenges on the way of development in the realistic application of these materials in the biomedical field. It also provides wonderful peculiar advantages to the next generation research by their progression in the chemicals, electronics, and approximately all sensible field of health care applications. In the present time, CNTs and their derivatives present genera of nanomaterials that are employed as biomaterials in the area of biomedicine and therapeutics. Furthermore, nanotechnology has been performed as a landmark in solving the restrictions and regulations. Therefore, size in nano-range along with the structural, electrical, and mechanical, properties of CNTs establish them as multifunctional nanocarriers, establishing CNTs functional candidates in research and developments of disease treatment, drug and gene delivery devices. There are various serious issues including cancer therapy targeted delivery of DOX [158], gene and drug delivery to the brain, that can be resolved with the help of CNTs. Confirmed targeted-tumor delivery with CNT is an approach for complex obstructions in cancer therapy. Literature survey revealed that CNT and its derivatives have been successfully used as targeted delivery scaffolds for drugs and gene delivery to the brain. Although further investigations is still needed to prepare functionalized CNT as a carrier device for proficient targeted brain therapy. Hence with the increasing demand, continuous researches are requisite to design multifunctional CNTs for effective and efficient drug delivery, for tissue engineering and stem cell therapy. However, functionalizations of CNTs reduce side effects, improve the bioavailability and physicochemical properties, but an absolute careful investigation of such biomaterials is still required. In addition, comprehensive understanding and safety knowledge of underlying physiological mechanisms of CNT incorporation is also essential. All above-reviewed strategies and approaches are only beneficial when the utilization CNTs will be sensibly effective, safe, along with obeying all safety and quality measures. There are more probabilities that drug delivery system based CNT and its derivatives will substitute most of the traditional imaging and diagnostic system. Despite the functionalized CNTs being valuable for deducting the cytotoxic effects and in offering improved bioavailability, it is essential to have a complete toxicological study of this nanocarrier. So far, among the present advanced drug delivery system based on CNTs, none of them were clinically approved. There is yet a further possibility to substitute all the traditional biomolecular detection and diagnostic systems through excellent and superspecific CNT-based systems in the future. Therefore, it can be suggested that in the future, with continuous research towards the advancement in CNTS, its utilization could reach heights in the field of biomedical.

14.7 Conclusion

In summary, chemically functionalized CNTs with a tunable length and large surface area as well as exclusive physicochemical characteristics, make CNTs promising biomaterials in the field of biomedicine. The literature survey confirms that, the possibilities of CNTs as targeted drug delivery carriers have produced them capable materials for the medication and diagnosis of dangerous diseases such as CNS disorders, infectious diseases, and cancer. CNTs' distinctive mechanical features also make them the main analytical topics in drug and gene targeting and release. Study on the possible toxicity of CNTs is still ongoing, but CNTs have been extensively established as a safer choice in comparison to the other nano-materials for example virus carriers and quantum dots. Thus, over the last two decades, the incredible growth of nanotechnology-based approaches in biomedical CNTs based materials has provided novel techniques for preparing unique functional materials.

(i) Herein, a collection of reports have been summarized on different techniques and functionalized CNTs with the improved clinical investigation.
(ii) With these approaches, desired molecules are combined with CNTs for specific selective imaging, drug delivery, and other therapies.
(iii) As a new type of promising candidate, the biocompatibility towards bone, suitability for *in vivo* analyses, and toxicity of CNTs have been extensively reviewed on the basis of published reports.
(iv) The literature survey revealed that, the composition of CNTs with particular targeting moiety or ligand that permits to improve the medication effect of the system is another fascinating research.
(v) Additionally, the broadening of these practical techniques to the 2-dimensional forms of carbon specifically graphene is also nowadays a rapidly emerging area. Whereas the mission for exploring newly hybrid materials is constantly on and the study on CNT-based materials in distinct areas for instance doping is still open for discussion and investigations.

Important Websites

https://now.northropgrumman.com/carbon-nanotube-applications-in-daily-life/
https://www.nanowerk.com/nanotechnology/introduction/introduction_to_nanotechnology_22.php
https://www.zdnet.com/article/5-surprising-uses-for-carbon-nanotubes/
https://en.wikipedia.org/wiki/Potential_applications_of_carbon_nanotubes

References

1. Jain, N. and Tiwari, S., Biomedical application of carbon nanotubes (CNTs) in vulnerable parts of the body and its toxicity study: A state-of-the-art-review. *Mater. Today Proc.*, 46, 7608–7617, 2021.
2. Park, W., Shin, H., Choi, B., Rhim, W.-K., Na, K., Keun Han, D., Advanced hybrid nanomaterials for biomedical applications. *Prog. Mater. Sci.*, 114, 100686, 2020.
3. Deshmukh, M.A., Jeon, J.-Y., Ha, T.-J., Carbon nanotubes: An effective platform for biomedical electronics. *Biosens. Bioelectron.*, 150, 111919, 2020.
4. Abd Elkodous, M., El-Sayyad, G.S., Abdelrahman, I.Y., El-Batal, A.I., Therapeutic and diagnostic potential of nanomaterials for enhanced biomedical applications. *Colloids Surf. B Biointerfaces*, 180, 411–428, 2019.
5. Rasheed, T., Hassan, A.A., Kausar, F., Sher, F., Bilal, M., Iqbal, H.M.N., Carbon nanotubes assisted analytical detection – Sensing/delivery cues for environmental and biomedical monitoring. *TrAC Trends Anal. Chem.*, 132, 116066, 2020.
6. Anzar, N., Hasan, R., Tyagi, M., Yadav, N., Narang, J., Carbon nanotube - A review on synthesis, properties and plethora of applications in the field of biomedical science. *Sens. Int.*, 1, 100003, 2020.
7. Kesharwani, P., Gajbhiye, V., Jain, N.K., A review of nanocarriers for the delivery of small interfering RNA. *Biomaterials*, 33, 7138–7150, 2012.
8. Ji, S., Liu, C., Zhang, B., Yang, F., Xu, J., Long, J., Jin, C., Fu, D., Ni, Q., Yu, X., Carbon nanotubes in cancer diagnosis and therapy. *Biochim. Biophys. Acta - Rev. Cancer*, 1806, 29–35, 2010.
9. Iijima, S., Helical microtubules of graphitic carbon. *Nature*, 354, 56–58, 1991.
10. Wick, P., Clift, M.J.D., Rösslein, M., Rothen-Rutishauser, B., A brief summary of carbon nanotubes science and technology: A health and safety perspective. *ChemSusChem*, 4, 905–911, 2011.
11. Pantarotto, D., Singh, R., McCarthy, D., Erhardt, M., Briand, J.-P., Prato, M., Kostarelos, K., Bianco, A., Functionalized carbon nanotubes for plasmid DNA gene delivery. *Angew. Chem. Int. Ed.*, 43, 5242–5246, 2004.
12. Liu, Y., Wu, D.-C., Zhang, W.-D., Jiang, X., He, C.-B., Chung, T.S., Goh, S.H., Leong, K.W., Polyethylenimine-grafted multiwalled carbon nanotubes for secure noncovalent immobilization and efficient delivery of DNA. *Angew. Chem. Int. Ed.*, 44, 4782–4785, 2005.
13. Foldvari, M. and Bagonluri, M., Carbon nanotubes as functional excipients for nanomedicines: II. Drug delivery and biocompatibility issues. *Nanomed. Nanotechnol. Biol. Med.*, 4, 183–200, 2008.
14. Watts, P.C.P., Fearon, P.K., Hsu, W.K., Billingham, N.C., Kroto, H.W., Walton, D.R.M., Carbon nanotubes as polymer antioxidants. *J. Mater. Chem.*, 13, 491–495, 2003.
15. Dresselhaus, M.S., Dresselhaus, G., Saito, R., Jorio, A., Raman spectroscopy of carbon nanotubes. *Phys. Rep.*, 409, 47–99, 2005.

16. Lefebvre, J., Austing, D.G., Bond, J., Finnie, P., Photoluminescence imaging of suspended single-walled carbon nanotubes. *Nano Lett.*, 6, 1603–1608, 2006.
17. Murakami, Y., Einarsson, E., Edamura, T., Maruyama, S., Polarization dependence of the optical absorption of single-walled carbon nanotubes. *Phys. Rev. Lett.*, 94, 087402, 2005.
18. Prato, M., Kostarelos, K., Bianco, A., Functionalized carbon nanotubes in drug design and discovery. *Acc. Chem. Res.*, 41, 60–68, 2008.
19. Alshehri, R., Ilyas, A.M., Hasan, A., Arnaout, A., Ahmed, F., Memic, A., Carbon nanotubes in biomedical applications: Factors, mechanisms, and remedies of toxicity. *J. Med. Chem.*, 59, 8149–8167, 2016.
20. Paul, A., Hasan, A., Al Kindi, H., Gaharwar, A.K., Rao, V.T.S., Nikkhah, M., Shin, S.R., Krafft, D., Dokmeci, M.R., Shum-Tim, D., Khademhosseini, A., Injectable graphene oxide/hydrogel-based angiogenic gene delivery system for vasculogenesis and cardiac repair. *ACS Nano*, 8, 8050–8062, 2014.
21. Yang, W., Thordarson, P., Gooding, J.J., Ringer, S.P., Braet, F., Carbon nanotubes for biological and biomedical applications. *Nanotechnology*, 18, 412001, 2007.
22. Zhou, J., Li, J., Wu, D., Hong, C., CNT-based and msn-based organic/inorganic hybrid nanocomposites for biomedical applications, pp. 169–192, ACS Publications, Washington, DC, 2017.
23. Hwang, Y., Park, S.-H., Lee, J., Applications of functionalized carbon nanotubes for the therapy and diagnosis of cancer. *Polym. (Basel)*, 9, 13, 2017.
24. Lombardo, D., Kiselev, M.A., Caccamo, M.T., Smart nanoparticles for drug delivery application: Development of versatile nanocarrier platforms in biotechnology and nanomedicine. *J. Nanomater.*, 2019, 1–26, 2019.
25. Mohammadi, M., Arabi, L., Alibolandi, M., Doxorubicin-loaded composite nanogels for cancer treatment. *J. Control. Release*, 328, 171–191, 2020.
26. Beg, S., Rizwan, M., Sheikh, A.M., Hasnain, M.S., Anwer, K., Kohli, K., Advancement in carbon nanotubes: Basics, biomedical applications and toxicity. *J. Pharm. Pharmacol.*, 63, 141–163, 2011.
27. Ali-Boucetta, H., Al-Jamal, K.T., McCarthy, D., Prato, M., Bianco, A., Kostarelos, K., Multiwalled carbon nanotube–doxorubicin supramolecular complexes for cancer therapeutics. *Chem. Commun.*, 4, 459–461, 2008.
28. DeRosa, A.M., Greco, K., Rajamani, S., Sitharaman, B., Recent patents on single-walled carbon nanotubes for biomedical imaging, drug delivery and tissue regeneration. *Recent Pat. Biomed. Eng.*, 3, 86–94, 2010.
29. Ren, Z.F., Huang, Z.P., Xu, J.W., Wang, J.H., Bush, P., Siegal, M.P., Provencio, P.N., Synthesis of large arrays of well-aligned carbon nanotubes on glass. *Sci. (80)*, 282, 1105–1107, 1998.
30. Araga, R. and Sharma, C.S., One step direct synthesis of multiwalled carbon nanotubes from coconut shell derived charcoal. *Mater. Lett.*, 188, 205–207, 2017.

31. Yaghoubi, A. and Ramazani, A., Anticancer DOX delivery system based on CNTs: Functionalization, targeting and novel technologies. *J. Control. Release*, 327, 198–224, 2020.
32. Tasis, D., Tagmatarchis, N., Bianco, A., Prato, M., Chemistry of carbon nanotubes. *Chem. Rev.*, 106, 1105–1136, 2006.
33. Sun, Y.P., Fu, K., Lin, Y., Huang, W., Functionalized carbon nanotubes: Properties and applications. *Acc. Chem. Res.*, 35, 1096–1104, 2002.
34. Dai, H., Carbon nanotubes: Synthesis, integration, and properties. *Acc. Chem. Res.*, 35, 1035–1044, 2002.
35. Huang, W., Taylor, S., Fu, K., Lin, Y., Zhang, D., Hanks, T.W., Rao, A.M., Sun, Y.-P., Attaching proteins to carbon nanotubes via diimide-activated amidation. *Nano Lett.*, 2, 311–314, 2002.
36. Lordi, V., Yao, N., Wei, J., Method for supporting platinum on single-walled carbon nanotubes for a selective hydrogenation catalyst. *Chem. Mater.*, 13, 733–737, 2001.
37. Niyogi, S., Hamon, M.A., Hu, H., Zhao, B., Bhowmik, P., Sen, R., Itkis, M.E., Haddon, R.C., Chemistry of single-walled carbon nanotubes. *Acc. Chem. Res.*, 35, 1105–1113, 2002.
38. Dyke, C.A. and Tour, J.M., Overcoming the insolubility of carbon nanotubes through high degrees of sidewall functionalization. *Chem. - Eur. J.*, 10, 812–817, 2004.
39. Dinesh, B., Bianco, A., Ménard-Moyon, C., Designing multimodal carbon nanotubes by covalent multi-functionalization. *Nanoscale*, 8, 18596–18611, 2016.
40. Hirsch, A. and Vostrowsky, O., *Functional molecular nanostructures*, A.D. Schlüter, (Ed.), pp. 193–237, Springer, Berlin, Heidelberg, 2005.
41. Zhou, H., Mu, Q., Gao, N., Liu, A., Xing, Y., Gao, S., Zhang, Q., Qu, G., Chen, Y., Liu, G., Zhang, B., Yan, B., A nano-combinatorial library strategy for the discovery of nanotubes with reduced protein-binding, cytotoxicity, and immune response. *Nano Lett.*, 8, 859–865, 2008.
42. Star, A., Stoddart, J.F., Steuerman, D., Diehl, M., Boukai, A., Wong, E.W., Yang, X., Chung, S., Choi, H., Heath, J.R., Preparation and properties of polymer-wrapped single-walled carbon nanotubes. *Angew. Chem.*, 113, 1771–1775, 2001.
43. Wang, X.L., Oh, I.K., Lu, J., Ju, J., Lee, S., Biomimetic electro-active polymer based on sulfonated poly (styrene-b-ethylene-co-butylene-b-styrene). *Mater. Lett.*, 61, 5117–5120, 2007.
44. Curran, S., Davey, A.P., Coleman, J., Dalton, A., McCarthy, B., Maier, S., Drury, A., Gray, D., Brennan, M., Ryder, K., de la Chapelle, M.L., Journet, C., Bernier, P., Byrne, H.J., Carroll, D., Ajayan, P.M., Lefrant, S., Blau, W., Evolution and evaluation of the polymer/nanotube composite. *Synth. Met.*, 103, 2559–2562, 1999.

45. Xing, R., Yu, H., Liu, S., Zhang, W., Zhang, Q., Li, Z., Li, P., Antioxidant activity of differently regioselective chitosan sulfates *in vitro*. *Bioorg. Med. Chem.*, 13, 1387–1392, 2005.
46. Yue, G.Z., Qiu, Q., Gao, B., Cheng, Y., Zhang, J., Shimoda, H., Chang, S., Lu, J.P., Zhou, O., Generation of continuous and pulsed diagnostic imaging x-ray radiation using a carbon-nanotube-based field-emission cathode. *Appl. Phys. Lett.*, 81, 355–357, 2002.
47. Choi, J.H., Nguyen, F.T., Barone, P.W., Heller, D.A., Moll, A.E., Patel, D., Boppart, S.A., Strano, M.S., Multimodal biomedical imaging with asymmetric single-walled carbon nanotube/iron oxide nanoparticle complexes. *Nano Lett.*, 7, 861–867, 2007.
48. Shiral Fernando, K.A., Lin, Y., Sun, Y.-P., High aqueous solubility of functionalized single-walled carbon nanotubes. *Langmuir*, 20, 4777–4778, 2004.
49. Liu, Z., Cai, W., He, L., Nakayama, N., Chen, K., Sun, X., Chen, X., Dai, H., *In vivo* biodistribution and highly efficient tumour targeting of carbon nanotubes in mice. *Nat. Nanotechnol.*, 2, 47–52, 2007.
50. Liu, Z., Chen, K., Davis, C., Sherlock, S., Cao, Q., Chen, X., Dai, H., Drug delivery with carbon nanotubes for *in vivo* cancer treatment. *Cancer Res.*, 68, 6652–6660, 2008.
51. Didenko, V.V., Moore, V.C., Baskin, D.S., Smalley, R.E., Visualization of individual single-walled carbon nanotubes by fluorescent polymer wrapping. *Nano Lett.*, 5, 1563–1567, 2005.
52. Vaisman, L., Wagner, H.D., Marom, G., The role of surfactants in dispersion of carbon nanotubes. *Adv. Colloid Interface Sci.*, 128-130, 37–46, 2006.
53. Wang, S., Humphreys, E.S., Chung, S.-Y., Delduco, D.F., Lustig, S.R., Wang, H., Parker, K.N., Rizzo, N.W., Subramoney, S., Chiang, Y.M., Jagota, A., Peptides with selective affinity for carbon nanotubes. *Nat. Mater.*, 2, 196–200, 2003.
54. Dasgupta, A. and Das, D., Designer peptide amphiphiles: Self-assembly to applications. *Langmuir*, 35, 10704–10724, 2019.
55. Arnold, M.S., Guler, M.O., Hersam, M.C., Stupp, S.I., Encapsulation of carbon nanotubes by self-assembling peptide amphiphiles. *Langmuir*, 21, 4705–4709, 2005.
56. Liu, Y., Ghaffari, M., Zhao, R., Lin, J.-H., Lin, M., Zhang, Q.M., Enhanced electromechanical response of ionic polymer actuators by improving mechanical coupling between ions and polymer matrix. *Macromolecules*, 45, 5128–5133, 2012.
57. Singh, R.P., Das, M., Thakare, V., Jain, S., Functionalization density dependent toxicity of oxidized multiwalled carbon nanotubes in a murine macrophage cell line. *Chem. Res. Toxicol.*, 25, 2127–2137, 2012.
58. Jaurand, M.C.F., Renier, A., Daubriac, J., Mesothelioma: Do asbestos and carbon nanotubes pose the same health risk? *Part. Fibre Toxicol.*, 6, 16, 2009.
59. Jain, K.K., Advances in use of functionalized carbon nanotubes for drug design and discovery. *Expert Opin. Drug Discovery*, 7, 1029–1037, 2012.

60. Amenta, V. and Aschberger, K., Carbon nanotubes: Potential medical applications and safety concerns. *WIREs Nanomed. Nanobiotechnol.*, 7, 371–386, 2015.
61. Bussy, C. and Kostarelos, K., Carbon nanotubes in medicine and biology — Safety and toxicology. *Adv. Drug Deliv. Rev.*, 65, 2061–2062, 2013.
62. Sharmeen, S., Rahman, A.F.M.M., Lubna, M.M., Salem, K.S., Islam, R., Khan, M.A., Polyethylene glycol functionalized carbon nanotubes/gelatin-chitosan nanocomposite: An approach for significant drug release. *Bioact. Mater.*, 3, 236–244, 2018.
63. Wang, J., Sun, P., Bao, Y., Dou, B., Song, D., Li, Y., Vitamin E renders protection to PC12 cells against oxidative damage and apoptosis induced by single-walled carbon nanotubes. *Toxicol. Vitr.*, 26, 32–41, 2012.
64. Shvedova, A.A., Kapralov, A.A., Feng, W.H., Kisin, E.R., Kagan, V.E., Impaired clearance and enhanced pulmonary inflammatory/fibrotic response to carbon nanotubes in myeloperoxidase-deficient mice. *PLoS One*, 7, 30923, 2012.
65. Kagan, V.E., Konduru, N.V., Feng, W., Allen, B.L., Conroy, J., Shvedova, A.A., Carbon nanotubes degraded by neutrophil myeloperoxidase induce less pulmonary inflammation. *Nat. Nanotechnol.*, 5, 354–359, 2010.
66. Ge, C., Du, J., Zhao, L., Wang, L., Liu, Y., Li, D., Yang, Y., Zhou, R., Zhao, Y., Chai, Z., Chen, C., Binding of blood proteins to carbon nanotubes reduces cytotoxicity. *Proc. Natl. Acad. Sci.*, 108, 16968–16973, 2011.
67. Mishra, V., Kesharwani, P., Jain, N.K., Biomedical applications and toxicological aspects of functionalized carbon nanotubes. *Crit. Rev. Ther. Drug Carrier Syst.*, 35, 293–330, 2018.
68. Firme, C.P. and Bandaru, P.R., Toxicity issues in the application of carbon nanotubes to biological systems. *Nanomed. Nanotechnol. Biol. Med.*, 6, 245–256, 2010.
69. Qu, G., Bai, Y., Zhang, Y., Jia, Q., Zhang, W., Yan, B., The effect of multiwalled carbon nanotube agglomeration on their accumulation in and damage to organs in mice. *Carbon N. Y.*, 47, 2060–2069, 2009.
70. Liopo, A.V., Stewart, M.P., Hudson, J., Tour, J.M., Pappas, T.C., Biocompatibility of native and functionalized single-walled carbon nanotubes for neuronal interface. *J. Nanosci. Nanotechnol.*, 6, 1365–1374, 2006.
71. Colvin, V.L., The potential environmental impact of engineered nanomaterials. *Nat. Biotechnol.*, 21, 1166–1170, 2003.
72. He, Z., Jiang, R., Long, W., Huang, H., Liu, M., Chen, J., Deng, F., Zhou, N., Zhang, X., Wei, Y., The combination of Diels-Alder reaction and redox polymerization for preparation of functionalized CNTs for intracellular controlled drug delivery. *Mater. Sci. Eng. C*, 109, 110442, 2020.
73. Huczko, A., Lange, H., Całko, E., Grubek-Jaworska, H., Droszcz, P., Physiological testing of carbon nanotubes: Are they asbestos-like? *Fuller. Sci. Technol.*, 9, 251–254, 2001.
74. Huczko, A. and Lange, H., Carbon nanotubes: Experimental evidence for a null risk of skin irritation and allergy. *Fuller. Sci. Technol.*, 9, 247–250, 2001.

75. Lam, C.-W., Pulmonary Toxicity of single-wall carbon nanotubes in mice 7 and 90 days after intratracheal instillation. *Toxicol. Sci.*, 77, 126–134, 2003.
76. Warheit, D.B., Comparative pulmonary toxicity assessment of single-wall carbon nanotubes in rats. *Toxicol. Sci.*, 77, 117–125, 2003.
77. Maynard, A.D., Baron, P.A., Foley, M., Shvedova, A.A., Kisin, E.R., Castranova, V., Exposure to carbon nanotube material: Aerosol release during the handling of unrefined single-walled carbon nanotube material. *J. Toxicol. Environ. Health Part A*, 67, 87–107, 2004.
78. Yang, S.T., Wang, X., Jia, G., Gu, Y., Wang, T., Nie, H., Ge, C., Wang, H., Liu, Y., Long-term accumulation and low toxicity of single-walled carbon nanotubes in intravenously exposed mice. *Toxicol. Lett.*, 181, 182–189, 2008.
79. Yokoyama, A., Sato, Y., Nodasaka, Y., Yamamoto, S., Kawasaki, T., Shindoh, M., Kohgo, T., Akasaka, T., Uo, M., Watari, F., Tohji, K., Biological behavior of hat-stacked carbon nanofibers in the subcutaneous tissue in rats. *Nano Lett.*, 5, 157–161, 2005.
80. Wahab, R., Khan, F., A.Siddiqui, M., Ahmad, J., Saquib, Q., Al-Khedhairy, A.A., Cytotoxic assessment of liver cancer cells (HepG2) with raw, functionalized multiwalled carbon nanotubes and their comparison with nanohydroxyapatite. *J. King Saud Univ. - Sci.*, 33, 101444, 2021.
81. Kunzmann, A., Andersson, B., Thurnherr, T., Krug, H., Scheynius, A., Fadeel, B., Toxicology of engineered nanomaterials: Focus on biocompatibility, biodistribution and biodegradation. *Biochim. Biophys. Acta - Gen. Subj.*, 1810, 361–373, 2011.
82. Johnston, H.J., Hutchison, G.R., Christensen, F.M., Peters, S., Hankin, S., Aschberger, K., Stone, V., A critical review of the biological mechanisms underlying the *in vivo* and *in vitro* toxicity of carbon nanotubes: The contribution of physico-chemical characteristics. *Nanotoxicology*, 4, 207–246, 2010.
83. Setyan, A., Sauvain, J., Guillemin, M., Riediker, M., Demirdjian, B., Rossi, M.J., Probing functional groups at the gas–aerosol interface using heterogeneous titration reactions: A tool for predicting aerosol health effects? *ChemPhysChem*, 11, 3823–3835, 2010.
84. Guo, L., Hong, S., Lin, X., Xie, Z., Chen, G., An organically modified sol–gel membrane for detection of lead ion by using 2-hydroxy-1-naphthaldehydene-8-aminoquinoline as fluorescence probe. *Sens. Actuators B Chem.*, 130, 789–794, 2008.
85. Chen, J., Chen, S., Zhao, X., Kuznetsova, L.V., Wong, S.S., Ojima, I., Functionalized single-walled carbon nanotubes as rationally designed vehicles for tumor-targeted drug delivery. *J. Am. Chem. Soc.*, 130, 16778–16785, 2008.
86. Kang, S., Herzberg, M., Rodrigues, D.F., Elimelech, M., Antibacterial effects of carbon nanotubes: size does matter! *Langmuir*, 24, 6409–6413, 2008.

87. Kang, S., Pinault, M., Pfefferle, L.D., Elimelech, M., Single-walled carbon nanotubes exhibit strong antimicrobial activity. *Langmuir*, 23, 8670–8673, 2007.
88. Yandar, N., Pastorin, G., Prato, M., Bianco, A., Patarroyo, M.E., Manuel Lozano, J., Immunological profile of a Plasmodium vivax AMA-1 N-terminus peptide-carbon nanotube conjugate in an infected Plasmodium berghei mouse model. *Vaccine*, 26, 5864–5873, 2008.
89. Pantarotto, D., Partidos, C.D., Hoebeke, J., Brown, F., Kramer, E., Briand, J.-P., Muller, S., Prato, M., Bianco, A., Immunization with peptide-functionalized carbon nanotubes enhances virus-specific neutralizing antibody responses. *Chem. Biol.*, 10, 961–966, 2003.
90. Shi, H., Xia, T., Nel, A.E., Yeh, J.I., Part II: Coordinated biosensors – development of enhanced nanobiosensors for biological and medical applications. *Nanomedicine*, 2, 599–614, 2007.
91. Mahmoud, K.A. and Luong, J.H.T., Impedance method for detecting HIV-1 protease and screening for its inhibitors using ferrocene–peptide conjugate/Au nanoparticle/single-walled carbon nanotube modified electrode. *Anal. Chem.*, 80, 7056–7062, 2008.
92. Dastagir, T., Forzani, E.S., Zhang, R., Amlani, I., Nagahara, L.A., Tsui, R., Tao, N., Electrical detection of hepatitis C virus RNA on single wall carbon nanotube-field effect transistors. *Analyst*, 132, 738, 2007.
93. Yang, F., Murugan, R., Ramakrishna, S., Wang, X., Ma, Y.-X., Wang, S., Fabrication of nano-structured porous PLLA scaffold intended for nerve tissue engineering. *Biomaterials*, 25, 1891–1900, 2004.
94. Abidian, M.R., Kim, D.H., Martin, D.C., Conducting-polymer nanotubes for controlled drug release. *Adv. Mater.*, 18, 405–409, 2006.
95. Hu, H., Ni, Y., Montana, V., Haddon, R.C., Parpura, V., Chemically functionalized carbon nanotubes as substrates for neuronal growth. *Nano Lett.*, 4, 507–511, 2004.
96. Lovat, V., Pantarotto, D., Lagostena, L., Ballerini, L., Carbon nanotube substrates boost neuronal electrical signaling. *Nano Lett.*, 5, 1107–1110, 2005.
97. Sitharaman, B., Shi, X., Tran, L.A., Spicer, P.P., Rusakova, I., Wilson, L.J., Mikos, A.G., Injectable *in situ* cross-linkable nanocomposites of biodegradable polymers and carbon nanostructures for bone tissue engineering. *J. Biomater. Sci. Polym. Ed.*, 18, 655–671, 2007.
98. Keefer, E.W., Botterman, B.R., Romero, M.I., Rossi, A.F., Gross, G.W., Carbon nanotube coating improves neuronal recordings. *Nat. Nanotechnol.*, 3, 434–439, 2008.
99. Gulyaev, A.E., Gelperina, S.E., Skidan, I.N., Antropov, A.S., Kivman, G.Y., Kreuter, J., Significant transport of doxorubicin into the brain with polysorbate 80-coated nanoparticles. *Pharm. Res.*, 16, 1564–1569, 1999.
100. Kateb, B., Van Handel, M., Zhang, L., Bronikowski, M.J., Manohara, H., Badie, B., Internalization of MWCNTs by microglia: Possible application in immunotherapy of brain tumors. *Neuroimage*, 37, S9–S17, 2007.

101. VanHandel, M., Alizadeh, D., Zhang, L., Kateb, B., Bronikowski, M., Manohara, H., Badie, B., Selective uptake of multi-walled carbon nanotubes by tumor macrophages in a murine glioma model. *J. Neuroimmunol.*, 208, 3–9, 2009.
102. Karimi, M., Solati, N., Ghasemi, A., Hamblin, M.R., Carbon nanotubes part II: A remarkable carrier for drug and gene delivery. *Expert Opin. Drug Deliv.*, 12, 1089–1105, 2015.
103. Kam, N.W.S. and Dai, H., Carbon nanotubes as intracellular protein transporters: Generality and biological functionality. *J. Am. Chem. Soc.*, 127, 6021–6026, 2005.
104. Lacerda, L., Pastorin, G., Wu, W., Prato, M., Bianco, A., Kostarelos, K., Luminescence of functionalized carbon nanotubes as a tool to monitor bundle formation and dissociation in water: The effect of plasmid-DNA complexation. *Adv. Funct. Mater.*, 16, 1839–1846, 2006.
105. Singh, R., Pantarotto, D., McCarthy, D., Kostarelos, K., Binding and condensation of plasmid dna onto functionalized carbon nanotubes: Toward the construction of nanotube-based gene delivery vectors. *J. Am. Chem. Soc.*, 127, 4388–4396, 2005.
106. Kam, N.W.S., Liu, Z., Dai, H., Functionalization of carbon nanotubes via cleavable disulfide bonds for efficient intracellular delivery of siRNA and potent gene silencing. *J. Am. Chem. Soc.*, 127, 12492–12493, 2005.
107. Zhang, Z., Yang, X., Zhang, Y., Zeng, B., Wang, S., Zhu, T., Roden, R.B.S., Chen, Y., Yang, R., Delivery of telomerase reverse transcriptase small interfering RNA in complex with positively charged single-walled carbon nanotubes suppresses tumor growth. *Clin. Cancer Res.*, 12, 4933–4939, 2006.
108. Behnam, B., Shier, W.T., Nia, A.H., Abnous, K., Ramezani, M., Non-covalent functionalization of single-walled carbon nanotubes with modified polyethyleneimines for efficient gene delivery. *Int. J. Pharm.*, 454, 204–215, 2013.
109. Pan, B., Cui, D., Xu, P., Ozkan, C., Feng, G., Ozkan, M., Huang, T., Chu, B., Li, Q., He, R., Hu, G., Synthesis and characterization of polyamidoamine dendrimer-coated multi-walled carbon nanotubes and their application in gene delivery systems. *Nanotechnology*, 20, 125101, 2009.
110. Cifuentes-Rius, A., Boase, N.R.B., Font, I., Coronas, N., Ramos-Perez, V., Thurecht, K.J., Borrós, S., *In vivo* fate of carbon nanotubes with different physicochemical properties for gene delivery applications. *ACS Appl. Mater. Interfaces*, 9, 11461–11471, 2017.
111. Ali, Z., Abul-faraj, A., Li, L., Ghosh, N., Piatek, M., Mahfouz, M.M., Efficient virus-mediated genome editing in plants using the CRISPR/Cas9 system. *Mol. Plant*, 8, 1288–1291, 2015.
112. Geng, Y., Liu, M.Y., Li, J., Shi, X.M., Kim, J.K., Effects of surfactant treatment on mechanical and electrical properties of CNT/epoxy nanocomposites. *Compos. Part A Appl. Sci. Manuf.*, 39, 1876–1883, 2008.
113. Kharaziha, M., Shin, S.R., Nikkhah, M., Topkaya, S.N., Masoumi, N., Annabi, N., Dokmeci, M.R., Khademhosseini, A., Tough and flexible CNT–polymeric

hybrid scaffolds for engineering cardiac constructs. *Biomaterials*, 35, 7346–7354, 2014.

114. Shokri, S., Movahedi, B., Rafieinia, M., Salehi, H., A new approach to fabrication of Cs/BG/CNT nanocomposite scaffold towards bone tissue engineering and evaluation of its properties. *Appl. Surf. Sci.*, 357, 1758–1764, 2015.
115. Chandrasekhar, P., *Conducting polymers, fundamentals and applications*, pp. 61–64, Springer International Publishing, Cham, 2018.
116. Liu, P., Modifications of carbon nanotubes with polymers. *Eur. Polym. J.*, 41, 2693–2703, 2005.
117. Wang, Q., Yao, Q., Chang, J., Chen, L., Enhanced thermoelectric properties of CNT/PANI composite nanofibers by highly orienting the arrangement of polymer chains. *J. Mater. Chem.*, 22, 17612, 2012.
118. Ortega-Guerrero, A., Espinosa-Duran, J.M., Velasco-Medina, J., TRPV1 channel as a target for cancer therapy using CNT-based drug delivery systems. *Eur. Biophys. J.*, 45, 423–433, 2016.
119. Zhou, F., Xing, D., Ou, Z., Wu, B., Resasco, D.E., Chen, W.R., Cancer photothermal therapy in the near-infrared region by using single-walled carbon nanotubes. *J. Biomed. Opt.*, 14, 021009, 2009.
120. Taghdisi, S.M., Lavaee, P., Ramezani, M., Abnous, K., Reversible Targeting and controlled release delivery of daunorubicin to cancer cells by aptamer-wrapped carbon nanotubes. *Eur. J. Pharm. Biopharm.*, 77, 200–206, 2011.
121. Cortazar, P., Zhang, L., Untch, M., Mehta, K., Costantino, J.P., von Minckwitz, G., Pathological complete response and long-term clinical benefit in breast cancer: The CTNeoBC pooled analysis. *Lancet*, 384, 164–172, 2014.
122. Wen, S., Liu, H., Cai, H., Shen, M., Shi, X., Targeted and pH-responsive delivery of doxorubicin to cancer cells using multifunctional dendrimer-modified multi-walled carbon nanotubes. *Adv. Healthc. Mater.*, 2, 1267–1276, 2013.
123. Guo, C., Al-Jamal, W.T., Toma, F.M., Bianco, A., Prato, M., Al-Jamal, K.T., Kostarelos, K., Design of cationic multiwalled carbon nanotubes as efficient siRNA vectors for lung cancer xenograft eradication. *Bioconjug. Chem.*, 26, 1370–1379, 2015.
124. Wang, J.T.-W., Fabbro, C., Venturelli, E., Al-Jamal, K.T., The relationship between the diameter of chemically-functionalized multi-walled carbon nanotubes and their organ biodistribution profiles *in vivo*. *Biomaterials*, 35, 9517–9528, 2014.
125. Chou, C.-C., Hsiao, H.-Y., Hong, Q.-S., Chen, C.-H., Peng, Y.-W., Chen, H.-W., Yang, P.-C., Single-walled carbon nanotubes can induce pulmonary injury in mouse model. *Nano Lett.*, 8, 437–445, 2008.
126. Huang, Y., Liang, J., Chen, Y., The application of graphene based materials for actuators. *J. Mater. Chem.*, 22, 3671–3679, 2012.
127. Cheng, J., Meziani, M.J., Sun, Y.-P., Cheng, S.H., Poly(ethylene glycol)-conjugated multi-walled carbon nanotubes as an efficient drug carrier for

overcoming multidrug resistance. *Toxicol. Appl. Pharmacol.*, 250, 184–193, 2011.
128. Yang, Z., Zhang, Y., Yang, Y., Sun, L., Han, D., Li, H., Wang, C., Pharmacological and toxicological target organelles and safe use of single-walled carbon nanotubes as drug carriers in treating Alzheimer disease. *Nanomed. Nanotechnol. Biol. Med.*, 6, 427–441, 2010.
129. Khazaei, A., Organic functionalization of single-walled carbon nanotubes (SWCNTs) with some chemotherapeutic agents as a potential method for drug delivery. *Int. J. Nanomed.*, 5, 639, 2010.
130. Bardi, G., Nunes, A., Gherardini, L., Bates, K., Al-Jamal, K.T., Gaillard, C., Prato, M., Bianco, A., Pizzorusso, T., Kostarelos, K., Functionalized carbon nanotubes in the brain: Cellular internalization and neuroinflammatory responses. *PLoS One*, 8, 80964, 2013.
131. Costa, P.M., Bourgognon, M., Wang, J.T.-W., Al-Jamal, K.T., Functionalised carbon nanotubes: From intracellular uptake and cell-related toxicity to systemic brain delivery. *J. Control. Release*, 241, 200–219, 2016.
132. R., H., M., J., Haridoss, P., Sharma, C.P., Novel nano-cocoon like structures of polyethylene glycol–multiwalled carbon nanotubes for biomedical applications. *Nano-Struct. Nano-Objects*, 13, 30–35, 2018.
133. Akhtari, J., Faridnia, R., Kalani, H., Bastani, R., Fakhar, M., Rezvan, H., Beydokhti, A.K., Potent *in vitro* antileishmanial activity of a nanoformulation of cisplatin with carbon nanotubes against Leishmania major. *J. Glob. Antimicrob. Resist.*, 16, 11–16, 2019.
134. Schwengber, A., Prado, H.J., Zilli, D.A., Bonelli, P.R., Cukierman, A.L., Carbon nanotubes buckypapers for potential transdermal drug delivery. *Mater. Sci. Eng. C*, 57, 7–13, 2015.
135. Pastorin, G., Wu, W., Wieckowski, S., Briand, J.-P., Kostarelos, K., Prato, M., Bianco, A., Double functionalisation of carbon nanotubes for multimodal drug delivery. *Chem. Commun.*, 11, 1182, 2006.
136. Wu, W., Wieckowski, S., Pastorin, G., Benincasa, M., Klumpp, C., Briand, J.-P., Gennaro, R., Prato, M., Bianco, A., Targeted delivery of amphotericin B to cells by using functionalized carbon nanotubes. *Angew. Chem.*, 117, 6516–6520, 2005.
137. Moon, H.K., Lee, S.H., Choi, H.C., *In vivo* near-infrared mediated tumor destruction by photothermal effect of carbon nanotubes. *ACS Nano*, 3, 3707–3713, 2009.
138. Jafari, M., Heidari, D., Ebrahimnejad, P., Synthesizing and characterizing functionalized short multiwall carbon nanotubes with folate, magnetite and polyethylene glycol as multitargeted nanocarrier of anti-cancer drugs. *Iran. J. Pharm. Res.*, 15, 449–456, 2016.
139. Kamath, A., Laha, A., Pandiyan, S., Aswath, S., Vatti, A.K., Dey, P., Atomistic investigations of polymer-doxorubicin-CNT compatibility for targeted cancer treatment: A molecular dynamics study. *J. Mol. Liq.*, 348, 118005, 2022.

140. Mirsalari, H., Maleki, A., Raissi, H., Soltanabadi, A., Investigation of the pristine and functionalized carbon nanotubes as a delivery system for the anticancer drug dacarbazine: Drug encapsulation. *J. Pharm. Sci.*, 110, 2005–2016, 2021.
141. Wei, C., Dong, X., Zhang, Y., Liang, J., Yang, A., Zhu, D., Liu, T., Kong, D., Lv, F., Simultaneous fluorescence imaging monitoring of the programmed release of dual drugs from a hydrogel-carbon nanotube delivery system. *Sens. Actuators B Chem.*, 273, 264–275, 2018.
142. Wang, C., Zhang, L., Jiang, Y., Zhang, M., Liu, L., Ye, D., Zhang, S., Yang, Y., A DFT study on the high-density assembly of doxorubicin drug delivery by single-walled carbon nanotubes. *Phys. E Low-Dimens. Syst. Nanostruct.*, 134, 114892, 2021.
143. Lotfi, M., Morsali, A., Bozorgmehr, M.R., Comprehensive quantum chemical insight into the mechanistic understanding of the surface functionalization of carbon nanotube as a nanocarrier with cladribine anticancer drug. *Appl. Surf. Sci.*, 462, 720–729, 2018.
144. Wahab, R., Khan, F., Hwang, I., Al-Khedhairy, A.A., Cytotoxic, genetic and statistical analytical evaluation of functionalized CNTs with C2C12 cells. *Vacuum*, 152, 348–357, 2018.
145. Chen, K. and Mitra, S., Incorporation of functionalized carbon nanotubes into hydrophobic drug crystals for enhancing aqueous dissolution. *Colloids Surf. B Biointerfaces*, 173, 386–391, 2019.
146. Cao, Y., Huang, H.-Y., Chen, L.-Q., Du, H.-H., Cui, J.-H., Zhang, L.W., Lee, B.-J., Cao, Q.-R., Enhanced lysosomal escape of ph-responsive polyethylenimine–betaine functionalized carbon nanotube for the codelivery of survivin small interfering RNA and doxorubicin. *ACS Appl. Mater. Interfaces*, 11, 9763–9776, 2019.
147. Li, H., Sun, X., Li, Y., Li, B., Liang, C., Wang, H., Preparation and properties of carbon nanotube (Fe)/hydroxyapatite composite as magnetic targeted drug delivery carrier. *Mater. Sci. Eng. C*, 97, 222–229, 2019.
148. Saeednia, L., Yao, L., Cluff, K., Asmatulu, R., Sustained releasing of methotrexate from injectable and thermosensitive chitosan–carbon nanotube hybrid hydrogels effectively controls tumor cell growth. *ACS Omega*, 4, 4040–4048, 2019.
149. Xu, H., Li, L., Fan, G., Chu, X., DFT study of nanotubes as the drug delivery vehicles of Efavirenz. *Comput. Theor. Chem.*, 1131, 57–68, 2018.
150. Kamel, M., Raissi, H., Morsali, A., Shahabi, M., Assessment of the adsorption mechanism of Flutamide anticancer drug on the functionalized single-walled carbon nanotube surface as a drug delivery vehicle: An alternative theoretical approach based on DFT and MD. *Appl. Surf. Sci.*, 434, 492–503, 2018.
151. Zhang, P., Yi, W., Hou, J., Yoo, S., Jin, W., Yang, Q., A carbon nanotube-gemcitabine-lentinan three-component composite for chemo-photothermal synergistic therapy of cancer. *Int. J. Nanomed.*, 13, 3069–3080, 2018.

152. Biagiotti, G., Ligi, M.C., Fedeli, S., Pranzini, E., Gamberi, T., Cicchi, S., Paoli, P., Metformin salts with oxidized multiwalled carbon nanotubes: *In vitro* biological activity and inhibition of CNT internalization. *J. Drug Deliv. Sci. Technol.*, 47, 254–258, 2018.
153. Seo, Y., Hwang, J., Lee, E., Kim, Y.J., Lee, K., Park, C., Choi, Y., Jeon, H., Choi, J., Engineering copper nanoparticles synthesized on the surface of carbon nanotubes for anti-microbial and anti-biofilm applications. *Nanoscale*, 10, 15529–15544, 2018.
154. Kordzadeh, A., Amjad-Iranagh, S., Zarif, M., Modarress, H., Adsorption and encapsulation of the drug doxorubicin on covalent functionalized carbon nanotubes: A scrutinized study by using molecular dynamics simulation and quantum mechanics calculation. *J. Mol. Graph. Model.*, 88, 11–22, 2019.
155. Mallakpour, S. and Khodadadzadeh, L., Ultrasonic-assisted fabrication of starch/MWCNT-glucose nanocomposites for drug delivery. *Ultrason. Sonochem.*, 40, 402–409, 2018.
156. Kumari, M., Liu, C.-H., Wu, W.-C., Wang, C.-C., Gene delivery using layer-by-layer functionalized multi-walled carbon nanotubes: Design, characterization, cell line evaluation. *J. Mater. Sci.*, 56, 7022–7033, 2021.
157. Mazzaglia, A., Scala, A., Sortino, G., Zagami, R., Zhu, Y., Sciortino, M.T., Pennisi, R., Pizzo, M.M., Neri, G., Grassi, G., Piperno, A., Intracellular trafficking and therapeutic outcome of multiwalled carbon nanotubes modified with cyclodextrins and polyethylenimine. *Colloids Surf. B Biointerfaces*, 163, 55–63, 2018.
158. Pereira, I., Saleh, M., Nunes, C., Reis, S., Veiga, F., Paiva-Santos, A.C., Preclinical developments of natural-occurring halloysite clay nanotubes in cancer therapeutics. *Adv. Colloid Interface Sci.*, 291, 102406, 2021.

15

Role of Functionalized Carbon Nanotubes in Antimicrobial Activity: A Review

Monika Aggarwal[1], Samina Husain[2*] and Basant Kumar[1]

[1]Department of Physics, Maharaja Agrasen Institute of Technology (affiliated to GGSIPU), New Delhi, India
[2]Centre for Nanoscience and Nanotechnology, Jamia Millia Islamia, New Delhi, India

Abstract

CNTs one of the novel carbon materials with its unique structure has certainly drawn attention of the researchers across the globe. Currently, much of the research enthusiasm has been shown for drug delivery, gene delivery, vaccine delivery and immense biomedical applications are due to the high surface area of CNTs that is capable of absorbing or conjugating with several therapeutic molecules. However, toxicity of CNTs in both human and animal cells, their hydrophobic nature, biocompatibility, solubility and dispersibility of CNTs in both organic and aqueous vehicles is a major concern. The lateral parameters especially cause agglomeration or clustering thereby altering the properties of CNTs and hence so forth this can be overcome with the use of functionalized Carbon Nanotubes (f-CNTs). An insight to CNT surface chemistry modification, as the surfaces coated with CNTs depicts antimicrobial activities, would be taken up in the chapter. The chapter will focus on introduction and overview of f-CNTs in which specific conjugation of bioactive molecules takes place with a variety of functional or chemical groups thereby enhancing the specific targeting. The chapter will focus on the possible mechanisms of both functionalized SWNTs and MWNTs antimicrobial activities so that they can be explicitly used in various biomedical applications.

***Keywords*:** Antimicrobial activity, functionalized carbon nanotubes, single-wall CNTs, multi-wall CNTs

**Corresponding author*: shusain3@jmi.ac.in

Jeenat Aslam, Chaudhery Mustansar Hussain and Ruby Aslam (eds.) Functionalized Carbon Nanotubes for Biomedical Applications, (377–412) © 2023 Scrivener Publishing LLC

15.1 Introduction

Life exists in the universe only because the carbon atom possesses certain exceptional properties (James Jeans)

Carbon, derived from the Latin word "carbo", has been acknowledged for a variety of applications and is the key ingredient for 95 percent of known compounds. It is the basic building block with a remarkable and unique property of showing strong affinity for bonding with itself and with different atoms.

Materials research and technology have paved the way for exploring the relation between the structure and properties of a variety of diverse well-known materials as well as processes for altering the structure and properties by taking care of the total control over the parameters. As a consequence, we have come across many unique and fascinating results not only for carbon but for its allotropes too. The bulk carbon materials like diamond and graphite have been known for decades but introduction of quasi-low-dimensional allotropes of carbon like fullerenes, carbon nanotubes and the two-dimensional graphene [1] completely gathered attention worldwide and set new area for explicit interdisciplinary research in physics, chemistry, material sciences, biomedical sciences etc. The unparalleled properties of these dynamic materials over conventional materials have really urged the desire for more and more research from researchers and scientists across the globe.

15.2 Introduction to CNTs

Carbon nanotubes (CNTs) are one of the greatest discoveries by Ijima in 1991 [2]. An allotrope of graphite, it is highly stable and a promising material for a variety of technological applications and so a major reason for extensive research in this area. As compared to the parent graphite material, CNTs exhibit a range of many unexpected phenomenon thereby increasing the expectation of the researchers to further explore the materials to the fullest. Their outstanding properties *viz.* small size, imperfection in their structures, metallic or semi-conducting property, their high thermal conductivity of about 1900 $Wm^{-1}K^{-1}$ when compared to diamond is twice [3, 4], thermal stability is up to 2800°C and electrical conductivity is about 10^3 S cm^{-1} with 1000 times higher electric current capacity, high elastic modulus (>1 TPa), high mechanical strength than other materials (around 10–100 times greater than the strongest steel) [5] have captured interest worldwide.

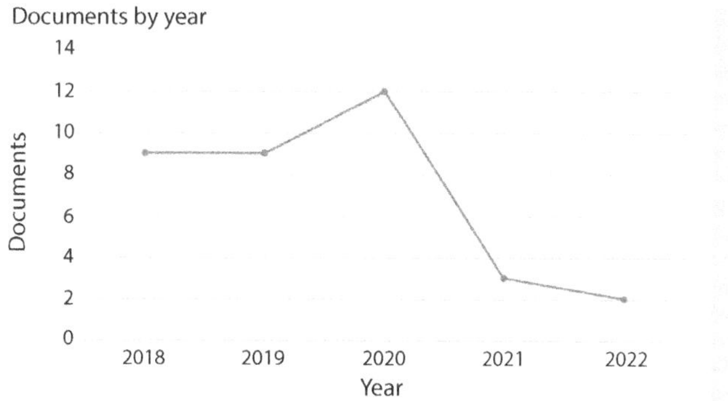

Figure 15.1 Last 5-year development in the field of CNTs/f-CNTs anti-microbial activity.

Figure (15.1) shows the last five-year development in the field of CNTs/f-CNTs-anti-microbial activity. The graph depicts the research publications on CNTs/f-CNTs for antimicrobial activity documented during last five years (2018–2022).

15.2.1 Classification of CNTs

Nanotubes can be broadly classified as:

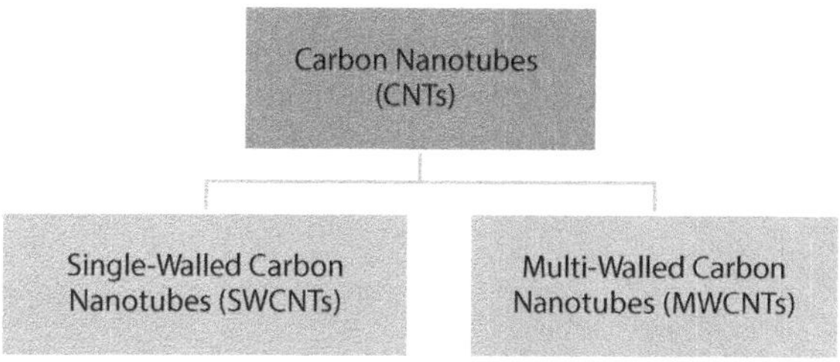

a) *Single-wall carbon nanotubes (SWCNTs)*
Its structure is like a seamless cylinder formed by wrapping up of one-atom thick layer of graphite which is called graphene (Figure 15.2(a)). With tube length being thousands of times larger and diameters ranging from 0.4 to 2–3 nm, SWNTs can be up to the order of cms. They are arranged hexagonally in bundles (Figure 15.2(b)) to form crystal-like structure [6] which exhibits unique electrical properties not shared by MWNT variants. These are less expensive to be produced and are good conductors, thus forming the basic part of electronics.

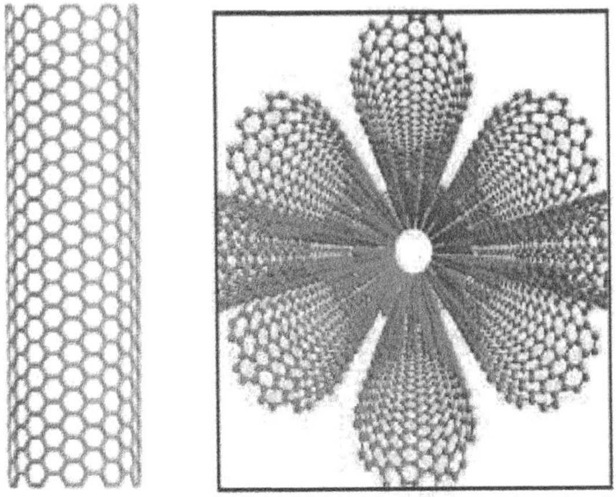

(a) A SWCNT **(b) Bundles of SWCNTs**

Figure 15.2 (a) A SWCNT (b) Bundles of SWCNTs. Reproduced with permission from [17].

b) *Multi-walled carbon nanotubes (MWCNTs)*
Multi-walled nanotubes (MWCNTs) are concentric layer of graphitic sheets rolled into a tube [7]. These are rolled around one common axis with a hollow centre (Figure 15.3(a, b)) with the interlayer spacing of ~0.34 nm close to that of graphite. The inner and outer diameter can vary from 0.4 nm up to a few nm and from 2 nm up to 20–30 nm respectively according to the number of layers. Due to the

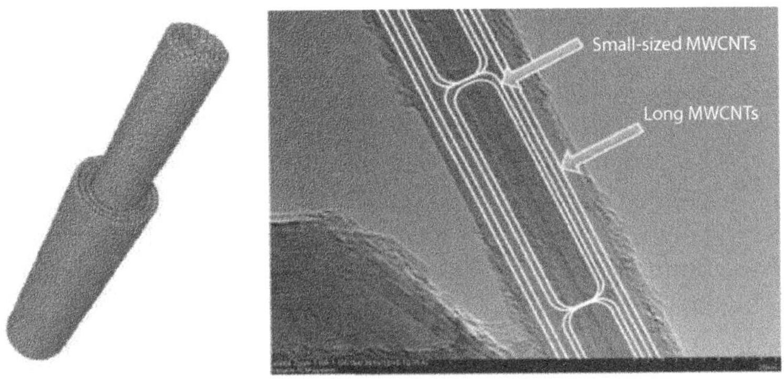

(a) MWCNT structure **(b) MWCNTs, as observed by HRTEM**

Figure 15.3 (a) MWCNT structure. (b) MWCNTs, as observed by HRTEM. Reproduced with permission from [2].

presence of pentagonal defects, nanotubes exhibit closed tips with their axial size ranging from 1 μm to a few cm. The structure of MWCNTs can be interpreted in terms of two models: the Russian Doll model wherein there are concentric sheets of graphite cylinders and the Parchment model wherein a single layer of graphite is rolled around itself to form a tube [8–10].

15.2.2 Structure of CNTs

Structural classification of these novel materials is an interesting challenge. Being crystalline in nature, many fundamentals and theories cannot be applied to these tubular structures using the conventional crystallography for three-dimensional solids.

Carbon nanotube (CNT) is tube of a few nanometers in diameter which is many microns long and made up of rolling of a sheet of hexagonal honeycomb carbon lattice (Figure 15.4). Few of its unique features include:

- Infinite length
- Presence of all the atoms on the surface
- Resemblance to graphene (single graphitic layer)

Though the binding is same in CNTs and graphite but CNTs are made up of rolling up of one or more sheets of graphene into a tubular structure. From the perpendicular view of the graphitic layers, we can observe honeycomb pattern of graphite. Turning back these patterns on top of themselves, edges can be joined keeping one end closed while the other to form a tube of graphite namely called as a nanotube (Figure 15.5).

CNTs popularly known as "wondrous material" possess unique, remarkable and extraordinary properties. The world-wide large-scale production of CNTs is due to the simplest chemical configuration with atomic covalent bonding and its ever-ending demand has fascinated us in all respect. CNTs can be synthesized on large scale with a variety of techniques. The oldest

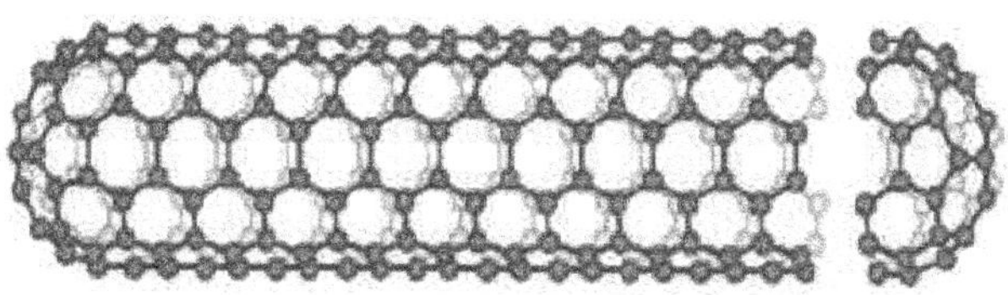

Figure 15.4 Structure of a CNT. Reproduced with permission from [17].

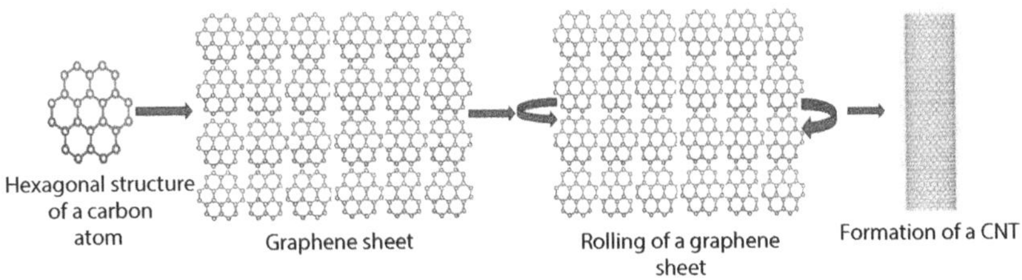

Figure 15.5 Schematic of formation of a CNT.

method known so far is the Arc-Discharge that was employed initially for the synthesis of carbon filaments and fibers. This method was used in the early 60s. R. Bacon used this technique for the synthesis of carbon fibers called whiskers, in nineties by Krätschmer and Huffman for the large-scale production of fullerenes, and later after improvisation, the technique was useful for the synthesis of MWCNTs and SWCNTs. Other techniques like Laser Ablation or Chemical Vapour Deposition (CVD) also marked their importance and relevance in producing CNTs. Economically Chemical Vapor Deposition (CVD) has proved to be one of the best methods because of its unique attributes: simple to operate, controlled process, energy efficient, simple raw materials used, high scalability, high yield and purity. Low temperature (<550°C) [11–13] growth of CNTs is still a challenge. The growth using CVD method depends largely on their diameter, alignment, purity, density, and orientation of CNTs along with factors like growth temperature and pressure. This is in contrast to the high temperature preparation techniques in case of Laser Ablation or Arc-Discharge. By controlling various parameters in the synthesis process of CVD, various studies were carried out for improvising the quality and the quantity of the as grown CNTs. Lot of efforts were also carried out to explore other new methods for CNT production besides Arc-Discharge, Laser Ablation and Chemical Vapor Deposition (CVD) but they were to no success. Still the greatest challenge is to produce pure CNTs in good quantity. It takes a whole lot of efforts and specialized equipment for synthesizing all the tubes with same width and length. The control over these two parameters is lost while considering the high-volume synthesis methods.

15.3 Overview on CNTs Functionalization

The uniqueness and remarkable properties of CNTs have drawn attention of the researchers across the globe for further utilization of its applications in various multidisciplinary fields. One of the ultimate examples of studying nanotubes lies in understanding its association and interaction with proteins, peptides, nucleic acids, polysaccharides etc. which helps in understanding the biological systems to a greater extent. The biocompatibility of CNTs can be understood only when we can overcome the problem of dispersion and dissolvability of the CNTs in water or any organic media. The hydrophobic nature and biocompatibility [14] of CNTs thus demands the need in modifying the surface properties of CNTs. One can easily overcome this problem by the use of proper functionalization wherein a CNT surface can be attached with a physical or a chemical attachment of functionalities. Doing so will not only improve the dispersibility and solubility of CNTs but also improves the interaction of CNTs with various solvents, polymers, nanoparticles etc. Functionalization in CNTs involves breaking up of C=C bonds through oxidation giving rise to carboxyl groups (sites for addition reactions) [15, 16]. The properties of functionalized CNTs (f-CNTs) are distinct and very much improved from the pristine CNTs and can be utilized accordingly. Its outstanding electrical, thermal, mechanical, magnetic, chemical, optical properties are ideal for their extraordinary behavior in various biomedical applications. Thus, a lot of research interest has been developed particularly in understanding its importance in various applications like drug delivery etc. Besides this their use is proved remarkable in a large number of biomedical tools like imaging agents, sensors etc.

However, improper functionalization of CNTs largely depends on the inertness of carbon atoms. The chemical reactivity is different for the end tips and sidewalls of the CNTs (Figure 15.6), due to the presence of pentagons and hexagons respectively. The end tips showing more reactivity and because of the regular graphene framework, functionalization at the sidewalls seems difficult. Conversion of sp^2 hybridization into sp^3 hybridized carbon atoms takes place as a result of addition reactions in carbon–carbon double bonds of hexagons leading to change in the geometry from a trigonal-planar structure to a tetrahedral geometry which is most favorable at the end tips of CNTs. An important factor that determines the physical or the chemical reactivity is the curvature of CNTs. In comparison to the planar graphene sheet, the end tips of CNTs are more reactive due to nonzero curvature of CNTs. As the curvature changes in CNTs, diameter reduces and as a result there is an increase in the binding energy of functional groups attached on the sidewall surface of CNTs. This nature has been known so far due to the theoretical studies carried out on the bond between the alkyl radicals and the sidewall of SWCNTs. In contrast, very less reactivity is shown by nanotubes towards addition reactions because of

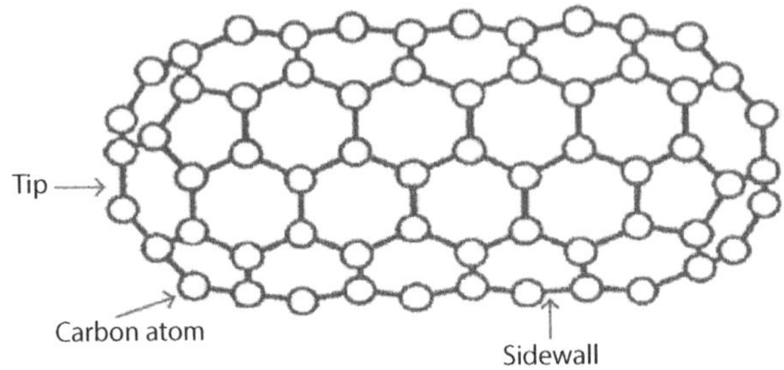

Figure 15.6 Carbon nanotube with closed ends. Reproduced with permission from [17].

the hollow curvature of the inner surface of a nanotube. A range of nanohybrids can be produced with special properties as CNTs can be considered as nanocontainers for the incoming guest molecules.

15.3.1 Types of Functionalization

There are two methods of functionalization: noncovalent bonding (physio-adsorption) and covalent (chemical bond formation).

a) *Non-covalent functionalization* (Physical modification) involves interaction of the CNT surface without altering the geometry and crystallinity of the graphene sheets [18]. Here a non-covalent interaction takes place between the small and large polymeric agents and the surface of the CNTs. The hydrophobic forces and π-π stacking interactions between the chains of the adsorbed molecules are responsible for the adsorption or wrapping of polynuclear aromatic compounds, surfactants, polymers, or biomolecules (Figure 15.7(a), (b)) on the surface of CNTs [19]. These forces help further in loading drugs into or onto CNTs [20]. On chemical treatment, a charge is developed on the nanotube surface which enables the adsorption of molecules on the surface via ionic interactions. Without disrupting the electronic structure and structure of CNTs, the surface of CNTs can be attached with chemical groups and so is one of the advantages of using non-covalent functionalization over covalent.
b) *Covalent functionalization* (Chemical modification) involves interaction by disturbing the system of graphene sheets. The disturbance is mainly on the basis of oxidative treatments

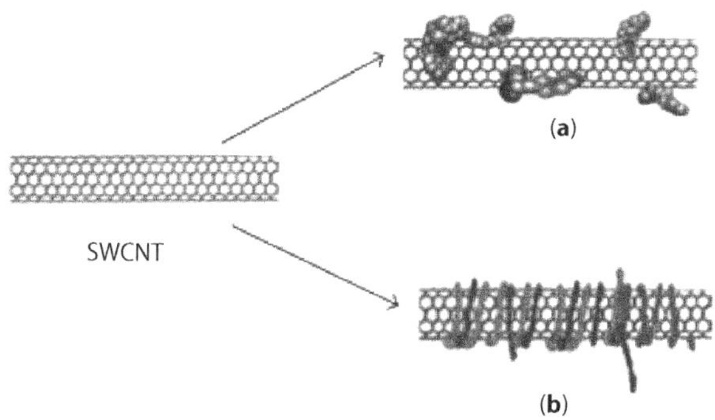

Figure 15.7 Non-covalent functionalization of CNTs with (a) surfactants such as protein adsorption and (b) polymers such as DNA wrapping. Reproduced with permission from [17].

or process like halogenation, addition etc. The functional groups can attach themselves to either the sidewall or to the tube ends. Usually, oxidation of CNTs (Figure 15.8(a)) gives rise to open tubes and can generate -COOH or -OH groups at the sidewall and the tube endings. Doing so enhances the amidation (Figure 15.8(b)), esterification, bromination and Dielsalder reaction essentially required for functionalization [21]. Haddon group have reported functionalization of CNTs using the carboxyl group for the first time with octadecylamine. At 70°C, the functionalization was carried after acylation with thionyl chloride in dimethylformamide for 24 hours [22]. The carboxylic acid groups at the open ends are generated at the defect point. The electronic properties get modify with presence of high surface defects and longer CNTs become short tubes, as drug carriers may need CNTs with different electronic properties [19]. Oxidation can take place via different acids and their combinations *viz.* HNO_3 + HCl, H_2SO_4 + HNO_3 + HCl, H_2SO_4 + HNO_3, etc. [23]. The -COOH or -OH groups can be used as chemical anchors for further derivatization. In spite of having similar structure of CNTs with graphene sheet, curvature of CNTs imparts addition reactions [24, 25]. As the curvature of sidewall increases exohedral chemical reactivity also increases. This is due to the non-alignment of π-orbitals and pyramidization of the sp^2-hybridized carbon atoms [26, 27].

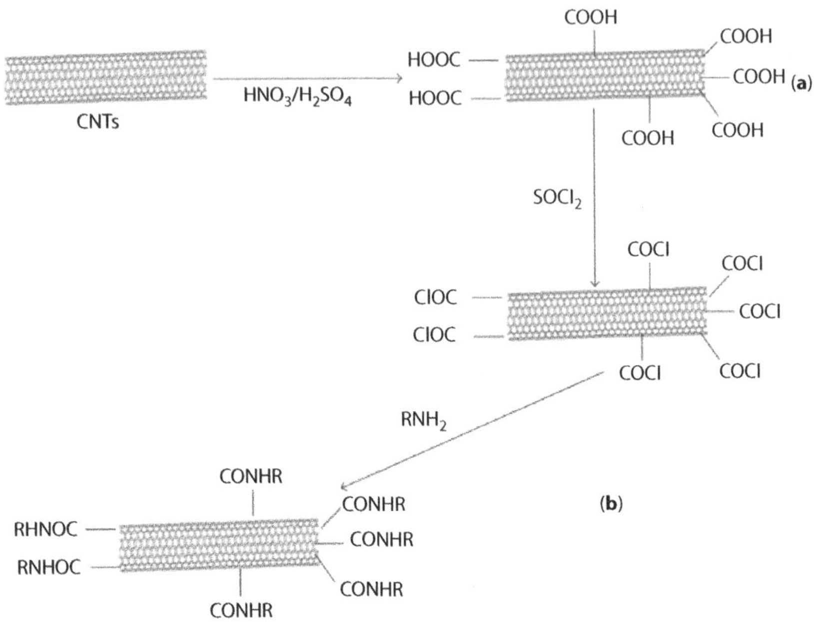

Figure 15.8 Covalent functionalization of CNTs by (a) oxidation reaction by strong acid and (b) further attaching hydrophilic molecules by amidation reactions. Reproduced with permission from [17].

Various defects can also arise on the CNT surface because of the interaction of the oxidation treatments with the chemically inert graphitic walls. High instability occurs due to the disruption of the hexagonal structure which leads to formation of reactive regions on the surface [28]. Organic solvents like chloroform, dichloromethane, aromatic solvents and CS_2 show dispersibility of these f-CNTs. Without disturbing mechanical and electrical properties [29] of CNTs, the covalent interactions thus cause an increase in the dispersibility of CNTs. This can further be enhanced significantly using polymers like sodium dodecyl sulfate (SDS), benzalkonium chloride etc. [21, 22].

Comparison of covalently functionalized (chemical modified) and non-covalently f-CNTs (physically modified) illustrates the higher loading of the former (52.52%) as compared to latter (38%). Studies also reveal that functionalization percentage of SWCNTs is more as compared to MWCNTs with the similar process due to the MWCNTs consisting of multi wall and larger outer diameter that leads to inaccessibility of their sidewalls; a comparative study on functionalizing SWCNTs and MWCNTs previously done so far so forth [30].

15.4 Anti-Microbial Activity of f-CNTs: Interaction and Action

f-CNTs enhance the surface activity of a CNT structure which may damage the cell wall and membrane of a particular microorganism (Figure 15.9). This is because when the surface of the CNT is attached with functional groups they become more mechanically and chemically robust. The length of the CNT nanotubes as well as their diameter are the prime factors which initially interact with the cell membrane leading to destruction of the cell wall and membrane. It is generally seen that shorter CNT nanotubes disrupts the cell wall of the microorganism. This is because shorter tubes tends to integrate themselves and longer tubes exhibits more bactericidal effects as they are able to encounter with more number of microbial cells. Also, diameter of tubes plays a vital in interaction with the cells. Smaller diameter CNT tubes are more likely to interact in large number with the cell surface while sidewall interaction of larger diameter tubes with cell membrane is mainly observed. Since CNTs are exceptional electrical conductors and are the best field emitters, the transfer of electrons from outer microbial surface of the cell wall can lead to enhanced cellular stress which ruptures cell wall eventually leading to the biological death of the microbial cells.

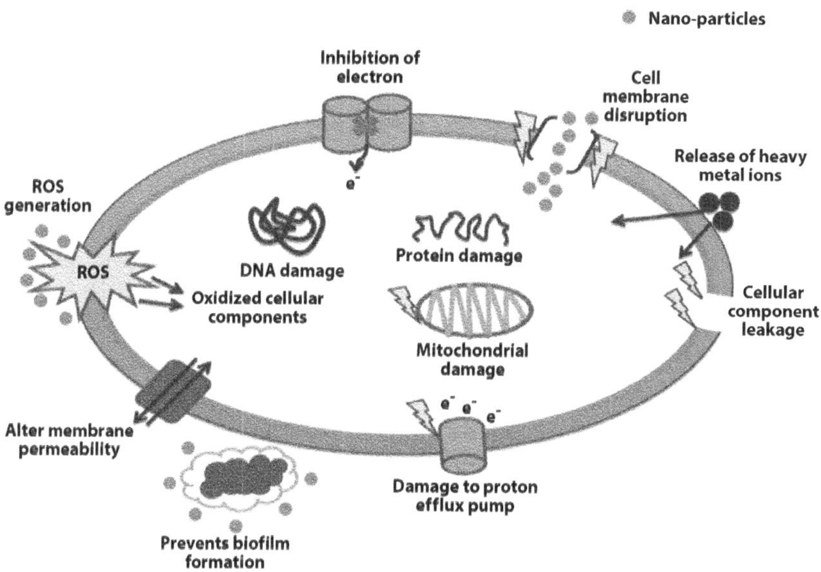

Figure 15.9 Schematic of anti-microbial activity of f-CNTs. Reproduced with permission from [30].

Also, some reactive oxygen species generated on the interaction of CNTs and microbial cells promotes disruption in their existing DNA structure.

15.5 Antifungal Activity of f-CNTs

Many authors have contributed in understanding the antifungal activity of f-CNTs. Zardini *et al.* [31] reported attaching functional groups like lysine and arginine for covalent functionalization of MWCNTs in a microwave assisted radiation. Using spectroscopic techniques like Fourier transform infrared and Raman spectroscopy, the chemical groups attached on the surface were analyzed for proving the phenomena of functionalization. Based on minimal inhibitory concentration (MIC) and radial diffusion assay various fungal species were tested for both the antifungal activities of functionalized and pristine MWCNTs. An outstanding and effectual antifungal activity of pristine MWCNTs, MWCNTs-lysine, and MWCNTs-arginine was observed at a concentration of 5 mg/ml using the radial diffusion assay (RDA) technique. The presence of lysine and arginine causes positive charge on MWCNTs thereby enhancing the interaction between MWCNTs and fungal membrane. Because of increase in the positive charge on arginine than lysine, the antifungal activity in case of MWCNTs-arginine is slightly more than that for MWCNTs-lysine. The studies on various fungi, including *Saccharomyces cerevisiae*, *Aspergillus niger*, *Candida albicans*, *Aspergillus fumigatus*, *Fusarium culmorum*, *Trichophyton rubrum*, *Microsporum canis*, *Trichophyton mentagrophytes*, *Penicillium chrysogenum*, and *Penicillium lilacinum* revealed the MIC results of the antifungal activity of MWCNTs-lysine to be increased manyfold in comparison to pristine MWCNTs. Similar results could be observed for MWCNTs-arginine. Both lysine and arginine helped in improving the antifungal properties of MWCNTs. In another study, Klumpp *et al.* [32] have used CNT functionalized with amphotericin B (AmB) which is considered as one of the best remedial antibiotic molecules handling large severe and chronic fungal infections. The use of such combination reduces toxicity which in turn is more hazardous if the drug amphotericin B is used alone. From the results it could be analyzed that using free AmB more than 40% of cells died by its very effect, but the antifungal activity increased when AmB is conjugated to the CNTs. Bianco *et al.* [33] have used on the other hand a variety of molecules in functionalization of MWCNTs. For functionalization, amphotericin B and fluorescein were used. The antibiotic when administered alone showed toxic effects and when conjugated with CNTs the antibiotic could be easily incorporated into the mammalian cells. Various pathogens including *Candida albicans*, *Cryptococcus neoformans* and *Candida parapsilosis* showed the antifungal activity for the combination AmB-CNT. The exploration of CNT conjugation with different molecules (f-CNTs) has extended our horizon in understanding the potential therapeutic applications. Various references and clinical fungal strains show [34] the remarkable antifungal activity of conjugates of f-CNTs-AMB in comparison to that of AMB sodium deoxycholate (AMBD) or AMB alone. Minimum inhibition concentration values displayed by AMB and AMBD are either comparable or less than that for f-CNT-AMB conjugates. Also, susceptibility towards f-CNT-AMB could be observed for AMBD-resistant *Candida* strains.

Since, 80% of the plant diseases are caused by a variety of pathogenic fungi. Wang *et al.* [35] have reported discussion on the growth of fungi and their propagation with respect to the chemistry of the surface of CNTs and their lengths. By substituting various functional groups (–OH, –COOH and $-NH_2$) on the surface of MWNTs, a study was carried out on the antifungal activity against Fusarium head blight or usually called as scab in wheat by the fungal plant pathogen *F. Graminearum*. The surface modified MWCNTs when compared to pristine MWCNTs, showed a remarkable increase in inhibition in spore elongation and germination. After the treatment with 500 $\mu g \cdot mL^{-1}$ MWCNTs-COOH, MWCNTs-OH, and MWCNTs-NH_2 separately, there was a tremendous reduction in spore length which was almost to half i.e., from 54.5 μm to 28.3, 27.4, and 29.5 μm respectively. With such kind of surface modification, the spore germination is restricted such that the germination rate of pristine MWCNTs becomes almost three times higher than that of conjugated or surface modified MWCNTs. This determines the superiority of surface-modified CNTs by inhibiting the germination rate thereby breaking the infection cycle.

In organic synthesis, by adopting the orthogonal approach one can have a better control on attaching active molecules either at the tips or the sidewalls of the CNTs. Wu *et al.* [36] reported this approach of orthogonal functionalization by using two different types of active molecules. The adoption of such type of method enables the linking of fluorescent probes to the CNTs which then can track the absorption or assimilation of the material used along with an antibiotic moiety acting as an active molecule. Fluorescein and amphotericin B (AmB), respectively are chosen for the same. Various groups have reported the use of AmB as an efficient antibiotic in treating chronic and severe fungal infections [37–39]. However, AmB when used alone is highly toxic to the mammalian cells [37] due to its very less solubility in water thereby forming aggregates [38]. Conjugation of such type of drugs with

CNTs leads to increase in the solubility of AmB in water, decreases aggregation, improves efficiency due to the internalization capacity of the CNTs and can modulate the antimicrobial activity (mammalian, bacterial, and fungal). Also, high doses of AmB is needed when used alone but once it is conjugated with the CNTs less amount would be needed. The extent of functionalization and dimension control can be readily achieved. The present work revealed the antifungal activity of CNT–AmB conjugates which can be administered into mammalian cells without showing any toxic effects. The collection strains of the three pathogenic fungi which might cause infections in humans namely *Candida parapsilosis* and Cryptococcus neoformans and clinical isolates *Candida albicans* showed antifungal activity for CNT-AmB conjugates. The conjugation of AmB took place with the covalent linked ammonium functionalized multi- and single-walled CNTs MWNT-AmB [37] and SWNT-AmB [38].

15.6 Antibacterial Activity of f-CNTs

15.6.1 For SWNTs

In comparison with other carbon nanomaterials SWNTs have proved to have enhanced antimicrobial activity [40]. The first investigation on the antimicrobial effect of both SWCNTs and MWCNTs was reported by Kang *et al.* [41]. They reported that highly purified or pristine SWNTs have a smaller size or diameter in comparison to MWNTs that leads to the rupturing of cell membrane giving a larger surface area [42] resulting in high toxicity against bacteria or in other words deactivation of bacteria. This proves their capability of showing higher antimicrobial behavior. Foo *et al.* [43] have reported on the other hand the functionalized SWNTs by an extract of the herb, *Hempedu bumi* so as to determine the antimicrobial activity against pathogens. In the earlier times, the herb was especially being used in Indian medicine. Based on the development of a zone of inhibition, they analyzed the effects or the biological activities of these conjugated SWNTs with *H. bumi* extract against three pathogens namely *Bacillus* sp. (a Gram-positive pathogen), *E. coli* (a Gram-negative pathogen), and *A. niger* (a pathogenic fungus). For the analysis part various techniques were adopted namely SEM, AFM, XPS, XRD and FTIR which gave the direct evidence of using such kind of conjugates in various biological applications. Chaudhari *et al.* [44] developed a unique approach using covalent attachment of antimicrobial peptide (TP359) to SWNTs-Ag exhibiting an extraordinary antibacterial effect. As compared to the plain TP359 (MIC: 7.8–1.9 μg/ml) and SWCNTs-Ag (MIC: 125–62.5 μg/ml), extraordinary antibacterial activity is shown by silver coated bio-conjugated SWCNTs due to the less concentration of the TP359 (0. 3–0. 08 μg/ml) and SWCNTs-Ag (7.8–1.9 μg/ml) in it so as to get the desired antibacterial properties against all bacterial pathogens. The IC50 values of plain SWCNTs-Ag IC50: ~31–35 μg/ml) was also large in comparison to functionalized SWCNTs-Ag (~4–5 μg/ml). The covalent functionalization of SWCNTs-Ag and TP359 was non-toxic and with high antibacterial activity when as compared to non-covalently functionalized SWCNTs-Ag. Arias *et al.* [45] have reported for the first time how the results are affected using different surface functionalizing groups (-OH, -COOH, and -NH_2) of SWNTS and MWNTs. The antimicrobial activity of Salmonella typhimurium (rod-shaped, Gram-negative), Bacillus subtilis (rod-shaped, Gram-positive spore-forming) and S. aureus (a sphere-shaped, Gram-positive) bacterial pathogen were investigated on various parameters including concentration, treatment time and buffer. The overall effectiveness of any reaction depends on the charge effects of the surrounding molecules. Positively charged CNTs-NH_2, negatively charged CNTs-COOH and neutrally charged CNTs-OH could be observed at the pHs of these buffers (~7.0). The conjugation of different charged surface groups on SWNTs and MWNTs can influence the bacterial-CNT interaction and cell-CNTs aggregation thereby affecting the net antimicrobial activity. Four different buffers *viz.* isotonic 0. 9% NaCl, DI water, 0.1M phosphate buffers saline (PBS) solution with pH 7.0 and a general bacterial growth broth namely brain heart infusion broth (BHI) was used to observe the buffer effect. The results showed that regardless of any shape of the cell, surface groups -OH and -COOH exhibited extremely strong antimicrobial activity on SWCNTs surface and 0.9% NaCl solution but no antimicrobial activity was found in PBS buffer and brain heart infusion broth. The antimicrobial activity of SWNTs with surface groups of -OH and -COOH increased with their concentration and treatment time. The antimicrobial activity of these two functionalized groups started at ~50 μg/mL in DI water or 0.9% NaCl solution and when the concentration rose to 200-250 μg/mL, 107 cfu/mL Salmonella cells got inactivated just within 15 min. At higher concentrations only, the antimicrobial activity of SWNTs-NH_2 took place. Even the concentrations till 500-875 μg/mL for any kind of buffer chosen were not able to justify major antimicrobial activity for MWNTs with surface groups of -OH, -COOH, and -NH_2. Studies on formation of cell-CNTs aggregates using fluorescence and electron scanning microscopes were carried out.

15.6.2 For MWCNTs

A lot of work has been carried out so far in case of functionalized Multiwalled CNTs (f-MWCNTs). Zardini *et al.* [46] have investigated functionalization Of Multiwalled CNTs (MWCNTs) with ethanolamine (EA) groups. Using microwave technique, functionalization of MWCNTs were carried out with mono-, di-, and tri-ethanolamine (MEA, DEA and TEA). Use of Fourier Transform Infrared (FTIR) and Thermogravimetric Analysis (TGA) confirmed the presence of various functional groups on the surface of MWCNTs and use of Transmission Electron Microscopy (TEM) analyzed the morphological variation. The functionalization of MWCNTs with mono-, di-, and tri-ethanolamine groups and pristine MWCNTs were tested for the antimicrobial activities for various pathogenic bacteria. Studies have been conducted for various Gram-negative bacteria (*Salmonella typhimurium*, *Pseudomonas aeroginosa*, *Klebsiella pneumoniae*, and *Escherichia coli*) and various Gram positive bacteria (*Bacillus subtilis*, *Bacillus cereus*, *Streptococcus pneumoniae*, and *Staphylococcus aureus*). Minimal Inhibitory Concentration (MIC) and Radial Diffusion Assay (RDA) results confirmed the least antibacterial activity of pristine MWCNT and the most for MWCNT-TEA followed by MWCNT-DEA and MWCNT-MEA. EA functionalized MWCNTs can prove to play an important role for the antibacterial activity. Antimicrobial activity [47] was investigated on thermoplastic polyurethane (TPU) nanofibers functionalized double-wall and multi-wall CNTs against various pathogenic bacteria (Gram-positive and Gram-negative) and various fungal strains. To obtain the electrospun nanofibers, the surfactant modified CNTs prepared in different concentrations were mixed in N, N-Dimethylformamide (DMF) solvent and 10 % solution of polymer (TPU) using a magnetic stirrer. This prepared solution was then passed through the electrospinning apparatus so as to obtain a highly stable dispersion. By taking various parameters into account, the use of such kind of apparatus for synthesis of f-CNTs/polymer nanofibers gives rise to increased distribution of diameter, decreased nanofiber diameter and much efficient alignment of the electrospun nanofibers. Thermoplastic polyurethane used here is a remarkable polymer with its intriguing properties of being transparent, resistant to abrasion, resistant to microorganism, withstanding very low temperatures, highly sturdy and having a good hydrolytic stability. Fourier transform infrared spectroscopy (FTIR) results confirmed the covalent attachment of the polyurethane polymer to the sidewalls of f-CNTs. The prepared TPU/f-CNTs fibers contain large antimicrobial activity especially TPU/f-MWCNTs nanofibers having more antimicrobial activity as compared to TPU/f-DWCNTs nanofibers. The treatment time and concentrations are important parameters on which the antimicrobial activity and cytotoxicity of TPU/f-CNTs nanofibers largely depends. Limited studies were carried on against *Klebsiella pneumoniae*. The electrospun nanofibers can be used as antimicrobial biomaterials due to its invigorating microbicidal and prolonged microbial growth inhibition properties. Zardini *et al.* [48] have reported functionalization of MWCNTs by arginine and lysine (amino acids) under microwave radiation. Since the cytotoxicity of MWCNTs is less as compared to SWCNTs in spite of knowing this fact that SWCNTs are most effective in showing antimicrobial properties, MWCNTs were being used for study. Functionalization of MWCNTs by arginine and lysine resulted in enhanced antibacterial activity of all the samples under study. Minimum inhibitory concentration (MIC) and radial diffusion assay (RDA) results showed least antibacterial activity of pristine MWCNTs as compared to lysine and arginine functionalized MWCNTs. Arginine functionalized MWCNTs exhibit greater antibacterial activity than lysine functionalized MWCNT. The two showed effective results against gram-negative bacteria (e.g., *Escherichia coli* and *Salmonella typhimurium*) and the resistant strain *Staphylococcus aureus*. The functionalized MWCNTs strongly interacts with the bacterial membranes thereby showing increased antibacterial activity. Ernst *et al.* [49] suggested that arginine and lysine groups of peptides act as electrostatic adsorbents of the negatively-charged membranes of the bacteria since both are positively-charged. With MWCNTs-Arg having more positive charge then MWCNTs-Lys, they show higher adsorption as compared to MWCNTs-Lys making arginine f-CNTs more effective (slightly higher) in showing antibacterial properties and lysine conjugated least. Dispersibility in CNTs is also increased due to amino acid functionalization. Neelgunda *et al.* [50] have reported the synthesis of CdS and Ag_2S QDs deposited functionalized MWCNTs with covalent conjugation of NH_2-terminated PAMAM dendrimer. These QDs (found in high density) were deposited *in situ* on PAMAM grafted MWCNTs. Various techniques like FTIR, powder XRD, elemental analysis, Raman spectroscopy, EDS and TEM were used for confirming the covalent grafting of PAMAM on MWCNTs and its conjugation with CdS and Ag_2S QDs. Such kind of nanohybrids have greatly enhanced the germicidal activity. An increased antibacterial activity was investigated against Escherichia coli, Staphylococcus aureus and *Pseudomonas aeruginosa* as compared with the activity of carbonylated MWCNTs, PAMAM dendrimer, PAMAM grafted MWCNTs, Ag_2S and CdS QDs. Both stability and water dispersibility is shown by f-MWCNTs and f-MWCNTs-QDs nanohybrids. In comparison to Gram-positive bacteria (S. aureus) the f-MWCNTs-QDs nanohybrids showed strong antimicrobial activity against Gram-negative bacteria (E. coli and P. aeruginosa). Gram-negative bacteria have thin membrane which can be easily disrupted but that of Gram-positive bacteria is thick and rigid and so more difficult for the dendrimer to disrupt it. The f-MWNTs have an edge over commercial antibiotics and stands as the best alternative for the antibacterial activities. Neima *et al.* [51] concluded the antibacterial activities of functionalized MWNTs against both gram-negative (E. Coli) and gram-positive bacteria (S. aureus). Firstly the MWNTs were purified and then they were functionalized. To study the efficacy of the f-MWNTs samples based on the processes of purification and oxidation and prepared by both the above treatments transmission electron microscope (TEM) and scanning electron microscope (SEM) equipped with energy dispersive X-ray spectroscopy were used. Surface analysis indicated the decrease in nanotube diameters and lengths after functionalization of MWNTs. Decrease in diameters can substantially cause damage to bacterial cell membrane through the cell-surface interaction while shorter length of the tube increases synergism between the bacteria and open ends of nanotubes. The antibacterial activity of f-MWNTs on E. coli and S. aureus cultures was confirmed by TEM. Due to the carbon particle size in nanoscale the antibacterial activity increases considerably due to the large surface area per unit volume. The inhibitory performance of f-MWNTs increased considerably reaching 85.2 and 57.1% for E. coli and S. aureus, respectively. These can prove to be ideal materials for medical purposes. Amiri *et al.* [52] reported the functionalization of MWNTs with lysine using microwave irradiation. The functionalization process is based on diazonium reaction with MWCNTs. TEM image showed smooth surface of pristine MWNT in comparison to an increase in the roughness for the functionalized MWCNTs. MIC results indicated that in comparison to pristine CNTs, enhanced activity of antimicrobial activity of CNT-Lysine against Gram-negative bacteria *E. coli*, *S. typhimurium*, and *K. pneumoniae* is observed. There is also an increase of 2.1, 1.84 and 1.80 times against Gram-positive bacteria *S. agalactiae*, *S. aureus*, and *S. dysgalactiae* respectively. The analysis of the results clearly signifies the effectiveness of CNT-lysine against gram negative bacteria. When lysine is attached to CNTs more positive charge increases due to these lysine groups and as a result there is more of electrostatic adsorption of bacteria membrane. Permeability increases across the membrane of bacteria, decreases the metabolic rate and kills the bacteria showing enhanced antibacterial activity. The use of Metal doped ZnO nanoparticles functionalized MWCNTs have attracted a widespread attention as antibacterial agent. The present study [53] reported functionalized pristine SWCNTs and MWCNTs with metal-doped ZnO nanoparticles to investigate the antimicrobial activity against Escherichia coli and Staphylococcus aureus using co-precipitation method. With extraordinary properties of ZnO being low cost, eco-friendly and high photocatalytic activity, it is a wide gap (3.37 eV) semi-conductor absorbing light only in the UV region. Using metals like silver (Ag) and gold (Au) which are highly stable and show good thermal/electrical conductivity can substantially increase the photocatalytic properties of ZnO in the visible region thereby enhancing the antimicrobial property. Using UV-vis spectrophotometer and TEM, the optical properties and shape and other morphological properties were studied respectively. Qualitative analysis of composition ensuring the uniform deposition of metal-doped ZnO nanoparticles on the surface of MWCNTs and SWCNTs was performed via energy dispersive spectroscopy (EDS). The results obtained revealed the fact that due to the thinner peptidoglycan layer of E. coli bacteria antimicrobial activity against such type of bacteria was greater than that of S. aureus. All the samples showed high bactericidal rate against two types of bacteria. The antimicrobial efficiency increases significantly by using Ag and Au as nanoparticles thereby enhancing the surface area of CNTs and thus increased interaction between the bacterial cell membrane and material. High photocatalytic activities of the metal-doped ZnO nanoparticles, large surface area of CNTs and high adsorption property has provided potential of such kind of nanohybrids to contribute as antimicrobial agents in biomedical applications. Gilbertson *et al.* [54] have reported the use of standard acid treatments for functionalizing MWNTs with oxygen. Further selective reduction takes place via annealing. For identifying various groups like COOH, C-OH and C=O groups on the surface of MWCNTs, CD-XPS is used. This is the basic elementary technique being used to vary the population of oxygen functional group. It is observed that there is a strong correlation between the population of the specific distribution of the oxygen group and variations in the electrochemical as well as the antimicrobial activities as verified by GSH and ORR oxidation experiment, respectively. For adding more of oxygen functional groups to the surface of CNTs and for complete minimization of the residual metal oxide catalyst strong oxidizing agents, such as nitric and sulfuric acid (HNO_3 and H_2SO_4), are required. The presence of C=O groups therefore greatly influences the electrochemical as well as the antimicrobial activities if compared with the surface COOH groups. So far, many materials have been used to observe the antimicrobial properties, Murugan *et al.* [55] have reported a MWCNTs hybrid by functionalizing generation 2 (G2) amphiphilic poly(propyleneimine) dendrimer (APPI), *viz.* MWCNTs-APPI and depositing silver (Ag) nanoparticles (AgNPs) on MWCNTs-APPI, *viz.* MWCNTs-APPI-AgNPs hybrid. Such kind of nanohybrids show its dispersibility in aqueous and organic solvents and can be best antibacterial agents. These nanohybrids functionalized with different groups *viz.* MWCNTs-COOH, MWCNTs-APPI, and MWCNTs-APPI-AgNPs were tested against *Escherichia coli* (*E. coli*, gram-negative), *Bacillus subtilis* (*B. subtilis*, gram-positive), and *Staphylococcus aureus* (*S. aureus*, gram-positive). Techniques like FTIR, zeta potential, SEM and HRTEM, EDS, thermogravimetric analysis, and Raman spectroscopy were used to quantify the covalent functionalization of APPI in MWNTs and amount of AgNPs deposited in MWCNTs-APPI. Thermal Gravimetric Analysis (TGA) confirmed that 67% amount of APPI was functionalized on MWNTs showing higher dispersibility in aqueous and organic solvents and non-aggregation of MWCNTs took place for at least 6 months showing their stable nature. Antimicrobial properties have really influenced the researchers worldwide for a variety of pathogens as antimicrobial resistance is subject of importance nowadays. Chaudhari *et al.* [44] have explored the synergistic effect of Ag-SWCNTs and antimicrobial peptides (APs) so as to form APs-functionalized SWCNTs-Ag. Initial step involves the carboxylation of silver coated single walled carbon nanotubes (SWCNTs-Ag) with Tri sodium citrate (TSC) followed by covalent functionalization with an antimicrobial peptide to get (FSWCNTs-Ag). Further SWCNTs-Ag were also non-covalently functionalized with antimicrobial peptides. An additional effect in terms of antibacterial activity is observed after functionalizing SWCNTs-Ag and antimicrobial peptides. As compared to the plain antimicrobial peptide and SWCNTs-Ag, much lower concentrations of antimicrobial peptides and SWCNTs-Ag in fSWCNTs-Ag are needed so as to achieve desired antibacterial properties. Hakami *et al.* [56] have reported the use of microwave radiation for functionalizing CNTs with 1-octadecanol groups (C18). This new approach eradicated Escherichia coli (E. coli) from water. Both as -received CNTs and f-CNTs were compared. The length and diameter of the as-received CNTs were characterized using FE-SEM and TEM. Using Fischer esterification, the surface modification/functionalization of CNTs could be realized using 1-octadecanol (C18) functional groups. The morphological/structural analysis was carried out using FTIR, TGA, and DSC. The effect of microwave radiation was observed both on as-received and f-CNTs and their efficacy was tested against destruction of pathogenic bacteria from water. The testing revealed minimal removal (3–5%) of E. coli bacteria for the as-received/unmodified CNTs used alone without using microwave radiation but contrary to this a high removal of the bacteria with microwave radiation. However, the highest antibacterial activity was observed for carbon-18f-CNTs synthesized via microwave radiation. They reported increased absorption rate of the microwave heat due to the presence of multiple chains of C18. This unique method can serve as one of the applications that can be employed as disinfection method. Tambur *et al.* [57] have studied antibacterial activity for Oxygen-fMWCNTs implanted onto silver nanoparticles in their pure phase. The antibacterial activity for Oxygen-functionalized Multi-Walled CNTs (O-MWCNTs) -Ag nanoparticles were tested against B. subtilis and S. aureus (Gram-Positive bacteria) and E. coli and P. aeruginosa (Gram-Negative bacteria). The former showed greater inhibition zone as compared to the latter. High antibacterial activity of Ag-MWCNTs against B. subtilis and S. aureus (Gram-Positive bacteria) was reported in the zone inhibition. Chaudhary *et al.* [58] have reported the use of pristine MWNTs and functionalized MWNTs [both CNT (OH-CNT) and (COOH-CNT)] containing tamarind gum hydrogels. FTIR spectroscopy, impedance analysis, and mechanical tester were used for studying the properties of hydrogels. The antibacterial analysis showed the efficacy of the CNTs with drug tigecycline enhanced the antibacterial activity in comparison to the CNTs injected without drug tigecycline. Shahriarya *et al.* [59] have reported heating the mixture of Graphene Oxide (GO), acid functionalized MWCNTs (f-MWCNTs) and $AgNO_3$ aqueous solution in the presence of sodium hydroxide to form silver reduced graphene oxide multiwalled CNTs (Ag-rGO-MWCNTs). This takes place without using any additional reducing agent and a surface modifier. Under alkaline conditions, GO and f-MWCNTS can be converted into rGO and MWCNTs respectively as they act as reducing agents. Various analysis techniques were carried *viz.* XRD, EDX and Ultraviolet-Visible spectroscopy (UV-Vis), taking Escherichia coli (E. coli) and Bacillus subtilis as bacteria model, the antimicrobial activity of Ag-rGO-MWCNTs was examined. The results were positive only for E. coli. Strong interaction of rGO, MWCNTs and silver (Si) nanoparticles resulted in enhanced antibacterial activity in comparison to Ag-rGO and Ag-MWCNTs. Youssef *et al.* [60] have reported crosslinked f-MWCNTs/PVA nanocomposites with citric acid with various concentrations. MWNTs lack water solubility, good surface chemistry, flexibility and formation of nano-composites themselves thereby restricting their use in several applications. For increasing its scope further, they mixed it with PVA that helps in formation of hydrophobic films and capturing of heavy metals, pesticides, bacteria and fungi from wastewater. Various techniques *viz.* FTIR, SEM and TEM were used to observe structural, morphological, shape and crystallinity respectively. The antimicrobial efficacy of f-MWCNTs/PVA nanocomposites film were analyzed against Pseudomonas aeruginosa (gram negative) and Staphylococcus aureus (gram positive) bacteria and *Aspargillus niger* and *Candida albicans* (fungi). Enhanced microbial activity was observed for high loadings of f-MWCNTs in the nanocomposites that increased the zone of inhibition and increased the antimicrobial activity and can be used as efficient candidate in removal of wastewater pollutants. S. Yallappa *et al.* [61] have studied the functionalization of CuNPs and AgNPs with MWCNTs via *T. arjuna* plant molecules extract using microwave irradiation. An increased biocide activity was visualized for the phytochemically functionalized Cu-MWCNTs and Ag-MWCNTs nanomaterials and the inhibitory activity for fungus was lesser as compared to bacteria. Sivaraj *et al.* [62] have reported the use of spray pyrolysis technique for the conjugation of silver (Si) substituted hydroxyapatite/fMWCNT (Ag-HA/f-MWCNT) with 316 L stainless steel (SS). With the help of well diffusion method antibacterial activity of the Ag substituted HA/f-MWCNT was studied against gram-positive and gram-negative bacteria. Different inhibitory level could be observed by varying the Si ions in Ag substituted HA/MWCNT on the antibacterial growth. By altering Ag wt% in Ag substituted hydroxyapatite/functionalized multiwall carbon nanotube (Ag-HA/f-MWCNT), one can observe change in the inhibitory zone. Sapalidis *et al.* [63] have reported the use of functionalized oxidized multi-walled nanotubes with quaternized hyperbranched polyethyleneimine (QPEI) to form hydrophilic hybrid material - ox-CNTs@QPEI via non-covalent interactions. Using solvent casting technique, nanocomposite films of PVA nanocomposite ox-CNTs@QPEI were prepared that was compatible with hydrophilic PVA. Many techniques were adopted including electron microscopy, XRD, TGA, DSC and UV–vis spectroscopy for studying the properties of the films. The techniques showed excellent dispersion of the f-CNTs in PVA matrix. Alone no antibacterial activity is possessed by QPEI but presence of QPEI helps in increasing the solubility of CNTs and forms positively charged hybrids. More the positive groups being functionalized on the surface of CNTs more will be the ability of killing bacteria as the interaction increases with the bacterial membranes resulting in increased diffusion across the membrane [48] and hence enhanced antibacterial properties. Krishnamoorty *et al.* [64] have investigated single-walled CNTs (SWCNTs), multiwalled CNTs (MWCNTs) and antioxidant Tannic Acid (TA)-functionalized SWCNTs and MWCNTs for the bactericidal cytotoxicity activity against *Escherichia coli*, *Shigella sonnei*, *Klebsiella pneumoniae*, and *Pseudomonas aeruginosa* (Gram-negative) bacteria. In comparison to naked CNTs no aggregation was observed in TA functionalized SWCNTs and MWCNTs. When treated under normal light in laboratory condition both SWCNTs and MWCNTs showed bactericidal activity as compared to TA-fCNTs showing no cytotoxicity to bacterial cells because the latter fails to generate reactive oxygen species (ROS) both in light and dark conditions. Amaku *et al.* [65] have reported the use of functionalized multiwalled CNTs (f-MWCNTs) using a plant extract from *Funtumia elastica* husk which helps in the reduction of $Ag+$ ions to attach themselves on the surface of CNTs. *Annona muricata* petals, *Funtumia elastica* husk and *Acaia xanthophloea* bark stem were used as plant biomasses or modifiers which are mixed with the product of plant waste/Ag/f-MWCNTs hybrid. Characterization of the novel biohybrid-nanocomposite was carried out using FTIR, powder XRD, HRTEM, Raman spectroscopy, FE-SEM, and TGA. The as-formed composites showed homogeneity with high stability against heat and increased surface area. The biohybrid-nanocomposites showed antibacterial activity in water against gram-negative *Escherichia coli* (ATCC 35218) and gram-positive *Staphylococcus aureus* (ATCC 219213). With proper optimization of temperature, the nanocomposites can be used as antimicrobial agents for water disinfection. Swathy *et al.* [66] have investigated the use of binary and ternary silver nanocatalysts for the fast heterogeneous catalytic reduction of p-nitrophenol so as to observe the solvent effects in aqueous media. Binary (BNC) and ternary (TNC) silver nanocatalysts utilized fMWCNTs and polythiophene fMWCNT nanocomposites as hosts for silver nanoparticles, respectively. The role played by each component of BNC and TNC nanocomposites was determined using control experiments. BNC shows non-stable dispersion in water as compared to the ternary nanocatalyst (TNC). They showed that it has more stability in water and stands appropriate antibacterial agent for Escherichia coli bacteria cultured in water-based nutrient. Yadav *et al.* [67] studied effective antimicrobial activity for alkyne-functionalized MWCNTs covalently attached to azide-moiety containing hyperbranched polyurethane using click chemistry reaction. Excellent dispersion property is shown by functionalized MWCNTs in the HBPU matrix due to covalent bonding with polymer chains resulting in high antimicrobial activity against gram-negative bacteria Escherichia coli (E. coli).

15.7 Commercial Application of Antimicrobial Activity of f-CNTs

As compared to bulk materials, nanomaterials have an increased surface area to volume ratio, which makes them more chemically reactive and also potentially more toxic materials. As hygiene in healthcare, sanitation and personal well-being is reaching a critical value, it is posing a major concern in the society. Many harmful microorganism such as bacteria have the ability to grow and multiply manyfold on numerous different types of surfaces and their bacterial strains are able to flourish despite the widespread use of antibiotics. As a result, resistance to antibiotic is increasing dramatically. This condition deteriorates when the surface retain even small amount of fluids which harbors not only a particular microorganism, but also provides enriched growth environment for a wider diversity of microorganisms. The surfaces that are frequently touched by many individuals throughout a day, thus potentially spreading harmful microbes throughout the population can include; doorknobs, soap dispensers, crosswalk buttons, handrails, support rails, phones, keyboards, computer mice, touchscreens, mobile phones, and the like, including many other commonly shared devices the medical device industry where bacterial infection and transfer is of great concern.

Recently, around 2500 patents have been filed/published and granted that assures that adding aCNT layer does not provide an adequate surface for the growth of microbial or bacterial microorganisms. Increasing application of CNTs in biomedical has compelled researchers all over the world to look into the biocompatibility and toxicity of CNTs so that they can be used for commercial applications. In general, it is necessary to have an accurate understanding of the pharmacological and toxicological properties of CNTs before using them in clinical applications. The optimal performance of CNTs for specific biomedical applications should be considered, whereas surface chemistry is designed to minimize toxicity. Parameters such as distribution, aggregation rate, and clearance pathways for CNT-based drug/gene delivery systems should also be considered.

15.8 Overview on Antimicrobial Activity of f-CNTs

Material	Method of functionalized CNTs	Antimicrobial activity
KGM, Gelatin (GL), Gallic acid (GA), Dopamine hydrochloride (DA), CNT, Tris (hydroxymethyl)-aminomethane (Tris), Sodium acetate, Monochloroacetic Acid, Glycerin, absolute ethanol [35]	Ultrasonicated CNTs in ultrapure water were mixed in a Tris buffer at pH 8.5 using magnetic stirring. Later, the infusion solution was stirred at 25°C for 32 h.	*S. aureus* *E. coli*
SWCNTs-Ag, Antimicrobial peptide TP359 of 5 amino acids, Trisodium citrate dihydrate (TSC), 1-Ethyl-3-(3-dimethy-laminopropyl) carbodiimide(EDS), N-Hydroxysuc-cinimide(NHS), sterile nuclease-free water [44]	SWCNTs-Ag carboxylated using Trisodium citrate dihydrate (TSC), SWCNTs-Ag were mixed with TSC at different ratios of TSC to SWCNTs-Ag (1 mg/ml) as 1:20 (1 part of TSC to 20 parts of SWCNTs-Ag), 1:10, 1:7.5 and 1:5. The mixture was subjected to continuous stirring at 37°C for 4 h.	*Staphylococcus aureus* *Streptococcus pyogenes* *Salmonella typhimurium* *E. coli*
CNTs, Thermoplastic polyurethane (TPU), N, N-Dimethyl-formamide (DMF) solvent, sodium dodecylbenzene sulfonate (SDBS) [47]	Purified CNTs dried in the oven at 60°C, treated with the suspension of 0.05 wt% Sodium dodecylbenzene Sulfonate (SDBS) solution. Ultrasonicated for 30 min. to obtain the f-CNTs.	*Pseudomonas aeruginosa* *Klebsiella pneumoniae* *Staphylococcus aureus* *Candida albicans*

(*Continued*)

Material	Method of functionalized CNTs	Antimicrobial activity
RawSCWNTs and MWCNTs, Silver nitrate, Zinc Acetate dehydrate, Oxalic Acid, Gold Chloride trihydrate, Sodium Citrate, Anhydrous Ethanol, H_2SO_4, HNO_3, Agar [53]	Raw SWCNTs treated with an acid mixture of H_2SO_4 and HNO_3, Ultrasonicated at 30°C for 4.0 mins, diluted with distilled water, vacuum-filtered through a cellulose nitrate membrane, and dried at 100°C for 12 h.	*E. coli* *Staphylococcus aureus*
Pristine MWCNTs, H_2SO_4 and HNO_3, Ethanol, Silver nitrate, Nitrogen stream, Deionized water, Nutrient broth, Agar–agar, Cotton swabs [57]	H_2SO_4 and HNO_3 functionalized P-MWCNTs. Stirred invariably stirring at 50°C for 8 h. Equal quantity of deionized water added and filtered after cooling to room temperature. Residue left is filtered and freeze-dried until neutral pH.	*Bacillus subtilis* *Staphylococcus aureus* *E. coli* *Pseudomonas aeruginosa*
Polyvinyl alcohol (PVA), MWCNTs, HCl, H_2SO_4, HNO_3, Citric acid, Thionyl chloride ($SOCl_2$), Tetrahydrofuran (THF) [60]	H_2SO_4 and HNO_3 functionalized MWCNTs sonicated at 60 Hz for 20 minutes. Mixtures is refluxed for 24 h at 60°C. Powdered MWCNT/Acid mixture is filtrated, dried at 110°C for 12h followed by raising the temperature to 350°C for 30 minutes. Dried powder of MWCNTs-COOH added to $SOCl_2$ solution in a new flask, sonicated for 1h and refluxed at 60°C for 24 h for substituting the COOH groups by COCl groups.	*Staphylococcus aureus* *Pseudomonas aeruginosa* *Candida albicans* *Aspargillus niger*

(*Continued*)

Material	Method of functionalized CNTs	Antimicrobial activity
Virgin CNT, OH-CNT and COOH-CNT, Tigecycline USP, Tamarind gum powder, Nutrient agar, Dialysis membrane, Double distilled water [58]	Purchased f-CNTs suspended in water, sonicated for 3 min, Tamarind powder stirred at room temperature 25°C, followed by heating at 70°C in a water bath for 5 mins. Kept half hour outside for gelation.	*S. aureus* *E. coli* *Acinetobacter baumannii*
T. arjuna plant, MWCNTs, $AgNO_3$, $(CuNO_3)_2$, 3-(4, 5-dimethyl-2-thiazolyl)-2, 5-diphenyl-2H-tetrazolium bromide (MTT), fetal calf serum (FCS), modified Eagles Medium (MEM), Glutamine, EDTA, Trypsin, Dimethyl sulfoxide (DMSO), methanol, ciprofloxacin, fluconazole, McFarland solution [61]	MWCNTs pretreated with nitric acid for few hours to remove any metal ions to impart functional groups such as – COOH.	*E. coli* *S. aureus* *S. typhi* *P. aeruginosa* *T. rubrum* *C. albicans* *C. indicum*
$(Ca(NO_3)_2{\cdot}4H_2O))$, $(NH_4)_2HPO_4$, $AgNO_3$, fMWCNTs, Surgical 316L SS, HNO_3, Acetone, Distilled water [62]	Purchased fMWCNTs used.	*S. aureus* *P. stuartii* *E. coli* *K. pneumoniae*

(*Continued*)

Material	Method of functionalized CNTs	Antimicrobial activity
Hyperbranched polyethyleneimine (PEI), partially hydrolyzed atactic PVA Mowiol 8–88, Glycidyl trimethylammonium chloride, Triethylamine, and Dialysis tubes, Oxidized MWCNTs, Ultra-pure water, PVA stock solution in water [63]	QPEI functionalized MWCNTs, mixture ultasonicated for 15 mins., unstirred further for 12 h at room temperature, left for ultracentrifugation followed by lyophilization giving rise to hybrid ox-CNTs@QPEI material.	*E. coli*
Annona muricata petals and *Funtumia elastica* husk, Acacia xanthophloea bark stem, MWCNTs, Glutaraldehyde, HCl, HNO_3 and $AgNO_3$ [65]	Oxidation on the surface of the MWCNTs using H_2SO_4:HNO_3 in a round-bottom flask for 5h at room temperature, followed by the neutralization of the mixture with 1 mol dm^{-3} NaOH, black suspension vacuum filtered, black residue obtained then washed with ethanol and acetone followed by oven-drying the fMWCNTs at 120°C for 12 h.	*E. coli* *Staphylococcus aureus*
Sodium borohydride, p-nitrophenol, p-aminophenol, acetone, glycerol (anhydrous), ethylene glycol, 1, 4-dioxane, Deionized water, a mixture of peptone, NaCl, and yeast extract, beef extract [66]	MWCNTs refluxed with HNO_3 at 100 °C.	*E. coli*

15.9 Future Scope

Although the field of CNTs, emerged as a boon in last 30 years for visualizing various applications but still the need to combat various bacterial, fungal and viral infections through surface modified CNTs stands much promising in enhancing the understanding of plethora of nanomaterials. The fabrication of surface modified CNTs on large scale with less toxicity is still challenging. The future scope of the current work lies on the scalability of f-CNTs to be cost effective, safety issues and tackling the toxicity profile. More of research and experimentation can be look forward for discovering variety of promising materials so as to update our understanding at the technological, biological and chemical front that can increase the scope of expansion of f-CNTs in various disciplines.

15.10 Conclusion

Though CNTs seem to be promising candidates but the need of surface modification or functionalization is essential so as to resolve various issues of toxicity, dispersion, biocompatibility, hydrophilicity etc. The hydrophilic nature of f-CNTs makes them more suitable for drug delivery over pristine or raw CNTs. Due to covalent or a non-covalent bonding, a variety of molecules or biologically active compounds can be conjugated with either SWCNTs or MWCNTs to overcome the barriers discussed above. Functionalization of CNTs can be either through acylation, amidation, oxidation, modified biomolecules or through surfactant and biopolymers. A lot of work has been carried to explain the antifungal activity against fungi like *Saccharomyces cerevisiae*, *Aspergillus niger*, *Candida albicans*, *Aspergillus fumigatus*, *Fusarium culmorum*, *Trichophyton rubrum*, *Microsporum canis*, *Trichophyton mentagrophytes*, *Penicillium chrysogenum* and *Penicillium lilacinum*, *Cryptococcus neoformans*, and *Candida parapsilosis* etc. using f-SWNTs or f-MWNTs. The pristine CNTs show less antifungal activity as compared to f-CNTs. However, a lot of work has being done to explain antibacterial activity as compared to antifungal activity. The antibacterial activity can be explained much due to the cell membrane damage in case of f-CNTs irrespective of toxicity issue in human cells thereby showing enhanced antibacterial activity of f-CNTs over pristine CNTs. Antibacterial activity against various gram-negative bacteria like *Salmonella typhimurium*, *S. aureus*, *Pseudomonas aeruginosa*, *Klebsiella pneumonia*, and *Escherichia coli* and gram-positive bacteria like *Bacillus subtilis*, *Bacillus cereus*, *Streptococcus pneumoniae*, and *Staphylococcus aureus* have been carried out by researchers worldwide. The capability of CNTs by appropriate surface modifications with their unique physico-chemical properties have proved them as excellent materials showing bactericidal and fungicidal activity. But still a more scope of research is there for exploring more of antifungal activity of f-CNTs.

Acknowledgement

The would like to thank all the contributors of the references 2, 17 and 30 mentioned in this chapter to give us valuable data in the form of their figures.

References

1. Ahmad, S., Mustonen, K., McLean, B., Jiang, H., Zhang, Q., Hussain, A., Kauppinen, E.I., Hybrid low-dimensional carbon allotropes formed in gas phase. *Adv. Funct. Mater.*, 30, 2005016, 2020.
2. Roslan, M.S., Abd Rahman, M.M., Jofri, M.H., Chaudary, K.T., Mohamad, A., Ali, J., Fullerene-to-MWCNT structural evolution synthesized by arc discharge plasma. *C*, 4, 58, 2018.
3. Maruyama, B. and Alam., K., Carbon nanotubes and nanofibers in composite materials. *SAMPE J.*, 38, 59, 2002.
4. Collins, P.G. and Avouris., P., Nanotubes for electronics –. *Sci. Am., Nat. Publishing Group, San Francisc.*, 283, 62, 2000.
5. Ahmadi, M., Zabihi, O., Masoomi, M., Naebe, M., Synergistic effect of MWCNTs functionalization on interfacial and mechanical properties of multi-scale UHMWPE fibre reinforced epoxy composites. *Compos. Sci. Technol.*, 134, 1, 2016.
6. Thess, A., Lee, R., Nikolaev, P., Dai, H., Petit, P., Robert, Crystalline ropes of metallic carbon nanotubes. *Science*, 273, 483, 1996.
7. Wei, H., Wang, Y., Guo, J., Qiu, B., Ding, D., Wei, S., Wujcik, E.K., Guo, Z., Synthesis of multifunctional carbon nanostructures, in: *Handbook of Carbon Nano Materials*, pp. 89–105, 2015.
8. Aydin, M. and Akins, D., Geometric and spectroscopic properties of carbon nanotubes and boron nitride nanotubes, in: *Electronic Properties of Carbon Nanotubes [Internet]*, Marulanda, J. M. (ed.), IntechOpen, London, 2011 [cited 2022 Aug 24]. Available from: https://www.intechopen.com/chapters/16845
9. Mehra, N.K., Mishra, V., Jain, N.K., Receptor based targeting of therapeutic. *Ther. Deliv.*, 4, 369, 2013.

10. Jain, N.K., Mishra, V., Mehra, N.K., Targeted drug delivery to macrophages. *Expert Opin. Drug Deliv.*, 10, 353, 2013.
11. Bower, C., Zhu, W., Jin, S.H., Zhou, O., Plasma-induced alignment of carbon nanotubes. *Appl. Phys. Lett.*, 77, 830, 2000.
12. Merkulov, V.I., Lowndes, D.H., Wei, Y.Y., Eres, G., Voelkl, E., Patterned growth of individual and multiple vertically aligned carbon nanofibers. *Appl. Phys. Lett.*, 76, 3555, 2000.
13. Teo, K.B.K., Chhowalla, M., Amaratunga, G.A.J., Milne, W.I., Hasko, D.G., Pirio, G., Pribat, D., Uniform patterned growth of carbon nanotubes without surface carbon. *Appl. Phys. Lett.*, 79, 1534, 2001.
14. Saliev, T., The advances in biomedical applications of carbon nanotubes. *J. Carbon Res.*, 5, 29, 2019.
15. Mehra, N.K., Mishra, V., Jain, N.K., A review of ligand tethered surface engineered carbon nanotubes. *Biomaterials*, 35, 1267, 2014.
16. Yao, Y., Ma, C., Wang, J., Qiao, W., Ling, L., Long, D., Rational design of high-surface-area carbon nanotube/microporous carbon core–shell nanocomposites for supercapacitor electrodes. *ACS Appl. Mater. Interfaces*, 7, 4817, 2015.
17. He, H., Pham-Huy, L.A., Dramou, P., Xiao, D., Zuo, P., Pham-Huy, C., Review article carbon nanotubes: Applications in pharmacy and medicine. *BioMed. Res. Int.*, 12, 2013.
18. Wang, Y. and Cao, G., Developments in nanostructured cathode materials for high-performance lithium-ion batteries. *Adv. Mater.*, 20, 2251, 2008.
19. Karousis, N., Tagmatarchis, N., Tasis, D., Current progress on the chemical modification of carbon nanotubes. *Chem. Rev.*, 110, 5366, 2010.
20. Mittal, G., Dhand, V., Rhee, K.Y., Park, S.-J., Lee, W.R., A review on carbon nanotubes and graphene as fillers in reinforced polymer nanocomposites. *J. Ind. Eng. Chem.*, 21, 11, 2015.
21. Cortazar, P., Zhang, L., Untch, M., Mehta, K., Costantino, J.P., Wolmark, N., Bonnefoi, H., Cameron, D., Gianni, L., Valagussa, P., Pathological complete response and long-term clinical benefit in breast cancer: The CTNeoBC pooled analysis. *Lancet*, 384, 164, 2014.
22. Hamon, M.A., Hu, H., Bhowmik, P., Niyogi, S., Zhao, B., Itkis, M.E., Haddon, R.C., End-group and defect analysis of soluble single-walled carbon nanotubes. *Chem. Phys. Lett.*, 347, 8, 2001.
23. Basheer, B.V., George, J.J., Siengchin, S., Parameswaranpillai, J., Polymer grafted carbon nanotubes—Synthesis, properties, and applications. A review. *Nano-Struct. Nano-Objects*, 22, 100429, 2020.
24. Hamon, M.A., Itkis, M.E., Niyogi, S., Alvaraez, T., Kuper, C., Menon, M., Haddon, R.C., Effect of rehybridization on the electronic structure of single-walled carbon nanotubes. *J. Am. Chem. Soc.*, 123, 11292, 2001.
25. Srivastava, D., Brenner, D.W., Schall, D.J., Ausman, K.D., Feng Yu, M., Ruoff, R.S., Predictions of enhanced chemical reactivity at regions of local conformational strain on carbon nanotubes: Kinky chemistry. *J. Phys. Chem. B*, 103, 4330, 1999.

26. Zhao, J., Lee, C.W., Han, X., Chen, F., Xu, Y., Huang, Y., Chan-Park, M.B., Chen, P., Li, L., Solution processable semi-conducting thin film transistors using single-walled carbon nanotubes chemically modified by organic radical initiators. *J. Chem. Commun.*, 46, 7182, 2009.
27. Gebhardt, B., Graupner, R., Hauke, F., Hirsch, A., A novel diameter-selective functionalization of SWCNTs with lithium alkynylides. *Eur. J. Org. Chem.*, 8, 1494, 2010.
28. Grimes, C.A. and Mor, G.K., *TiO2 nanotube arrays: Synthesis, properties, and applications*, pp. 1–48, Springer Science & Business Media, US, Boston, MA, 2009.
29. Vaisman, L., Wagner, H.D., Marom, G., The role of surfactants in dispersion of carbon nanotubes. *Adv. Colloid Interface Sci.*, 128, 37, 2006.
30. Mubeen, B., Ansar, A.N., Rasool, R., Ullah, I., Imam, S.S., Alshehri, S., Ghoneim, M.M., Alzarea, S.I., Nadeem, M.S., Kazmi, I., Nanotechnology as a novel approach in combating microbes providing an alternative to antibiotics. *Antibiotics*, 10, 1473, 2021.
31. Zare-Zardini, H., Amiri, A., Shanbedi, M., Memarpoor-Yazdi, M., Asoodeh, A., Studying of antifungal activity of functionalized multiwalled carbon nanotubes by microwave-assisted technique. *Surf. Interface Anal.*, 45, 751, 2013.
32. Klumpp, C., Kostarelos, K., Prato, M., Bianco, A., Functionalized carbon nanotubes as emerging nanovectors for the delivery of therapeutics. *Biochim. Biophys. Acta*, 1758, 404, 2006.
33. Bianco, A., Kostarelos, K., Prato, M., Applications of carbon nanotubes in drug delivery. *Curr. Opin. Chem. Biol.*, 9, 674, 2005.
34. Jain, K.K., Advances in use of functionalized carbon nanotubes for drug design and discovery. *Expert Opin. Drug Discovery*, 7, 1, 2012.
35. Xiuping, W., Zilin, Z., Fangfang, C., Surface modification of carbon nanotubes with an enhanced antifungal activity for the control of plant fungal pathogen. *Materials*, 10, 1375, 2017.
36. Wu, W., Wieckowski, S., Pastorin, G., Benincasa, M., Klumpp, C., Briand, J.P., Bianco, A., Targeted delivery of amphotericin B to cells by using functionalized carbon nanotubes. *Angew. Chem.*, 117, 6516, 2005.
37. Zotchev, S.B., Polyene macrolide antibiotics and their applications in human therapy. *Curr. Med. Chem.*, 10, 211, 2003.
38. Szlinder-Richert, J., Cybulska, B., Grzybowska, J., Bolard, J., Borowski, E., Interaction of amphotericin B and its low toxic derivative, N-methyl-N-D-fructosyl amphotericin B methyl ester, with fungal, mammalian and bacterial cells measured by the energy transfer method. *IL Farmaco*, 59, 289, 2004.
39. Zumbuehl, A., Jeannerat, D., Martin, S.E., Sohrmann, M., Stano, P., Vigassy, T., Carreira, E.M., An amphotericin B–fluorescein conjugate as a powerful probe for biochemical studies of the membrane. *Angew. Chem. Int. Ed.*, 43, 5181, 2004.

40. Oyelami, A.O. and Semple, K.T., Impact of carbon nanomaterials on microbial activity in soil. *Soil Biol. Biochem.*, 86, 172, 2015.
41. Kang, S., Pinault, M., Pfefferle, L.D., Elimelech, M., Single-walled carbon nanotubes exhibit strong antimicrobial activity. *Langmuir*, 23, 8670, 2007.
42. Kang, S., Herzberg, M., Rodrigues, D.F., Elimelech, M., Antibacterial effects of carbon nanotubes: Size does matter. *Langmuir*, 24, 6409, 2008.
43. Foo, M.E., Anbu, P., Gopinath, S.C., Lakshmipriya, T., Lee, C.G., Yun, H.S., Yaakub, A.R.W., Antimicrobial activity of functionalized single-walled carbon nanotube with herbal extract of Hempedu bumi. *Surf. Interface Anal.*, 50, 354, 2008.
44. Chaudhari, A.A., Kate, K., Dennis, V., Singh, S.R., Owen, D.R., Palazzo, C., Pillai, S.R., A novel covalent approach to bio-conjugate silver coated single walled carbon nanotubes with antimicrobial peptide. *J. Nanobiotechnol.*, 14, 1, 2016.
45. Arias, L.R. and Yang, L., Inactivation of bacterial pathogens by carbon nanotubes in suspensions. *Langmuir*, 25, 3003, 2009.
46. Zardini, H.Z., Davarpanah, M., Shanbedi, M., Amiri, A., Maghrebi, M., Ebrahimi, L., Microbial toxicity of ethanolamines-multiwalled carbon nanotubes. *J. Biomed. Mater. Res. Part A*, 102, 1774, 2014.
47. Mansab Ali, S., Phelim Voon Chen., Y., Eng Hwa, W., Investigation of antimicrobial activity and cytotoxicity of synthesized surfactant-modified carbon nanotubes/polyurethane electrospun nanofibers. *Nano-Struct. Nano-Objects*, 24, 100612, 2020.
48. Zardini, H.Z., Amiri, A., Shanbedi, M., Maghrebi, M., Baniadam, M., Enhanced antibacterial activity of amino acids-functionalized multi walled carbon nanotubes by a simple method. *Colloids Surf. B: Biointerfaces*, 92, 196, 2012.
49. Ernst, W.A., Thoma-Uszynski, S., Teitelbaum, R., Ko, C., Hanson, D.A., Clayberger, C., Modlin, R.L., Granulysin, a T cell product kills bacteria by altering membrane permeability. *J. Immunol.*, 165, 7102, 2000.
50. Neelgund, G.M., Oki, A., Luo, Z., Antimicrobial activity of CdS and Ag2S quantum dots immobilized on poly(amidoamine) grafted carbon nanotubes. *Colloids Surf. B: Biointerfaces*, 100, 215, 2012.
51. Abo-Neima, S.E., Motaweh, H.A., Elsehly, E.M., Antimicrobial activity of functionalised carbon nanotubes against pathogenic microorganisms. *IET Nanobiotechnol.*, 14, 457, 2020.
52. Amiri, A., Zardini, H.Z., Shanbedi, M., Maghrebi, M., Baniadam, M., Tolueinia, B., Efficient method for functionalization of carbon nanotubes by lysine and improved antimicrobial activity and water-dispersion. *Mater. Lett.*, 72, 153, 2012.
53. Mohammed, M.K., Ahmed, D.S., Mohammad, M.R., Studying antimicrobial activity of carbon nanotubes decorated with metal-doped ZnO hybrid materials. *Mater. Res. Express*, 6, 055404, 2019.

54. Gilbertson, L.M., Goodwin Jr., D.G., Taylor, A.D., Pfefferle, L., Zimmerman, J.B., Toward tailored functional design of multi-walled carbon nanotubes (MWNTs): Electrochemical and antimicrobial activity enhancement via oxidation and selective reduction. *Environ. Sci. Technol.*, 48, 5938, 2014.
55. Murugan, E. and Vimala, G., Effective functionalization of multiwalled carbon nanotube with amphiphilic poly(propyleneimine) dendrimer carrying silver nanoparticles for better dispersability and antimicrobial activity. *J. Colloid Interface Sci.*, 357, 354, 2011.
56. Al-Hakami, S.M., Khalil, A.B., Laoui, T., Atieh, M.A., Fast disinfection of *Escherichia coli* bacteria using carbon nanotubes interaction with microwave radiation. *Bioinorg. Chem. Appl.*, 1, 2013.
57. Tambur, P., Bhagawan, D., Siva Kumari, B., Reddy Kasa, R., A facile synthesis of implantation of silver nanoparticles on oxygen-functionalized multi-walled carbon nanotubes: Structural and antibacterial activity. *SN Appl. Sci.*, 2013, 2, 981, Article ID 458943, 2020.
58. Choudhary, B., Paul, S.R., Nayak, S.K., Singh, V.K., Anis, A., Pal, K., Understanding the effect of functionalized carbon nanotubes on the properties of tamarind gum hydrogels. *Polym. Bull.*, 75, 4929, 2018.
59. Shahriary, L., Nair, R., Sabharwal, S., Athawale, A.A., One step synthesis of Ag-reduced graphene oxide multiwalled carbon nanotubes for enhanced antibacterial activities. *New J. Chem.*, 39, 4583, 2015.
60. Youssef, A.M., El-Naggar, M.E., Malhat, F.M., El Sharkawi, H.M., Efficient removal of pesticides and heavy metals from wastewater and the antimicrobial activity of f-MWCNTs/PVA nanocomposite film. *J. Cleaner Prod.*, 206, 315, 2019.
61. Yallappa, S., Manjanna, J., Dhananjaya, B.L., Vishwanatha, U., Ravishankar, B., Gururaj, H., Niranjana, P., Hungund, B.S., Phytochemically functionalized Cu and Ag nanoparticles embedded in MWCNTs for enhanced antimicrobial and anticancer properties. *Nano-Micro Lett.*, 8, 120, 2016.
62. Sivaraj, D. and Vijayalakshmi, K., Enhanced antibacterial and corrosion resistance properties of Ag substituted hydroxyapatite/functionalized multiwall carbon nanotube nanocomposite coating on 316L stainless steel for biomedical application. *Ultrason. Sonochem.*, 59, 104730, 2019.
63. Sapalidis, A., Sideratou, Z., Panagiotaki, K.N., Sakellis, E., Kouvelos, E.P., Papageorgiou, S., Katsaros, F., Fabrication of antibacterial poly (vinyl alcohol) nanocomposite films containing dendritic polymer functionalized multi-walled carbon nanotubes. *Front. Mater.*, 5, 11, 2018.
64. Rajavel, K., Gomathi, R., Manian, S., Rajendra Kumar, R.T., *In vitro* bacterial cytotoxicity of CNTs: Reactive oxygen species mediate cell damage edges over direct physical puncturing. *Langmuir*, 30, 592, 2014.
65. Amaku, J.F. and Martincigh, B.S., Synthesis and characterization of nanocomposites containing silver nanoparticle – decorated multiwalled carbon nanotubes for water disinfection. *Waste and Biomass Valorization*, 13: 149-172, 2022.

66. Swathy, T.S., Jinish Antony, M., George, N., Active solvent hydrogen-Enhanced p-Nitrophenol reduction using heterogeneous silver nanocatalysts@surface-functionalized multiwalled carbon nanotubes. *Ind. Eng. Chem. Res. 2021*, 60, 7050, 2021.
67. Yadav, S.K., Mahapatra, S.S., Cho, J.W., Synthesis of mechanically robust antimicrobial nanocomposites by click coupling of hyperbranched polyurethane and carbon nanotubes. *Polymer*, 53, 2023, 2012.

Index

Printed and bound by CPI Group (UK) Ltd, Croydon, CR0 4YY
17/09/2026
14964275-0007